MOSFETs: Properties, Preparations and Performance

MOSFETs: Properties, Preparations and Performance

NOAH T. ANDRE AND LUCAS M. SIMON
EDITORS

Nova Science Publishers, Inc.
New York

For permission to use material from this book please contact us:
Telephone 631-231-7269; Fax 631-231-8175
Web Site: http://www.novapublishers.com

LIBRARY OF CONGRESS CATALOGING-IN-PUBLICATION DATA

MOSFETS : properties, preparations to performance / Noah T. Andre and Lucas M. Simoná, Editors.
p. cm.
Includes bibliographical references and index.
ISBN 978-1-60456-762-5 (hardcover : alk. paper)
1. Metal oxide semiconductor field-effect transistors. I. Andre, Noah T. II. Simon, Lucas M.
TK7871.95.M665 2008
621.3815'284--dc22 2008023105

Published by Nova Science Publishers, Inc. ✢ New York

CONTENTS

PREFACE

The metal–oxide–semiconductor field-effect transistor (MOSFET, MOS-FET, or MOS FET) is a device used to amplify or switch electronic signals. It is by far the most common field-effect transistor in both digital and analog circuits. The MOSFET is composed of a channel of n-type or p-type semiconductor material, and is accordingly called an NMOSFET or a PMOSFET (also commonly nMOSFET, pMOSFET. The width of the channel, which determines how well the device conducts, is controlled by an electrode called the gate, separated from channel by a thin layer of oxide insulation. The insulation keeps current from flowing between the gate and channel. MOSFETs are useful for high-speed switching applications and also on integrated circuits in computers.

The conventional MOSFET with its drift-diffusion driven current transport mechanism dominated the silicon semiconductor technology for many decades. Its functionality is well understood and it was easy to scale down the device along the ITRS road map.

While technological problems were the ones to solve for further shrinking in the past, now physical barriers inhibit the shrinking process. Furthermore the conventional MOSFET suffers from inherent fundamental limits in some fields of application like high temperature electronics. Future concepts like the FINFET or the SONFET (Silicon-On-Nothing Field Effect Transistor) use three-dimensional channel engineering in order to achieve fully depleted channel regions leading to a large I_{ON}/I_{OFF} ratio and suppression of short channel effects.

Another promising way out of these problems is the investigation of field effect devices using current transport mechanisms other than drift-diffusion. Remarkable candidates in this field are the Tunneling Field Effect Transistor (TFET), the vertical Impact Ionization MOSFET (vertical I-MOS) and the lateral Impact Ionization MOSFET (lateral I-MOS). By using tunneling, impact ionization or body charging effects these devices overcome barriers like the temperature limit in silicon, the subthreshold slope limitation to kT/q or the radiation sensitivity of the conventional MOSFET.

Chapter 1 will discuss the work done on these devices over the past years and their advantages as well as disadvantages compared to the classic approach. Their chances and possible fields of application will be outlined as well as the physics which is behind their outstanding performance.

A 3D Monte Carlo simulator for modelling transport in nano-MOSFETs, based on tetrahedral elements, is described in detail. The basic physical models used to simulate carrier transport in semiconductor devices and how these models can be implemented within the

Finite Element method, are presented in Chapter 2. In the case of ensemble Monte Carlo simulations, these details include the particle-mesh coupling generalised for arbitrary meshes, treatment of contacts and integration of the equations of motion. The simulator is parallel and uses a domain decomposition strategy for the solution of the Poisson equation whereas a random distribution of particles among processors is used to parallelize the Monte Carlo. A very good scalability of the 3D simulator has been obtained thanks to the Finite Element solver, which is the most computationally intensive stage in self-consistent simulations.

Development of cryogenic power electronic system has attracted much attention recently due to the substantially better performances of semiconductor power devices such as power MOSFETs at cryogenic temperature, e.g., higher efficiency and reduced weight and size[1-4]. For instance, at low temperatures, power MOSFETs demonstrate much lower on-resistance and higher switching speed resulting from increased carrier mobility and saturation velocity [5-7]. At lower temperatures, the thermal conductivities of the device itself and the substrate materials in the packaging improve drastically, which lead to better and simpler thermal management and better reliability. Recent advances in $YBa_2Cu_3O_x$-based coated conductor development are enabling long lengths of Second Generation High Temperature Superconductor (HTS) tapes to operate above liquid nitrogen temperature (>77K) in high magnetic fields. Adapting power electronics control systems for low-temperature operation in combination with HTS devices (such as HTS motors or generators) is a natural step that provides synergistic benefits such as higher power density, further reductions in size, reduced loads on the cryogenic system due to fewer connections to outside world, higher reliability and improved performance. Being able to operate power MOSFETs at cryogenic temperatures is also very beneficial for deep space exploration spacecrafts since radioisotope heating unit (RHU) will no longer be needed to maintain these electronics within the tolerable temperature range (normally near room temperature on earth) [8], thereby reducing system size and weight. In Chapter 3, the present status of CPE is discussed in order to provide an outlook on emerging device technologies that are gaining in interest. Advanced power electronic packaging/interconnect methods that adapt well to cryogenics and help further improve system performance is discussed. Development of cryogenic integrated power package is discussed in detail. Furthermore, understanding how the structure of the device unit-cell affects the cryogenic performance of a Si power MOSFETis an important step toward optimizing of the device for cryogenic operations. Physically based numerical simulations of the Si power DMOS are performed in order to analyze the effects of device structure on their performances at cryogenic temperatures. The simulation results suggest that the device optimized for room temperature operation can be further optimized at cryogenic temperatures. In the end of the chapter, the possibility of utilizing superjunction power MOSFET for cryogenic operation is also discussed based on experimental testing results.

The problem of redesign and optimization of semiconductor devices is revised from a different perspective from what has been done in the past, namely by using the formalism of doping sensitivity functions. The doping sensitivity functions are introduced in this chapter as the set of operators on the adjoint space of the space of square integrable functions generated by all possible doping configurations. On this functional space we are able to construct analytical and numerical algorithms for the redesign and optimization of semiconductor devices and circuits subject to technological or practical constraints. These constraints are taken into consideration by using the Lagrange multipliers technique. Our redesign technique relies on first-order perturbations, for which reason is computationally very efficient and requires only the solution of linear systems of equations. The technique is applied the design of random doping fluctuation resistant structures (fault-tolerant devices) and to the optimization of semiconductor devices and circuits. The design is performed by starting with an initial device structure and applying an iterative multidimensional maximization procedure to "readjust" the doping configuration and device structure in order to obtain the desired final properties for the device. The technique presented in Chapter 4 is universal in the sense that it can be applied to any semiconductor device such as field-effect transistors or heterojunctions devices and can be used in the framework of any transport model. It requires only the solution of a few linear sparse systems of equations, for which reason is computationally very efficient. Numerical results related to the suppression of random doping induced fluctuations in a 35-nm channel length transistor and suppression of the effects of the drain-induced barrier lowering (DIBL) on the value of the threshold voltage are presented for a 30 nm channel length MOSFET device. It is found that by carefully designing the doping profiles in these devices the effects of the random doping concentration on threshold voltage can be decreased by almost one order of magnitude, while the effects of the DIBL can be reduced at least 50%.

With the increasing of integrated circuit density, the transistor size (channel length and width, gate oxide thickness) must decrease drastically. The reduction of the MOS structure size implies a good knowledge of the physical degradation mechanism and the reliability of the devices when the gate oxide thicknesses are lower than 3 nm. The ultra-thin gate oxide reliability is then an essential factor in CMOS technologies. As actually, no alternative and operative solution can replace the Si/SiO_2 technology, that reaches its physical limits, those studies are important so that the CMOS device development is even interesting. It is then necessary to improve our knowledge on the degradation mechanisms which occur at operating voltage. When the oxide thickness becomes smaller than 3 nm, the gate current becomes measurable when the substrate is in depletion. This current increases with the decrease of the oxide thickness. After electrical stresses, this gate current, called LVSILC (Low Voltage Stress Induced Leakage Current) increases. It does not affect the device and circuit functionality because it is observed in depletion or weak inversion regimes and not at usual operating voltages. The low voltage gate current in ultra-thin oxide metal–oxide–semiconductor devices is very sensitive to electrical stresses, it can be used to monitor the oxide degradation, to predict the charge to breakdown and then to predict the lifetime of the device. For various oxide thicknesses ranging from 1.2 to 4 nm and for P and N-MOSFETs, we have realized an experimental database on the LVSILC evolution, as a function of different stress conditions (voltages, temperatures, etc...). This large database provides a coherent understanding for describing the reliability of ultra-thin SiO_2 gate oxide under a constant voltage tunneling stress for positive and negative bias voltages. The authors have shown in Chapter 5 that the voltage dependence of the time to breakdown follows a power law behavior rather than an exponential law. From the LVSILC measurements, they have shown that it can be possible to deduce the time to breakdown as a function of the stress voltage, to predict the device lifetime and to project reliability limits.

Metal-oxide semiconductor field effect transistor (MOSFET) dosimeter has been used to measure and monitor radiation dose for mega-voltage (MV) photon and electron beams in radiation therapy. The dosimeter has a very small physical size (active region of about $0.2 \times 0.2\ mm^2$), and the recorded dose can be read immediately after or during the treatment using a remote reader setup. This simplifies the cumbersome dosimetric procedure and minimizes the time of quality assurance in the treatment. Recently, the properties, preparations and performances of the MOSFET dosimeter are studied extensively in different aspects such as angular and energy dependence, lifespan and sensitivity. Monte Carlo simulation is also employed to predict dosimetric results under different conditions of measurements, and provides valuable information to improve the design of the dosimeter. In Chapter 6, some medical applications using the MOSFET dosimeter to measure the MV photon and electron beams were presented. Moreover, the properties and performances of the MOSFET dosimeter in determining the electron backscatter from the lead were studied using Monte Carlo simulations. The accuracy of using MOSFET dosimeter to predict the electron backscatter from the lead was also evaluated in this chapter.

For photon beam study in radiation therapy, surface doses for the 6 and 15 MV prostate intensity modulated radiation therapy (IMRT) plans were measured, and compared to those for typical prostate four-field box (4FB) plans. MOSFET dosimeters were used and placed on a transverse contour line along a Rando Phantom surface at the central beam axis in the measurement. The aims of measurement are to investigate how: (1) the moving multi-leaf collimator (MLC) leaves contribute to the surface dose during IMRT, and (2) the effects of both photon beam energy and number of beam used in the IMRT plan on the surface dose. Results showed that with the same number of beam in the IMRT plan, the 6 MV plan gave more surface dose than that of the 15 MV. However, when the number of beam was increased in the plan, the surface dose difference between the 6 and 15 MV photon beams became less. It was also found that the increase of surface dose in the IMRT plan compared to 4FB was predominantly due to the number of beam, and the calculated monitor units in the plans.

For electron beam study in radiation therapy, the electron backscatter, produced by clinical electron beams, was studied using the MOSFET dosimeter and Monte Carlo simulation (EGSnrc-based code). Electron backscatter factor (EBF), defined as the ratio of dose at the tissue-lead interface with and without the presence of backscatter, was calculated using the Monte Carlo simulation, and verified with the MOSFET measurement. The dependences of the (1) electron beam energy, (2) depth of the lead shield, (3) beam angle, and (4) aluminium sheet used to absorb electron backscatter, on the EBF were studied. It was found that for particular depth of lead shield, EBF decreases with increase of electron beam energy (4 – 16 MeV). In addition, within the electron practical range, R_p, and particular beam energy, EBF increases with the depth of lead. When the electron beam angle increases from 0 to 5 deg, EBF decreases only slightly (< 4%) for all beam energies. For low electron beam energy (4 MeV), 2 mm aluminium was found to effectively reduce the electron backscatter by 31%. While the electron beam energy increases, less electron backscatter were reduced by the same thickness of aluminium (only about 6% for the 16 MeV).

The performance of the MOSFET dosimeter in determining the electron backscatter from the lead was evaluated. The EBF was predicted by Monte Carlo simulation using the GEANT4-based code. Clinical electron beams with energies of 4, 6, 9 and 12 MeV were used in the simulations. It was found that with presence of the MOSFET body, EBF was underestimated to about 2% for the low electron beam energy of 4 MeV. However, such deviation decreases to 0.9% when higher electron beam energy of 12 MeV was used. This is because the small MOSFET dosimeter, mainly made of epoxy and silicon, not only attenuated the electron fluence from the upstream direction, but the electron backscatter generated by the lead underneath the dosimeter. Corresponding measurements were carried out and uncertainties compared to Monte Carlo results were within ± 2%. The electron energy spectra with and without the lead and MOSFET body were also determined by Monte Carlo simulations. It was found that the presence of the MOSFET body widened up the electron energy spectrum in the dosimetric volume. Moreover, no matter the MOSFET body is present or not in the simulation, deviations of electron energy spectra with and without the lead decrease with an increase of the electron beam energy. It is concluded that the MOSFET dosimeter performed well for measuring the electron backscatter from the lead using electron beams. The uncertainty of EBF is within 2% according to the Monte Carlo simulation, and decreases when higher electron beam energy is used (from 4 to 12 MeV).

Sustaining functional MOSFET device scaling encounters severe technological challenges as optical lithography and conventional transistor structure are approaching their physical limits. To break lithographic and structural barriers for continuous device scaling down to sub-10 nm half pitch, non-conventional transistors, fabrication processes, and their device physics have been extensively studied. In Chapter 7, recent research progress made in surrounding gate based nanowire and nanodot MOSFETs and a vertical integrated-gate CMOS (VIG CMOS) device is presented.

Vertical surrounding-gate MOSFET, compared with double-gate and tri-gate structures, proffers simultaneous gains in transistor density and device performance by consuming less space with better gate control and reduced leakage current. Device physics and modeling of surrounding-gate MOSFET are discussed, and the analytical solutions to the simplified Poisson's equation in both cylindrical and spherical coordinates are given to accurately calculate the electric potential and inversion charge concentration in the ultra-thin bodies. The stochastic characteristics of the device parameters of nanodot and nanowire MOSFETs are studied and several important process control issues are addressed.

For logic applications, a vertical integrated-gate CMOS (VIG CMOS) device is proposed to integrate vertical surrounding-gate PMOS and NMOS together to significantly increase the transistor density and improve the process control capability. Based on the VIG CMOS structure, novel NOR and NAND circuit architectures as well as their interconnect schemes are invented. Device physics of VIG CMOS is examined and we find that solving Poisson's equation in the domain of a Si shell under the single surrounding-gate control faces challenges related to device's boundary conditions and needs more research effort.

Chapter 8 presents one of the most attractive devices for implementing nanoscale planar MOSFET because of its excellent characteristics such as ideal subthreshold swing, high transconductance, and reasonable short-channel effect. A kind of new sub-50 nm N channel self-aligned double gate MOSFETs has been studied by solving coupled Poisson-Schrödinger equations with a finite element method. A systematic simulation-based study has been presented on the quantum effects and the gate leakage current. The simulation results indicate that the deviation from the classical model becomes more important as the gate oxide, gate length and Fin channel width becomes thinner and the Fin channel doping increases. Gate tunneling current density is reduced with the body thickness decreasing. The excessive scaling increases the gate current below Fin thickness of 5nm. The gate current can be dramatically reduced beyond $10^{17}cm^{-3}$ with increasing the Fin body doping. Quantum mechanical simulation results are compared with the results from the classical approach indicating that quantum mechanical simulation is essential for the realistic optimization of the double gate structure.

Recent advances in GaN-based MOS-HFET have opened the way for their practical applications in high power and high frequency electronic devices. Undoped GaN-based MOS-

HFETs with atomic-layer-deposited Al_2O_3 gate dielectrics are fabricated. Using a two-dimensional numerical calculation, we report the results of self-heating simulations of the GaN-based MOS-HFETs, including hot electron and quantum effects. The simulated electrical characteristics are in good agreement with the reported experimental data. The effect of the gate and source/drain extension lengths on both the output performance and self-heating is discussed in detail, allowing for device optimization. The dissipated Joule electric power causes the self-heating effects, which lead to negative differential output conductance. Our results have demonstrated that the hot electrons make a negligible contribution to the negative differential output conductance in our long channel MOS-HFETs. In order to investigate their joint interactions to the MOS-HFET's operation, the different static interface trap and charge densities created at the AlGaN/Al_2O_3 interface are considered in the output characteristics. Results show that the presence of the interface charges and traps are directly responsible for the observed current collapse and device switching in the GaN-based MOS-HFETs.

As explained in Chapter 9, there are two types of FinFETs: SOI FinFETs and bulk FinFETs. SOI FinFETs have an excellent scalability and high performance, but require expensive SOI wafer and has heat dissipation problems. To solve these problems, bulk FinFETs was devised and demonstrated firstly by the author's group. Since the bulk FinFETs have also excellent scalability and high performance, they can be applied to future high speed logics, scaled memory cells like sub-50 nm DRAM and flash memory cells. First, SOI and bulk FinFETs were compared in terms of scalability and heat dissipation. Then the author's group focused on the bulk FinFETs by covering from fabrication to threshold voltage modeling. The bulk FinFETs have no back-bias effect differently from conventional planar channel MOSFETs. For the fabrication of the bulk FinFETs, we explained key fabrication steps and resultant structures with many drawings and pictures. The first 40 nm bulk FinFET was briefly introduced with device structure and basic I-V characteristics. SEM views of the bulk FinFETs fabricated at industry were shown. The bulk FinFETs showed much better DIBL than that of conventional planar channel devices, and also shown very small back-bias effect. Key device characteristics such as threshold voltage, maximum transconductance, subthreshold swing, and saturation drain current were compared with those of the control devices by changing fin body width. Especially, p-type bulk FinFET shown much better key characteristics compared to those of planar channel p-type MOSFETs. The threshold voltage (V_{th}) behaviors of the bulk FinFETs were modeled based on charge-sharing. Threshold voltage modeling of the double/triple-gate bulk FinFETs was performed by considering the potential lowering at the channel surface of fin body at $V_{GS}=V_{th}$ condition. With decreasing gate length and/or fin width, the surface band-bending (theoretically $2\psi_B$) at $V_{GS}=V_{th}$ in fully depleted bulk FinFETs decreases since actual surface potential decreases at the V_{th} condition. Based on the $2\psi_B$ correction, charge-sharing length (x_h) and corner factor (α_c) were extracted. The V_{th} behaviors of bulk FinFETs were modeled based on the surface potential lowering, 3-D charge-sharing, narrow-width effect, and corner factor. The V_{th} model predicted well the V_{th} behavior with fin body width, body doping concentration, gate length, gate height, and corner shape of the fin body. The compact models predicted accurately V_{th} of the devices with the gate length up to 20 nm and the fin body width up to 5 nm.

In Chapter 10 the most important flicker noise models that have been proposed in the literature are analytically presented and discussed. The available flicker noise simulation models in software packages like SPICE and HSPICE and SPECTRE are also examined and selection criteria in relation to the type of MOSFET and the operating regime are proposed. The impact of scaling down technologies is discussed and extra analysis is presented in relation to the type of the bias in CMOS transistors. In particular the switched biasing effect in the MOSFET low frequency noise performance is presented. Additionally, low frequency noise measurement set-ups are presented and measurement techniques are suggested.

The experimental conditions allowing the extraction of reliable SILC characteristics in n-channel MOSFET's with ultrathin oxides are discussed first in Chapter 11. These characteristics are often studied using the relative variations in the gate current density. This approach is critically reviewed. It is shown that the variations (instead of the relative variations) in the gate current density, allow the extraction of reliable SILC properties and information on the SILC mechanism to be obtained. These curves are used to discuss the model proposed in [1]. In this paper and at negative bias, the gate electrons tunnel from the metal Fermi level into Si-SiO_2 interface states from where they recombine with the substrate majority carriers. It is shown that the SILC mechanism in those devices cannot be that one. Then, results confirming that conclusion and allowing information on the actual SILC mechanism to be obtained are discussed. This is done through the comparison between the trap density relative variations extracted from the SILC and from charge pumping, and through the characteristics of the traps required in $I_g(V_g)$ curves simulation with regard to those measured usually in MOS devices. The information obtained on the actual SILC mechanism is finally gathered.

In Chapter 12, to investigate the impact of the gate oxide degradation and breakdown (BD) on CMOS circuits functionality, an accurate description of the electrical characteristics of the stressed devices is presented, which can be included in circuit simulators. The proposed picture considers, on the one hand, the variation of the channel current and, on the other, the increase of the gate current due to the oxide degradation and BD and is able to account for different levels of oxide damage. The effect of gate oxide degradation and BD on simple analog (current mirror) and digital (RS latches, SRAM memory cells, CMOS inverters) circuits has been evaluated.

Chapter 13 describes a mixed-signal VLSI design utilizing the controlled conductance properties of MOSFETs, for tunable filters, low power amplifiers, and analog-mixed signal processing. The programmable linear conductance is realized by the pair of MOSFETs in the triode region, and it produces the voltage-controlled tunable function or the analog computing element. It demonstrates the application areas of neural synaptic functions, electronic neurons with biologically plausible characteristics, and tunable analogue complex filters for RF communication. The programmable conductance by MOSFETs is analyzed by the experimentation of prepared MOSFETs and the SPICE simulation.

The amplifier based on the proposed MOSFET conductance is employed to enforce the tunable filters for RF applications. It has the tuning range of +/- 50% and the bandwidth over 1 GHz is exhibited. For applications to analog computing, the design of an analog multiplier and biologically plausible neuromorphic devices are illustrated by utilizing MOSFET's programmable conductance.

In: MOSFETs: Properties, Preparations and Performance ISBN: 978-1-60456-762-5
Editors: N. T. Andre et al. pp. 1-59

Chapter 1

NOVEL DEVICE CONCEPTS TO OVERCOME MOSFET LIMITS

Ulrich Abelein and Ignaz Eisele
Bundeswehr University Munich, Institute of Physics, Werner-Heisenberg-Weg 39, 85577 Neubiberg, Germany

Abstract

The conventional MOSFET with its drift-diffusion driven current transport mechanism dominated the silicon semiconductor technology for many decades. Its functionality is well understood and it was easy to scale down the device along the ITRS road map.

While technological problems were the ones to solve for further shrinking in the past, now physical barriers inhibit the shrinking process. Furthermore the conventional MOSFET suffers from inherent fundamental limits in some fields of application like high temperature electronics. Future concepts like the FINFET or the SONFET (Silicon-On-Nothing Field Effect Transistor) use three-dimensional channel engineering in order to achieve fully depleted channel regions leading to a large I_{ON}/I_{OFF} ratio and suppression of short channel effects.

Another promising way out of these problems is the investigation of field effect devices using current transport mechanisms other than drift-diffusion. Remarkable candidates in this field are the Tunneling Field Effect Transistor (TFET), the vertical Impact Ionization MOSFET (vertical I-MOS) and the lateral Impact Ionization MOSFET (lateral I-MOS). By using tunneling, impact ionization or body charging effects these devices overcome barriers like the temperature limit in silicon, the subthreshold slope limitation to kT/q or the radiation sensitivity of the conventional MOSFET.

This chapter will discuss the work done on these devices over the past years and their advantages as well as disadvantages compared to the classic approach. Their chances and possible fields of application will be outlined as well as the physics which is behind their outstanding performance.

I. Introduction

What we know today as "Moore's Law" was the driving force of innovation in semiconductor technologies over the last 40 years. Gordon Moore predicted in 1965 that the number of transistors in an integrated circuit would continue to grow exponentially for at least 10 years [1]. Today we know that the time of exponential increase in complexity is still continuing. While Intel's 4004 microprocessor was fabricated 1971 using a 10 micron technology to integrate 2,300 transistors on a chip, the Quadcore Xeon® processor presented in 2007 consists of 840 million transistors, produced in a 45 nm technology. It is easy to see from these numbers that a tremendous effort on reducing the feature size of devices integrated in modern circuits was necessary to keep this development going on and on. But what was the reason for this procedure over many decades? Why didn't we start with 90 nm? Is this the core of Moore's Law?

The answer is obvious. Forty years ago we did not have the technology necessary to fabricate structures as small as 0.1 micron and below in silicon. So the information technology age started with 10-micron-structures. But what made it a "law" to shrink? Moore was not talking of a law when he published his article in 1965. It was an observation and a forecast with a very important second part, which is often not mentioned today. It is not just the number of transistors that increases. It is the number of transistors that can be produced at minimum costs that increases. This means more complexity for the same or less money. So Moore's Law described the way to achieve enormous growth rates in semiconductor industries. And as it was the way to grow, the whole semiconductor industry followed this prediction and gave it the status of a law. After four decades this led to an impressive world market volume for semiconductors of 247.7 billion US dollars in 2007 [2].

The device which dominates semiconductor technology is the silicon-based MOSFET. While the first transistor discovered in 1947 was a point-contact transistor in germanium [3, 4] no other device was as successful as the field effect transistor using silicon and silicon dioxide, which was demonstrated 13 years later [5, 6]. There are several reasons for this success: First of all, silicon is a comparatively cheap semiconductor, non-toxic and easy to produce. Second, it forms a natural oxide, SiO_2, which has excellent electrical properties and is ideal to form a MOS capacitance. Furthermore a MOSFET is a device with very good scaling capabilities. Finally, and this might be the most important reason, the use of p- and n-channel MOSFET for integrated circuits opened the door to CMOS, a technology which allowed a dramatic reduction in static power consumption as under ideal conditions only during the switching of the logical gates is a significant current flowing.

During the whole era of the MOSFET, mostly technological problems had to be solved to increase the complexity of the circuits by shrinking the device. New methods of lithography were developed, the methods to achieve the doping of source, drain and channel region changed and the process flow was optimized. But the device stayed the same for a long time. It just became smaller and smaller.

But with the continuous downscaling the MOSFET enters regions where physical limits become dominant. The first signs of these limits are known as short channel effects [7, pp. 469]. By improving the MOSFET design, for example with the introduction of pocket implants, the reduction of feature size was continued. Now in 2008 we approach a point which seems to be the end of scaling the MOSFET. Effects like tunneling, diffusion limited subthreshold slope or statistical fluctuations in dopant concentrations limit the functionality of sub-50 nm MOSFETs.

This chapter will introduce first the fundamental physical limits of the MOSFET when scaled down into the sub-50 nm regime. Furthermore, we will have a look at a field of application which at all times was not easy to handle for the MOSFET, namely high temperature environments above 200 °C. Also the physics limiting the device under these circumstances will be explained.

As all these limitations have their origin in the operating principle of the MOSFET, the only solution to overcome them is the design of a new device which is free of these physical boundaries. We will have a look at five alternative concepts, all recently presented and with different advantages compared to the conventional MOSFET: the FINFET, the SONFET, the p-i-n and the n-i-p-i-n based I-MOS and the TFET. Their device physics and how they overcome the MOSFET's limits will be described and explained.

II. Drift Diffusion Driven Transistors

When talking about the limits of the conventional drift diffusion driven MOSFET concept (figure 1 shows a schematic cross section of the device) one should not get the idea that the device will suddenly fail at a certain point and lose all its functionality. The device properties will rather degrade with ongoing shrinking or the step to next technology node is inhibited by technological barriers (i.e. there is no technology to realize the next node or the technology to realize the next node is too expensive). The device development community invested a lot of work to overcome these difficulties and keep the concept of the drift diffusion driven MOSFET. This subsection will provide an overview of the concept and technological limits and what efforts have been made to overcome them.

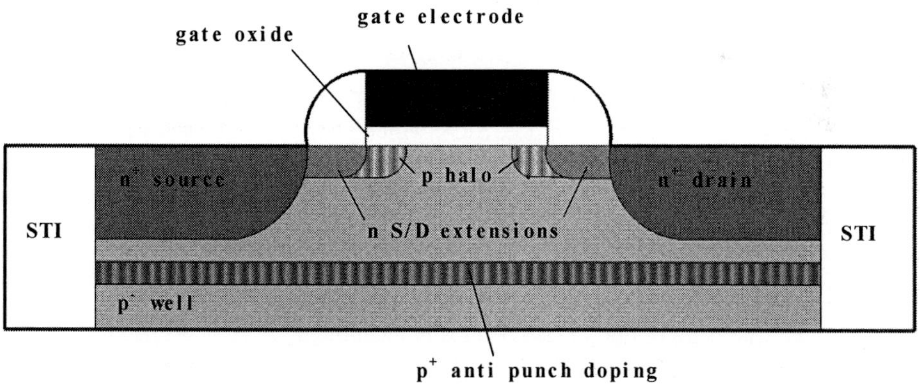

Figure 1. Schematic cross section of a conventional short channel MOSFET.

A. Concept Inherent Limits

1. Limitation of the Subthreshold Slope S

The carriers to form the channel of a MOSFET are provided by diffusion. Therefore the switching behavior strongly depends on temperature and is limited by the diffusion velocity of the charge carries. Furthermore this limitation determines how efficiently a change in gate voltage influences the formation of the channel. This results in a rising static leakage current, if the threshold voltage and the drive current of the device are technological requirements and the limitation of the slope inhibits a complete turn-off of the device. A better understanding of this correlation can be derived from the calculation of the theoretical limit of the slope.

For a gate voltage V_G below the threshold voltage V_T the MOSFET is working in the so called subthreshold region with its channel being in weak inversion. As the dependency of the drain current I_D on V_G in this region describes how good the transistor can be switched off, it is especially important with respect to the total power consumption. The correlation between I_D and V_G is described by the subthreshold slope S which is defined as

$$S = \left(\frac{d(\log I_D)}{dV_G} \right)^{-1} \quad (1)$$

S is commonly given in mV/dec and is the value the gate voltage has to be increased to gain an increase of the drain current by one order of magnitude. In weak inversion the current transport is dominated by diffusion and so the minimum of S depends on temperature. It can be calculated from the equation for the drain current in the subthreshold region which is in case of an n-channel MOSFET [10]:

$$I_D = \mu_n \cdot \frac{W}{L} \cdot \frac{\varepsilon_{Si} \cdot t_{ox} \cdot C_{ox}}{\varepsilon_{ox} \cdot L_D} \cdot \frac{k^2 \cdot T^2}{q^2} \cdot \frac{n_i^2}{N_A^2} \cdot \left(1 - e^{\frac{-V_D \cdot q}{k \cdot T}} \right) \cdot e^{\frac{\Phi_s \cdot q}{k \cdot T}} \cdot \sqrt{\frac{\Phi_s \cdot q}{k \cdot T} - 1}^{\,-1} \quad (2)$$

μ_n is the electron mobility, W and L are gate length and width, ε_{Si} is the silicon permittivity[1], t_{ox}, C_{ox} and ε_{ox} are thickness, capacitance per unit area and permittivity[2] of the gate oxide, L_D is the Debye length[3], k is the Boltzmann constant, T is the absolute temperature, q is the elementary charge, n_i is the intrinsic charge carrier density and N_A the acceptor density . Φ_s is the surface potential which depends on the gate voltage by the equation [8]:

$$\Phi_S = (V_G - V_{FB}) - \frac{2 \cdot \varepsilon_{Si}^2 \cdot t_{ox}^2 \cdot k \cdot T}{\varepsilon_{ox}^2 \cdot L_D^2 \cdot q} \left(\sqrt{1 + \frac{\varepsilon_{ox}^2 \cdot L_D^2}{\varepsilon_{Si}^2 \cdot t_{ox}^2} \left(V_G \cdot \frac{q}{k \cdot T} - V_{FB} \cdot \frac{q}{k \cdot T} - 1 \right)} - 1 \right) \quad (3)$$

V_{FB} is the flatband voltage. For drain voltages >3kT/q the drain current becomes independent of V_D. Therefore under these conditions the logarithm of the drain current can be calculated by:

$$\ln(I_D) = \ln\left(\mu_n \cdot \frac{W}{L} \cdot \frac{\varepsilon_{Si} \cdot t_{ox} \cdot C_{ox}}{\varepsilon_{ox} \cdot L_D} \cdot \frac{k^2 \cdot T^2}{q^2} \cdot \frac{n_i^2}{N_A^2} \right) + \frac{\Phi_s \cdot q}{k \cdot T} - \frac{1}{2} \ln\left(\frac{k \cdot T}{\Phi_s \cdot q} - 1 \right) \quad (4)$$

---

[1] $\varepsilon_{Si} = \varepsilon_{r,Si} \cdot \varepsilon_0 = 11{,}9 \cdot 8.8542 \cdot 10^{-12} CV^{-1}m^{-1}$

[2] $\varepsilon_{ox} = \varepsilon_{r,SiO_2} \cdot \varepsilon_0 = 3{,}9 \cdot 8.8542 \cdot 10^{-12} CV^{-1}m^{-1}$

[3] The Debey length is defined as $L_D \equiv \sqrt{\frac{kT \cdot \varepsilon_{Si}}{q \cdot N_A}}$ in p doped silicon [7, p. 437].

Furthermore the derivative of the surface potential with respect to the voltage applied to a MOS capacitance [7, p. 383] is given by

$$\frac{d\Phi_s}{dV} = \frac{C_{ox}}{C_{ox} + C_D} \tag{5}$$

with C_D the depletion capacitance, depending on the surface potential [8] by the equation

$$C_D = \frac{\varepsilon_{Si}}{L_D} \cdot \frac{1}{\sqrt{2\left(\frac{q}{kT}\Phi_s - 1\right)}} \tag{6}$$

From equation (6) we can obtain an expression for the ratio of oxide capacitance and depletion layer capacitance:

$$\frac{C_D}{C_{ox}} = \frac{d_{ox} \cdot \varepsilon_{Si}}{\varepsilon_{ox} \cdot L_D} \cdot \frac{1}{\sqrt{2\left(\frac{q}{kT}\Phi_s - 1\right)}} \tag{7}$$

Now we can calculate the subthreshold slope S form (1) by substituting (4) and use (5) to calculate the derivative and finally substituting (7):

$$S = \left(\frac{d(\log I_D)}{dV_G}\right)^{-1}$$

$$= \ln 10 \left(\frac{d(\ln I_D)}{dV_G}\right)^{-1}$$

$$= \ln 10 \cdot \left(\frac{q}{kT} \cdot \frac{C_{ox}}{C_{ox} + C_D} - \frac{1}{2} \cdot \frac{1}{\frac{\Phi_s \cdot q}{kT} - 1} \cdot \frac{q}{kT} \cdot \frac{C_{ox}}{C_{ox} + C_D}\right)^{-1}$$

$$= \frac{kT}{q} \cdot \ln 10 \cdot \left[\frac{C_{ox}}{C_{ox} + C_D} \cdot \left(1 - \frac{1}{2\left(\frac{\Phi_s \cdot q}{kT} - 1\right)}\right)\right]^{-1}$$

$$= \frac{kT}{q} \cdot \ln 10 \cdot \left(1 + \frac{C_D}{C_{ox}}\right) \cdot \left(1 - \frac{C_D^2}{C_{ox}^2} \cdot \frac{\varepsilon_{ox}^2 \cdot L_D^2}{d_{ox}^2 \cdot \varepsilon_{Si}^2}\right)^{-1} \tag{8}$$

As $(C_D^2 \cdot \varepsilon_{ox}^2 \cdot L_D^2)/(C_{ox}^2 \cdot d_{ox}^2 \cdot \varepsilon_{Si}^2)$ is close to zero in a MOSFET [7, p. 447] we can approximate the slope S by:

$$S \approx \frac{kT}{q} \cdot \ln 10 \cdot \left(1 + \frac{C_D}{C_{ox}}\right) \tag{9}$$

The theoretical limit for S is reached when $C_D \ll C_{ox}$ and is 60 mV/dec at room temperature [9]. Interface traps are not considered in this calculation but lead to a rising slope as they cause a capacitance C_{it} parallel to the depletion capacitance. Therefore in state of the art MOSFETs the typical slope achieved is around 80-100 mV/dec [10].

An increase of the static leakage current results from this limitation when the threshold voltage V_T is reduced during the shrinking procedure while an increasing ON current is desired according to the ITRS roadmap. Starting from a given drain current at the threshold voltage and a constant drain voltage, which will be the supply voltage according to the roadmap, the resulting leakage current at $V_G = 0$ V can be calculated by:

$$I_{LEAK} = I_D(V_G = V_T) \cdot e^{-(V_T/S)}\Big|_{V_D=\text{const.}} \tag{10}$$

It is easy to see from (10) that the static leakage current will increase exponentially with a linear decrease of V_T if S is not reduced by the same factor. As the limitation of the slope is strictly connected to the drift diffusion driven current transport in a conventional MOSFET there are no chances to overcome this limit without changing the device concept completely.

2. Short Channel Effects (SCE)

Short channel effects (SCE) summarize a group of degradation mechanisms of the electrical device properties which became relevant when the MOSFET was scaled below 1.0 µm channel length. The most difficult to control effects which are furthermore of importance for alternative device concepts, too, are drain induced barrier lowering (DIBL) [11], gate induced drain leakage (GIDL) [12] V_T roll-off [15] and hot electron effects (which will be discussed in more detail in section III.C). All of them were described as early as in the 1970s and the 1980s respectively but are still of very much interested for the research activities of the electron device community.

DIBL results from the influence of the drain source field on the barrier height without any gate bias. Similar to the punch-through effect the drain sided space charge region influences the OFF state current of the transistor but in contrast to punch-through not via the bulk silicon but directly in the channel region of the device. For short channel devices already a low drain source voltage is sufficient to lower the barrier in the OFF state and lead to an decreasing threshold voltage and possibly with respect to the limitation of the subthreshold slope to an increasing static leakage current. Figure 2 shows the influence of the drain voltage on the energy bands of a short channel MOSFET.

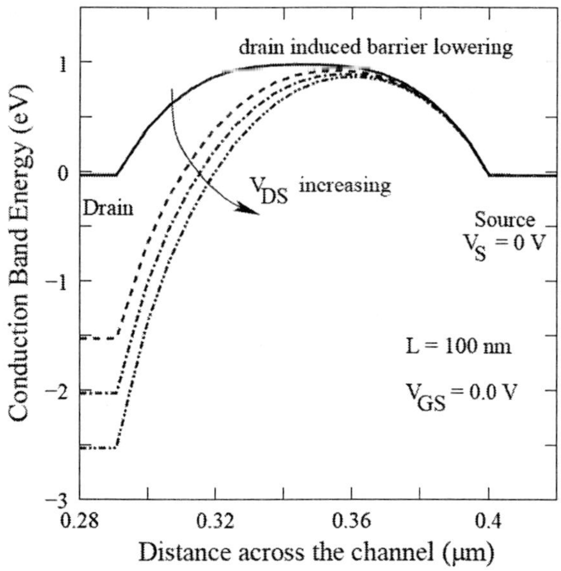

Figure 2. Influence of increasing drain source voltage on barrier height: simulated conduction band energy at the Si/SiO_2 interface (flatband conditions). [13]

A possibility to reduce the DIBL would be an increase of the channel doping. But the maximum channel doping is limited: First it directly increases the threshold voltage, which reduces I_{ON}. For extremely high doping levels the gate field will no longer be sufficient to bring the channel region in inversion. Second too high channel doping levels will lead to a rising leakage current due to Zener tunneling between the S/D regions and the channel (this effect will be described in more detail in section III.A). The answer on the DIBL driven rising leakage current were local halo or pocket implants [14]. These implantation regions are placed in the channel region next to source and drain. They reduce the influence of the drain field on the channel barrier and still grant gate control of the barrier height as the threshold voltage is determined by the V_T control doping in the region close to the gate oxide-channel interface.

The GIDL effect does not influence the switching behavior or the ON state in any case but only occurs in the OFF state of the transistor which means in an integrated circuit supply voltage potential at the drain and ground potential at the gate for an n-channel MOSFET. The reason for GIDL are overlap capacities between gate and drain caused by diffusion of drain extension below the gate. The potential difference between drain and gate leads to the formation of a deep-depletion region in the overlap region. Due to the extreme band bending valence band electrons from the substrate can tunnel into the conduction band of the drain extension region and are collected at drain.

V_T roll-off results from the lowering of the intrinsic barrier height even without any voltages applied. Figure 3 shows the change in the conduction band energy of a n-channel MOSFET when the channel length is reduced from 100 to 20 nm. The maximum of the barrier is first reduced from a plateau to a point and then the total barrier height is drastically decreasing which leads to a reduction of the threshold voltage. On the one hand this results in an rising static leakage current. On the other hand small variations in the channel length lead to significant changes in the threshold voltage which can lead to a large standard deviation of the threshold voltages of the transistors integrated on a chip.

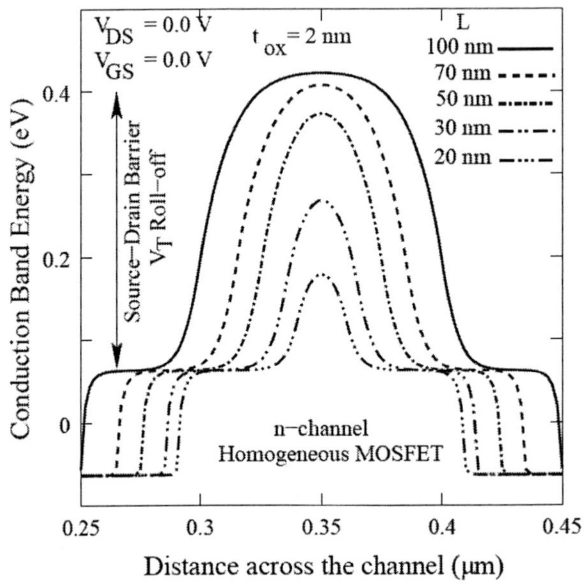

Figure 3. Simulated conduction band energy (flat band conditions) for a n-channel MOSFET with various channel lengths. [13]

3. RF Limitation

An important issue which should not be disregarded when talking about limitations of MOSFETs is their usefulness for RF applications as they cover a huge part of the field of applications for integrated circuits.

Mainly two parameters are important for high frequency operation of MOSFETs: the cutoff frequency f_T and the maximum frequency of oscillation f_{max}. While at f_T the small signal current gain decreases to unity (i. e. 0 dB), f_{max} is the frequency where the unity power gain rolls off to 0 dB. For most applications f_{max} is the more critical parameter as it determines the change from an active to a passive device.

The limitation in use with high frequencies is mostly due to (parasitic) capacities and the limitation of the low field charge carrier mobility. For high frequencies the charge carriers can no longer follow the change in bias and the current and power gain will drop down.

f_T scales very well with reduction of the channel length and thereby the gate capacitance [16, 17]. The dependences of f_{max} on the device design are more complex. It is mainly influenced, beside by the cutoff frequency, by the ratio of gate-drain to gate-source capacities and the output conductance of the device [16]. Both lead to a decrease in f_{max} when increasing, which is the case for sub-100 nm MOSFETs.

A detailed insight in the RF performance and its dependency on device scaling is given by *Dambrine et al.* [16] and *Liou and Schwierz* [17]. *Liou and Schwierz* also point out the improvement of RF performance by using fully depleted devices which are not limited by a depletion capacitance.

B. Technological Limits in MOSFET Fabrication

Besides effects originating from the device concept the downscaling of the MOSFET has met technological barriers all the time during the past decades. Most of them have been overcome or were at least delayed with huge expense in device fabrication. We want to have a short look at the most fundamental ones which have to be taken into account for any device.

1. Fringing Fields

The influence of fringing fields was not a major issue until the introduction of high-k dielectrics. To avoid direct tunneling through thin gate oxides (see section “Tunneling leakage currents”) gate dielectrics with high values of their relative permittivity ε were proposed. The idea behind was to achieve the same coupling between the gate electrode and the channel with a larger thickness of the dielectric layers thanks of its ε being larger than that of SiO_2, which was used as gate dielectric so far. Most of the investigated materials provide a relative permittivity not only larger than 3.9 (which corresponds to SiO_2) but also larger than 11.9 (which corresponds to Si). Furthermore the use of high-k[4] materials leads (for very high relative permittivities) to a thickness of the gate dielectric in the range of the channel length. [18]

This causes two different effects which degrade the device properties [19-21]:

The fringing fields between gate and source/drain will increase. This on the one hand induces a electrical field from source to channel and decreases the source-channel-barrier height. On the other hand the increased field from gate to source and drain decrease the gate control on the channel.

The drain-gate overlap together with the thicker high-k dielectric material leads to an increased electrical field between drain and channel compared to SiO_2 as dielectric. The electric field originating from the drain reaches further into the channel, which increases short channel effects caused by this field like DIBL, degradation of subhthreshold slope, and reduction of threshold voltage.

These effects limit the possibilities to support the downscaling process by the use of high-k materials and should be taken into account when designing high-k gate stacks.

2. Statistical Fluctuations of Dopant Concentrations

The next issue we want to have a look on in this section deals with the usefulness of the device in an integrated circuit combined with millions of other devices. For a proper functionality of the integrated circuit it is essential for the electrical parameters of the devices to be in a certain well defined range. Variations in threshold voltage or drain current will result in variations of delay times or even inhibit switching of a certain MOSFET. Furthermore the prediction of OFF currents and therefore static power consumption depends on detailed knowledge about the threshold voltage. And last but not least especially SRAMs need a very precise threshold matching for desired operation.

---

[4] The term “high-k” is widespread in use instead of “high-ε”. In this chapter we use consequently ε as abbreviation for the permittivity and k for the Boltzman constant. However, as “high-k” is commonly used term for

With aggressive downscaling of feature size the intrinsic variations of device properties, especially threshold voltage, became a certain issue. Even devices being next to each other on the same chip may vary by more than 100 mV in V_T. The reason why this problem arose is the strongly decreasing absolute number of doping atoms in the channel region with ongoing scaling of device dimensions. The importance of this topic to semiconductor manufacturing can be seen from the amount of work, which was done over the last years to investigate and describe the influence of statistical dopant fluctuations on MOSFET properties [22-31].

Although the doping level was increased compared to former technology nodes the doping in the depletion region of a state-of-the art MOSFET is only determined by a few hundred doping atoms. The doping in the inversion layer is even far less than 100 doping atoms. With so few atoms determining the electrical properties of the channel region variations in both the absolute number and the positioning of the dopant ions will cause variations of the threshold voltage [29, 30]. IBM presented measurement data obtained from 3,481 identical n-channel MOSFETs fabricated in a 90 nm technology and laid out close together on the same chip [30]. The determination of the threshold voltages resulted in a comparatively wide Gaussian distribution with a maximum at 0.26 V and a standard deviation σ_{VT} of 25.6 mV. Furthermore experimental results proved that σ_{VT} is reciprocally proportional to the square root of the effective gate area. This result is in good correlation to an approximation suggested by *Asenov et. al* [31] giving σ_{VT} by the equation:

$$\sigma_{VT} = 3.19 \cdot 10^{-8} \left(\frac{d_{ox} \cdot N_{Ch}^{0.4}}{\sqrt{L_{eff} \cdot W_{eff}}} [V] \right) \tag{11}$$

were d_{ox}, L_{eff} and W_{eff} are oxide thickness, effective channel length and width, and N_{Ch} the average channel doping.

As the channel volume will further decrease and doping will rise during the miniaturization process, the accelerating increase in threshold voltage variation might become a crucial topic for some applications long before the ultimate scaling limits will be reached. All of the alternative approaches discussed in this chapter differ from the conventional MOSFET, beside other properties, in the doping profiles. This has in all cases of course a huge impact on the statistical variations of device properties.

3. Lithography

In planar device technology the further development of lithography turned out to be the bottle neck when scaling the device dimensions below the wavelength of light sources used for exposure. Optical lithography is the dominant technology used for MOSFET fabrication. The use of wavelengths of 193 nm or even 157 nm combined with advanced mask technologies like phase shifting masks allows to be still in the business even for the 45 nm technology node [101].

A detailed discussion of the challenges in lithography development would go beyond the scope of this chapter. As lithography is a very complex but purely technological limit to

dielectrics having a permittivity larger than SiO_2 this term is used in this case instead of the consequently more correct term "high-ε".

further shrinking and affects all kind of devices we will not deal in detail with this issue here. An excellent overview of optical lithography is given by *Fuller* [32]. *Saitou* [33] and *Silverman* [34] give a very deep insight in the two most investigated alternative technologies, X-ray and E-beam lithography. These would already provide the possibility to realize 10 nm structures in silicon. But so far they do not play an important role in semiconductor mass production as they require extremely high financial investments. If future developments can reduce the costs for these technologies they might become interesting for further shrinking of minimum feature size.

C. Advanced Drift-Diffusion MOSFET Concepts

When facing barriers during the shrinking process all improvements we have seen so far did not touch the fundamental structure of the device. The doping profiles became more complicated and new fabrication techniques were developed but the basic design did not change very much. Device concepts which are taking not only the dopings but also the geometrical shape of the channel region in all three dimensions into account to improve the electrical properties are going a step ahead. We want to have a look at the two most investigated of these advanced drift diffusion driven MOSFETs.

1. FinFET

a. Device Conecpt

Dealing with short channel effects so called multi-gate devices were one of the most investigated suggestions over the last two decades to solve the above described problem. The probably best known device is the so called FinFET. Since it was introduced in 1999 by *Huang et al.*[37] nearly all big semiconductor companies like Hitachi, IBM, Intel (calling it Tri-Gate MOSFET), AMD, Texas Instruments, Infineon Technologies, IMEC or TSM have been working on the device beside several university groups [35-66]. Since 2006 the ITRS officially expects the FinFET to be introduced in fabrication for the 32 nm node [67]. Figure 4 shows a schematic drawing of the FinFET.

The idea of the FinFET is based on the Fully Depleted Lean-Channel Transistor (DELTA), introduced by *Hisamoto et al.* [35, 36] already in 1990. But the big run on the concept did not start until the beginning of the 21st century. The increasing importance of short channel effects in the 90 nm node accelerated the investigations on this alternative concept. The transistor itself consists of a silicon fin (typical height H_{fin}=50 nm, typical width W_{fin}<20 nm), giving the device its name, which acts as channel region, and source and drain regions first realized in highly doped poly silicon [37,38]. Because these source and drain regions are much thicker than the fin height the device is often called to be "quasi planar". This makes it easy to integrate the device in an existing CMOS process flow including passivation and metal layers by adopting an existing mask set to the FinFET's special requirements [52].

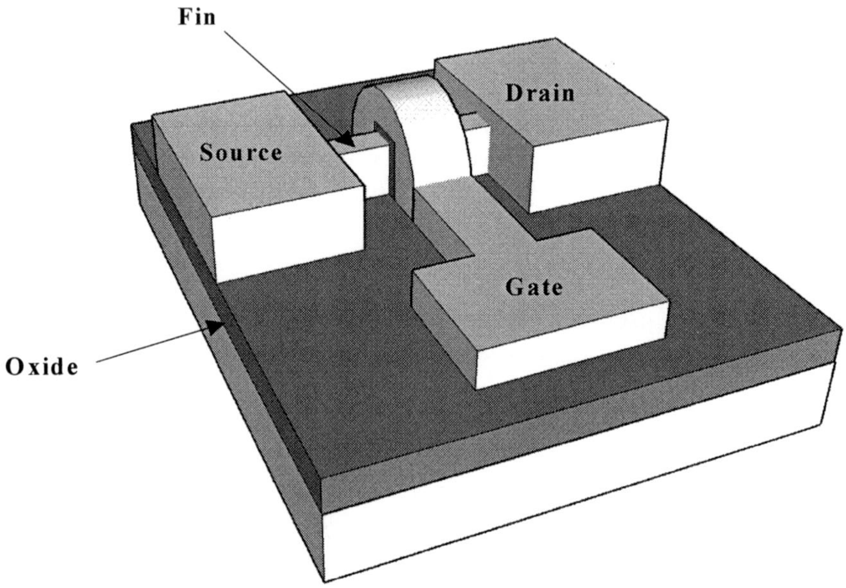

Figure 4. Schematic drawing of the FinFET.

The great advantage of the FinFET is its ultra thin channel, which can be fully depleted and because of the surrounding gate the barrier can be controlled much more efficiently than in a conventional MOSFET. A typically process flow for the production of FinFETs for the 22 nm node can be described as follows [59, 63, 64]:

The fins are fabricated from an SOI substrate to provide electrical isolation between the devices. The thickness of the device layer is reduced to the desired fin height by thermal oxidation of the spare silicon. Although a higher fin would be advantageous with respect to the maximum drive current practically a typical fin height of 50 nm is aspired. If the fin is higher the device cannot be produced in a quasi planar design. Channel doping can be achieved from defined doping of the device layer and implantation for opposite doping.

The fabrication of the fin is the most critical part of the process flow. The fin has to be very thin to achieve full depletion which means for the 22 nm node 15 nm or less. Furthermore with respect to variations in device parameters the fins should be very uniformly with very low sidewall roughness [64]. Finally multi fin devices are state of the art rising the drive current by connecting source and drain regions via several fins in parallel. Therefore the parallel fins should be as close together as possible to keep the area necessary for the device layout as low as possible.

Sidewall pattern transfer (SWT) today is commonly used to produce the thin and narrow fins [40, 47, 50, 52, 59]. This technique allows the fabrication of very thin lines by defining them through spacers deposited on the sidewalls of a patterned sacrificial layer. First a hardmask layer is deposited on the device layer, typical silicon nitride (Si_3N_4), optionally with a thin thermal oxide interface layer to relieve stress between the nitride mask and the silicon. For the definition of the fins a sacrificial SiGe layer is deposited using CVD[5]. By electron

[5] CVD: chemical vapour deposition

beam or laser lithography prior to reactive ion etching the SiGe layer is patterned to be used to support the spacers. PSG[6] is deposited and etched anisotropicelly to define the spacers on the sidewalls of the SiGe patterns. After removing the SiGe (e. g. using $H_20:H_20_2:NH_4OH$ 5:1:1 at 75 °C [40]) the PSG is used as hardmask to fabricate the fins by plasma etching. Finally PSG is removed while silicon nitride remains on the fins. The Si_3N_4 acts as CMP[7] stopper for the gate poly silicon planarization, what is necessary if the gate should be produced using SWT. Furthermore high fields at the corners of the fins are reduced and the fin is protected during the gate etching process [59].

First step of the gate stack fabrication is the growth of thermal oxide or nitrided oxide. Second highly doped poly silicon is deposited all over the wafer and planarized by CMP to the height of the fins. A thin layer of highly doped poly silicon is deposited on the planarized surface to provide electrical contact between the gates on both sides of the fin. Now using SWT the gate is defined with a pattern positioning lines in a 90° angle with respect to the fins. Before growing silicon to form source and drain the gate has to be protected by isolating spacers on both sides. Selective growth of crystalline silicon on the fins is used to fabricate source and drain and finally deposition of highly doped poly silicon ensures good contacts. For further improvement of drain and source contact resistance implantation and silicidation was suggested prior to selective epitaxial growth and showed good results [64].

Earlier reported gate, source and drain fabrication process flows consisted of deposition of poly silicon on the fin and opening a trench by anisotropic etching which was filled up by the gate material [38]. The disadvantage of this method is the inaccuracy of lithography to define the gate length. The most significant variations in V_T result from variations in gate length and fin width [64]. SWT results in the lowest possible variation in line width [59], which means the lowest possible variation in fin width when used for fin fabrication and the lowest possible variation in gate length when used for gate fabrication.

SWT furthermore increases the fin density by a factor of 2 for constant minimum feature size as a fin is masked at both sides of each sacrificial SiGe line. Theoretically this results in a factor of 4 current gain compared to planar MOSFET fabricated on the same area (2 gates on 2 fins consume the same chip space as a conventional MOSFET gate if both were fabricated using minimum feature size to define the structures).

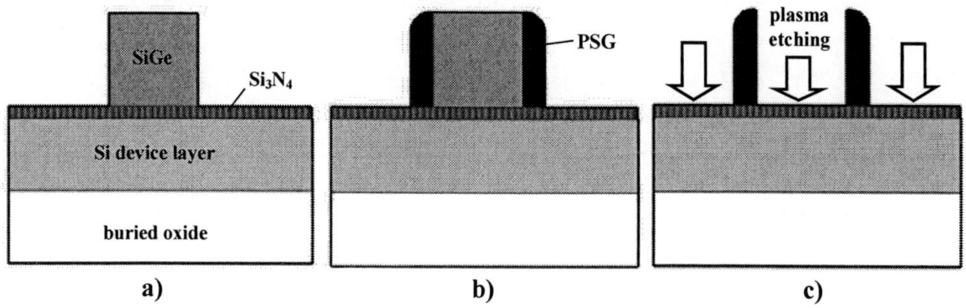

Figure 5. Continued on next page.

[6] PSG: phosphosilicate glass
[7] CMP: chemical mechanical polishing

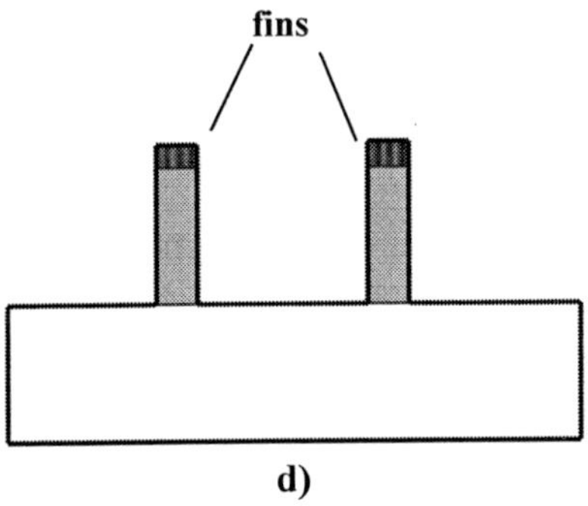

Figure 5. Schematic illustration of the SWT process flow: a) On a SOI with silicon nitride layer the sacrificial SiGe layer is deposited and patternd. b) In the next step PSG is deposited on the sidewalls of the SiGe structures. c) The SiGe is removed and the remaining PSG is used as mask d) to define the fins.

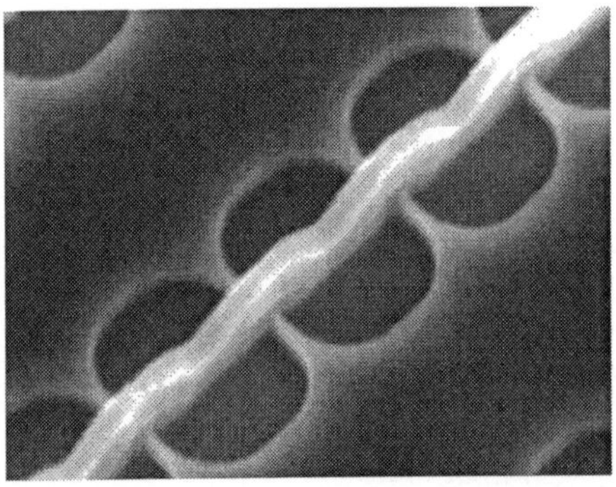

Figure 6. SEM image of mulitgate FinFET, Source: Infineon Technologies, © 2007.

b. Properties and Performance

The unique properties of this device derive from its three-dimensional channel structures. The gate surrounding the channel on three sides together with the very low thickness of the fin allows an effective gate control of the potential within the complete channel volume.

Operating as a fully depleted MOSFET the FinFET shows excellent suppression of short channel effects. First of all punch-through will obviously not occur in this structure as the punch-through-pass via the bulk is missing. Furthermore the excellent gate control of the electric fields in the channel region inhibits drain induced barrier lowering (DIBL) even for comparatively low channel doping and short channels. Necessary for a good DIBL suppression is a significant high L/W_{fin} ratio to ensure the domination of the gate field over the drain source field. Experimental results from different groups demonstrated that $L/W_{fin}=2$ is a good tradeoff between shrinking and short channel effect suppression. The results are compared in table 1.

Table I. Comparison of FinFET Parameters

	Ref. [40]		Ref. [42]	Ref. [44]	Ref. [45]		Ref. [53][8]
Gate length L [nm]	25	80	100	100	10	100	10
Fin width W_{fin} [nm]	26	26	38	25	21	21	10
Supply voltage [V]	1.0	1.0	1.5	1.25	1.2	1.2	1.0
DIBL [mV/V]	180	30	57	25	71	11	80
S [mV/dec]	80	65	68	70	125	64	75

Due to the excellent gate control of the potential in the channel region these low DIBL values can be realized with comparatively low substrate doping levels. Typical levels of doping in the channel are between 10^{16} and 10^{17} cm^{-3} compared to $5\cdot10^{18}$ cm^{-3} necessary in sub-50 nm conventional MOSFETs [69]. As easy obtainable from equation (12) a decreased channel doping leads to a decreased standard deviation of the threshold voltage. In principle even the use of undoped channels with setting V_T only by work function engineering is possible [70]. This would make variations of the device parameters independent of dopant fluctuations as long as unintentional impurities are well controlled [65]. Furthermore the low channel doping excellently prevents leakage currents caused by Zener tunneling.

For the thin and fully depleted body of the device the depletion capacitance can no longer be calculated according to (6). Because the current transport mechanism is the same as in a conventional MOSFET, the current is still diffusion limited. The theoretical limit for the slope as given by equation (9) is still 60 mV/dec at room temperature. Because the channel thickness is much smaller than the depletion layer width (which actually means "fully depleted") the depletion capacitance becomes very small and the slope is decreased according to equation (9) [36]. Furthermore the capacitive coupling between the two gates on both sides of the fin dominates the subthreshold behavior of the device compared to the influence of the depleted film capacitance [68]. This results in experimentally verified subthreshold slopes close to the theoretical limit. Detailed values, which were obtained from different prototypes, are cited in table I.

Furthermore the source drain tunneling leakage current is proportional to the cross sectional channel area. This means a FinFET fabricated using a fin width of 10 nm is expected to have the same source drain tunneling leakage current as an conventional SOI MOSFET with equivalent gate width if the SOI MOSFET is build on a 5 nm device layer. In this regard the FinFET provides an advantage as fins of 10 nm width are much more reproducible in fabrication using SWT than SOI substrates with a 5 nm device layer. Furthermore for ultra thin double gate devices simulations predict a reduced tunneling current due to reduced drain induced barrier thinning [21]. This is not a major issue as it does not become valid before shrinking the body thickness below 3 nm but should be taken into account, too.

The FinFET structure can also be used to enhance the total output current by combining multiple fins to one transistor. A FinFET produced using SWT with the sacrificial pattern defined in minimum feature size puts two fins on the same chip space as one conventional MOSFET channel in minimum feature size. As the fin height defines the width of one gate

[8] The FinFET presented in this work is not built on a classical fin with a rectangular cross section but on a silicon nanowire. Therefore the fin width given here is the diameter of this wire.

and this structure consists of totally four gates a multiplication of a factor of 4 – 8 can be achieved, depending on technology node and fin height.

State of the art FinFETs in the 22 nm node offer an off state leakage current as low as 100 nA/µm combined with a drive current of 960 µA/µm as recently reported by Toshiba [64]. This hits the ITRS requirements for low power applications up to 2020 [67].

2. Silicon-on-Nothing MOSFET (SON MOSFET)

a. Device Concept

The silicon-on-nothing (SON) MOSFET (sometimes only refered as SONFET) was first introduced in 1999 by *Jurczak et al.* [71]. Its approach is similar to the FinFET by reducing short channel effects and improving device performance by the use of an ultra thin fully depleted body. Since its first introduction different groups investigated and improved the device [71 – 95]. The dominating part of this work focuses on two different approaches, a lateral and a vertical one. Both define the thickness of the transistor body by epitaxial growth in contrast to the FinFET were the body thickness is commonly defined by lithography and spacer formation.

The lateral SON MOSFET concept is the earlier reported one. The development of this device aimed at an SOI transistor with an ultra thin channel on an ultra thin buried oxide. Simulations proved that a thick buried oxide reduced the positive influence on short channel effects of an SOI transistor for short channels in the sub-50 nm regime even when the device layer was thin enough to achieve full depletion. The electrostatic coupling between source and drain through a thick buried oxide layer leads to a equipotential distribution which is comparable to DIBL conditions in a conventional short channel device [73, 82]. This is not the case if the buried oxide is thin, which means in the range of 20 nm.

The following two sub sections describe the fabrication of both the lateral and the vertical SON MOSFET using a sacrificial strained SiGe layer. There were other methods reported like the Empty Space in Silicon (ESS) technique [74], the Si layer transfer over a pre-etched cavity [90] or void formation by He^+ implantation [91]. But the use of sacrificial layers is the most widespread method and therefore this chapter will focus on this technique.

b. Lateral SON MOSFET Fabrication

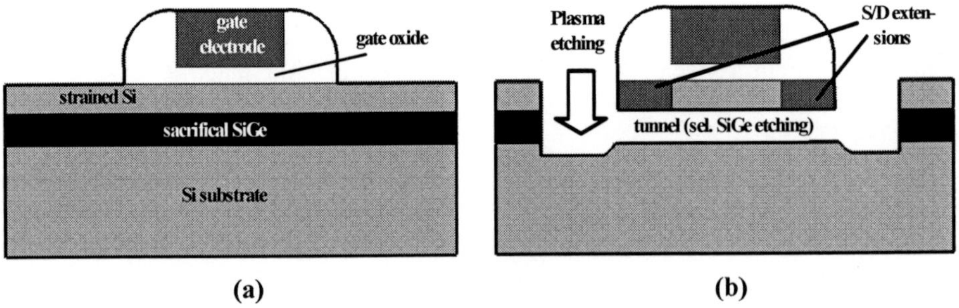

Figure 7. Continued on next page.

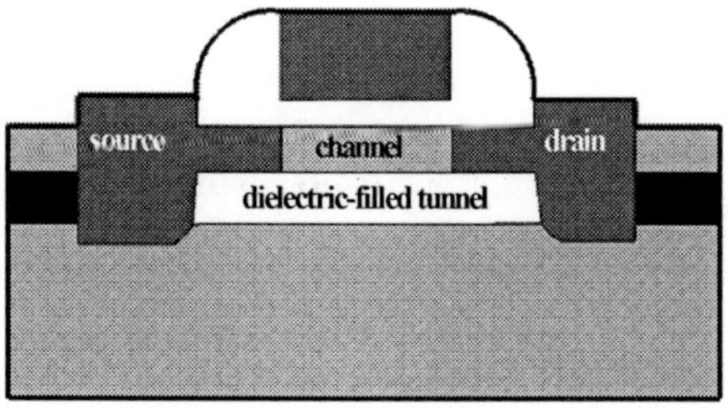

(c)

Figure 7. Fabrication flow of a lateral SON MOSFET: a) A conventional CMOS gate stack is fabricated on the Si/SiGe/Si-substrate. b) After fabrication of S/D extension by ion implantation the source and drain region are opend using plasma etching. The sacrifical SiGe layer is removed by selectiv etching. c) The remaining tunnel is filled with a dielectric and source and drain are fabircated using slective epitaxy.

The fabrication process is quite similar to a conventional MOSFET in most steps beside the formation of the thin buried oxide. Very detailed information on the process flow is provided by *Jurczak et al.*[73] and *Monfray et al.*[82]. Starting from bulk Si wafers STI[9] is prepared for later device isolation. A layer of strained SiGe, typically 20 nm, is grown epitaxially on these substrates followed by a thin layer, typically <15 nm, of Si. This silicon layer will be the channel of the device later.

On this substrate a conventional CMOS gate stack is fabricated consisting of gate dielectric, electrode and spacers. After the implantation of the source and drain extensions anisotropic etching, commonly plasma etching, is used to open trenches in the source and drain regions. The depth of these trenches is below the buried SiGe layer so that this layer is accessible. Next selective wet chemical etching or plasma etching is used to remove the sacrificial layer so that a silicon channel "on nothing" is formed. The empty tunnel below the channel can be left open with air as dielectric or can be filled by any solid dielectric. Mostly described is the use of an oxide-nitride-oxide sandwich to fill the tunnel. After a cleaning step selective epitaxial growth is used to form the source and drain regions. Doping of source and drain can either be done in-situ or by ion implantation. The backend processing is equivalent to a conventional MOSFET. Figure 7 shows a schematic overview of the fabrication process.

c. Vertical SON MOSFET Fabrication

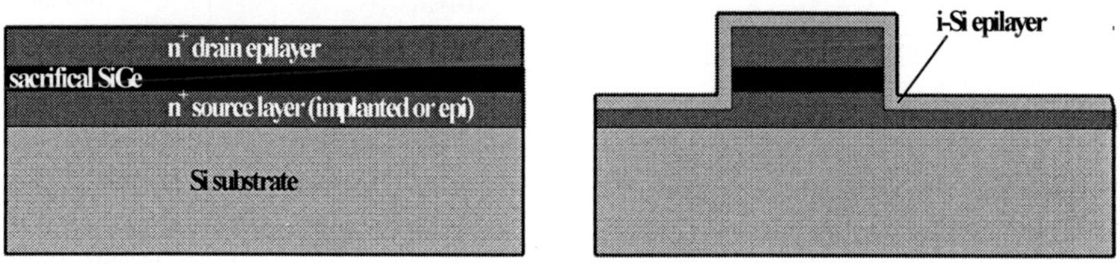

a) b)

Figure 8. Continued on next page.

[9] STI: shallow trench isolation

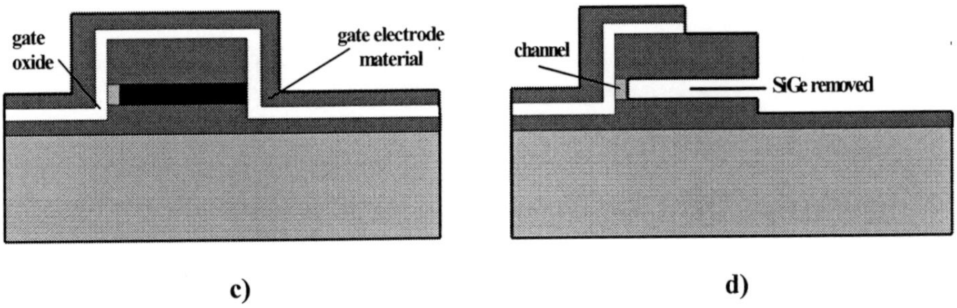

Figure 8. Fabrication flow of a vertical n-channel SON MOSFET: a) A n^+-Si/SiGe/n^+-Si layer stack is fabricated. b) This layer stack is patterned and the mesa is overgrown with a thin i-Si layer. c) Gate oxide is grown on this layer and the gate electrode material is deposited. Diffusion during high temperature step leads to an n-doping of the i-Si layer beside the part on the SiGe, which will form the channel. d) After patterning the gate the SiGe is removed using selective etching.

The vertical SON MOSFET concept was first presented by *Schulze et al.* [85] in 2004. It is based on the vertical transistor concept developed at the Bundeswehr University Munich, Germany, in 1993 by *Gossner et al.* [96]. The basic idea is to control the channel length by means of epitaxy and not lithography which makes it much easier to aim at channel lengths below 50 nm as epitaxy is well controllable in the range of nm.

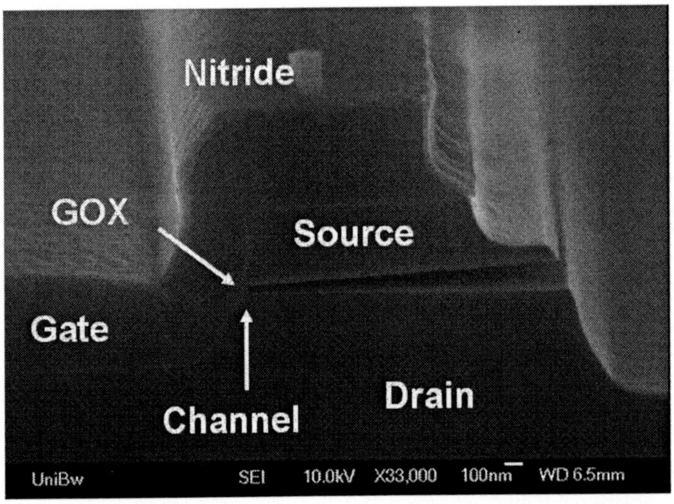

Figure 9. SEM Image of a vertical 30 nm SON MOSFET presented by *Höllt et al.* [94]

This technology can be used for the fabrication of fully depleted SON MOSFETs based on the selective etching of SiGe. Therefore on a conventional p-doped substrate the drain regions are defined by fabricating a SiO_2 or Si_3N_4 hardmask to cover the area beside the drain and setting the drain doping by ion implantation or diffusion from a phosphorus spin-on glass layer. After removal of mask and optional doping glass layer the sacrificial SiGe layer is grown on top followed by an in-situ n-doped silicon layer which will be used to form the source. From this layer stack mesas are fabricated by reactive ion etching. On this structure a intrinsic silicon layer is grown by CVD. By an RTP[10] step this layer is doped by diffusion of dopant atoms from source and drain except the region grown on the SiGe layer so that an undoped channel region is remaining. An RTP gate oxide is grown on top of this layer. Finally poly silicon is deposited by CVD to form the gate electrode. The gate stack is fabricated using an additional hard mask and wet chemical etching to remove poly silicon and gate oxide. After protecting the gate stack with a Si_3N_4 passivation layer the sacrificial SiGe layer is removed by wet chemical etching and the gap is filled with an RTP oxide like it is done with the lateral version. Figures 8 and 9 show the schematic fabrication flow of a vertical SON MOSFET and a SEM[11] image of the resulting device, respectively.

d. Properties and Performance

Similar to the FinFET the device takes advantage of the fully depleted channel region. This leads to excellent suppression of SCE and DIBL, improves high frequency behavior and makes the device interesting for radiation-hard applications.

The lateral type, which was presented first, is also the one with more experimental data being published. So far a channel length as short as 38 nm with a 9 nm thick channel was demonstrated [78]. Of more technological relevance than this record small prototype device is the demonstration of a complete CMOS node in SON technology including fully operational SRAM cells with 65 nm NMOS gate length and 55 nm PMOS respectively. These were presented 2004 by *Monfray et al.* [86]. As expected these devices show a very good suppression of SCE and a DIBL of less than 80 mV/V. Drive current and especially OFF state leakage current make them competitive to conventional SOI MOSFETs. A detailed overview of the device parameters is given in table II.

On the vertical SON MOSFET there is not as much work published as on the lateral one which is not astonishing as its first introduction was more than four years later. Simulations predict an excellent SCE suppression and high frequency operation at $f_T > 100$ GHz for a 30 nm device [87] but the first prototype at these geometrical dimension was not published until 2008 [94]. These devices suffered from parasitic pn-diodes formed between source and drain by remaining SiGe of the sacrificial layer after etching. The leakage current through these diodes degrades the device performance, which is the main reason for the comparatively bad parameters given in table II. Simulations indeed predict a drive current of 1980 µA/µm, a DIBL of 73 mV/V and a subthreshold slope as low as 75 mV/dec [94].

Furthermore the vertical SON MOSFET offers a new possibility of three-dimensional integration of a CMOS inverter. This approach was proposed 2005 by *Schulze* [95] including a process flow for its realization. A midgap metal is used to form the gate electrode which is put into the etched tunnel between source and drain. The disadvantage of this integration technique will be the overlap capacities between gate-source and gate-drain. But so far this issue was not further investigated. Figure 10 shows a schematic drawing of such an 3-D SON MOSFET inverter.

Beside these two SON MOSFET concepts other approaches were presented to realize "real" silicon-on-nothing channels [74, 90]. Their main advantage compared to SOI technologies is a better suppression of self-heating effects but the channel layers are not as

---

[10] RTP: rapid thermal processing

[11] SEM: scanning electron microscopy

thin as they can be for channels grown epitaxially on sacrificial layers. Table II gives their electrical parameters for comparison.

Table II. Comparison of SONFET Parameters

	Ref. [74]		Ref. [90]	Ref. [86]		Ref. [94]
	nMOS	pMOS		nMOS	pMOS	
Gate length L [nm]	250	250	2000	65	55	30
Channel thickness [nm]	350	350	100	17	17	9
Supply voltage [V]	1.5	1.5	1.0	1.2	1.2	1.1
Drive current [μA/μm]	320	105	450	1100	350	30
Off state leakage [nA/μm]	0.01	0.01	0.02	138	20	10
DIBL [mV/V]	-	-	68	80	80	-

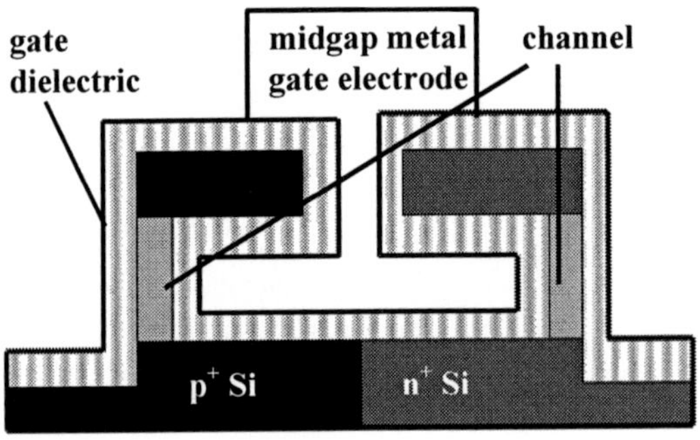

Figure 10. Schematic drawing of the 3D vertical SON MOSFET CMOS inverter according to [95]. The left part of the structure is the p-channel SON MOSFET, the right part is the n-channel SON MOSFET. The common gate electrode forms the inverter input.

III. Fundamental Physical Limits of MOSFET Operation

Beside technological limitations and barriers of the drift-diffusion driven carrier transport concept there are fundamental physical limits which can not overcome by improving the device design. Their understanding is very important as they might affect any device which controls a current flow by controlling a charge carrier barrier.

A. Tunneling Leakage Currents

Tunneling leakage currents occur at several points in the MOSFET and are becoming more important with shrinking the device dimensions. Tunneling is a quantum mechanical phenomenon allowing a charge carrier to pass a barrier although its energy is not sufficient in the classical approach. The resulting tunneling current depends on barrier width and height as well as on the applied voltage. Most important for semiconductor devices is tunneling through an insulating layer and band-to-band tunneling in the semiconductor. Both phenomena have been intensively investigated and analytical models were published already in the 1960s [97-99].

Most important in today's CMOS technology is the gate oxide leakage current. In modern devices using a 1.2 nm gate oxide current densities of 100 A/cm^2 [100] are exceeded. This is responsible for a significant part of the total power consumption. The use of high-k gate dielectrics is the common approach to deal with this problem and Intel was the first to introduce Hf-oxide in a microprocessor production flow in 2007 [101]. As this solution is again improving technology without changing the concept we will not deal with it in this chapter.

The tunneling mechanism, which sets the ultimate limit to MOSFET scaling is the direct tunneling of electrons from source to drain [102-105]. The typical behavior of a MOSFET is the control of the potential barrier height between source and drain by the gate voltage. If the distance between source and drain is rather short, electrons can tunnel directly through this barrier and the current flow through the channel region of the device is no longer controllable by the gate voltage. Simulations predict the total failure of transistor operation for a channel length <10 nm [102, 103]. In 2000 *Kawaura et al.*[102] were first able to measure the source-drain-leakage current in a sub-10 nm MOSFET and proved its significant influence of this current starting at 10 nm and exceeding the thermal current at 5 nm gate length.

A third tunneling junction that limits the functionality of a conventional MOSFET is band-to-band tunneling at reverse biased highly doped p-n-junctions (so called Zener tunneling). This leads to a shortcut-like behavior of this structure which is actually intended to act as barrier for charge carriers. To understand this behavior, which begins at doping levels of 10^{19} cm^{-3}, a closer look at the energy band diagrams like shown in figure 11 is helpful.

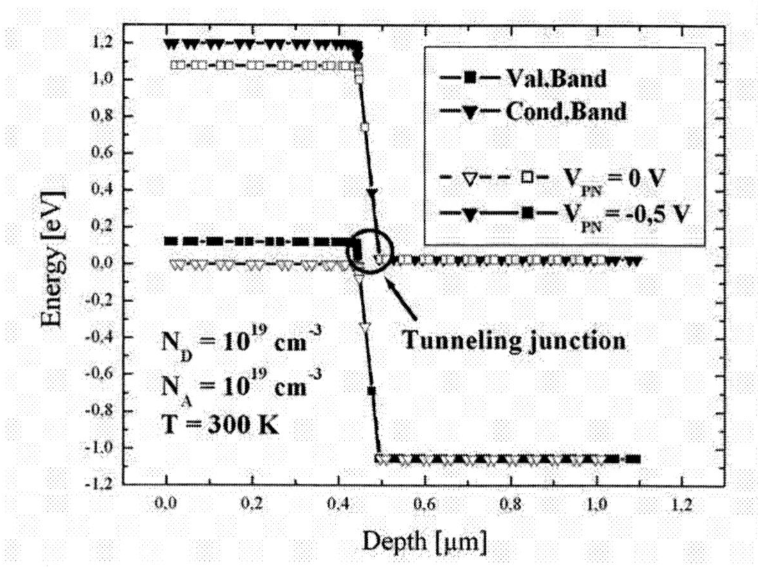

Figure 11. Simulated energy band diagrams of a highly doped pn-junction without biasing and reverse biased. Only a small voltage results in the formation of a tunneling junction causing this doping structure to behave like a shortcut instead of a charge carrier barrier.

The valence band edge energy on the p-doped side and the conduction band edge energy on the n-doped side only differ by a few 10 mV for high doping levels as obtained from ATLAS [107] simulation results shown in figure 11. Applying a small voltage in reverse bias leads to equal energy levels in both the conduction band on the n-doped side and the valence band on the p-doped side. Due to the very close distance electrons can tunnel from the p- to the n-doped side making such a doping profile useless to act as barrier. This means a limitation for the maximum doping levels for pn-junctions in the MOSFET, e. g. the source-channel or drain-channel junction. On the one hand it affects of course the possible range of scaling the doping and on the other hand this is a reason for a very low range of possibilities to optimize a MOSFET for high temperature applications as will be discussed in the following sub section.

B. High Temperature pn-Barrier Lowering

Disregarding any scaling issues a MOSFET fabricated in silicon in general is not very suitable for high temperature applications. This is caused by the temperature dependence of the potential barrier formed by a pn-junction. The barrier height is determined by the difference in the Fermi levels of p and n type silicon. The explicit value of the Fermi level difference can be obtained from the equation:

$$W_{Fn} - W_{Fp} = W_g - kT \cdot \ln \frac{N_C \cdot N_V}{N_D \cdot N_A} \tag{11}$$

W_{Fn} and W_{Fp} are the Fermi levels in n- and p-type silicon respectively, W_g is the band gap, N_C and N_V are the density of states in the conduction and valence band, respectively, and N_D and N_A are the donor and acceptor concentrations in n- and p-type silicon. Equation (11) can be easily calculated from the equation for the difference between Fermi level and valence band energy in p-type silicon and the difference between conduction band energy and Fermi level in n-type silicon [7, p. 27].

We see from (11) that the difference decreases with rising temperature as kT, N_C and N_V are increased with rising temperature. Furthermore this effect is less pronounced for higher doping levels N_D and N_A. Therefore the first barrier to be expected for thermal breakdown in CMOS technology is the one with the lowest levels of doping, i. e. the well-substrate barrier (see figure 12). Figure 14 presents ATLAS simulations of the junction barrier with typical doping levels of 10^{15} to 10^{16} cm^{-3}. At 200 °C the barrier is almost completely vanished what will lead to a rather high bulk leakage current.

The demand for a better p- and n-channel device isolation can be fulfilled by the use of SOI[12] substrates instead of a n-well doping as illustrated by figure 13. In this case no well-substrate junction exists, and one could now think of a rather high channel doping to achieve a good temperature stability. But beside other problems like quite high threshold voltage, a channel doping of 10^{19} cm^{-3} will lead to a high source-drain leakage current due to band-to-band tunneling as we have already seen in the previous subsection.

[12] SOI: silicon on insulator

All of this problems do not begin at a certain point of shrinking the device but are inherent in the conventional MOSFET concept because of its direct pn-junctions. Without fundamental changes in the device structure we can only adopt the device in a very small parameter range to high temperature applications.

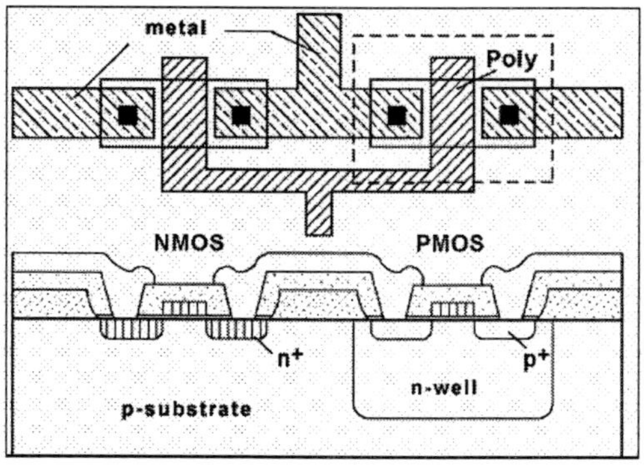

Figure 12. Side and top view of a conventional CMOS inverter structure with typical doping profiles.

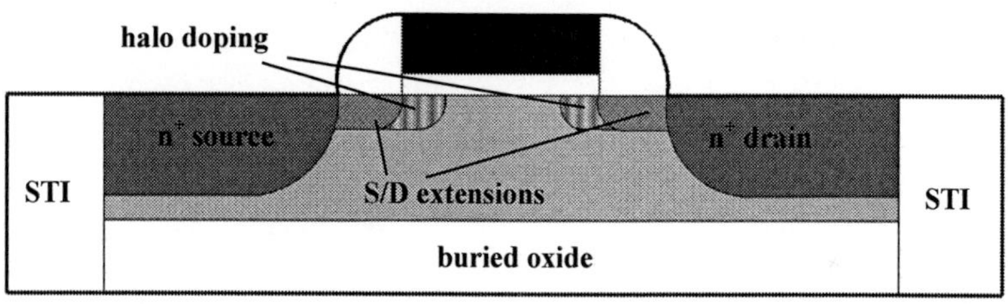

Figure 13. Schematic drawing of a MOSFET realzied on a SOI substrate. The burried oxide and the shallow trench isolation (STI) is used instead of the well doping.

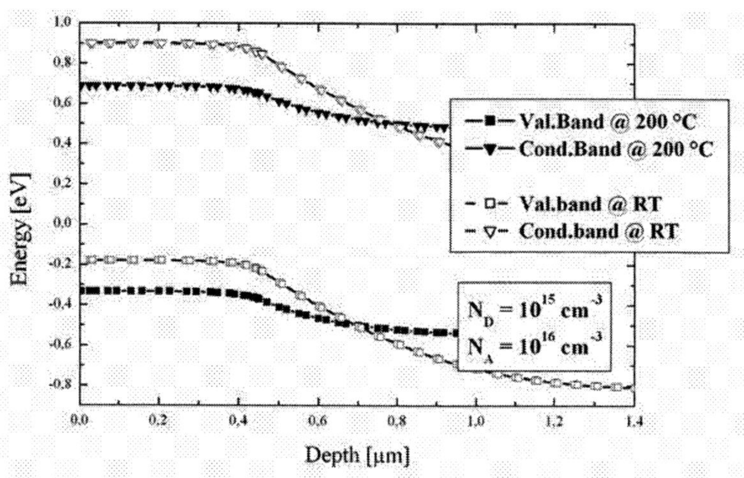

Figure 14. Simulated band energy diagrams with typical doping levels for a well-substrate pn-junction. Simulated at room temperature (RT) and at 200 °C. For high temperatures the barrier is almoust competely vanished.

C. Impact Ionization Effects

1. Avalanche Breakdown

Besides Zener tunneling (see section III.A) avalanche breakdown is the dominant breakdown mechanism of MOSFETs in the off state and reverse biased pn-junction. Avalanche breakdown is induced by impact ionization and occurs due to the exponential increase in the multiplication factor of charge carriers with increasing electric field between source and drain [109] and across the pn-junction respectively [108].

Breakdown actually occurs when the multiplication factor tends towards infinity. Impact ionization, especially by electrons, which show significant higher ionization rates than holes in silicon [110], begins at much lower field strength. This leads to a rising leakage current and possibly to damage of gate oxide by hot electron injection. The avalanche breakdown field in silicon was determined to approximately 300 kV/cm [7, p. 850].

Short channel devices show a disproportionately low breakdown voltage. In a long channel NMOS device the holes generated near the drain via impact ionization flow towards the body contact of the transistor while the electrons drift towards the drain region. The same in principle happens in a short channel device. But as the mobility of holes is lower than of electrons they remain in the channel region longer than the electrons. In case of a short channel device this leads to a earlier breakdown by a drain-source-barrier lowering induced by this hole charge or even by a bipolar latch-up effect [111].

2. The Kink Effect

Charging of the channel region by carriers created by impact ionization is also responsible for a hot carrier phenomenon called kink effect. The kink effect only occurs in devices with a floating body and was first described for SOS[13] devices [112, 113]. As the impact ionization rate in silicon is higher for electrons than for holes [110] the effect is more important for n-channel devices but occurs in p-channel SOI transistors, too [114]. Therefore the physics described for a NMOS device here also applies to a PMOSFET just with the roles of the charge carriers changed:

For a drain source voltage high enough to achieve impact ionization at the drain side of the channel region of the transistor (but far below the avalanche breakdown voltage) electron-hole pairs are generated. The electrons immediately flow towards the drain while the holes remain in the channel region. They accumulate at the buried oxide near the drain and lower the threshold voltage. This leads to an abrupt increase in the drain current and a "kink" in the I_D-V_D characteristics.

Depending on device geometry and doping a second kink can occur at higher drain-source voltage but still below avalanche breakdown. This is due to a parasitic bipolar transistor. When the drain voltage is increased beyond the first kink the hole current will increase until the body-source diode is in forward bias, which is the case for a body potential of approximately 0.7 V. At this point the body potential will not change any longer but electron injection from the source will increase further with drain voltage leading to an increase in the drain current.

[13] SOS: silicon on sapphire

Both effects are shown in the qualitative I_D-V_D characteristics shown in figure 15.

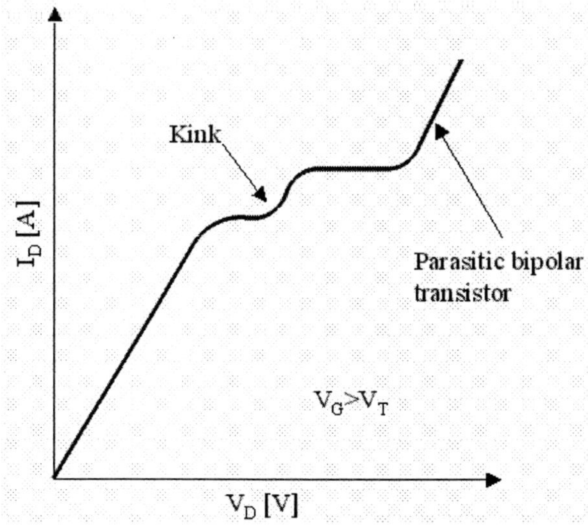

Figure 15. Observation of the Kink effect in the I_D-V_D characteristics of a SOI MOSFET.

D. Heisenberg Limit

An active electronic device needs at least one energy barrier to control charge carrier flow. The device becomes useless if passing of this barrier by electrons or holes is no longer controllable. Such a barrier is defined by its height and its width. Below a certain barrier width the Heisenberg uncertainty principle leads to a uncontrollable tunneling of electrons through the barrier.

Theoretical approaches, which followed this idea to determine the ultimate minimum dimensions of an electronic device predicted a minimum barrier width of 1.4 nm [115,116]. As we have explained in section III.A, in case of a conventional MOSFET the leakage current caused by direct source-drain tunneling sets a limit at approximately 5 nm gate length. Therefore one should have the existence of the Heisenberg uncertainty principle as ultimate limit in mind but it is not of practical relevance for MOSFET scaling.

IV. Novel Device Concepts

Several approaches have been proposed to overcome the MOSFET limits. Beside the already described improvements of the drift diffusion driven MOSFET completely different field effect transistors concepts were introduced. All devices described here take one of the serious deficits of the conventional MOSFET and turn it into an advantage.

A. p-i-n I-MOS

1. Device Concept

The (p-i-n structured) Impact Ionization MOSFET (p-i-n I-MOS) was first introduced in 2002 by *Gopalakrishnan et al.* [117] followed by publications on improvement, modeling, and device physics by several groups [118-140]. The device concept focuses on overcoming the subthreshold slope limitation of the conventional MOSFET. Therefore, instead of drift-diffusion, avalanche breakdown is used as current transport mechanism.

The device is structured as a p-i-n diode with a gate stack on part of the intrinsic region next to the drain side. For a n-channel device, drain is formed by the n-region and for a p-channel device by the p-region of the diode. By applying an appropriate gate voltage a channel of the same charge carrier type like the drain is formed. Thereby the length of the intrinsic zone and the breakdown voltage in reverse bias can be modulated.

For the desired operation as a transistor with steep subthreshold slope the device is reverse biased at a drain source voltage which is below the breakdown voltage of the intrinsic p-i-n diode but well above the modulated breakdown voltage. Under these conditions the reduction of the effective length of the intrinsic region leads to an increase of the electric field between source and drain above avalanche multiplication field strength. This leads to a amplification effect of drain current and a very sharp turn on behavior.

The fabrication of the p-i-n I-MOS (see figure 16) is commonly done in form of a lateral design by self-aligned source and drain fabrication by ion implantation after gate stack formation. Therefore this device is in some publications referred to as "lateral I-MOS" in contrast to the n-i-p-i-n I-MOS which uses a different working principle and is reported so far only vertically (see section IV.B).

After gate stack fabrication first the intrinsic and source regions are protected by photo resist and the drain region is defined by ion implantation. In the next step drain and intrinsic region are covered by photo resist (i. e. a kind of photo resist spacer is formed next to the gate stack to define the length of the intrinsic region) and the source doping is implanted. Contacts are fabricated identically to a conventional MOSFET. Compared to a CMOS process there is just one additional mask and implantation step necessary to realize the opposite doping of source and drain.

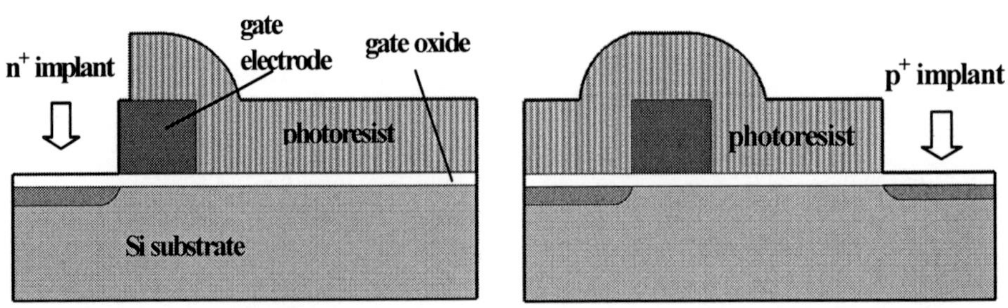

a) b)

Figure 16. Continued on next page.

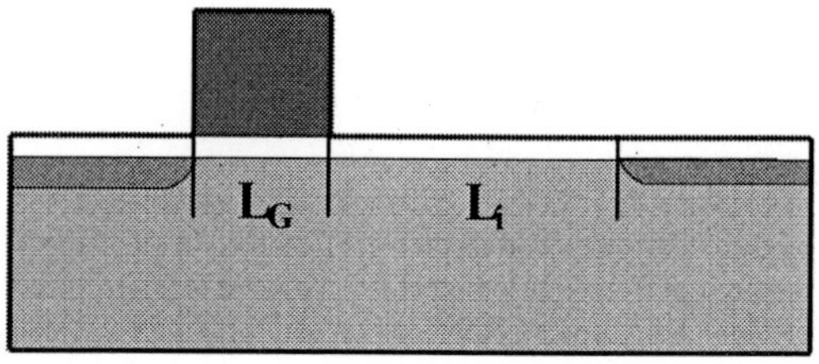

c)

Figure 16. Schematic process flow for fabrication of a classical p-channel p-i-n I-MOS: a) After gate oxide growth and gate electrode depostion and pattering the n-region is implanted. b) A photoresist spacer is used to define the intrinsic region and the p-region is implanted. c) The gate defines the gate length L_G and the photoresist spacer defines the length of the intrinsic region L_i.

Alternatively *Choi et al.* [118, 119] proposed a process flow where an elevated drain region was realized by means of epitaxy (see figure 17). Therefore a highly doped layer is grown on the substrate, a mesa with a height approximately equivalent to the epi layer thickness is etched and the gate stack is formed next to this mesa. The intrinsic region is then defined by a TEOS sidewall spacer and the source is again fabricated by ion implantation and RTA[14] (which is also necessary to achieve a diffusion of the drain doping down to the channel level). The advantage of this process flow is the abdication of one implantation step but on the other hand side it needs a very well controlled etching process.

Unfortunately the use of avalanche breakdown as switching mechanism leads inescapably to hot charge carriers in the channel region as they are necessary to achieve impact ionization. The injection of these carries into the gate oxide leads to a shift of the threshold voltage after only five switching cycles of the device [121]. *Toh et al.* [122] suggested a device design were the impact ionization region is realized in SiGe and "L" shaped as shown in figure 18. This leads to a hot electron distribution concentrated at the source side with strongly decreasing energy towards the gate oxide. Simulations predict an extremely reduced hot carrier degradation of this device. Although a experimental proof was not reported so far.

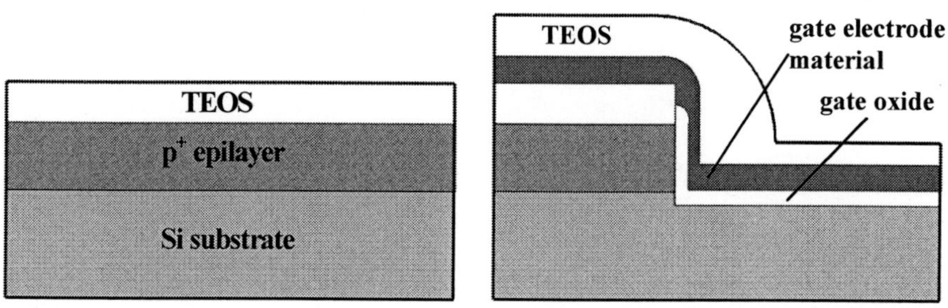

a) b)

[14] RTA: rapid thermal annealing

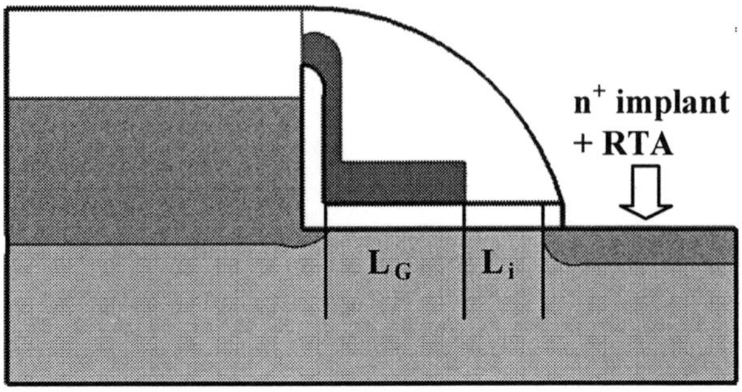

c)

Figure 17. Schematic process flow of the self-aligned p-channel p-i-n I-MOS with elevated drain: a) A p^+ layer is grown epitaxially and a TEOS oxide is depostied on top. b) After patterning the gate oxide is grown, the gate electrode material is deposited and a TEOS spacer is formed. c) After patterning the gate the n^+ region is fabricated using implantation. The RTA step after implantation leads to the necessary diffuson of p-dopants down on channel level.

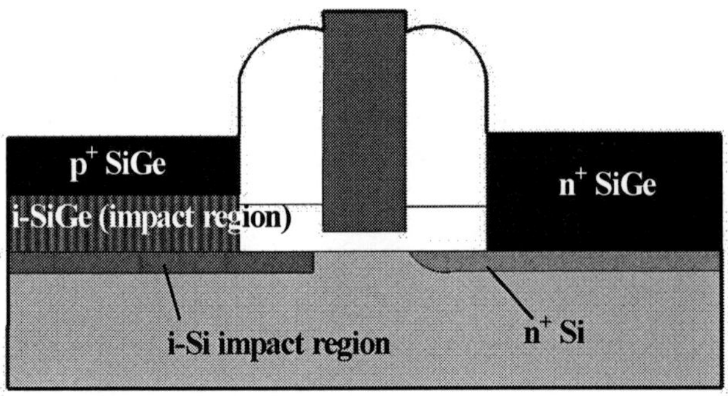

Figure 18. I-MOS with L-shaped impact ionization region and SiGe source/drain regions grown by selective epitaxy.

2. Properties and Performance

The use of avalanche breakdown as turn-on mechanism results in a very steep subthreshold slope. Once the field in the intrinsic region rises above the critical limit of approximately 10^5 V/cm [7, p. 850] the device switches from off to on state at once. Figure 19 shows one of the earliest reported I_D-V_G characteristics published by *Gopalakrishnan et al.* [121] in 2005. The demonstrated slope was approximately 10 mV/dec and so far values as low as 2 mV/dec were experimentally verified [129] (see also table III). Figure 19 also shows the already described effect of hot carrier injection into the gate oxide. After injection fixed charges degrade the superior device performance and can only be removed by a high temperature annealing step.

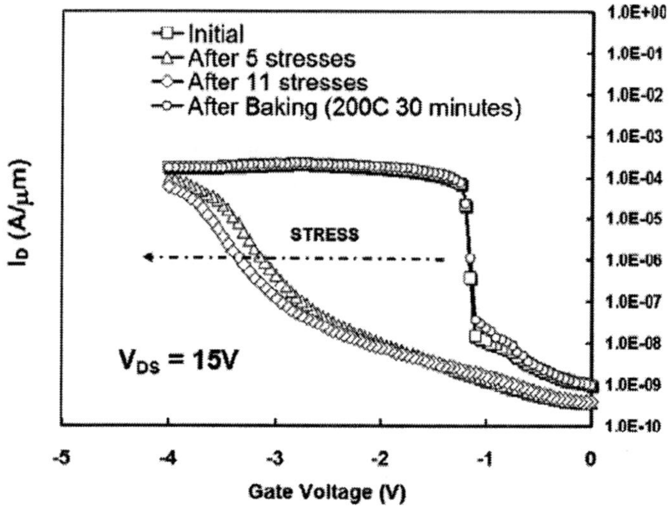

Figure 19. Experimental I_D-V_G characteristics of a p-i-n I-MOS at room temperature. L_i=300 nm, L_G= 100 nm, t_{ox}= 10...20 nm. The device was fabricated on SOI with t_{Si} = 200 nm. [121], ©2005 IEEE.

One major drawback of the concept can easily be derived from the working principle, and is evident in the characteristics shown in figure 19. To achieve an electric field high enough to cause avalanche breakdown a very high drain source voltage is necessary, typically around 10 V or more. This conflicts with the main motivation for the development of this device, the reduction of the subthreshold slope to reduce the threshold voltage without rising static leakage current. But the reason for reducing the threshold voltage is the reduction of the supply voltage during the scaling process. With a supply voltage of more then 10 V there is no demand for an threshold voltage of 0.4 V.

There are two promising approaches which have been followed to reduce the input voltage of a p-i-n I-MOS based logic:

- Reducing the electrical field necessary for breakdown
- Moving to a biasing scheme different from those of conventional CMOS

The first approach can be realized by reducing the band gap in the impact ionization region. This can either be done by using new materials such as SiGe or by modulating the silicon band gap by the use of strained silicon. Both approaches have been intensively studied by *Toh et al.* [126, 130, 133-135]. However the effect on the drain source voltage necessary for I-MOS operation is much smaller than the effect of shrinking the device. While a decrease of L_i from 60 to 40 nm results in a decrease of breakdown voltage from 8.8 V to 6.3 V the use of SiGe and therefore a narrower band gap in the impact ionization region leads only to a decrease from 8.8 V to 8.0 V for L_i kept at 60 nm (with L_G=60 nm in both cases) [135, 139]. More relevant is the change in the ON current by the use of narrow-band gap material. *Toh et al.* showed an increase of the ON current by a factor of 2 for strained silicon and by a factor of 3 for SiGe [135].

The lowest published breakdown voltage was achieved by a combination of band gap engineering with a three dimensional channel pattering. Impact Ionization FinFETs (I-

FinFET) were reported to work at 4.75 V with the tradeoff of a more complicated fabrication process [138]. These difficulties in integration of I-FinFETs into the conventional CMOS process flow are also the reason why there is such few works published on these devices and why we do not focus on them.

The second approach does not affect the absolute breakdown voltage of the device but focuses on its usability in an integrated circuit. *Choi et al.* suggested a biasing scheme different from that of a conventional MOSFET [124]. While in the latter case source and bulk are connected to ground and drain and gate are variable biased, the suggested scheme for the I-MOS only connects the bulk to ground and sets source to a fixed voltage, in case of a n-channel device a negative voltage and in case of a p-channel device to a positive voltage. This reduces drain induced leakage currents and improves the switching behavior of the device. Furthermore it is suitable for an I-MOS inverter structure [125]. Therefore this biasing scheme is commonly used for all the experimental and simulated data achieved in recent work [126-140].

Besides the difficulty to meet the road map requirements with an operating voltage of more than 10 V, high leakage currents are another critical problem arising from these high drain source voltages. Especially for bulk devices leakage currents reach unacceptable high values in the range of μA and degrade the I_{ON}/I_{OFF} ratio down to one order of magnitude [118]. Therefore the use of SOI substrates is preferable. It allows easy isolation between the devices by either shallow trench isolation or by using a mesa structure [129]. The significant reduction of the bulk leakage current allows I_{ON}/I_{OFF} ratios > 10^5 to hit the roadmap requirements (see also table III).

The demand on a significant I_{ON}/I_{OFF} ratio also sets an upper limit to the drain source voltage even when using SOI substrates. The reverse biased pin diode leakage current rises with the applied voltage and decreases the dynamic switching range. On the other hand the switching speed of the p-i-in I-MOS strongly depends on the desired I_{ON}/I_{OFF} ratio. According *to Shen et al.* [132] the delay time τ_i necessary to achieve an increase in current by one order of magnitude is given by:

$$\tau_i = 2.3 \cdot \frac{L_i}{2 \cdot (r_{impac} \cdot L_i - 1) \cdot v_{sat}} \qquad (12)$$

where r_{impac} is the average ionization rate and v_{sat} is the saturation velocity of charge carriers. For typical values of an I-MOS transistor this results in time constants in the ps range. For a desired I_{ON}/I_{OFF} ratio of 10^5 the total intrinsic delay time amounts to $5 \cdot \tau_i$. This is far away from the ITRS roadmap requirements for both high performance and low power technology, which for the 32 nm node are 0.39 ps and 0.73 ps, respectively. This let the p-i-n I-MOS not appear very suitable for logic applications, especially as conventional MOSFETs hit the ITRS requirements. In simulations of the dynamic behavior of a three stage ring oscillator it also is evident that the delay time per stage is 12 ps [125] compared to experimentally shown 5.1 ps of ring oscillators fabricated by Intel in their 45 nm technology [101].

Table III. Comparison of p-i-n I-MOS parameters

	Ref. [118]	Ref. [121]	Ref. [129]	Ref. [135]		Ref. [139]
Impact region material	Si	Si	Si	Strained Si	$Si_{0.75}Ge_{0.25}$	Si
Substrate	Bulk	SOI	SOI	Bulk	Bulk	Bulk
S [mV/dec]	7.2	10	2	3.56	3.28	5.88
I_{ON} [A/µm]	$5{\cdot}10^{-6}$	10^{-4}	$3{\cdot}10^{-5}$	$2.5{\cdot}10^{-4}$	$4{\cdot}10^{-4}$	$2{\cdot}10^{-4}$
I_{OFF} [A/µm]	$5{\cdot}10^{-5}$	10^{-9}	10^{-10}	10^{-7}	$1.5{\cdot}10^{-7}$	10^{-8}
L_G [nm]	130	100	1000	60	60	60
L_i [nm]	60	300	400	60	60	40
V_{DS} [V]	6.75	15	18	8.8	8.8	6.75

B. n-i-p-i-n I-MOS[15]

1. Device Concept

A second device which overcomes the kT/q limit of the conventional MOSFET is the n-i-p-i-n based vertical[16] I-MOS [144-148]. Impact ionization also plays an important role for the electrical behavior of this device but in a quite different way compared to the lateral I-MOS. Besides a low subthreshold slope it offers superior high temperature stability and is far less sensitive to statistical fluctuations of doping concentrations in the channel region compared to a conventional short channel MOSFET.

The device is based on a so called triangular barrier diode. The doping profile of this structure is of the type n-i-p-i-n (and p-i-n-i-p respectively) with a p^+-region of only 3 nm in width embedded between the intrinsic regions[17]. As the realization of such a narrow highly doped region is hardly possible by means of lithography these structures are commonly realized vertically as an epitaxially grown layer stack. This doping profile results in the formation of a triangular shaped band energy diagrams and therefore of a triangular shaped potential barrier as visualized in figure 20.

A doping profile of this kind was first used by *Hansch, Rao and Eisele* [141, 142] to realize a transistor by adding a gate stack to the sidewall of a mesa etched from the layer stack. The resulting device was called Planar Doped Barrier MOSFET (PDBFET) due to the channel doping being concentrated in a plane between the intrinsic zones. These devices did not show a subthreshold behavior being fundamentally different from that of a conventional MOSFET. The reason was the comparatively moderate delta layer doping of approximately 10^{18} cm^{-3}.

[15] The description of the operating modes given in this section refer to a n-i-p-i-n structure. For a p-channel device which would be based on a p-i-n-i-p structure the carrier types and biasing must be change visa verse. So far only the n-channel device was demonstrated experimentally.

[16] In contrast to the p-i-n I-MOS this device was so far only realized in a vertical design.

[17] Really "intrinsic" regions are impossible to realize for technological reasons. In this section we assume always an unintentional doping of 5 10^{15} cm^{-3} n-type when talking about intrinsic silicon.

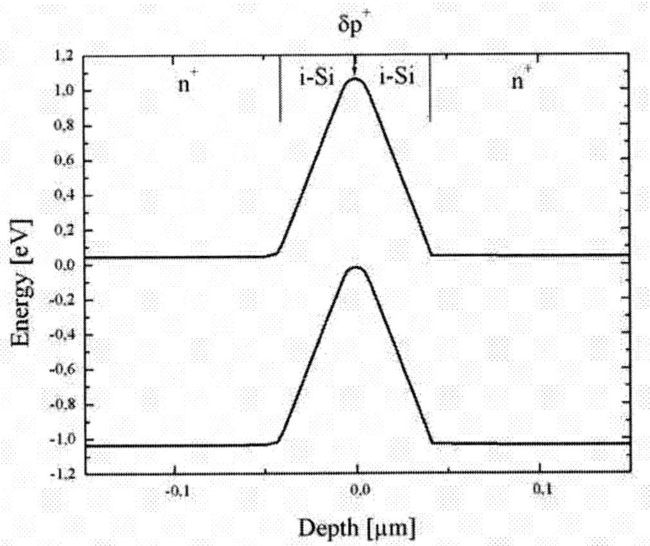

Figure 20. Simulated energy band levels of a triangular barrier diode. The n region doping is N_D=5 10^{18} cm^{-3}, the δp^+ layer doping is N_A=2 10^{19} cm^{-3}, thickness of the delta layer is 3 nm. Length of the intrinsic zones is 40 nm, unintentional doping 5 10^{15} cm^{-3} n-type. To achieve more realistic band structures a diffusion of dopants at 800 °C for 10 minutes was simulated to add a typical thermal budget. From [147] with permission of Elsevier.

The development of the vertical I-MOS out of the PDBFET was possible by achieving electrical active dopant concentrations of more than one order of magnitude higher in the delta layer even after outdiffusion due to high temperature treatment during transistor fabrication [144, 147]. These devices no longer operate like short channel MOSFETs over the complete bias range but show three clearly distinguishable operating modes depending on the drain source voltage. Due to the high doping the band edge at the delta layer, i. e. the barrier height, is not significantly affected by the drain potential. The potential drop between drain and source rather occurs on the drain side and leads to an electric field in the drain side intrinsic region. The barrier can be controlled by the gate voltage. The electric field strength in the drain side intrinsic regions determines the operating mode of the transistor.

- Conventional MOSFET mode:

For low drain source voltages the device is reported to behave like a conventional MOSFET. The high delta layer doping suppresses SCE very well and leads to excellent low OFF state leakage currents. Figure 21 illustrates the band energy diagrams for this conditions in ON and OFF state.

- Impact Ionization mode:

With increasing voltage the field in the drain side intrinsic region rises and finally reaches a value where impact ionization occurs. In contrast to the p-i-n based I-MOS the carrier multiplication is not used to amplify the current directly. Rather the electrons generated during this process move towards the drain while the holes accumulate in the floating body of the device at the delta layer similar to the kink effect in thin body SOI devices [112]. This results in a dynamic reduction of the threshold voltage during switching the device on.

Therefore the I_D-V_{GS} characteristics will show a very steep subthreshold slope. Figure 22 shows the band diagrams for these biasing conditions and the localization and absolute value of impact ionization for on and off state in this operating mode.

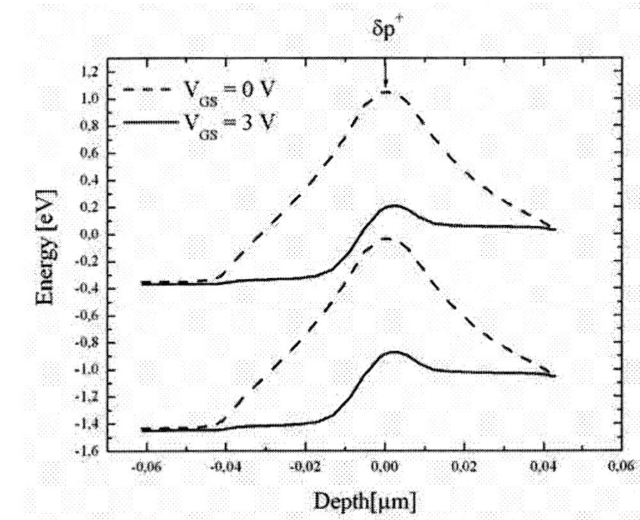

Figure 21. Simulated energy band levels of a vertical I-MOS in conventional MOSFEY mode (V_{DS} = 0.4 V). Then n region doping is N_D=5 10^{18} cm^{-3}, the δp^+ layer doping is N_A=2 10^{19} cm^{-3}, thickness of the delta layer is 3 nm. Length of the intrinsic zones is 40nm, unintentional doping 5 10^{15} cm^{-3} n-type. From [147] with permission of Elsevier.

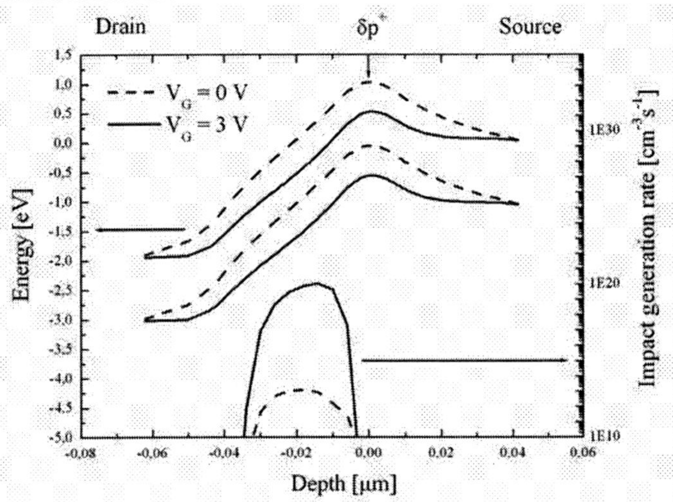

Figure 22. Simulated energy band levels of a vertical I-MOS in conventional MOSFEY mode (V_{DS} = 2.0 V). Then n region doping is N_D=5 10^{18} cm^{-3}, the δp^+ layer doping is N_A=2 10^{19} cm^{-3}, thickness of the delta layer is 3 nm. Length of the intrinsic zones is 40nm, unintentional doping 5 10^{15} cm^{-3} n-type. From [147] with permission of Elsevier.

As only a few holes are necessary to lower the barrier and rise the drain current a much lower impact ionization rate and therefore a much lower voltage is sufficient compared to the p-i-n based I-MOS. Figure 22 shows simulation results predicting impact ionization rates around 10^{20} $cm^{-3}s^{-1}$ while the lateral I-MOS exceeds 10^{30} $cm^{-3}s^{-1}$ [122]. For a device with 40 nm length of the as grown intrinsic regions the transition between conventional and impact ionization mode takes place at 1.25 V [147].

A possible and so far not experimentally investigated disadvantage of the operation in impact ionization mode is the discharging of the floating body which happens through the body-source diode. From the kink effect it is known that this discharging process proceeds very slow once the body potential has dropped below the threshold voltage of this diode. This might lead to higher leakage currents at RF operation.

- Bipolar mode:

If the drain source voltage is increased further a point is reached where the hole current generated by impact ionization is sufficient to switch on the parasitic bulk npn bipolar transistor being part of this structure. This leads to an additional increase in the drain current. As the bipolar current is not gate controlled the device will latch if the drain source voltage is chosen too high. In the transition region between impact ionization and bipolar mode this amplification mechanism can be used to achieve an even steeper subthreshold slope but with an visible hysteresis in the I_D-V_{GS} characteristics.

All experimental realizations [145–148] described so far are fabricated from a MBE[18] grown layer stack. From these layer stack mesas are fabricated by reactive ion etching (RIE). RTO[19] is commonly used to grow a gate oxide on the sidewalls of these mesas to keep the thermal budget low. After depositing the gate electrode material, which is commonly formed by highly doped poly silicon, the gate stack is patterned and the device is passivated by deposition of silicon nitride. Opening of contact holes and metal contact formation completes the process flow. Figure 23 shows the schematic drawing of the resulting device.

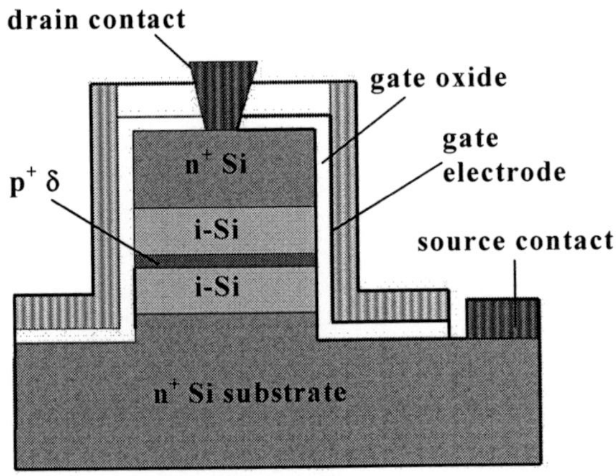

Figure 23. Schematic drawing of the vertical n-i-p-i-n based I-MOS.

[18] MBE: molecular beam epitaxy
[19] RTO: rapid thermal oxidation

2. Properties and Performance

Although the device concept was published just two years ago detailed experimental results on the electrical behavior are avaliable. The unique property of dynamic V_T reduction and therefore decrease of the subthreshold slope below the kT/q limit is visible in both the I_D-V_{GS} and I_D-V_{DS} characteristics. Figure 24 shows the I_D-V_{GS} characteristics for different drain source voltages. The transistion from conventional MOSFET mode to impact ionitaion mode is clearly visible. The subthreshold slope decreases with increasing drain source voltage because the impact ionization rate increases with drain source voltage. Therefore the shift in threshold voltage is enhancend which leads to a steeper subthreshold slope.

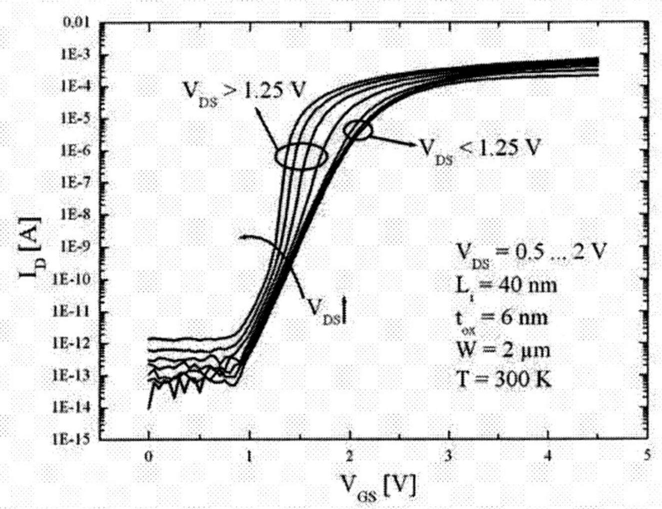

Figure 24. I_D-V_{GS} characteristics of the vertical I-MOS for various drain source voltages at room temperature. From [147] with permission of Elsevier.

In the I_D-V_{DS} characteristics the transition between the operating modes results in an increase of the drain current for constant V_{GS}. This becomes eminantly visible if the device is operated at a gate voltage, which is below the threshold voltage in conventional MOSFET mode. The resulting characteristics are plotted in figure 25. At around 1.25 V the transition to the impact ionization mode takes place lowering the the threshold voltage below the applied gate voltage of 1.5 V, and leading to a significant ON current. In conventional MOSFET mode the threshold voltage of such a device is around 1.8 V [148]. A further increase of drain source voltage leads to the transition to bipolar mode with a further increase of the drain current at around 2.4 V.

As already explained in the concept-section the operation deep in the bipolar regime will lead to a loss of gate control over switching behavior. However, the choice of a drain source voltage in the transition region between impact ionization and bipolar mode can be used to achieve a record low subthreshold slope of 1.06 mV/dec at room temperature but with a visible hystersis. This can be easier achieved if the delta layer doping has been broadened due to diffusion leading to a wider but lower barrier. It results in a transition region between impact ionization and bipolar mode over a wider drain source voltage range[147]. Depending on the needs of the application the device can be optimized to achieve a hysteresis free I_D-V_{GS} characteristics with a very low slope S (figure 26) or a reduction of the slope by almost a factor of 4 but with a visible hysteresis (figure 27).

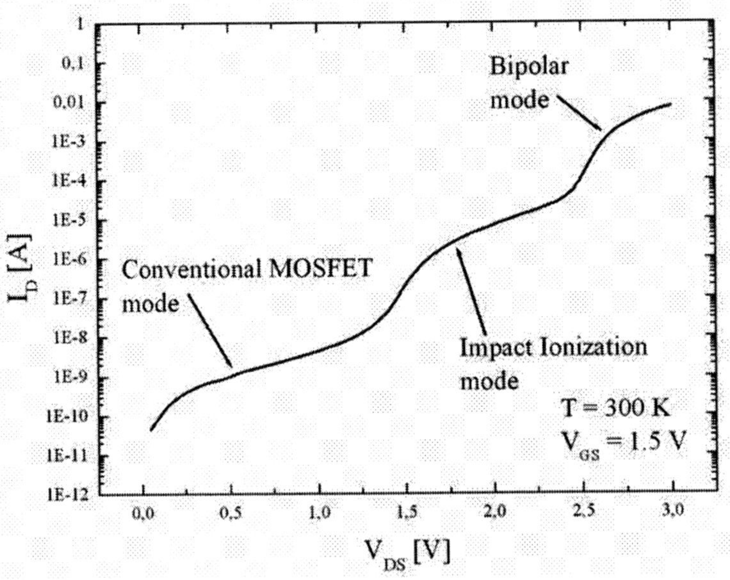

Figure 25. I_D-V_{DS} characteristics of the vertical I-MOS at room temperature.

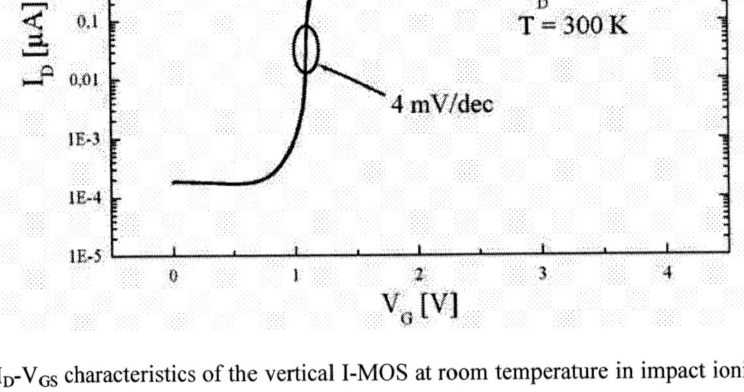

Figure 26. I_D-V_{GS} characteristics of the vertical I-MOS at room temperature in impact ionization mode. From [147] with permission of Elsevier.

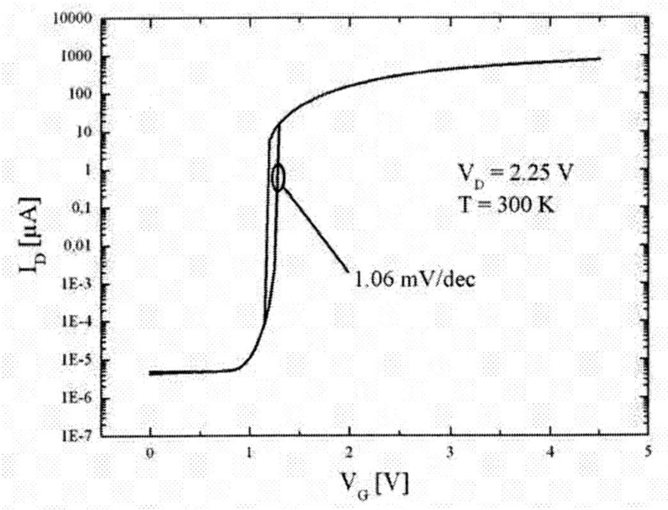

Figure 27. I_D-V_{GS} characteristics of the vertical I-MOS at room temperature in bipolar mode with gate control. From [147] with permission of Elsevier.

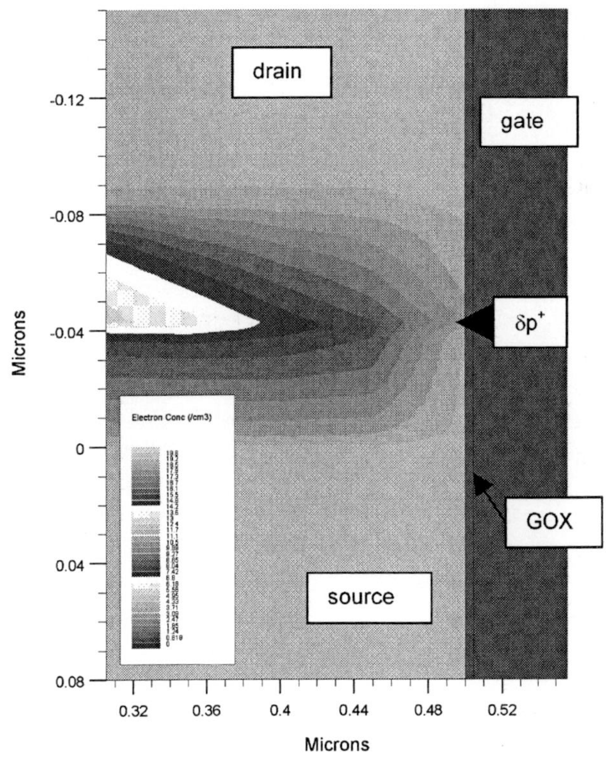

Figure 28. 2D ATLAS [107] simulation of the electron distribution in a vertial I-MOS in impact ionization mode ($V_D = 2$ V, $V_G = 2$ V). The n region doping is N_D=5 10^{18} $cm^{-3,}$ the δp^+ layer doping is N_A=2 10^{19} cm^{-3} thickness of the delta layer is 3nm. Length of the intrinsic zones is 40 nm, unintentional doping 5 10^{15} cm^{-3} n-type. From [147] with permission of Elsevier.

In contrast to the p-i-n I-MOS the n-i-p-i-n I-MOS shows an outstanding reduction of hot electron damage. While hot carrier injection into the gate oxide degrades the characteristics of the p-i-n device after only a few switching cycles [121] (see figure 19) the n-i-p-i-n device shows no tendency to change its electrical properties [146]. The reason for this stability is a phenomenon already described by *Hansch et al.*[141] for the first PDBFET realitzation: only at the delta layer the channel is confined to the gate oxide. Especially in the drain side intrinsic region where impact ionization takes place the electrons spread out in the bulk which was proven by 2D ATLAS simulations (see figure 28). In combination with the lower voltage sufficient for device operation this effects leads to an efficient suppression of hot electron damage. An experimental study of this prediction showed stable electrial properties during a 160-hour long time measurement with 50,000 switching cycles. Also there was no evidence of beginning degradation at the end of this experiment. The threshold voltage as well as ON and OFF currents remained almost unchanged [146].

The specific doping profile where the highly doped layers are separated by intrinsic regions provides another advantage of the device as it allows much higher doping levels compared to a device with direct pn-junctions of a conventional MOSFET. Section III.B already described why MOSFET functionality is limited at high temperatures due to pn barrier lowering and that higher doping reduces the influence of temperature on the potential barrier. On the other hand in section III.A it was explained why a significant tunneling current will flow through a reverse biased pn-junction if the doping levels are chosen too high. By separating the highly doped regions by intrinsic zones the doping levels can be set much higher without any tunneling leakage currents. Experimental results still showed an I_{ON}/I_{OFF} ratio of 10^5 at 250 °C and a change in threshold voltage by only 0.4 mV/K in conventional MOSFET mode [148]. This makes the n-i-p-i-n based I-MOS a very promising candidate for high-temperature applications.

Another important advantage of the device is its insensitivity to statistical fluctuations of the absolute number of dopant atoms. This is an effect directly achieved from the very high doping in the delta layer. Although the as grown layer thickness is only 3 nm the amount of doping atoms is more than two orders of magnitude higher than in a conventional MOSFET's channel region. While only a few ten dopant atoms determine the doping level there, a few hundred are present in the channel region of a vertical I-MOS. Furthermore these are spatially more concentrated, which does not allow a significant influence of their position on the electrical properties of the device.

C. TFET

1. Device Concept

While the previous presented devices take advantage of impact ionization the tunneling field effect transistor uses another parasitic phenomenon of the conventional MOSFET for device operation instead of avoiding it: quantum mechanical tunneling. Since the first experimental observation of interband tunneling in a semiconductor device by *Esaki* [149] in 1957 several structures were discussed. An overview of these early approaches is given by *Sze* [7, pp. 513].

The most interesting concept with respect to its compatibility to the CMOS process flow is the p-i-n based tunnel field effect transistor (TFET). A gated p-i-n diode as semiconductor device was first proposed by *Quinn et al.* [150] in 1978, not as a transistor but to investigate the properties of the 2D electron inversion channel. The first realization of a transistor in silicon based on this structure was published by *Reddick and Amaratunga* [154] in 1995. Meanwhile wide range of work on silicon based[20] TFETs was reported [155–183].

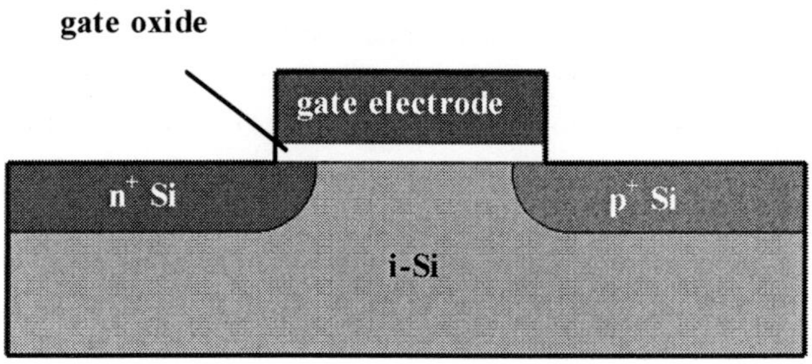

Figure 29. Schematic drawing of the TFET.

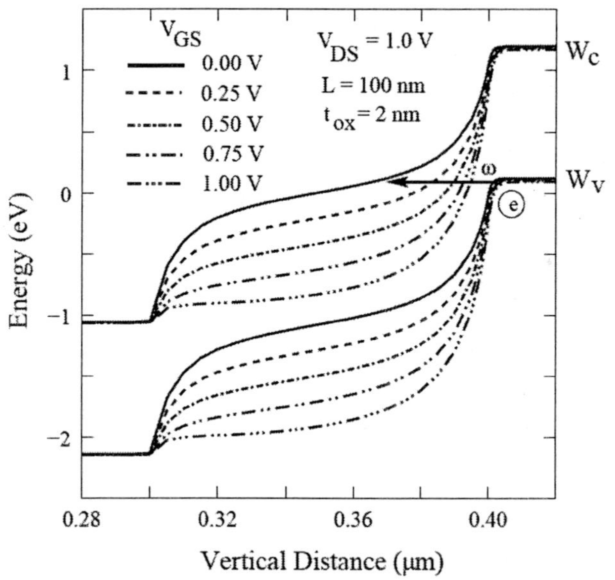

Figure 30. Simulated band diagrams for different gate source voltages in n-channel operation of a TFET with 100 nm intrinsic region. Source and drain doping are set to 10^{20} cm^{-3}, unintentional doping of the intrinsic zone is 10^{16} cm^{-3} n-type. [13]

The device is structured as a gated p-i-n diode with the gate covering the complete intrinsic region in contrast to the p-i-n based I-MOS (see figure 29). For operating as a transistor the diode has to be reverse biased and a heavy doping has to be chosen for the p and n region. Under these conditions applying a positive bias to the gate electrode leads to the formation of a electron inversion channel in the intrinsic region and results in gate controlled band-to-band tunneling of electrons from the valence band in the p^+ region into the

[20] The first realization of a tunneling field effect transitor ever was reported by *Baba* [151] and *Uemura* [152] in a III-V compound semiconductor. Furthermore also tunneling transistor based on carbon nanotubes [153] were

conduction band of the channel (figure 30 shows the band diagrams for these conditions and the dependence of the tunneling width ω on the gate source voltage). The same device can be operated as p-channel transistor by applying a negative gate bias. In this case electrons tunnel from the valence band of the inversion channel into the conduction band of the heavily doped n^+ region.

The realization of a lateral device can be done by implantation or diffusion based fabrication of the p+ and n+ region equivalent to the fabrication of a conventional MOSFET with an additional mask to cover the opposite doped region (similar to the lateral I-MOS but just without any spacer as the complete intrinsic region is covered by the gate). However, to achieve good tunnel probabilities a doping profile as sharp as possible is desired [172]. Therefore the formation of the highly doped regions by epitaxy is more desirable.

One approach which was not realized so far is the use of selective epitaxial growth of these regions in a lateral design [180, pp. 142]. For this purpose after formation of the gate stack the later p-doped region of the transistor is protected by a SiO_2 hardmask while on the other side TMAH is used for etching a well which is in sequence filled up by selectively grown highly n-doped silicon. The same sequence is then used for the formation of the p-doped region.

Another approach has already been realized in 2000 by *Hansch et al.* [157] and was investigated further by both experimental and theoretical studies [13, 158,162,164-167, 171-173, 180] in the following years. It is based on the fabrication of a vertical device by a process flow similar to those used for a n-i-p-i-n I-MOS. The doping profile of this vertical device is realized by epitaxial growth of intrinsic silicon on a highly doped substrate and a highly doped region (of the opposite doping type than the substrate) on top. The transistor is then fabricated by etching mesas from these layer stack, growing a gate oxide and depositing the gate electrode material. After patterning the gate the devices are passivated and electrical contacts are fabricated by passivation opening and metal deposition. *Born* [179] gives a very good overview of the fabrication process and the special difficulties and tasks.

2. Properties and Performance

In a conventional MOSFET the drain source current is diffusion driven in the subthreshold region and drift determined in the linear region of the device characteristics. In contrast the drain source current of a TFET is driven by quantum mechanical tunneling for any bias condition. Therefore I_{DS} can be described by only one equation. Kane's model [14] proved to provide quite accurate results over a wide range of biasing [167]. More detailed modeling approaches have been investigated to obtain the full device characteristics via simulation [173]. This chapter will not focus on these approaches, which are extremely useful for further device improvement, but not necessary to understand the principles of operation of this device. *Bhuwalka et al.* [167] derived from Kane's model for constant V_{DS} the equation for the drain current:

$$I_{DS} = A_{Kane} \cdot D^2 \cdot W_g^{-1/2} \cdot V_{GS}^2 \cdot e^{-B_{Kane} \cdot W_g^{3/2} \cdot (V_{GS} \cdot D)^{-1}} \tag{13}$$

demonstrated. Both are only of minor usfullness as add-on or replacment for the conventional silicon based MOSFET in CMOS technology. Therefore we will focus on silicon based TFETs in this chapter.

A_{Kane} and B_{Kane} are in general functions of the effective mass[21] but in good approximation constants. D is a function of V_{DS} and device parameters and therefore constant for constant V_{DS}.

Equation (13) lead to two interesting aspects of the TFET:

First as the equation is valid for the whole range of V_{GS} there is no threshold voltage like in the conventional MOSFET. Therefore the threshold voltage of a TFET is commonly defined using the constant current method. This means the device is defined as being on for a certain drain source current and the gate source voltage necessary to reach is this current is the threshold voltage. This of course means the threshold voltage depends on V_{DS}, too.

Second the only temperature dependent parameter in equation (13) is the band gap W_g. But the change in the band gap with temperature is very small. This means the device characteristics are only very weakly temperature dependent. This recommends the TFET for high temperature applications [172]. Furthermore as the current is not diffusion limited, the slope is not limited to kT/q like in the subthreshold region of the conventional MOSFET. Therefore a slope lower than 60 mV/dec at room temperature is possible but it strongly depends on oxide thickness and sharpness of doping profiles [165]. Therefore the experimental verification was not shown until 2007 when *Choi et al.* [177] reported a TFET in silicon with a slope of 52.8 mV/dec.

A further aspect emphasizes the use for high temperature applications. The height of the potential barrier of a TFET is determined by the doping levels in the n^+ and p^+ region. They are supposed to be as high as possible to form a good tunneling junction. This makes their Fermi level very insensitive to temperature variations and therefore results in an excellent high temperature OFF state performance of the device. Experimental data by *Born et al.* [172] showed a change of less than one order of magnitude in the OFF current of the device over a range of 340 K. Furthermore simulations of a device with ideal doping profiles (source and drain doping 10^{20} cm^{-3}) predict a off state leakage current in the range of pA/µm at 450 K for a device with a 70 nm intrinsic region.

Due to the band diagram and the high doping levels the TFET is also less sensitive to short channel effects and provides excellent scaling properties. In contrast to the conventional MOSFET the drain potential has almost no influence on the barrier height and only a weak influence on the tunnel width in the OFF state. This is clearly visible from the band diagrams for different drain source voltages without any gate bias as shown in figure 31. This results in a very low OFF state leakage current in the pA/µm range [160, 172, 175, 177]. Furthermore due to the opposite doping of source and drain no punch through is possible in this structure.

The most relevant leakage current mechanism influenced by scaling is direct tunneling. Simulations predict this effect not becoming visible before scaling the device down to 25 nm. At this point exponential increase of the static leakage current with decreasing gate length begins but the total OFF state current is still far below the ITRS roadmap requirements and far below what can be expected from a state of the art MOSFET [166]. Therefore the device an be scaled down to the theoretical limit of direct drain source tunneling around 10 nm (see

---

21 $A_{Kane} = \frac{e^2 \cdot m_0^{1/2}}{18\pi \cdot \hbar^2}$; $B_{Kane} = \frac{\pi \cdot m_0^{1/2}}{2e \cdot \hbar}$; with m_0 being the effective mass of electrons and $\hbar$ Planck's constant

section III.A) and is expected to show better performance with respect to static leakage currents than any conventional MOSFET structure.

This excellent scalability is supported by the intrinsic channel region. It makes the device, similar to the already discussed fully depleted intrinsic channel MOSFET concepts (FinFET and SONFET), independent from statistical fluctuations of dopant atom positions and concentration in the channel region.

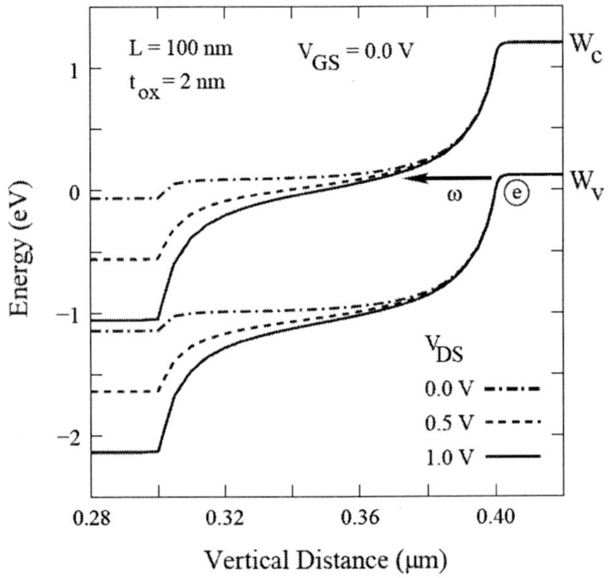

Figure 31. Simulated band diagrams for different drain source voltages of a TFET with 100 nm intrinsic region. Source and drain doping are set to 10^{20} cm^{-3}, unintentional doping of the intrinsic zone is 10^{16} cm^{-3} n-type. [13]

The main drawback of a TFET realized in silicon with SiO_2 as gate dielectric is its very low ON current. The drain source current is determined by V_{GS}, V_{DS} and the band gap W_g. *Bhuwalka et al.* [165] demonstrated by 2D device simulations that the dependency of the current on increasing V_{DS} is very weak and saturates very early. An increase of the gate field by increasing V_{GS} in contrast leads to a exponential increase in the current. However, the limiting factor is the band gap, which leads to a limitation of the drain current for a silicon based TFET with SiO_2 as gate dielectric to about 10 μA/μm which is far below the ITRS roadmap requirements.

This corresponds to all experimental data reported on TFET devices where the ON current usually was even below this theoretical limit due to imperfect doping profiles [160, 172, 175, 177]. Reports on experimental realizations of TFETs in silicon with SiO_2 or nitrided SiO_2 as gate dielectric showing drive currents of more than 100 μA/μm [161, 163, 168-170] were corrected by the authors meanwhile. Due to an unwanted n-doping in the tunneling region at the p-side of the device a significant drift-diffusion current caused these high drain source currents[176]. The theoretical and practical limitation in drive currents for silicon/silicon-dioxide based TFETs is nowadays commonly accepted.

To overcome this crucial limit of the TFET two improvements of the device structure were suggested: the use of a SiGe heterojunction in the tunneling region [164] and the use of a high-k gate dielectric [174].

The use of a thin highly doped SiGe region at the source of the transistor was first proposed and intensively investigated by *Bhuwalka et al.* [164-167]. In SiGe the valence band level is higher compared to silicon which results not only in a reduced band gap but also directly in a narrower tunneling width ω_{SiGe} for the same biasing conditions (see illustration in figure 32). This enhances the tunneling current further.

Because I_D-V_{GS} characteristics show an exponential current increase with increasing gate voltage, the threshold voltage (which is determined by the constant current method for a TFET) is reduced with the introduction of SiGe, too.

For a germanium mole fraction of 0.5 2D device simulations predict a drive current as high as 850 µA/µm and a threshold voltage of 0.3 V for 100 nm devices [165]. The tradeoff for this achievement is a rising static leakage current. But thanks to a very low initial OFF state diffusion current even for high germanium mole fractions OFF currents can be expected to be in the nA/µm range, which would still hit the roadmap requirements [166].

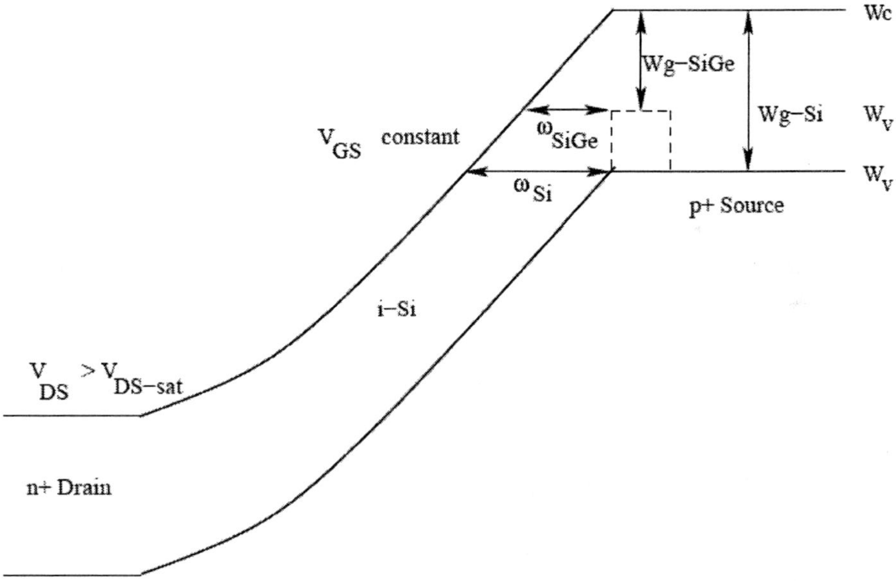

Figure 32. Schematic illustration of the band diagrams with and without SiGe delta layer at the source end of the TFET. Clearly visible is that not only the band gap is reduced but also the tunneling width for the same biasing conditions. [13]

The second approach addressing the problem of low drive currents was first detailed investigated by *Boucart and Ionescu* [174, 178, 179]. They proposed a double gate TFET (which means the use of a FinFET structure but with tunneling transistor specific doping) with high-k gate dielectric. The potential in the thin channel region can completely by controlled by the gate. As known from FinFET devices the complete channel volume can be in inversion although the fin width is more than twice the inversion layer thickness of a single gate transistor. By using high-k gate dielectric this effect is even enhanced. This leads to the formation of a larger cross section of the tunneling junction and a stronger band bending, which reduces tunneling width. Both effects lead to a rising current.

Toh et al. [181] investigated the influence of the fin width on the maximum achievable ON current with the use of HfO_2 as gate dielectric by 2D device simulations. With increasing width the current first strongly increases up to 500 μA/μm as long as the volume is limiting the tunneling junction. Then a slower increase follows until the gate coupling is lost. This point at a fin width of 20 nm determines the maximum current which is approximately 600 μA/μm. For wider fins the current decreases again.

The next consequent step was done also by *Toh et al.* [183] with combining FinFET approach, high-k dielectric and a SiGe heterojunction at the source. For these investigations a complete SiGe source region was assumed and not only a thin SiGe layer between source and channel as done by *Bhuwalka et al.* [164-167]. By 2D device simulations a drive current of 1.6 mA/μm for a 50 nm device was predicted with an OFF state leakage current being in the pA/μm range. A critical issue to achieve this high current is the position of the gate edge with respect to the end of the source region. *Toh et al.* determined a sharp increase in the drive current when the gate edge is close to the p^+-SiGe/i-Si interface with a maximum when the gate edge is positioned with 2 nm overlap. Obviously under these conditions the lowest tunneling width is achieved as the tunneling junction is formed right at the SiGe/Si interface.

For both approaches, SiGe heterojunction in the tunneling region and high-k gate dielectric, no experimental results have been reported to prove the theoretical estimations. Nevertheless the TFET seems to have the potential as a candidate for improving the performance of integrated circuits further than the conventional MOSFET can do.

V. Conclusion

The ongoing downscaling of the conventional MOSFET will soon come to an end. Technological barriers, cost aspects and fundamental physical limits will make future propagation to the next technology nodes difficult or even impossible. A change in the device concept is the most likely way to further increase complexity and circuit performance.

A closer look at the three alternative concepts, which were described in this chapter, identifies the most promising candidate for the post-MOSFET era:

The p-i-n based I-MOS has the drawback of the necessary high supply voltage. This makes a co-integration with conventional CMOS parts, e.g. memory or I/O components, hardly possible. A wide range of applications, especially mobile electronics, are difficult to realize with operating voltages of 10 V or more. Furthermore the RF behavior of the device is inferior compared to a conventional MOSFET. Therefore the p-i-n based I-MOS will not replace the MOSFET.

The n-i-p-i-n based I-MOS is more compatible to CMOS requirements. Low OFF state leakage, operating voltage in the range of CMOS technology, and excellent high temperature characteristics make the device suitable to overcome certain MOSFET limits. But two aspects inhibit the realization of future integrated CMOS circuits using the n-i-p-i-n I-MOS: First the vertical design of the device is not very suitable for large scale integration. Second a vertical transistor will always have larger overlap capacities than a lateral one, which was fabricated using a self aligning gate technique. This decreases possible switching speed and results in poor RF performance.

However, the ultra sharp switching behavior and its superior leakage current suppression makes the vertical I-MOS an interesting candidate for some special applications in sensing technologies. Single transistors are commonly used there as transducers or amplifiers to turn any physical or chemical effect (like the surface potential change of a sensing layer due to adsorption of gas molecules) in a measurable electrical signal. In this area large-scale integration is not the task. Rather well defined electrical behavior of the transistor over a wide range of temperature or low power consumption for mobile applications are the requirements to be fulfilled. In this field the vertical I-MOS is superior to any conventional MOSFET.

The concept which will most likely find its way in CMOS technology is the TFET. When realized as double gate transistor in silicon with high-k gate dielectric it has all advantages of the MOSFET but overcomes the limits of MOSFET downscaling. Due to the p-i-n structure the barrier is stable down to 10 nm, the leakage current is very low, and the device does not suffer from V_T roll-off. Dopant fluctuations are not an issue as the channel region is intrinsic. The technology to realize the double gate device does already exist. FinFET structures are very well investigated and are expected to be used for CMOS in the 32 nm node already. The fabrication of a TFET instead of a MOSFET is possible by just one additional mask to define the two different source and drain dopings.

Taking all this into account the TFET offers the solution to continue the process of rising complexity and performance of integrated circuits with decreasing costs per transistor. Finally, direct source drain tunneling will also set the limit for scaling this device. From today's point of view we can only speculate about the development of silicon technology beyond this point. Probably scaling will come to an end or we will move to a completely different concept.

In any case Moore's Law will be pursued as long as we can realize both parts: shrinking the device and reducing the costs. It is not impossible that economics instead of physics will stop the scaling process.

References

[1] G. Moore, "Cramming more components onto integrated circuits", *Electronics*, vol. 38(8), 1965.

[2] Semiconductor Industries Association, "Global Chip Sales Hit Record 247.7 Billion in 2006", Press Release 02/02/2007, San Jose, CA, USA.

[3] Bardeen, J., Brattain, W., "The Transistor, a Semiconductor Triode", *Physical Review,* vol. 74, pp. 230–231, 1948.

[4] Bardeen, J., Brattain, W., "Three-Electrode Circuit Element Utilizing Semiconducitve Material", US Patent No. 2,524,035, 1950.

[5] Kahng, D., Atalla, M. M., "Silicon-Silicon Dioxide Field Induced Surface Devices", *IRE-AIEE Solid-State Device Research Conference*, Carnegie Institute of Technology, Pittsburgh, PN, USA, 1960.

[6] Kahng, D., "Electric Field Controlled Semiconductor Device", US Patent No. 3,102,230, 1963.

[7] Sze, S. M., "Physics of Semiconductor Devices", 2nd edition, New York, 1981.

[8] Brews, J., "A Charge-Sheet Model of the MOSFET", *Solid State Electronics*, vol. 21, pp. 345-355, 1978.

[9] Brews, J., "Subhreshold Behaviour of Uniformly and Nonuniformly Doped Long-Channel MOSFET", *IEEE Transaction on Electron Devices*, vol. 26(9), pp. 1282-1291, 1979.

[10] Bai, P., Auth, C., Balakrishnan, S., Bost, M., Brain, R., Chikarmane, V., Heussner, R., Hussein, M., Hwang, J., Ingerly, D., James, R., Jeong, J., Kenyon, C., Lee, E., Lee, S.-H., Lindert, N., Liu, M., Ma, Z., Marieb, T., Murthy, A., Nagisetty, R., Natarajan, S., Neirynck, J., Ott, A., Parker, C., Sebastian, J., Shaheed, R., Sivakumar, S., Steigerwald, J., Tyagi, S., Weber, C., Woolery, B., Yeoh, A., Zhang, K., Bohr, M., '"A 65nm logic technology featuring 35nm gate lengths, enhanced channel strain, 8 Cu interconnect layers, low-k ILD and 0.57 μm^2 SRAM cell", *Technical Digest of the IEEE International Electron Device Meeting 2004*, San Fransico, CA, pp. 657-660, 2004.

[11] Troutman, R., "VLSI Limitations from Drain-Induced Barrier Lowering", *IEEE Transactions on Electron Devices*, vol. 26(4), pp. 461-469, 1979.

[12] Chen, J., Chan, T., Chen, I., Ko, P., Hu, C., "Subbreakdown Drain Leakage Current in MOSFET", *IEEE Electron Device Letters,* vol. 8(11), 1987.

[13] Bhuwalka, K.K., "Novel Tunneling Devices for Future CMOS Technologies", PhD Thesis, Bundeswehr University Munich, 2006.

[14] Ogura, S., Codella, C., Rovedo, N, Shepard, J., Riseman, J., "A Half Micron MOSFET Using Double Implanted LDD", *Technical Digest of the IEEE International Electron Devices Meeting 1982*, San Fransisco, CA, pp. 718-722, 1982.

[15] Troutman, R., Fortino, A., "Simple Model for Threshold Voltage in a Short-Channel IGFET", *IEEE Transactions on Electron Devices*, vol. 24(10), pp. 1266-1268, 1977.

[16] Dambrine, G., Raynaud, C., Lederer, D., Dehan, M., Rozeaux, O., Vanmackelberg, M., Danneville, F., Lepilliet, S., Raskin, J.-P., "What are the Limiting Parameters of Deep-Submicron MOSFETs for High Frequency Applications?", *IEEE Electron Device Letters,* vol. 24(3), pp. 189-191, 2003.

[17] Liou, J., Schwierz, F., "RF MOSFET: recent advances, current status and future trends", *Solid State Electronics*, vol. 47, pp. 1881-1895, 2003.

[18] Yeap, G., Krishnan, S., Lin, M.-R., "Fringing-induced barrier lowering (FIBL) in sub-100nm MOSFETs with high-K gate dielectrics", *IEEE Electronics Letters*, vol. 34(11), pp. 1150-1152, 1998.

[19] Cheng, B., Cao, M., Rao, R., Inani, A., Voorde, P., Greene, W., Stork, J., Yu, Z., Zeitzoff, P., Woo, J., "The Impact of High-κ Gate Dielectrics and Metal Gate Electrodes on Sub-100 nm MOSFET's", *IEEE Transactions on Electron Devices*, vol. 46(7), pp. 1537-1544, 1999.

[20] Mohapatra, N., Desai, M., Narendra, S, Rao, R., "The Effect of High-K Gate Dielectrics on Deep Submicrometer CMOS Device and Circuit Performance", *IEEE Transactions on Electron Devices*, vol. 49(5), pp. 826-831, 2002.

[21] Mohapatra, N., Desai, M., Rao, R., "Detailed Analysis of FIBL in MOS Transistors with High-K Gate Dielectrics", *Proceedings of the 16th International Converence on VLSI Design*, New Delhi, India, pp. 99-104, 2003.

[22] Mizuno, T., Okamura, J., Toriumi, A., "Experimental Study of Threshold Voltage Fluctuation Due to Statistical Variation of Channel Dopant Number in MOSFETs", *IEEE Transactions on Electron Devices*, vol. 41(11), pp. 2216-2221, 1994.

[23] Taur, Y., Mii, Y.-J., Frank, D., Wong, H.-S., Buchanan, D., Wind, S., Rishton, S., Sai-Halasz, G, Nowak, E., "CMOS scaling in to the 21st century: 0.1 µm and beyond", *IBM Journal of Research and Development*, vol. 39(1/2), pp. 245-260, 1995.

[24] Stolk, P., Klaassen, D., "The Effect of Statistical Dopant Fluctuations on MOS Device Performance", *Technical Digest of the IEEE International Electron Devices Meeting 1996,* San Fransisco, CA, pp. 627-630, 1996.

[25] Stolk, P., Widdershoven, F., Klaassen, D., "Modeling Statistical Dopant Fluctuations in MOS Transistiors", *IEEE Transactions on Electron Devices*, vol. 45(9), pp. 1960-1971, 1998.

[26] Wong, H.-S., Taur, Y., Frank, D., "Discrete Random Dopant Distribution Effectsin Nanometer-Scale MOSFETs", *Microelectronics Reliability*, vol. 38(9), pp. 1447-1456, 1998.

[27] Frank, D., Taur, Y., Ieong, M., Wong, H.-S., "Monte Carlo Modeling of Threshold Variation Due to Dopant Fluctuations", *Digest of Technical Papers of the VLSI Symposium on VLSI Circuits* 1999, pp. 171-172, 1999.

[28] Hane, M., Ikezawa, T., Ezaki, T., "Atomistic 3D Process/Device Simulation Considering Gate Line-Edge Roughness and Poly-Si Random Crystal Orientation Effects", *Technical Digest of the IEEE International Electron Device Meeting 2003*, Washington, DC, pp. 9.5.1- 9.5.4, 2003.

[29] Frank, D., Dennard, R., Nowak, E., Solomon, P., Taur, Y., Wong, H.-S., "Device Scaling Limits of Si MOSFETs and Their Application Dependencies", *Proceedings of the IEEE*, vol. 89(3), pp. 259-287, 2001.

[30] Bernstein, K., Frank, D., Gattiker, A., Haensch, W., Ji, B., Nassif, S., Nowak, E., Pearson, D., Rohrer, N., "High-performance CMOS variability in the 65-nm regime and beyond", *IBM Journal of Research and Development*, vol. 50(4/5), pp. 433-449, 2006.

[31] Asenov, A., Brown, A., Davis, J., Kaya, S., Slavcheva, G., "Simulation of Intrinsic Parameter Fluctuations in Decananometer and Nanometer Scale MOSFETs", *IEEE Transactions on Electron Devices*, vol. 50(9), pp. 1837-1852, 2003.

[32] Fuller, G., "Optical Lithography", in: Nishi, Y., Doering, R., "Handbook of Semiconductor Manufacturing Technology", pp. 461-497, Marcel Dekker, New York, NY, 2000.

[33] Silverman, J., "X-Ray Lithography", in: Nishi, Y., Doering, R., "Handbook of Semiconductor Manufacturing Technology", pp. 543-570, Marcel Dekker, New York, NY, 2000.

[34] Saitou, N., "Electron-Beam Lithography", in: Nishi, Y., Doering, R., "Handbook of Semiconductor Manufacturing Technology", pp. 571-587, Marcel Dekker, New York, NY, 2000.

[35] Hisamoto, D., Kaga, T., Kawamoto, Y., Takeda, E., "A Fully Depleted Lean-Channel Transistor (DELTA) – A Novel Vertical Ultrathin SOI MOSFET", *IEEE Electron Device Letters,* vol. 11(1), pp. 36-38, 1990.

[36] Hisamoto, D., Kaga, T, Takeda, E., "Impact of the Vertical SOI "DELTA" Structure on Planar Device Technology", *IEEE Transactions on Electron Devices*, vol. 38(6), pp. 1419-1424, 1991.

[37] Huang, X., Lee, W.-C., Kuo, C., Hisamoto, D., Chang, L., Kedzierski, J., Anderson, E., Takeuchi, H, Choi, Y.-K., Asano, K., Subramanian, V., King, T.-J., Bokor, J., Hu, C.,

"Sub 50-nm FinFET: PMOS", *Technical Digest of the IEEE International Electron Device Meeting 1999*, Washington, DC, pp. 67-70, 1999.

[38] Hisamoto, D., Lee, W.-C., Kedzierski, J., Takeuchi, H., Asano, K., Kuo, C, Anderson, E., King, T.-J., Bokor, J., Hu, C., "FinFET – A Self-Aligned Double-Gate MOSFET Scalable to 20 nm", *IEEE Transactions on Electron Devices*, vol. 47(12), pp. 2320-2325, 2000.

[39] Tang, S., Chang, L., Lindert, N., Choi, Y.-K., Lee, W.-C., Huang, X., Subramanian, V., Bokor, J., King, T.-J., Hu, C., "FinFET – A Quasi-Planar Double-Gate FET", *Digest of Technical Papers, IEEE International Solid-State Circuits Conference 2001*, San Fransisco, CA, pp. 118-119, 2001.

[40] Choi, Y.-K., Lindert, N., Xuan, P., Tang, S., Ha, D., Anderson, E., King, T.-J., Bokor, J., Hu, C., "Sub-20 nm CMOS FinFET Technologies", *Technical Digest of the IEEE International Electron Devices Meeting 2001,* Washington, DC, pp. 19.1.1-19.1.4, 2001.

[41] Kedzierski, J., Fried, D., Nowak, E., Kanarsky, T., Rankin, J., Hanafi, H., Natzle, W., Boyd, D., Zhang, Y., Roy, R., Newbury, J., Yu, C., Yang, Q., Saunders, P., Willets, C., Johnson, A., Cole, S., Young, H., Carpenter, N., Rakowski, D., Rainey, B., Cottrell, P., Ieong, M., Wong, H.-S., "High-performance symmetric-gate and CMOS-compatible V_t asymmetric-gate FinFET devices" , *Technical Digest of the IEEE International Electron Devices Meeting 2001*, Washington, DC, pp. 437-440, 2001.

[42] Fried, D., Johnson, A., Nowak, E., Rankin, J., Willets, C., "A Sub-40nm Body Thickness N-type FinFET", *Conference Digest Device Research Conference 2001,* Notre Dame, IN, pp. 24-25, 2001.

[43] Huang, X., Lee, W.-C., Kuo, C., Hisamoto, D., Chang, L, Kedzierski, J., Anderson, E., Takeuchi, H., Choi, Y.-K., Asano, K., Subramanian, V., King, T.-J., Bokor, J., Hu, C., "Sub-50 nm P-Channel FinFET", *IEEE Transaction on Electron Devices*, vol. 48(5), pp, 880-886, 2001.

[44] Kedzierski, J., Nowak, E., Kanarsky, T., Zhang, Y., Boyd, D., Carruthers, R., Cabral, C., Amos, R., Lavoie, C., Roy, R., Newbury, J., Sullivan, E., Benedict, J., Saunders, P., Wong, K., Canaperi, D., Krishnan, M., Lee, K.-L., Rainey, B.A., Fried, D., Cottrell, P., Wong, H.-S.P., Ieong, M., Haensch, W., "Metal-gate FinFET and fully-depleted SOI devices using total gate silicidation", ", *Technical Digest of the IEEE International Electron Device Meeting 2002*, San Fransisco, CA, pp. 247-250, 2002.

[45] Yu, B., Chang, L, Ahmed, S., Wang, H., Bell, S., Yang, C.-Y., Tabery, C., Ho, C., Xiang, Q., King, T.-J., Bokor, J., Hu, C., Lin, M.-R., Kyser, D., "FinFET Scaling to 10nm Gate Length", *Technical Digest of the IEEE International Electron Device Meeting 2002*, San Fransisco, CA, pp. 251-254, 2002.

[46] Nowak, E., Rainey, B., Fried, D., Kedzierski, J., Ieong, M., Leipold, M., Wright, J., Breitwisch, M., "A Functional FinFET-DGCMOS SRAM Cell", *Technical Digest of the IEEE International Electron Device Meeting 2002*, San Fransisco, CA, pp. 411-414, 2002.

[47] Choi, Y.-K., King, T.-J., Hu, C., "Spacer FinFET: nanoscale double-gate CMOS technology for the terabit era", *Solid-State Electronics*, vol. 46, pp. 1595-1601, 2002.

[48] Ieong, M., Wong, H.-S., Nowak, E., Kedzierski, J., Jones, E., "High Performance Double-Gate Device Technology Challenges and Opportunities", *Proceedings of the 3rd*

International Symposium on Quality Electronic Design, San Jose, CA, pp. 492-495, 2002.

[49] Doyle, B., Datta, S., Doczy, M., Hareland, S., Jin, B., Kavalieros, J., Linton, T., Murthy, A., Rios, R., Chau, R., "High Performance Fully-Depleted Tri-Gate CMOS Transistor", *IEEE Electron Device Letters*, vol. 24(4), pp. 263-265, 2003.

[50] Kedzierski, J., Ieong, M., Nowak, E., Kanarsky, T., Zhang, Y., Roy, R., Boyd, D., Fried, D., Wong, H.-S., "Extension and Source/Drain Design for High-Performance FinFET Devices", IEEE Transactions on Electron Devices, vol. 50(4), pp. 952-958, 2003.

[51] Fried, D., Duster, J., Kornegay, K., "Improved Independent Gate N-Type FinFET Fabrication and Characterization", *IEEE Electron Device Letters*, vol. 24(9), pp. 592-594, 2003.

[52] Ludwig, T., Aller, I., Gernhoefer, V., Keinert, J., Nowak, E., Joshi, R., Mueller, A., Tomaschko, S., "FinFET Technology for Future Microprocessors", *Proceedings of the IEEE International SOI Conference* 2003, Newport Beach, CA, pp. 33-34, 2003.

[53] Fu-Liang Yang, Di-Hong Lee, Hou-Yu Chen, Chang-Yun Chang, Sheng-Da Liu, Cheng-Chuan Huang, Tang-Xuan Chung, Hung-Wei Chen, Chien-Chao Huang, Yi-Hsuan Liu, Chung-Cheng Wu, Chi-Chun Chen, Shih-Chang Chen, Ying-Tsung Chen, Ying-Ho Chen, Chih-Jian Chen, Bor-Wen Chan, Peng-Fu Hsu, Jyu-Horng Shieh, Han-Jan Tao, Yee-Chia Yeo, Yiming Li, Jam-Wem Lee, Pu Chen, Mong-Song Liang, Chenming Hu, "5nm-Gate Nanowire FinFET", *2004 Symposium on VLSI Technology, Digest of Technical Papers, Honolulu*, HI, pp. 196-197, 2004.

[54] Tsui, B.-Y., Lin, C.-P., "A Novel 25-nm Modified Schottky-Barrier FinFET With High Performance", *IEEE Electron Device Letters*, vol. 25(6), pp. 430-432, 2004.

[55] Collaert, N., Dixit, A., Goodwin, M., Anil, K., Rooyackers, R., Degroote, B., Leunissen, L, Veloso, A., Jonckheere, R., De Meyer, K., Jurczak, M., Biesemans, S., "A Functional 41-Stage Ring Oscillator Using Scaled FinFET Devices With 25-nm Gate Lengths and 10-nm Fin Widths Applicable for the 45-nm CMOS Node", *IEEE Electron Device Letters*, vol. 25(8), pp. 568-570, 2004.

[56] Trivedi, V., Fossum, J., "Source/Drain-Doping Engineering for Optimal Nanoscale FinFET Design", *Proceedings of the IEEE International SOI Conference 2004,* Charlston, SC, pp. 192-194, 2004.

[57] Guo, Z., Balasubramanian, S., Zlatanovici, R, King, T.-J., Nikolic, B., "FinFET-Based SRAM Design", *Proceedings of the 2005 International Symposium on Low Power Electronics and Design, San Diego*, CA, pp. 2-7, 2005.

[58] Collaert, N., Brus, S., De Keersgieter, A., Dixit, A., Ferain, I., Goodwin, M., Kottantharayil, A., Rooyackers, R., Verheyen, P., Yim, Y., Zimmerman, P., Beckx, S., Degroote, B., Demand, M., Kim, M., Kunnen, E., Locorotondo, S., Mannaert, G., Neuilly, F., Shamiryan, D., Baerts, C., Ercken, M., Laidlcr, D., Leys, F., Loo, R., Lisoni, J., Snow, J., Vos, R., Boullart, W., Pollentier, I., De Gendt, S., De Meyer, K., Jurczak, M., Biescmans, S., "Integration challenges for multi-gate devices", *Proceedings of the* 2005 *International Conference on Integrated Circuit Design and Technology*, Austin, TX, pp. 187-194, 2005.

[59] Kaneko, A., Yagishita, A., Yahashi, K., Kubota, T., Omura, M., Matsuo, K., Mizushima, I., Okano, K., Kawasaki, H., Inaba, S., Izumida, T., Kanemura, T., Aoki, N., Ishimaru, K., Ishiuchi, H., Suguro, K., Eguchi, K., Tsunashima, Y.,"Sidewall

Transfer Process and Selective Gate Sidewall Spacer Formation Technology fur Sub-15nm FinFET with Elevated Source/Drain Extension", *Technical Digest of the IEEE International Electron Devices Meeting 2005*, Washington, DC, pp. 844-847, 2005.

[60] Byun, H.-S., Lee, W.-S., Lee, J.-W., Lee, K.-H., Park, Y.-K., Kong, J.-T., "3-Dimensional Analysis on the GIDL Current of Body-tied Triple Gate FinFET", *Proceedings of the International Conference on Simulation of Semiconductor Processes and Devices 2006*, Monterey, CA, pp. 267-270, 2006.

[61] Kavalieros, J., Doyle, B., Datta, S., Dewey, G., Doczy, M., Jin, B., Lionberger, D., Metz, M., Rachmady, W., Radosavljevic, M., Shah, U., Zelick, N., Chau, R., "Tri-Gate Transistor Architecture with High-k Gate Dielectrics, Metal Gates and Strain Engineering", *2006 Symposium on VLSI Technology, Digest of Technical Papers*, Honolulu, HI, pp. 50-51, 2006.

[62] Nawaz, M., Molzer, W., Haibach, P., Landgraf, E., Roesner, W., Staedele, M., Luyken, H., Gencer, A., "Validation of 30 nm process simulation using 3D TCAD for FinFET devices", *Semiconductor Science and Technology*, vol. 21, pp. 1111-1120, 2006.

[63] Yagashita, A., "FinFET SRAM Process Technology for *hp*32 nm node and beyond", Proceedings of the 2007 *International Conference on Integrated Circuit Design and Technology,* Austin, TX, pp. 1-4, 2007.

[64] Kawasaki, H., Kaneko, A., Yagishita, A., Okano, K., Izumida, T., Kanemura, T., Suguro, K., Eguchi, K., Tsunashima, Y., Inaba, S., Aoki, N., Ishimaru, K., Toyoshima, Y., "FinFET Process and Integration Technology for High Performance LSI in 22 nm node and beyond", *Extended Abstracts of the 7th International Workshop on Junction Technology,* Kyoto, Japan, pp. 3-7, 2007.

[65] Chiang, M.-H., Lin, J.-N., Kim, K., Chuang, C.-T., "Random Dopant Fluctuation in Limited-Width FinFET Technologies", *IEEE Transactions on Electron Devices*, vol. 54(8), pp. 2055-2060, 2007.

[66] Sachid, A., Manoj, C., Sharma, D., Rao, V., "Gate Fringe-Induced Barrier Lowering in Underlap FinFET Structures and Its Optimization", *IEEE Electron Device Letters*, vol. 29(1), pp. 128-130, 2008.

[67] *International Technology Roadmap for Semiconductors, 2006 Update, Process Integration, Devices, and Structures*, http:/www.itrs.net/, 01/30/2008.

[68] Wouters, D, Colinge, J.-P., Maes, H., "Subthreshold Slope in Thin-Film SOI MOSFETs", *IEEE Transactions on Electron Devices*, vol. 37(9), pp. 2022-2033, 1990.

[69] Nayfeh, H., Leitz, C., Pitera, A., Fitzgerald, E., Hoyt, J., Antoniadis, D., "Influence of High Channel Doping on the Inversion Layer Electron Mobility in Strained Silicon n-MOSFETs", *IEEE Electron Device Letters*, vol. 24(4), pp. 248-250, 2003.

[70] Hu, C., King, T.-J., Subramanian, V., Chang, L, Huang, X., Choi, Y.-K., Kedzierski, J., Lindert, N., Bokor, J., Lee, W.-C., "FinFET Transistor Structures Having a Double Gate Channel Extending Vertically from a Substrate and Methods of Manufacture", US Patent No. 6,413,802, 2002.

[71] Jurczak, M.,Skotnicki, T., Paoli, M., Tormen, B., Regolini, J.-L., Morin, C., Schiltz, A., Martin, J., Pantel, R., Galvier, J., "SON (Silicon On Nothing) – A New Device Architecture for the ULSI Era", 1999 Symposium on VLSI Technology, *Digest of Technical Papers*, Kyoto, Japan, pp. 29-30, 1999.

[72] Mizushima, I., Sato, T., Taniguchi, S., Tsunashima, Y., "Empty-space-in-silicon technique for fabricating a silicon-on-nothing structure", *Applied Physics Letters*, vol. 77(20), pp. 3290-3292, 2000.

[73] Jurczak, M., Skotnicki, T., Paoli, M., Tormen, B., Martins, J., Regolini, J.-L., Dutartre, D., Ribot, P., Lenoble, D., Pantel, R., Monfray, S., "Silicon-on-Nothing (SON) – an Innovative Process for Advanced CMOS", *IEEE Transactions on Electron Devices*, vol. 47(11), pp. 2179-2187, 2000.

[74] Sato, T., Nii, H., Hatano, M., Takenaka, K., Hayashi, H., Ishigo, K., Hirano, T., Ida, K., Aoki, N., Ohguro, T., Ino, K., Mizushima, I., Tsunashima, Y., "SON (Silicon on Nothing) MOSFET using ESS (Empty Space in Silicon) technique for SoC applications", *Technical Digest of the IEEE International Electron Devices Meeting 2001*, Washington, DC, pp. 809-812, 2001.

[75] Monfray, S., Skotnicki, T., Morand, Y., Descombes, S., Paoli, M., Ribot, P., Talbot, A., Dutartre, D., Leverd, F., Lefriec, Y., Pantel, R., Haond, M., Renaud, D., Nier, M.-E., Vizioz, C., Louis, D., Buffet, N., "First 80nm SON (Silicon-On-Nothing) MOSFETs with perfect morphology and high electrical performance", *Technical Digest of the IEEE International Electron Devices Meeting 2001*, Washington, DC, pp. 645-648, 2001.

[76] Monfray, S., Skotnicki, T., Tavel, B., Morand, Y., Decombes, S., Talbot, A., Dutartre, D., Jenny, C., Mazoyer, P., Palla, R., Leverd, F., Le Friec, Y., Pantel, R., Haond, M., Charbuillet, C., Vizioz, C., Louis, D., Bueffet, N., "SON (Silicon-On-Nothing) P-MOSFETs with totally silicided ($CoSi_2$) Polysilicon on 5nm-thick Si-films: The simplest way to integration of Metal Gates on thin FD channels", *Technical Digest of the IEEE International Electron Device Meeting 2002*, San Fransisco, CA, pp. 263-266, 2002.

[77] Monfray, S., Skotnicki, T., Morand, Y., Descombes, S., Coronel, P., Mazoyer, P., Harrison, S., Ribot, P., Talbot, A., Dutartre, D., Haond, M., Palla, R., Le Friec, Y., Leverd, F., Nier, M.-E., Vizioz, C., Louis, D., "50nm – Gate All Around (GAA) – Silicon On Nothing (SON) – Devices: A Simple Way to Co-integration of GAA Transistors within bulk MOSFET process", *2002 Symposium on VLSI Technology, Digest of Technical Papers*, Honolulu, HI, pp. 108-109, 2002.

[78] Monfray, S., Skotnicki, T., Morand, Y., Descombes, S., Talbot, A., Dutartre, D., Leverd, F., Le Friec, Y., Palla, R., Pantel, R., Haond, M., Nier, M.-E., Vizioz, C., Louis, D., "Highly performant 38nm SON (Silcion-On-Nothing) P-MOSFETs with 9nm-thick channels", *Proceedings of the IEEE International SOI Conference 2002*, Williamsburg, VI, 2002.

[79] Pretet, J., Cristoloveanu, S., Skotnicki, T., Monfray, S., "Short Channel and Back-Gate Coupling Effects in Silicon-On-Nothing (SON) MOSFETs", *Proceedings of the IEEE International SOI Conference 2003*, Newport Beach, CA. 2003.

[80] Harrison, S., Coronel, P., Leverd, F., Cerutti, R., Palla, R., Delille, D., Borel, S., Jullian, S., Pantel, R., Decombes, S., Dutartre, D., Morand, Y., Samson, M., Lenoble, D., Talbot, A., Vilarret, A., Monfray, S., Mazoyer, P., Bustos, J., Brut, H., Cros, A., Munteanu, D., Autran, J.-L., Skotnicki, T., "Highly Performant Double Gate MOSFET realized with SON process", *Technical Digest of the IEEE International Electron Devices Meeting 2003*, Washington, DC, pp. 449-452, 2003.

[81] Pretet, J., Monfray, S., Cristoloveanu, S, Skotnicki, T., "Silicon-on-Nothing MOSFETs: Performance, Short-Channel Effects, and Backgate Coupling", *IEEE Transactions on Electron Devices*, vol. 51(2), pp. 240-245, 2004.

[82] Monfray, S., Skotnicki, T., Fenouillet-Beranger, C., Carriere, N., Chanemougame, D., Morand, Y., Descombes, S., Talbot, A., Dutartre, D., Jenny, C., Mazoyer, P., Palla, R., Leverd, F., Le Friec, Y., Pantel, R., Borel, S., Louis, D., Buffet, N., "Emerging silicon-on-nothing (SON) devices technology", *Solid State Electronics*, vol. 48, pp. 887-895, 2004.

[83] Harrison, S., Munteanu, D., Autran, J.-L., Cros, A., Cerutti, R., Skotnicki, T., "Electrical characterization and modelling of high-performance SON DG MOSFETs", *Proceedings of the 34th European Solid State Device Research Conference*, Leuven, Belgium, pp. 373-376, 2004.

[84] Chanemougame, D., Poncet, A., Monfray, S., Souifi, A., Bourdon, H., Talbot, A., Leverd, F., Delille, D., Skotnicki, T., "2D Poisson-Schrödinger simulations in Ultra-Thin Silicon-On-Nothing devices: Quantum effects impact evaluation", *Proceedings of the IEEE International SOI Conference 2004,* Charlston, SC, pp. 77-78, 2004.

[85] Schulze, J., Eisele, I., Thompson, P.E., Jernigan, G., Suligoj, T., "Silicon-Germanium-Based Combined MBE and CVD Processing for Vertical "Silicon-On-Nothing" (SON) Device Technology", in: Electrochemical Society, "SiGe: materials, processing, and devices", pp. 719-729, Pennington, NJ, 2004.

[86] Monfray, S., Chanemougame, D., Borel, S., Talbot, A., Leverd, F., Planes, N., Delille, D., Dutartre, D., Palla, R., Morand, Y., Descombes, S., Samson, M.-P., Vulliet, N., Sparks, T., Vandooren, A., Skotnicki, T., "SON (Silicon-On-Nothing) technological CMOS Platform: Highly performant devices and SRAM cells", Technical Digest of the *IEEE International Electron Devices Meeting 2004*, San Francisco, CA, pp. 635-638, 2004.

[87] Thompson, P.E., Jernigan, G., Schulze, J., Eisele, I., Suligoj, T., "Vertical SiGe-based silicon-on-nothing (SON) technology for sub-30 nm MOS devices", *Materials Science in Semiconductor Processing*, vol. 8, pp. 51-57, 2005.

[88] Lee, J.-J., Hsu, S., "Silicon-On-Nothing Fabrication Process", US Patent No. 7,078,298, 2006.

[89] Svilicic, B., Jovanovic, V., Suligoj, T., "Vertical Silicon-on-Nothing FET: Threshold Voltage Calculation Using Compact Capacitance Model", *International Semiconductor Device Research Symposium 2007,* College Park, MD, 2007.

[90] Kilchytska, V., Chung, T., Olbrechts, ., Vovk, Y., Raskin, J.-P., Flandre, D., "Electrical characterization of true Silicon-On-Nothing MOSFETs fabricated by Si layer transfer over a pre-etched cavity", *Solid State Electronics*, vol. 51, pp. 1238-1244, 2007.

[91] Peng, B., Yu, T, Yu, F., "A Novel Process of Silicon-on-Nothing MOSFETs with Double Implantation", *Proceedings of the 2007 IEEE International Conference on Integration Technology,* Shenzhen, China, pp. 237-240, 2007.

[92] Kasturi, P., Saxena, M., Gupta, M., Gupta, R. S., "Dual Material Double-Layer Gate Stack SON MOSFET: A Novel Architecture for Enhanced Analog Performance – Part I: Impact of Gate Metal Workfunction Engineering", *IEEE Transactions on Electron Devices*, vol. 55(1), pp. 372-381, 2008.

[93] Kasturi, P., Saxena, M., Gupta, M., Gupta, R. S., "Dual Material Double-Layer Gate Stack SON MOSFET: A Novel Architecture for Enhanced Analog Performance – Part

II: Impact of Gate-Dielectric Material Engineering", *IEEE Transactions on Electron Devices*, vol. 55(1), pp. 382-387, 2008.

[94] Höllt, L., Schulze, J., Eisele, I. Suligoj, T., Jovanovic, V., Thompson, P.E., "First sub-30nm vertical SONFET", *31st International Convention MIPRO*, Opatija, Croatia, in press, 2008.

[95] Schulze, J., "Konzepte Silizium-basierter MOS-Bauelemente", Springer Verlag, Berlin, Germany, 2005.

[96] Gossner, H., Kiunke, W., Eisele, I., Risch, L., Hofmann, K., Treichler, R., Cerva, H., „Vertical Si-MOSFETs with Channel Lenghts Down to 45nm", *Extended Abstracts of the 1993 International Conference on Solid State Devices and Materials*, Chiba, Japan, pp. 422-424, 1993.

[97] Simmons, J., "Generalized Formula for the Electric Tunnel Effect between Similar Electrodes Separated by a Thin Insulating Film", *Journal of Applied Physics*, vol. 34(6), pp. 1793-1803, 1963.

[98] Kane, E., "Zener Tunneling in Semiconductors", *Journal of Physics and Chemistry of Solids*, vol. 12, pp. 181-188, 1959.

[99] Kane, E., "Theory of Tunneling", *Journal of Applied Physics*, vol. 32(1), pp. 83-91, 1961.

[100] Thompson, S., Chau, R., Ghani, T., Mistry, S, Bohr, M., "In Search of 'Forever,' Continued Transistor Scaling One New Material at a Time", *IEEE Transactions on Semiconductor Manufacturing*, vol. 18(1), pp. 26-36, 2005.

[101] Mistry, K., Allen, C., Auth, C., Beattie, B., Bergstrom, D., Bost, M., Brazier, M., Buehler, M., Cappellani, A., Chau, R., Choi, C.-H., Ding, G., Fischer, K., Ghani, T., Grover, R., Han, W., Hanken, D., Hattendorf, M., He, J., Hicks, J., Huessner, R., Ingerly, D., Jain, P., James, R., Jong, L., Joshi, S., Kenyon, C., Kuhn, K., Lee, K., Liu, H., Maiz, J., McIntyre, B., Moon, P., Neirynck, J., Pae, S., Parker, C., Parsons, D., Prasad, C., Pipes, L., Prince, M., Ranade, P., Reynolds, T., Sandford, J., Shifren, L., Sebastian, J., Seiple, J., Simon, D., Sivakumar, S., Smith, P., Thomas, C., Troeger, T., Vandervoorn, P., Williams, S., Zawadzki, K., "A 45nm Logic Technology with High-k+Metal Gate Transistors, Strained Silicon, 9 Cu Interconnect Layers, 193nm Dry Patterning, and 100% Pb-free Packaging," *Technical Digest of the IEEE International Electron Devices Meeting 2007*, Washington, DC, pp. 247-250, 2007.

[102] Kawaura, H., Sakamoto, T., Baba, T., "Oberservation of source-to-drain tunneling current in 8 nm gate electrically variable shallow junction metall-oxide-semiconductor field-effect transistors", Applied Physics Letters, vol. 76(25), pp. 3810-3812, 2000.

[103] Naveh, Y., Likharev, K., "Modeling of 10-nm-Scale Ballistic MOSFET's", *IEEE Electron Device Letters*, vol. 21(5), pp. 242-244, 2000.

[104] Wang, J., Lundstrom, M., "Does source-to-drain tunneling limit the ultimate scaling of MOSFETs?", *Technical Digest of the IEEE International Electron Device Meeting 2002*, San Fransisco, CA, pp. 707-710, 2002.

[105] Kawaura, H., "The Scaling Limit of MOSFETs due to Direct Source-Drain Tunneling", in: Oda, S., Ferry, D., "Silicon Nanoelectronics" , pp. 65-87, CRC Press Taylor & Francis Group, Boca Raton, FL, 2006.

[106] Lundstorm, M., "Device Physics at the Scaling Limit: What Matters?", Technical Digest of the IEEE International Electron Device Meeting 2003, Washington, DC, pp. 789-792, 2003.

[107] ATLAS User Manual, 2-D Device Simulation Software, SILVACO International, Santa Clara, CA,1997.

[108] McKay, K., "Avalanche Breakdown in Silicon", *Physical Review*, vol. 94(4), pp. 877-884, 1954.

[109] Young, K., Burns, J., "Avalanche-Induced Drain-Source Breakdown in Silicon-on-Insulator n-MOSFET"s", *IEEE Transactions on Electron Devices*, vol. 35(4), pp. 426-431, 1988.

[110] Miller, S., "Ionization Rates for Holes and Electrons in Silicon", *Physical Review*, vol. 105(4), pp. 1246-1249, 1957.

[111] Eitan, B., Frohman-Berntchkowsky, D., "Surface Conduction in Short-Channel MOS Devices as a Limitation to VLSI Scaling", *IEEE Transactions on Electron Devices*, vol. 29(2), pp. 254-265, 1982.

[112] Tihanyi, J., Schlötterer, H., "Influence of the Floating Substrate Potential on the Characteristics of ESFI MOS Transistors", *Solid-State Electronics*, vol. 18., pp. 309-314, 1975.

[113] Tihanyi, J., Schlötterer, H., "Properties of ESFI MOS Transistors Due to the Floating Substrate and the Finite Volume", *IEEE Transactions on Electron Devices*, vol. 22(11), pp. 1017-1023, 1975.

[114] Lyu, J.-S., Kang, S.-W., Lee, C.-C., "Substrate Floating Effects of P-Channel SOI MOSFETs", *Proceedings of the 1990 IEEE SOS/SOI Technology Conference*, Key West, FL, pp. 87-88, 1990.

[115] Zhirnov, V., Cavin, R., Hutchby, J., Bourianoff, G., "Limits to Binary Logich Switching Scaling – A Gedanken Model", *Proceedings of the IEEE*, vol. 91(11), pp. 1934-1939, 2003.

[116] Cavin, R., Zhirnov, V., "Future Devices for Information Processing", *Proceedings of the 35th European Solid State Device Research Conference*, Grenoble, France, pp. 7-12, 2005.

[117] Gopalakrishnan, K., Griffin, P., Plummer, J., "I-MOS: A Novel Semiconductor Device with a Subthreshold Slope lower than kT/q*", Technical Digest of the IEEE International Electron Device Meeting 2002*, San Fransisco, CA, pp. 289-292, 2002.

[118] Choi, W., Choi, B., Woo, D.-S., Lee, J., Park, B.-G., "A New Fabrication Method for Self-aligned Nanoscale I-MOS (Impact-Ionization MOS)", *Conference Digest 62nd Device Research Conference*, Notre Dame, IN, pp. 211-212, 2004.

[119] Choi, W., Song, J., Choi, B., Lee, J., Park, Y., Park, B.-G., "80nm Self-Aligned Complementary I-MOS Using Double Sidewall Spacer and Elevated Drain Structure and Its Application to Amplifiers with High Linearity", *Technical Digest of the IEEE International Electron Device Meeting 2004*, San Fransico, CA, pp. 203-206, 2004.

[120] Gopalakrishnan, K., Griffin, P., Plummer, J., "Impact Ionization MOS (I-MOS)-Part I: Device and Circuit Simulations", *IEEE Transactions on Electron Devices*, vol. 52(1), pp. 69-76, 2005.

[121] Gopalakrishnan, K., Woo, R., Jungemann, C., Griffin, P., Plummer, J., "Impact Ionization MOS (I-MOS)-Part II: Experimental Results", *IEEE Transactions on Electron Devices,* vol. 52(1), pp. 77-84, 2005.

[122] Toh, E.-H., Wang, G., Lo, G.-Q., Balasubramanian, N., Tung, C.-H., Benistant, F., Chan, L., Samudra, G., Yeo, Y.-C., "A Nove CMOS Compatible L-Shaped Impact-

Ionization MOS (LI-MOS) Transistor", *Technical Digest of the IEEE International Electron Device Meeting 2005*, Washington, DC, pp. 951-954, 2005.

[123] Choi, W., Song, J., Lee, J., Park, Y., Park, B.-G., "100-nm n-/p-Channel I-MOS Using a Novel Self-Aligned Structure", *IEEE Electron Device Letters*, vol. 26(4), pp. 261-263, 2005.

[124] Choi, W., Song, J., Lee, J., Park, Y., Park, B.-G., "A Novel Biasing Scheme for I-MOS (Impact Ionization MOS) Devices", *IEEE Transactions on Nanotechnology*, vol. 4(3), pp. 322-325, 2005.

[125] Mayer, F., Le Royer, C., Le Carval, G., Clavelier, L., Deleonibus, S., "Static and Dynamic TCAD Analysis of I-MOS Performance: From the Single Device to the Circuit", *IEEE Transactions on Electron Devices*, vol. 53(8), pp. 1852-1857, 2006.

[126] Toh, E.-H., Wang, G., Chan, L, L, G.-Q., Samudra, G., Yeo, Y.-C., "I-MOS Transistor With an Elevated Silicon-Germanium Impact-Ionization Region for Band gap Engineering", *IEEE Electron Device Letters*, vol. 27(12), pp. 975-977, 2006.

[127] Choi, W., Song, J., Lee, J., Park, B.-G., "Effect of Source Extension Junction Depth and Substrate Doping Concentration on I-MOS Device Characteristics", *IEEE Transactions on Electron Devices*, vol. 53(5), pp. 1282-1285, 2006.

[128] Charbuillet, C., Dubois, E., Monfray, S., Bouillon, P., Skotnicki, T., "Fabrication and Analysis of CMOS Fully-Compatible High Conductance Impact-Ionization MOS (I-MOS) Transistors", *Proceedings of the 36th European Solid State Device Research Conference,* Montreux, Switzerland, pp. 299-302, 2006.

[129] Mayer, F., Le Royer, C., Le Carval, G., Tabone, C., Clavelier, L, Deleonibus, S., "Co-integration of 2mV/dec Subthreshold Slope Impact Ionization MOS (I-MOS) with CMOS", *Proceedings of the 36th European Solid State Device Research Conference*, Montreux, Switzerland, pp. 303-306, 2006.

[130] Toh, E.-H., Wang, G., Chan, L, Samudra, G., Yeo, Y.-C., "Reduction of impact-ionization threshold energies for performance enhancement of complementary impact-ionization metal-oxide-semiconductor transistors", *Applied Physics Letters*, **91**, 153501, 2007.

[131] Choi, W., Lee, J., Park, B.-G., "Integration Process of Impact-Ionization Metal-Oxide-Semiconductor Devices with Tunneling Field-Effect-Transistors and Metal-Oxide-Semiconductor Field-Effect Transistors", *Japanese Journal of Applied Physics*, vol. 46(1), pp. 122-124, 2007.

[132] Shen, C., Lin, J.-Q., Toh, E.-H., Chang, K.-F., Bai, P., Heng, C.-H., Samudra, G., Yeo, Y.-C., "On the Performance Limit of Impact-Ionization Transistors", *Technical Digest of the IEEE International Electron Device Meeting 2007*, Washington, DC, pp. 117-120, 2007.

[133] Toh, E.-H., Wang, G., Chan, L., Lo, G.-Q., Sylvester, D., Heng, C.-H., Samudra, G., Yeo, Y.-C., "A Complementary-I-MOS Technology Featuring SiGe Channel and I-region For Enhancement of Impact-ionization, Breakdown Voltage, and Performance", *Proceedings of the 37th European Solid State Device Research Conference,* Munich, Germany, pp. 295-298, 2007.

[134] Toh, E.-H., Wang, G., Lo, G.-Q., Choy, S.-F., Chan, L., Samudra, G., Yeo, Y.-C., "A Strained N-channel Impact-ionization MOS (I-MOS) Transistor with Elevated Silicon-Carbon Source/Drain for Performance Enhancement", *Proceedings of the 2007*

International Symposium on VLSI Technology, Systems and Appliccations, Hsinchu, Taiwan, pp. 1-2, 2007.

[135] Toh, E.-H., Wang, G., Chan, L., Lo, G.-Q., Samudra, G., Yeo, Y.-C., "Strain and Materials Engineering for the I-MOS Transistor With an Elevated Impact-Ionization Region", *IEEE Transactions on Electron Devices*, vol. 54(10), pp. 2778-2785, 2007.

[136] Mayer, F., Le Royer, C., Le Carval, G., Tabone, C., Clavelier, L, Delenoibus, S., "Comparative Study of the fabricated and simulated Impact Ionization MOS (IMOS)", *Sold-State Electronics*, vol. 51, pp. 579-584, 2007.

[137] Mayer, F., Le Royer, C., Le Carval, G., Clavelier, L, Deleonibus, S., "Experimental and TCAD Investigation of the Two Components of the Impact Ionization MOSFET (IMOS) Switching", *IEEE Electron Device Letters,* vol. 28(7), pp. 619-621, 2007.

[138] Toh, E.-H., Wang, G., Zhu, M., Shen, C., Chan, L., Lo, G.-Q., Tung, C.-H., Sylvester, D., Heng, C.-H., Samudra, G., Yeo, Y.-C., "Impact Ionization Nanowire Transistor with Multiple-Gates, Silicon-Germanium Impact Ionization Region, and Sub-5 mV/decade Subthreshold Slope", *Technical Digest of the IEEE International Electron Device Meeting 2007*, Washington, DC, pp. 195-198, 2007.

[139] Toh, E.-H., Wang, G., Chan, L., Samudra, G., Yeo, Y.-C., "A Double-Spacer I-MOS Transistor With Shallow Source Junction and Lightly Doped Drain for Reduced Operating Voltage and Enhanced Device Performance", *IEEE Electron Device Letters*, vol. 29(2), pp. 189-191, 2008.

[140] Toh, E.-H., Wang, G., Chan, L., Samudra, G., Yeo, Y.-C., "Simulation and design of a germanium L-shaped impact-ionization MOS transistor", *Semiconductor Science and Technology*, **23**, 015012, 2008.

[141] Hansch, W., Rao, R., Eisele, I., "Tha Planar-Doped-Barrier-FET: MOSFET Overcomes Conventional Limitations", ", *Proceedings of the 27th European Solid State Device Research Conference*, Stuttgart, Germany, pp. 624-627. 1997.

[142] Rao, R., Hansch, W., Eisele, I., "Simulation, Fabrication and Characterization of High Performance Planar-Doped-Barrier Sub 100 nm Channel MOSFETs", *Technical Digest of the IEEE International Electron Device Meeting 1997*, Washington, DC, pp. 811-814, 1997.

[143] Hansch, W., Rao, R., Fink, C., Kaesen, R., Eisele, I., "Electric field tailoring in MBE-grown vertical sub-100 nm MOSFETs", *Thin Solid Films*, vol. 321, pp. 206-214, 1998.

[144] Born, M., Abelein, U., Bhuwalka, K., Schindler, M., Schmidt, M., Ludsteck, A., Schulze, J., Eisele, I., "Sub-50 nm high performance PDBFET with impact ionization", *Thin Solid Films*, vol. 508, pp. 323-325, 2006.

[145] Abelein, U., Born, M., Bhuwalka, K., Schindler, M., Schmidt, M., Sulima, T., Eisele, I., "A Novel Vertical Impact Ionisation MOSFET (I-MOS) Concept", *Proceedings of the 25th International Conference on Microelectronics MIEL 2006*, Belgrade, Serbia and Montenegro, pp. 121-123, 2006.

[146] Abelein, U., Born, M., Bhuwalka, K., Schindler, M., Schlosser, M., Sulima, T., Eisele, I., "Improved Reliability by Reduction of Hot-Electron Damage in the Vertical Impact-Ionization MOSFET (I-MOS)", *IEEE Electron Device Letters*, vol. 28(1), pp. 65-67, 2007.

[147] Abelein, U., Assmuth, A., Iskra, P., Schindler, M., Sulima, T., Eisele, I., "Doping profile dependence of the vertical impact ionization MOSFET's (I-MOS) performance", *Solid-State Electronics*, vol. 51, pp. 1405-1411, 2007.

[148] Abelein, U., Assmuth, A., Iskra, P., Reinl, M., Schlosser, M., Sulima, T., Eisele, I., "Vertical 40 nm Impact Ionization MOSFET (I-MOS) for High Temperature Applications", *Proceedings of the 26th International Conference on Microelectronics MIEL 2008*, Nis, Serbia and Montenegro, in press, 2008.

[149] Esaki, L., "New Phenomenon in Narrow Germanium p-n Junctions", *Physical Review*, **109**, pp. 603-604, 1958.

[150] Quinn, J., Kawamoto, G., McCombe, B., "Subband Spetroscopy by Surface Channel Tunneling", *Surface Science*, vol. 73, pp. 190-196, 1978.

[151] Baba. T., "Proposal for surface tunnel transistors", *Japanese Journal of Applied Physics*, vol. 31, pp. L445-L457, 1992.

[152] Uemura, T., Baba., T., "Characterization of depletion-type surface tunnel transistor", *Japanese Journal of Applied Physics*, vol. 31, pp. L1727-L1729, 1992.

[153] Knoch, J., Appenzeller, J., "A novel concept for field-effect transistors – the tunneling carbon nanotube FET", Conference Digest Device Research Conference 2004, Santa Barbara, CA, pp. 153-156, 2005.

[154] Reddick, W., Amaratunga, G., "Silicon Surface tunnel transistor", *Applied Physics Letters*, vol. 67(4), pp. 494-496, 1995.

[155] Koga, J., Toriumi, A., "Negative differential conductance in three-terminal silicon tunneling device", *Applied Physics Letters*, vol. 69(10), pp. 1435-1437, 1996.

[156] Koga, J., Toriumi, A., "Three-Terminal Silicon Surface Junction Tunneling Device for Room Temperature Operation", *IEEE Electron Device Letters*, vol. 20(10), pp. 529-531, 1999.

[157] Hansch, W., Fink, C., Schulze, J., Eisele, I., "A vertical MOS-gated Esaki tunneling transistor in silicon", *Thin Solid Films*, vol. 369, pp. 387-389, 2000.

[158] Bhuwalka, K., Sedlmaier, S., Ludsteck, A., Tolksdorf, C., Schulze, J., Eisele, I., "Vertical Tunnel Field-Effect Transistor", *IEEE Transactions on Electron Devices*, vol. 51(2), pp. 279-282, 2004.

[159] Nirschl, T., Wang, P.-F., Hansch, W., Schmitt-Landsiedel, D., "The Tunneling Field Effect Transistor (TFET): The Temperature Dependence, the Simulation Model, and its Application", *Proceedings of the 2004 International Symposium on Circuits and Systems*, Vancouver, Canada, vol. 3, pp. 713-716, 2004.

[160] Wang, P.-F., Hilsenbeck, K., Nirschl, T., Oswald, M., Stepper, C., Weis, M., Schmitt-Landsiedel, D., Hansch, W., "Complementary tunneling transistor for low power application", *Solid State Electronics*, vol. 48, pp. 2281-2286, 2004.

[161] Nirschl, T., Henzler, S., Pacha, C., Wang, P.-F., Hansch, W., Georgakos, G., Schmitt-Landsiedel, D., "The Tunneling Field Effect Transistor (TFET) used in a Singel-Event-Upset (SEU) insensitive 6 transistor SRAM cell in ultra-low voltage applications", 4th *IEEE Conference on Nanotechnology*, Munich, Germany, pp. 402-404, 2004.

[162] Bhuwalka, K., Schulze, J., Eisele, I., "Vertical Tunnel Field-Effect Transistor with Band gap Modulation and Workfunction Engineering", *Proceedings of the 34th European Solid State Device Research Conference,* Leuven, Belgium, pp. 241-244, 2004.

[163] Nirschl, T., Wang, P.-F., Weber, C., Seldmeir, J., Heinrich, R., Kakoschke, R., Schrüfer, K., Holz, J., Pacha, C., Schulz, T., Ostermayr, M., Olbrich, A., Georgakos, G., Ruderer, E., Hansch, W. , Schmitt-Landsiedel, D., "The Tunneling Field Effect Transistor (TFET) as an Add-on for Ultra-Low-Voltage Analog and Digital Processes",

Technical Digest of the IEEE International Electron Device Meeting 2004, San Fransico, CA, pp. 195-198, 2004.

[164] Bhuwalka, K., Schulze, J., Eisele, I., “Scaling Issues of n-channel Vertical Tunnel FET with δp^+ SiGe layer”, *Conference Digest Device Research Conference 2004*, Notre Dame, IN, pp. 215-216, 2004.

[165] Bhuwalka, K., Schulze, J., Eisele, I., “Performance Enhancement of Vertical Tunnel Field-Effect Transistor with SiGe in the δp^+ Layer”, *Japanese Journal of Applied Physics*, vol. 43(7A), pp. 4073-4078, 2004.

[166] Bhuwalka, K., Schulze, J., Eisele, I., “Scaling the Vertical Tunnel FET With Tunnel Band gap Modulation and Gate Workfunction Engineering”, *IEEE Transactions on Electron Devices*, vol. 52(5), pp.909-917, 2005.

[167] Bhuwalka, K., Schulze, J., Eisele, I., “A Simulation Approach to Optimize the Electrical Parameters of a Vertical Tunnel FET”, *IEEE Transactions on Electron Devices,* vol. 52(7), pp. 1541-1547, 2005.

[168] Nirschl, T., Schaper, U., Einfeld, J., Henzler, S., Sterkel, M., Singer, J., Fulde, M., Hansch, W., Georgakos, G., Schmitt-Landsiedel, D., “Impact of Mask Alignment on the Tunneling Field Effect Transistor (TFET)”, *Proceedings of the IEEE 2005 International Conference on Microelectronic Test Structures*, Leuven, Belgium, pp. 43-46, 2005.

[169] Nirschl, T., Henzler, S., fische,r J., Bargagli-Stoffi, A., Fulde, M., Sterkel, M., Teichmann, P., Schaper, U., Einfeld, J., Linnenbank, C., Sedlmeir, J., Weber, C., Heinrich, R., Ostermayr, M., Olbrich, A., Dobler, B., Ruderer, E., Kakoschke, R., Schrüfer, K., Georgakos, G., Hansch, W., Schmitt-Landsiedel, D., “The 65nm Tunneling Field Effect Transistor (TFET) 0.68 μm^2 6T Memory Cell and Multi-V_{th} Device”, *Proceedings of the 35th European Solid State Device Research Conference*, Grenoble, France, pp. 173-176, 2005.

[170] Nirschl, T., Henzler, S., Fischer, J., Fulde, M., Bargagli-Stoffi, A., Sterkel, M., Sedlmeir, J., Weber, C., Heinrich, R., Schaper, U., Einfeld, J., Neubert, R., Feldmann, U., Stahrenberg, K., Ruderer, E., Georgakos, G., Huber, A., Kakoschke, R., Hansch, W., Schmitt-Landsiedel, D., “Scaling properties of the tunneling field effect transistor (TFET): Device and circuit”, *Solid State Electronics*, vol. 50, pp. 44-51, 2006.

[171] Sterkel, M., Wang, P.-F., Nirschl, T., Fabel, B., Bhuwalka, K., Schulze, J., Eisele, I., Schmitt-Landsiedel, D., Hansch, W., “Characteristics and optimisation of vertical and planar tunneling-FETs”, *Journal of Physics: Conference Series – 2nd Conference on Microelectronics and Nanotechnology*, pp. 15-18, 2005.

[172] Born, M., Bhuwalka, K., Schindler, M., Abelein, U., Schmidt, M., Sulima, T., Eisele, I., “Tunnel FET: A CMOS Device for High Temperature Applications”, *Proceedings of the 25th International Conference on Microelectronics MIEL 2006,* Belgrade, Serbia and Montenegro, pp. 124-127, 2006.

[173] Heigl, A., Wachutka, G., “Simulation of Advanced Tunneling Devices”, 6th *International Conference on Advanced Semiconductor Devices and Microsystems*, Bratislava, Slovakia, pp. 121-124, 2006.

[174] Boucart, K., Ionescu, A., “Double Gate Tunnel FET with ultrathin silicon body and high-k gate dielectric”, *Proceedings of the 36th European Solid State Device Research Conference,* Montreux, Switzerland, pp. 383-386, 2006.

[175] Choi, W., Lee, J., Park, B.-G., "Novel Tunneling Devices with Multi-Functionality", *Japanese Journal of Applied Physics*, vol. 46(4B), pp. 2622-2625, 2007.

[176] Nirschl, T., Weis M., Fulde, M., Schmitt-Landsiedel, D., "Correction to "Revision of Tunneling Field-Effect Transistor in Standard CMOS Technologies"", *IEEE Electron Device Letters*, vol. 28(4), p. 315, 2007.

[177] Choi, W., Park, B.-G., Lee, J., Liu, T.-J., "Tunneling Field-Effect Transistor (TFETs) With Subthreshold Slope (SS) Less Than 60 mV/dec", *IEEE Electron Device Letters*, vol. 28(8), pp. 743-745, 2007.

[178] Boucart, K., Ionescu, A., "Double-Gate Tunnel FET With High-κ Gate Dielectric", *IEEE Transactions on Electron Devices*, vol. 54(7), pp. 1725-1733, 2007.

[179] Boucart, K., Ionescu, A., "Length scaling of the Double Gate Tunnel FET with a high-K gate dielectric", *Solid State Electronics*, vol. 51, pp. 1500-1507, 2007.

[180] Born, M., "Vertical Gate Controlled Tunnel Transistors in Si and SiGe", 1st edition, Göttingen, Germany, 2007.

[181] Toh, E.H., Wang, G., Samudra, G., Yeo, Y.-C., "Device physics and design of double-gate tunneling field-effect transistor by silicon film thickness optimization", *Applied Physics Letters*, vol. 90, 263507, 2007.

[182] Vershulst, A., Vandenberghe, W., Maex, K., Groeseneken, G., "Tunnel field-effect transistor without gate-drain overlap", *Applied Physics Letters*, vol. 91, 053102, 2007.

[183] Toh, E.-H., Wang, G., Chan, L., Samudra, G., Yeo, Y.-C., "Device physics and guiding principles for the design of double-gate tunneling field-effect transistor with silicon-germanium source heterojunction", *Applied Physics Letters*, vol. 91, 243505, 2007.

Reviewed by: PD Dr. habil. Joerg Schulze
Siemens AG
Corporate Technology
Virtual Design
CT PP 2
Otto-Hahn-Ring 6
81739 Munich
Germany

In: MOSFETs: Properties, Preparations and Performance ISBN 978-1-60456-762-5
Editors: N.T. Andre et al. pp. 61-84

Chapter 2

EFFICIENT PARALLEL MONTE CARLO SIMULATIONS USING FINITE ELEMENT TETRAHEDRAL MESHES FOR NOVEL THIN-BODY MOSFET ARCHITECTURES

***Manuel Aldegunde*[1], *Antonio J. García-Loureiro*[1], *Natalia Seoane*[1] *and Karol Kalna*[2]**
[1]Departamento de Electrónica y Computación,
Universidad de Santiago de Compostela,
15782 Santiago de Compostela, Spain
[2]Department of Electronics and Electrical Engineering,
University of Glasgow, Glasgow G12 8LT, United Kingdom

Abstract

A 3D Monte Carlo simulator for modelling transport in nano-MOSFETs, based on tetrahedral elements, is described in detail. The basic physical models used to simulate carrier transport in semiconductor devices and how these models can be implemented within the Finite Element method, are presented. In the case of ensemble Monte Carlo simulations, these details include the particle-mesh coupling generalised for arbitrary meshes, treatment of contacts and integration of the equations of motion. The simulator is parallel and uses a domain decomposition strategy for the solution of the Poisson equation whereas a random distribution of particles among processors is used to parallelise the Monte Carlo. A very good scalability of the 3D simulator has been obtained thanks to the Finite Element solver, which is the most computationally intensive stage in self-consistent simulations.

1. Introduction

Monte Carlo (MC) methods have been widely used to simulate the behaviour of semiconductor devices [1–4]. They become decisive to obtain proper on-state characteristics of the transistors since they can account for non equilibrium effects which cannot be accounted for in simpler approaches as drift-diffusion. However, they are computationally very intensive and their use in very large 3D problems (in real space) such as studies of fluctuation effects is often prohibitive, or relatively strong restrictions has to be applied [5,6]. Therefore, a speed-up of the simulation process in order to save computational time through the use of parallel computers is highly desirable [7,8].

Self-consistent simulations are especially expensive, where Poisson equation has to be solved in a 3D mesh at every time step. The size of the time step depends on the material and doping characteristics of the device, but it can be as small as 0.05 fs [9], resulting in thousands of iterations for each bias point for simulations in the order of ps. The need for a frequent solution of the Poisson equation becomes the main bottleneck in simulation which can be tackled using an optimised device mesh. Moreover, novel thin-body architectures are expected to deliver the required on-current in the scaling of semiconductor transistors into nanometre dimensions. However, some of these novel devices do not have planar geometry [10,11]. Therefore, the need of an optimal mesh and the discretisation of complex geometry call for the use of non-rectangular meshes to describe the simulation domain [12].

We have developed a parallel 3D finite element MC simulator based on tetrahedral elements [13]. To assist for a start of the ensemble MC simulation close to steady state, the nonlinear Poisson and current continuity equations are solved using the Finite Element Method (FEM) at the pre-processing stage of each MC simulation. The parallel simulator has been developed for distributed-memory computers [14]. It employs a Multiple Instruction-Multiple Data strategy (MIMD) [15] under the Single Program-Multiple Data paradigm (SPMD) [15] which is achieved by using the Message Passing Interface (MPI) standard library [16]. We have chosen the MPI library due to its availability on many computer systems which guarantees the portability of the code [17].

The chapter is organised as follows. Firstly, Section 2 presents a brief description of the physical models. Section 3 describes the Finite Element (FE) solver and its parallelisation. Section 4 is dedicated to the MC module and Section 5 to the parallelisation of the global system of equations. Section 6 addresses some issues that must be considered when tetrahedral elements are used in particle MC simulations: treatment of self-forces and respective calculation of the electric field. Finally, Section 7 presents some numerical results and we finish in Section 8 with some concluding remarks.

2. Physical Model

To simulate the transport of carriers in semiconductor devices, it is required to calculate the electrostatic potential ψ due to a charge distribution ρ in the device. This is made through the solution of the Poisson equation:

$$\nabla(\varepsilon\nabla\psi) = -\rho.$$

In a semiconductor device, charge density can be decomposed into four contributions [18]:

$$-\rho = q(n - p - N_D^+ + N_A^-),$$

where q denotes the electron charge, n is the concentration of electrons, p is the concentration of holes, N_D^+ is the concentration of ionised donor impurities and N_A^- is the concentration of ionized acceptor impurities.

The carrier transport description will be made within a framework of the Boltzmann Transport Equation (BTE) [18]. There are several methods to solve it with different degrees of approximation. The simplest one, which we will use to initialise the simulation for each bias point, is the Drift-Diffusion (DD) approximation [19]. This solution is a starting point for a MC method.

In the following subsections, both transport models, the DD approach and the MC technique, will be described. We will restrict ourselves to the simulation of n-type MOSFETs, where majority carriers are electrons. In other words, we will not consider the equations for the hole transport. Note that if one needs to consider a hole transport for p-type MOSFETs, the MC technique is very similar. However, care has to be taken to replace the quantities related to electrons (e.g. band structure, scattering rates, etc.) with the ones related to holes. Furthermore, we will also consider the total generation-recombination term to be zero since the applied bias will not exceed conditions leading to an excessive creation of holes in the device.

2.1. The Drift-Diffusion Model

In stationary state, if we only consider one type of charge carrier (electrons), we must solve two coupled equations: the Poisson equation and the electron continuity equation. If, as stated above, we do not consider the generation-recombination of carriers, we have:

$$-\nabla(\varepsilon\nabla\psi) + q[n - p - N_D^+ + N_A^-] = 0, \tag{1}$$

$$\nabla \mathbf{J}_n = 0. \tag{2}$$

To obtain an expression for $\mathbf{J}_n$ we must make several assumptions [18]. In the DD approximation, introducing the quasi-Fermi level for electrons ϕ_n, we obtain the following expressions [18,20]:

$$\mathbf{J}_n = -q\mu_n n(\psi, \phi_n)\nabla\phi_n, \tag{3}$$

where μ_n is the mobility of the electrons and is material dependent. Finally, we have to consider the carrier density n as a function of the quasi-Fermi level and the electrostatic potential given by:

$$n = \sum_j \int_{E_{c_j}}^{E_{sup_j}} f_e(E) g_j(E) dE, \tag{4}$$

where the sums are over all conduction bands. E_{sup_j} and E_{c_j} are the maximum and minimum energy levels of the band j, respectively. The function $f_e(E)$ is the occupation probability density and $g_i(E)$ the density of states of the band i. This expression can be rewritten as [18]:

$$n = n_{ien} \exp\left(\frac{\psi - \phi_n}{V_T}\right), \tag{5}$$

where $V_T = \frac{KT}{q}$, being K the Boltzmann constant and T the temperature. The value of the effective intrinsic concentration of electrons n_{ien} can be calculated to include degeneracy effects of the semiconductor, variations of the parameters with the composition or the existence of several bands or valleys, including non parabolicity effects in that bands.

These parameters can be functions of the electrostatic potential and the quasi-Fermi potential. Since these magnitudes are the unknowns of the problem, an iterative solution process is established in order to guarantee coherent results. Applying Fermi-Dirac statistics, the electron concentration in a single parabolic conduction band approximation can be obtained as:

$$n = N_c F_{1/2}(\eta_c), \tag{6}$$

where N_c is the effective density of states in the conduction band E_c, $F_{1/2}$ is the Fermi-Dirac integral of order $1/2$, and η_c is given by

$$\eta_c = \frac{q\phi_n - E_c}{KT}. \tag{7}$$

2.2. The Boltzmann Transport Equation

The previously described DD model, which can be regarded as further approximation to the BTE [18], assumes that carrier transport is at equilibrium (a change of state of the system is slower than the energy relaxation in the system) and that the electron temperature is the same as the lattice temperature. However, with the scaling of devices, non-equilibrium processes start to play a crucial role in the behaviour of the device.

In order to study non-equilibrium statistics of classical gas, Ludwig Boltzmann, introduced [21] the famous BTE which describes the time evolution of the distribution function f due to binary interparticle collisions [22]. The fundamental quantity in the BTE is the distribution function $f(\mathbf{r}, \mathbf{k}, t)$, which gives the probability of finding a particle at time t with position $\mathbf{r}$ and k-vector $\mathbf{k}$. In essence the BTE is the equation which describes its evolution in the phase space [1]:

$$\frac{\partial f}{\partial t} + \mathbf{v} \cdot \nabla_{\mathbf{r}} f + \frac{q}{\hbar} \nabla_{\mathbf{r}} \psi \cdot \nabla_{\mathbf{k}} f = \left(\frac{\partial f}{\partial t} \right)_C. \tag{8}$$

Note that the BTE can be also derived from the first principles (from Schrödinger equation assuming a specific Hamiltonian) [23] within a framework of Boltzmann transport theory. The RHS of Eq. (8) represents the variation of the distribution function caused by collisions (C). It is given by

$$\left(\frac{\partial f}{\partial t} \right)_C = \int [f(\mathbf{r}, \mathbf{k}', t) P(\mathbf{k}', \mathbf{k})(1 - f(\mathbf{r}, \mathbf{k}, t)) - f(\mathbf{r}, \mathbf{k}, t) P(\mathbf{k}, \mathbf{k}')(1 - f(\mathbf{r}, \mathbf{k}', t))] d\mathbf{k}', \tag{9}$$

where $P(\mathbf{k}, \mathbf{k}')$ is the probability per unit time of the transition of an electron from an initial state $\mathbf{k}$ to a final state $\mathbf{k}'$. The $(1 - f)$ factors come from the Pauli exclusion principle.

The transition probability per unit time is obtained for each considered scattering mechanism using the Fermi's golden rule [1,2]:

$$P(\mathbf{k}, \mathbf{k}') = \frac{2\pi}{\hbar} | \langle \mathbf{k}' | H' | \mathbf{k} \rangle |^2 \delta[\varepsilon(\mathbf{k}) - \varepsilon(\mathbf{k}')], \tag{10}$$

where $\varepsilon(\mathbf{k}')$ represents the energy of the whole system and H' the Hamiltonian of the perturbation, which depends on the particular scattering mechanism.

The collision term on the right-hand side of Eq. (8) is a summation over individual scattering events. It implies that the collision duration τ_c must be much smaller than any other time scale in the problem [24]. In other words, the collisions are regarded as instantaneous processes which interrupt free flight. That means a weak electron-phonon and electron-electron coupling during their mutual interaction. Validity of the BTE requires:

$$\tau_c \ll \tau_L, \tag{11}$$

where the transit time τ_L is defined as the ratio $l/\bar{v}$, where l is the carrier mean free path and $\bar{v}$ is typical carrier velocity. Under the condition (11), one expects an instantaneous response of the distribution function to fields and scattering and the distribution should then evolve, in a Markovian sense, to the local equation of motion. But this model fails when the energy absorbed by the electron from an external field becomes comparable with the width of the energy band.

The electron distribution function $f(\mathbf{r}, \mathbf{k}, t)$ represents essentially the diagonal elements of the electron density matrix $\rho(\mathbf{r}, \mathbf{k}, t)$ in momentum space. This construction makes practical sense only if: i) the momentum states $|\mathbf{r}, \mathbf{k}\rangle$ are good approximations to the true electron states, ii) the distribution function satisfies a closed equation of motion and iii) correct boundary conditions for f can be constructed. The first condition is actually not restrictive, since we can always choose a better set of states. Moreover, the wave vector of the electron is defined by the centre of gravity of the wave packet $|\mathbf{r}, \mathbf{k}\rangle$ what implies that the extension of the wave packet in $\mathbf{k}$-space must be small compared with the dimension of the Brillouin zone. The second condition is more difficult to fulfil. If the Liouville-von Neumann equation [25] is formally projected into equations for f, it will also involve the dynamics of phonons. If these equations are solved for the off-diagonal elements in terms of f, the equations for f can be closed only in the integro-differential form. However, if the electrons are not independent as in the case when the phonon dynamics affects the dynamics of electrons, the equations for f cannot be closed and have to be solved iteratively.

3. Finite Element Discretisation, Solver and Parallelisation

In this section we present the discretisation of the equations of the DD model using the FEM and the parallel algorithms implemented to solve the resulting system of equations.

3.1. Finite Element Solver

The physical model is composed by two non linear equations (1 and 2) with two unknowns, the electrostatic potential ψ and the quasi-Fermi level for the electrons ϕ_n. This system of equations cannot be solved directly, being necessary the use of numerical methods. In our simulation process, they are discretised using the finite element method (FEM) [26] on an unstructured tetrahedral mesh, which allows the simulation of complex domains with great flexibility. At the pre-processing the simulation domain representing the device is triangulated into tetrahedrons using our in-house mesh generator [27]. Then, using the program METIS [28], the solution domain is partitioned into sub-domains and each assigned to an individual processor. The same program is subsequently used to achieve an improved ordering of the nodes of each sub-domain.

3.2. Discretisation of the Poisson Equation

The scaled Poisson equation in stationary state for a material with a relative permittivity ε_r is given by [20]:

$$\varepsilon_r \lambda^2 \nabla^2 \psi = (n - p - C), \qquad x \in \Omega \tag{12}$$

where $\lambda = \sqrt{\frac{\varepsilon_0 V_T}{q C_0 l_0}}$, ε_0 is the permittivity of free space, V_T is the thermal voltage, q is the electron charge, C_0 and l_0 are the scaled values for concentration and length respectively. The equation (12) is subjected to the following boundary conditions:

$$\left.\frac{\partial \psi}{\partial \nu}\right|_{\partial \Omega_N} = 0 \tag{13}$$

$$\psi\Big|_{\partial \Omega_D} = \psi_D\Big|_{\partial \Omega_D}. \tag{14}$$

In addition, if an interface charge Q_I between materials is considered, the boundary condition used in that interface is modified as:

$$\varepsilon_a \left.\frac{\partial \psi}{\partial \nu}\right|_{\partial \Omega_{I,a}} - \varepsilon_b \left.\frac{\partial \psi}{\partial \nu}\right|_{\partial \Omega_{I,b}} = Q_I. \tag{15}$$

Using the Ritz-Galerkin approximation and applying the FEM based on tetrahedral elements with piecewise linear shape functions [26], we can obtain the following weak formulation of the Poisson equation [20]:

$$\lambda^2 \sum_{j=1}^{K} \int_{\Omega_e} (\nabla \widehat{\varphi_j})^t \nabla \widehat{\varphi}_i |J_e| d\Omega_e \psi_j + \lambda^2 \int_{\partial \Omega_I} \frac{Q_I}{\varepsilon_0} \widehat{\varphi}_i dS$$
$$+ \sum_{j=1}^{K} \int_{\Omega_e} (n_j - C_j) \widehat{\varphi_j} \widehat{\varphi}_i |J_e| d\Omega_e = 0$$
$$\forall i = 1, \ldots, K. \tag{16}$$

3.3. Discretisation of the Continuity Equation

The equation which has to be discretised is:

$$\nabla \mathbf{J}_n = 0 \tag{17}$$
$$\forall x \in \Omega, \qquad t \geq 0$$

where

$$\mathbf{J}_n = -q \mu_n n \nabla (\phi_n). \tag{18}$$

The boundary conditions for Eqs. (17) and (18) are

$$n|_{\partial \Omega_D} = n_{eq} \tag{19}$$

$$\left.\frac{\partial n}{\partial \nu}\right|_{\partial \Omega_N} = 0 \tag{20}$$

In principle, it would be possible to use the method described in the previous section for the discretisation of these equations. However, the behaviour of electron concentrations would require a very fine discretisation not affordable for the solution of the whole problem. One of the most important methods to discretize these equations has been originally proposed by Scharfetter and Gummel [29]. It was developed for a one-dimensional device using a finite difference approach and later extended for multi-dimensional devices and finite element methods.

The procedure used to discretise the continuity equation is as follows. Let G_e be the baricenter of an element Ω_e and $\psi_\triangle(x)$ the electrostatic potential defined on the finite element. We can then approximate the current density in the element Ω_e using a mobility in the centre of the element, μ_{n,G_e}, instead of μ_n, and piecewise functions defined on the domain instead of ϕ_n [20].

$$\mathbf{J}_{n,G_e} = \mu_{n,G_e} n_{ien_\triangle,G_T} \exp(\psi_\triangle(P_0)) J_e^{-t} \\ B_T(\psi_\triangle) J_e^t(\nabla \exp(-\phi_{n,\triangle}(x)))|_\triangle, \qquad \forall x \in \Omega_e \tag{21}$$

where J_e is the Jacobian matrix of the transformation from the element Ω_e to the master element, $\{P_0, \ldots, P_d\}$ are the vertices of the element Ω_e, $\psi(P_i)$ is the electrostatic potential on the i-th vertex and $B_T(\psi)$ is the following diagonal matrix:

$$B_T(\psi) = \begin{pmatrix} B(\psi(P_0) - \psi(P_1)) & \cdots & 0 \\ \vdots & \ddots & \vdots \\ 0 & \cdots & B(\psi(P_0) - \psi(P_d)) \end{pmatrix} \tag{22}$$

where B is the Bernoulli function defined as [18]:

$$B(z) = \begin{cases} \frac{z}{(e^z - 1)} & \text{if } |z| \geq 10^{-4} \\ 1 - \frac{z}{2}\left(1 + \frac{z}{6}\right)\left(1 - \frac{z^2}{60}\right) & \text{if } |z| < 10^{-4} \end{cases} \tag{23}$$

The electron continuity equation in terms of the current density defined in the elements is given by:

$$\sum_{\Omega_e} \nabla \mathbf{J}_{n,Ge} = 0. \tag{24}$$

Applying the FEM, the following discretised version of the Eqs. (21), (22), (23) and (24) is obtained [30]:

$$\sum_{\Omega_e} \int_{\Omega_e} \mu_{n,G_e} n_{ien_\triangle,G_T} \exp(\psi_\triangle(P_0)) \nabla \varphi_j J_e^{-t} \\ B_T(\psi_\triangle) J_e^t \nabla(\exp(-\phi_{ni})) \nabla \varphi_i d\Omega_e = 0, \qquad \forall i = 1, \ldots, K \tag{25}$$

3.4. Parallel Linear Solver

To solve the system of equations (16) and (25), the Poisson and current continuity equations are decoupled using Gummel methods and linearised using Newton's algorithm. The

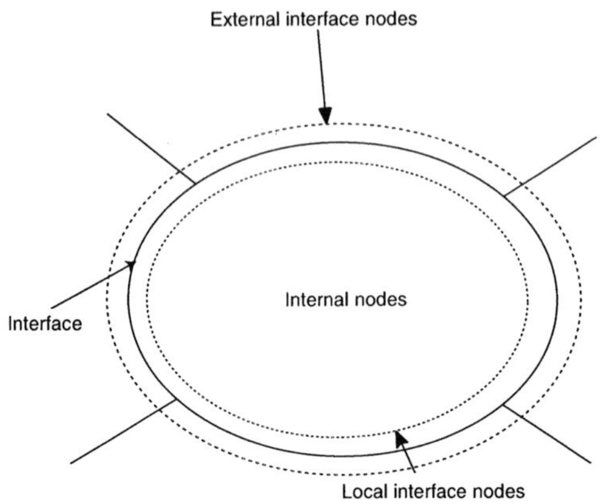

Figure 1. Schematic representation of the nodes in a subdomain.

resulting linear system of equations is solved in parallel using domain decomposition techniques [31] as follows. The solution domain Ω is partitioned in p subdomains Ω_i as

$$\Omega = \bigcup_{i=1}^{p} \Omega_i \tag{26}$$

In essence, the domain decomposition methods attempt to solve the problem on the entire domain Ω by concurrent solutions on each subdomain Ω_i.

A subdomain of the physical solution domain is illustrated in Fig. 1. Each node belonging to a subdomain is an unknown of the whole problem. We have to distinguish between three types of unknowns: (i) interior nodes, those that are coupled only with local nodes, (ii) local interface nodes, those coupled with external nodes as well as local nodes, and (iii) external interface nodes, which are those nodes in other subdomains coupled with local nodes. The nodes are labelled according to their subdomains, first the internal nodes and then both the local and external interface nodes. As a result, the linear system associated with the problem has the following structure:

$$\begin{pmatrix} B_1 & & & & E_1 \\ & B_2 & & & E_2 \\ & & \cdot & & \\ & & & \cdot & \\ & & & B_p & E_p \\ F_1 & F_2 & & F_p & C \end{pmatrix} \begin{pmatrix} x_1 \\ x_2 \\ \cdot \\ \cdot \\ x_s \\ y \end{pmatrix} = \begin{pmatrix} f_1 \\ f_2 \\ \cdot \\ \cdot \\ f_s \\ g \end{pmatrix} \tag{27}$$

where, for the subdomain Ω_i, y represents the vector of all interface unknowns, B_i represents the equations of internal nodes, C the equations of interface nodes, E_i the subdomain to the interface coupling seen from the subdomains and F_i represents the interface to the subdomain coupling seen from the interface nodes.

The METIS program is then used to partition the mesh into subdomains. The same program was subsequently used to relabel the nodes in the subdomains in order to obtain a more suitable rearrangement to reduce the bandwidth of the matrix. When the mesh is finally split to the optimal subdomains, the linear system (27) is solved with parallel sparse iterative solvers library (PSPARSLIB) [32] modified to take advantage of the reordering of the matrices. A great advantage of this library is its optimisation for various multicomputers. Among many domain decomposition techniques supported within this library, the best result was obtained when using the additive Schwarz technique [33].

Additive Schwarz technique

This technique is similar to a block-Jacobi iteration and consists of updating all the new components from the same residual. The basic additive Schwarz iteration can be described as follows:

1.– Obtain $y_{i,ext}$
2.– Compute a local residual $r_i = (b - Ax)_i$
3.– Solve $A_i\delta_i = r_i$
4.– Update solution $x_i = x_i + \delta_i$

The linear system $A_i\delta_i = r_i$ is solved using an standard ILUT preconditioner combined with an FGMRES solver [34].

4. Monte Carlo Method for the Solution of the BTE

MC methods [35] are widely used in various fields of physics including nuclear physics, solid-state physics and statistical physics. When applied to carrier transport in solids, MC provides exact numerical solution of the BTE without necessity to solve this equation directly. In other words, the distribution function $f(\mathbf{r}, \mathbf{k}, t)$ obtained from the MC satisfies the BTE.

This method is by far the more popular simulation technique because it enables us to obtain a solution of the BTE using relatively simple and very effective program algorithms, when compared with direct numerical techniques. Simultaneously, MC provides satisfactory microscopic interpretation of simulated processes. The essence of MC method is to simulate motion of the ensemble of carriers in the **k** space as well as in the **r** space. The motion of each carrier is governed by classical equations of motion and by stochastic collisions with various perturbations (for example, phonons, ions or other carriers). These collisions cause instantaneous transitions between unperturbed Bloch states, with transition probabilities given by quantum mechanics in the approximation of Fermi's golden rule. Using these probabilities, the carrier free-flight time, the scattering channel and final states after scattering can be generated by random numbers. During the simulation any **k**, **r** or t dependent physical quantity can be calculated.

The starting point of the MC program (see Fig. 2) is the definition of the physical system of interest including the material parameters and the values of physical quantities, such as lattice temperature T_0 and electric field **E** [36]. At this level, we also define the parameters that control the simulation: the duration of each subhistory and the time step dt, the desired precision of the result, and so on. The next step is a preliminary calculation of each scattering rate as a function of electron energy. The choice of the dispersion relation $\varepsilon(\mathbf{k})$ usually depends on the simulated transport problem. We have decided to use an analytical

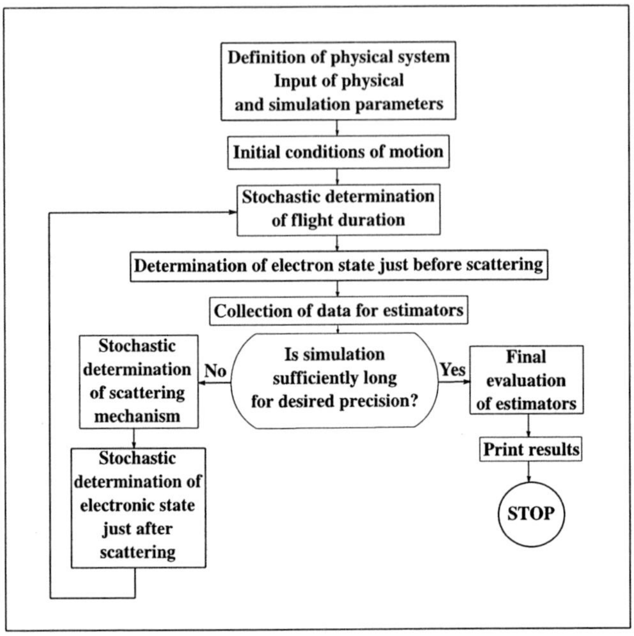

Figure 2. Flowchart of a typical Monte Carlo program.

bandstructure, which is a good compromise between accuracy and efficiency, since our intention is to model the carrier transport in nanoscaled MOSFETs. A typical supply voltage is expected to be smaller than 1 V and we do not intend to investigate any breakdown device mechanisms which can occur at very large electric fields. Therefore, a nonparabolic dispersion law $\sum \hbar^2 k_\alpha^2/(2m_\alpha) = \varepsilon(1+\alpha\varepsilon), \alpha = x, y, z$, is used taking into account an anisotropic shape of the material valleys.

The subsequent step is the generation of the flight duration (Fig. 2). The electron wave vector $\mathbf{k}$ changes continuously during a free flight because of the applied field. Thus, if $\lambda[\mathbf{k}(t)]$ is the scattering probability for an electron in the state $\mathbf{k}$ during the small time interval dt then the probability that the electron, which already suffered a scattering event at time $t = 0$, has not yet suffered further scattering after time t is

$$\exp\left[-\int_0^t dt'\ \lambda[\mathbf{k}(t')]\right], \tag{28}$$

which gives the probability that the interval $(0, t)$ does not contain a scattering. Consequently, the probability $P(t)$ that the electron will suffer its next collision during the time interval dt at a time t is

$$P(t)dt = \lambda[\mathbf{k}(t)]\ \exp\left[-\int_0^t dt'\ \lambda[\mathbf{k}(t')]\right] dt. \tag{29}$$

After some mathematical manipulations, the free-flight time t can be generated from the equation

$$r = \int_0^t dt'\ P(t'), \tag{30}$$

where r is a random number between 0 and 1.

Once the electron free flight is terminated, a scattering mechanism has to be selected. The weight of the i-th scattering mechanism (when n scattering mechanisms are present) is given by a probability

$$P_i(\mathbf{k}) = \frac{\lambda_i(\mathbf{k})}{\lambda(\mathbf{k})}, \qquad \lambda(\mathbf{k}) = \sum_{i=1}^{n} \lambda_i(\mathbf{k}). \tag{31}$$

A random number between 0 and 1 is generated and the following inequalities are tested:

$$\sum_{i=1}^{j-1} \frac{\lambda_i(\mathbf{k})}{\lambda(\mathbf{k})} < r < \sum_{i=1}^{j} \frac{\lambda_i(\mathbf{k})}{\lambda(\mathbf{k})}, \qquad j = 1, \ldots, n. \tag{32}$$

We will accept the j-th mechanism if the j-th inequality is fulfilled. It should be noted that the discussed selection of the free flight time and the scattering channel can be simplified by introducing the self-scattering λ_0 [37,38].

The next step is the choice of the state after scattering. Once the scattering mechanism that caused the end of the electron free flight has been determined, the new state after scattering of the electron, $\mathbf{k}'$, must be chosen as the final state of the scattering event. If the free flight ended with a self-scattering, $\mathbf{k}'$, must be taken as equal to $\mathbf{k}$, the state before scattering. When, in contrast, a true scattering occurred, then $\mathbf{k}'$ must be generated stochastically according to the differential cross section of that particular scattering mechanism.

The scattering mechanisms for Si included in the simulation are acoustic phonon scattering, nonpolar intravalley optical scattering, nonpolar intervalley optical scattering (g- and f-processes), nonpolar optical intervalley scattering (X-L and X-Γ transitions), ionized impurity scattering and interface roughness scattering.

5. Parallelisation of the Global System

In Section 3. we have described how the Poisson equation is parallelised using a domain decomposition strategy. However, this is not, in general, the best way to share the load of the MC engine. With this approximation we would have to perform load balancing since the regions with the highest electron density have the highest number of particles to simulate. Furthermore, they have more scattering events (ionised impurity scattering in regions of high doping and interface roughness scattering in the inversion layer) and, as a consequence, more computational weight. Although a partitioning algorithm weighted using the electron density profile could help to overcome this difficulty, this is well beyond the scope of this work. With this in mind, we have chosen a different strategy to parallelise the MC engine in the simulation. The main idea is to replicate the whole mesh and simulate an equal number of particles in each processor. Although this requires the replication of the mesh a number of times, this will not increase significantly the use of the total memory required by the program since even for a relatively big mesh it would only be a few MB of memory. The other drawback of the strategy is the higher number of communications, but we believe that this is compensated by their regularity and the balancing of the computations. This can be clearly seen, for example, in an extreme case. During the initial iteration of movement, some particles may reach the boundary of their domain. In order for their flight to be completed, communication processes are required to send their properties to the processor which is simulating the domain into which they are transferring. After the communication, a new iteration of the flight is required for those particles which have changed domain in order to finish the time step. However, after this second iteration, some of these particles may again change to a new domain, necessitating additional communications. This process may take several iterations: communications of particle properties to different processors and free flight generations. Although each iteration will be faster than the previous ones, with a high number of processors it may impose a bottleneck. This could be alleviated with specific partitions, but they also slow down the parallel Poisson solver.

In the implemented scheme, free flights are computed completely in parallel, requiring the first communications to be executed for the boundary conditions in the contacts. In this step, a reduction must be done in every tetrahedron of the contact to evaluate the condition of thermal equilibrium [39]. The next communication is required to share the local values of the electron density needed to solve the Poisson equation. Taking advantage of the linearity of the charge assignment process, this can be done with a single reduction step. The electrostatic potential must be sent to every processor after the resolution of the Poisson equation to update the electric fields of the whole mesh. Once the electric field is obtained, the next iteration would be ready to be started.

6. Use of Tetrahedral Element Meshes

The use of tetrahedral elements meshes in the simulation requires the generalisation of the procedures used for cartesian finite difference meshes. These procedures include charge assignment, electric field interpolation, modelling of contacts and particle free flights. In the following, we will briefly address these issues for tetrahedral meshes.

6.1. Charge Assignment and Electric Field Interpolation

The impact of self-forces in particle simulations have been extensively studied in the past [40, 41] and various methods have been proposed to minimise them, but most of these works have achieved a satisfactory result only for orthogonal meshes [41]. From a mathematical point of view, self-forces have been proved to be identically zero for uniform orthogonal meshes if the charge assignment and the force interpolation schemes are the same [40]. However, it must be pointed out that this is not a necessary condition as shown in [41]. In real devices, we have boundary conditions and material inhomogeneities, so the force on a particle which is alone in the device does not have to be zero. However, although the lack of spatial accuracy to resolve the effect of this inhomogeneities will also lead to incorrect forces on the particle, this could be, in principle, avoided by the use of a finer mesh, whereas the impact of self-force cannot.

Hockney and Eastwood [40] demonstrated that the use of the same interpolation scheme for the charge and for the calculation of the electric field may lead to a zero self-force. For this reason, we have decided to adopt the same interpolation scheme for both variables. The most intuive choice is the use of the same interpolant for the finite element solver. We have chosen a linear interpolation which in the case of tetrahedra corresponds to the volume

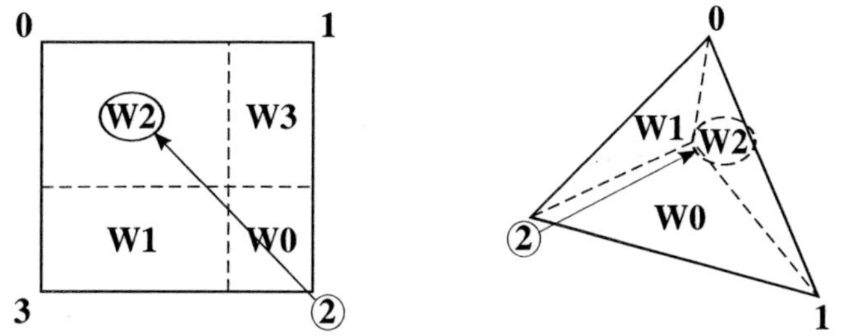

Figure 3. Geometrical interpretation of charge assignment schemes for a Cartesian element (left) and a simplex element (right) in 2D. Wi is the weight associated to node i and corresponds to the area of the subelement opposite to the node. Subelements are defined a point which would be the position of the charge.

coordinates [26]. Fig. 3 compares, in 2D, this interpolation and the standard cloud-in-cell interpolation used in cartesian grids, showing its similar geometrical meaning.

Once the charge has been assigned to the particles, the Poisson equation is solved to obtain the electrostatic potential on the nodes of the mesh. From the potential we can obtain the electric field in every element of the mesh. However, this is not enough, in general, to maintain the zero self-force [41]. In fact, to fulfil the requirement stated above (the use of the same interpolation function for charge density and electric field), we need the electric field value on the nodes of the mesh. For this purpose, we have interpolated the value at the nodes from the value in their neighbouring tetrahedra. Numerical experiments using analytical potential profiles mimicking those in MOSFETs, show that the error is below 5%. The next step to evaluate the quality of the interpolation in our problem is the ability to reproduce the symmetry of the analytical profile, since it will also be an important factor to contributing to the self-forces [40, 41]. To check our interpolation scheme we have solved the Poisson equation in a domain with charge density different from zero in only one node and with boundary conditions far enough so that they do not affect the potential profile near the charged node. From the resulting potential profile we calculate the electric field. We repeat this for the four different nodes of one element. What we have to check is that the electric field in node j due to the charge density in node i is opposite to the electric field in node i due to the charge density in node j. If we write the resulting data in a matrix, then it should be antisymmetric in the ideal case. This matrix would give us the electric field in a node i of the tetrahedron due to the four charges in the following way: $E_i^a = \sum_j G_{ij}^a \rho_j$, with $a = x, y, z$. We have obtained the following normalised (to the maximum value) matrix G^x:

$$G^x \simeq \begin{pmatrix} 4.1 \cdot 10^{-8} & -0.22826 & -0.32240 & -0.99933 \\ 0.22827 & 1.1 \cdot 10^{-5} & -0.16331 & -0.51641 \\ 0.32241 & 0.16333 & 1.1 \cdot 10^{-5} & -0.52093 \\ 1.00000 & 0.51709 & 0.52161 & 6.8 \cdot 10^{-4} \end{pmatrix}$$

The property of anti-symmetry is satisfied by this approximation with an accuracy of almost 0.1%.

6.2. Charge Injection at the Contacts

Other process, which must be adapted to meshes for general geometries, is the injection of particles in the ohmic contacts. First, the implementation of the velocity-weighted model for a particle injection, as described by González and Pardo [39], has to be modified for tetrahedra elements. Tetrahedra with only one or two nodes on the contact do not have a contact surface so particle injection to surface can not be done. The solution is to consider charge neutrality globally in the contact. The number of particles that need to be injected is calculated through the difference between the doping charge in the volume of the contacts (volume of every tetrahedron with some node on the contact) and the charge in terms of number of particles in the same volume. Then, the required particles are injected on the surface of the contact according to the local doping in the surface element. To do this we have implemented the same method used to select the scattering mechanism. We assign a probability proportional to its area and to its doping to every triangle on the contact. Then, a random number is generated and compared to the cumulative normalised $area \cdot doping$ product of the triangles. The particle is injected in the first triangle with cumulative $area \cdot doping$ product greater than the random number,

$$r < \sum_{i=1}^{j} \frac{a_i C_i}{AC}, \qquad j = 1, \ldots, n$$

where a_i is the area of triangle i, AC is the total $area \cdot doping$ product of the contact, $AC = \sum_{i=1}^{n} a_i C_i$, and n the number of triangles of the contact. Once the triangle has been selected, a random position is chosen on the triangle,

$$\begin{cases} \mathbf{r} = \mathbf{r}_0 + r_1 \cdot \mathbf{r}_{01} + r_2 \cdot \mathbf{r}_{02} \\ r_1 + r_2 \leq 1 \\ r_1, r_2 \geq 0 \end{cases}$$

where $\mathbf{r}_0$, $\mathbf{r}_1$ and $\mathbf{r}_2$ are the positions of the nodes of the triangle; r_0 and r_1 two random numbers and $\mathbf{r}_{ij}$ the vector $\mathbf{r}_j - \mathbf{r}_i$.

6.3. Integration of the Equations of Motion

The movement of particles due to an external force (electric field $\mathbf{E}$) is governed by the equations of motion. A new position of a particle can be determined by their integration which is, in principle, mesh independent. The integration becomes rather complex when the particle movement intersects a boundary of the mesh element. First, we will show the equations to be solved and later the difficulties associated with the use of tetrahedral elements.

The equations to be solved for the free flights of the electrons in the semiclassical model are:

$$\mathbf{v} = \frac{d\mathbf{r}}{dt} = \frac{1}{\hbar} \nabla_{\mathbf{k}} \varepsilon(\mathbf{k}) \tag{33}$$

and

$$\frac{d\mathbf{k}}{dt} = \frac{-e\mathbf{E}}{\hbar}, \tag{34}$$

where $\varepsilon(\mathbf{k})$ is given by the band structure of the material. We are using an analytical non-parabolic anisotropic model for the bands (see Section 4.),

$$\varepsilon(\mathbf{k})(1+\alpha\varepsilon(\mathbf{k})) = \frac{\hbar^2}{2}\mathbf{k}\mathbf{M}^{-1}\mathbf{k} = \gamma = \frac{\hbar^2|\mathbf{k}^s|^2}{2m^*}, \tag{35}$$

where $\mathbf{M}^{-1}$ is the inverse effective mass tensor. We have introduced the effective mass m^* with the use of the Herring-Vogt transformation [42]. For example, the transformation for the X valleys is as follows:

$$k_a^s = \sqrt{\frac{m^*}{m_a}}k_a, \qquad a = x, y, z. \tag{36}$$

The symmetry of the inverse effective mass tensor leads to the following equation for the position of the particle:

$$\frac{d\mathbf{r}}{dt} = \frac{\hbar\mathbf{M}^{-1}\mathbf{k}}{\sqrt{1+4\alpha\gamma}} = \frac{\hbar\mathbf{k}^*}{\sqrt{1+4\alpha\gamma}}, \tag{37}$$

Again, for the X valley, we have

$$k_a^* = \frac{k_a}{m_a}, \qquad a = x, y, z. \tag{38}$$

Since the electric field must be constant in the free flight between two consecutive solutions of the Poisson equation to avoid self-forces,

$$\mathbf{k}^*(t) = \mathbf{k}^*(0) - \frac{q\mathbf{E}^*}{\hbar}t \tag{39}$$

and

$$\mathbf{r}(t) = \mathbf{r}(0) + \hbar\int_0^t \frac{\mathbf{k}^*(0) - \frac{q\mathbf{E}^*}{\hbar}t}{\sqrt{1+4\alpha\gamma(t)}}dt. \tag{40}$$

The time dependency of γ is given by

$$\gamma = \hbar^2\frac{\mathbf{k}^s\cdot\mathbf{k}^s}{2m^*} = \frac{\hbar^2}{2m^*}\left[|\mathbf{k}^s(0)|^2 - 2\frac{q}{\hbar}\mathbf{E}^s\cdot\mathbf{k}^s(0)t + \frac{q^2}{\hbar^2}|\mathbf{E}^s|^2t^2\right]. \tag{41}$$

At this point we may make an approximation by letting γ constant during the free flights. We will come back later to this approximation. If we decide to include the time dependency of γ , we have to solve the following integral:

$$\int_0^t \frac{a+bt}{\sqrt{c+dt+et^2}}dt, \tag{42}$$

where a, b, c, d and e are constants. This integral can be solved analytically,

$$\begin{aligned}\int_0^t \frac{a+bt}{\sqrt{c+dt+et^2}}dt &= \frac{b}{e}\sqrt{c+dt+et^2}+ \\ &+ \left(\frac{a}{\sqrt{e}} - \frac{bd}{2e\sqrt{e}}\right) arcsenh\frac{2et+d}{\sqrt{4ec-d^2}} - \\ &- \frac{b}{e}\sqrt{c} - \left(\frac{a}{\sqrt{e}} - \frac{bd}{2e\sqrt{e}}\right) arcsenh\frac{d}{\sqrt{4ec-d^2}}.\end{aligned} \tag{43}$$

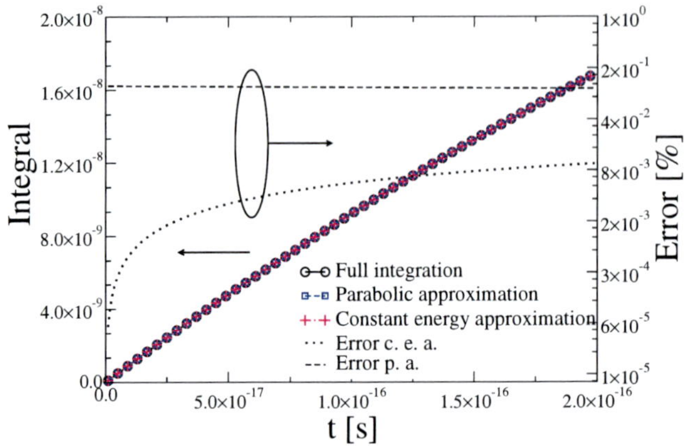

Figure 4. Value of the integral involved in the calculation of the trajectories over a typical time step of 0.2 fs. The parabolic and the constant γ approximations and the error made with them are also shown.

Solving this equation to find the position of the particle after the free flight can be too expensive. As we said above, we can approximate the trajectory of the particle by using a constant γ during the free flight. In this case the integral is

$$\int_0^t \frac{a+bt}{\sqrt{1+4\alpha\gamma}}dt = \frac{at+\frac{bt^2}{2}}{\sqrt{1+4\alpha\gamma}} \tag{44}$$

and the trajectory is a parabola:

$$\mathbf{r}(t) = \mathbf{r}(0) + \frac{\hbar\mathbf{k}^*(0)}{\sqrt{1+4\alpha\gamma}}t - \frac{q\mathbf{E}^*}{2\sqrt{1+4\alpha\gamma}}t^2. \tag{45}$$

This is equivalent to the result obtained using a parabolic band approximation with mass modulated by the energy of the electron. In Fig. 4 we show the value of the integral for the three cases: non-parabolic band, non-parabolic band with constant γ during the free flight and parabolic band approximations. We also show the error when using the two later approximations when their result is compared with the first one for a given initial condition. The values were calculated for the initial conditions $\mathbf{k} = (10^8, 0, 0)$ $[m^{-1}]$ and $\mathbf{E}^* = (10^8, 0, 0)$ $[V/m]$. However, the difference between both approximation may increase for different sets of initial conditions, e.g., with lower electric fields found, for example, in source/drain regions.

After we have introduced the equations of trajectories, we can discuss their treatment for tetrahedral elements. The first of the points to discuss is related to the identification of the element enclosing the particle. This task is both more accurate and faster in cartesian meshes, since it only involves comparisons of the coordinates of the particle with the coordinates of the planes of the element. However, in tetrahedral meshes, the comparisons cannot be done in such straightforward way. We have two main options: transform the coordinates to a system where the comparison is simple or make comparisons of relative positions of a point and four planes. These two methods involve more operations when compare to cartesian meshes, which result in a slower and less accurate determination. The

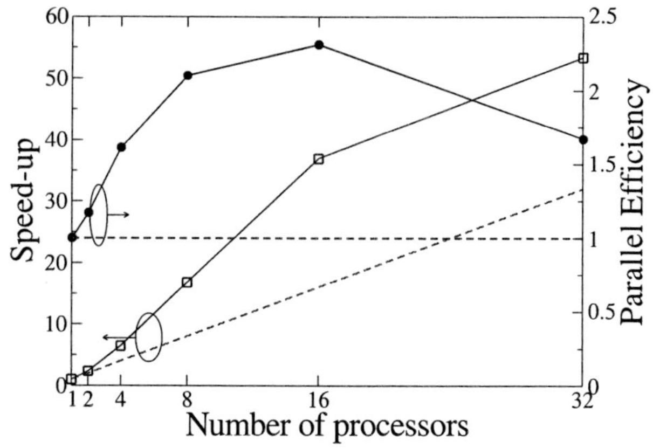

Figure 5. Speedup (crosses) and parallel efficiency (circles) of one time step. We also show the ideal values (dashed lines) for comparison.

next point involves the determination of time when the trajectory of the particle crosses the element boundary. In the case of tetrahedral meshes, we have to check four boundary planes instead of six. This check is more time consuming for tetrahedral meshes since it involves the complete three dimensional trajectory, whereas in cartesian meshes the three components of the trajectory can be decoupled. We have to solve the following equation for t:

$$Ax(t) + By(t) + Cz(t) + D = 0, \tag{46}$$

where A, B, C and D are the coefficients which define the plane of a given face of the element. In cartesian meshes, the six equations to solve are $x = x_0$, $x = x_1$, $y = y_0$, $y = y_1$, $z = z_0$ and $z = z_1$. However, in tetrahedral meshes, we have to solve Eq. (46) for four general planes. The main difference is that having to solve for general planes leads to a higher uncertainty. A tolerance in the position of the particle becomes almost compulsory to avoid later problems in the determination of the element enclosing the particle, which, as commented above, has some degree of numerical uncertainty in tetrahedral meshes. This depends on the approximation we have used to move the particle. The tolerance has to be larger when we use the full solution of the equations of motion for the non-parabolic energy band, since it requires the use of numerical methods to compute the solution to Eq. (46).

7. Numerical Results

In this section we show some results obtained with the presented FEM MC simulator. Firstly, we show the scalability of the simulator with the number of processors. This study is done for a small period of simulated time in equilibrium. After this study, we show results obtained after real simulations of the I_D-V_G curve of a DG MOSFET.

7.1. Parallel Performance

The developed MC code has been tested by simulating a double gate (DG) MOSFET with a channel length of 4 nm, a body thickness of 3.3 nm, a gate dielectric stack of 0.54 nm and a source/drain doping of 1×10^{20} cm^{-3}. The parallel performance of the code has been investigated over a small period of 20 fs. All the results presented here were obtained on an HP Superdome cluster with 128 Itanium 2 1.5 GHz processors.

The scalability of the program can be characterised by the speedup (S) or the parallel efficiency (PE), defined for p processors as:

$$S(p) = \frac{t_1}{t_p} \qquad PE(p) = \frac{t_1}{t_p p} \tag{47}$$

where t_1 and t_p are the times employed running the simulation on one and p processors, respectively. Fig. 5 shows their dependence with the number of processors. We can see that their values exceed the linear scaling limit. This effect is due to the superefficiency of the Poisson solver and it starts to vanish for a high number of processors when operations corresponding to non scaling stages (e.g., the electric field update) begin to be of the same order of magnitude than the scaling operations (the solution of linear system or the particle free flights). The performance limit of the superlinear behaviour is intrinsic to our parallelisation strategy and cannot be avoided. The limit will mainly depend on the size of the employed mesh for a given number of processors.

Table 1 shows the execution times of different parts of the simulation process in the ensemble Monte Carlo method. We obtain a very good scalability for the particle free flight times being limited by the contact treatment, where communications and sequential code are required in order to maintain the equilibrium. The other limitation is the update of the electric field, which requires a constant time independent of a number of processors. These two limitations do not affect the scalability for a low number of processors, but will affect the scalability for a high number of processors, when these times are of the same order of magnitude or larger than those required for MC free flights or for the solution of the Poisson equation. This effect can be alleviated if we calculate $\mathbf{E}$ in parallel and then communicate among assigned processors. The communication times for passing electrostatic potential show that this could speed up the simulation. However, one has take into account that it could slow down the simulation in systems with a slow interprocessor communication.

7.2. Numerical Examples

To test our ensemble MC code in real simulations, we have chosen a larger DG transistor, the 10 nm gate length MOSFET with a body thickness of 6.1 nm and a source/drain doping of 1×10^{20} cm^{-3}.

Fig. 6 shows the drift electron velocity as a function of applied electric field in bulk Si in $\langle 100 \rangle$ and $\langle 111 \rangle$ directions compared to experimental data. This justifies our transport model for electrons in Si.

The particle positions in the device are initialised using a rejection technique to reproduce the electron density obtained from a drift-diffusion solution for each bias point. Fig. 7 shows an example of the initial distribution for equilibrium conditions. Average electron concentration and velocity are also shown in Figs. 8 and 9, respectively.

Table 1. Mean simulation times (s) for one time step ($t_{iteration}$), linear Poisson solver ($t_{Poisson}$), particle flights (t_{flight}), treatment of contacts ($t_{contacts}$), reduction of electron concentration ($t_{reduction}$), communication of potentials ($t_{comun-poten}$) and electric field update ($\mathbf{E}_{update}$) for different number of processors (n_p)

n_p	$t_{iteration}$	$t_{Poisson}$	t_{flight}	$t_{contacts}$	$t_{reduction}$	$t_{comun-poten}$	$\mathbf{E}_{update}$
1	27.50	24.9	0.68	0.35	0.013	0.019	0.207
2	11.74	11.1	0.33	0.20	0.010	0.007	0.204
4	4.27	3.87	0.173	0.146	0.0085	0.0054	0.203
8	1.64	1.08	0.102	0.212	0.0188	0.0050	0.198
16	0.745	0.386	0.0420	0.101	0.005	0.0042	0.200
32	0.5152	0.186	0.0209	0.079	0.009	0.006	0.206

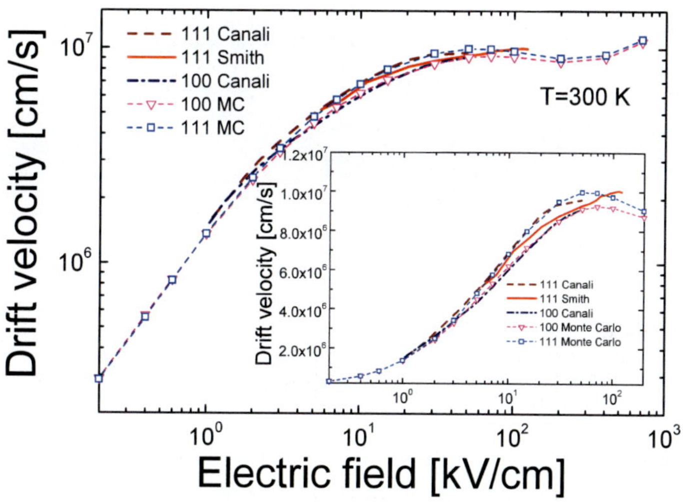

Figure 6. Average electron velocity as a function of applied electric field for bulk Si in $\langle 100 \rangle$ and $\langle 111 \rangle$ crystallographic directions compared with indicated experimental data. The inset shows the same data on a linear scale.

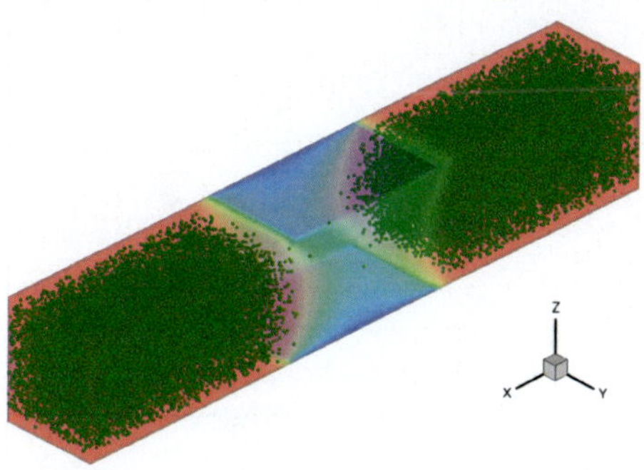

Figure 7. Particles and electrostatic potential in the 10 nm DG MOSFET after particle initialisation.

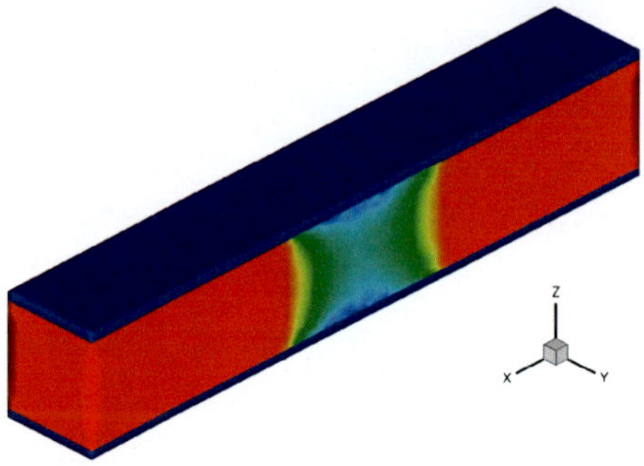

Figure 8. Average electron density during a simulation at a bias of $V_D = 0.05$ V, $V_G = 0.4$ V.

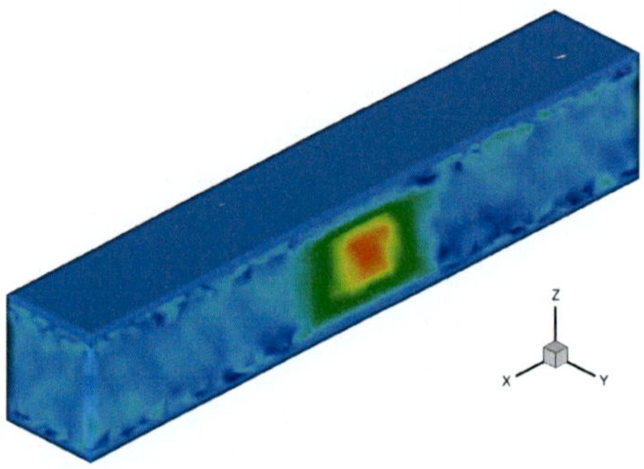

Figure 9. Average x-velocity in the device during a simulation at a bias of $V_D = 0.05$ V, $V_G = 0.4$ V.

As we mentioned in Section 6., the way in which the electron density is obtained from the charge density will affect the Poisson equation. Fig. 10 shows a comparison between the I_D-V_G characteristics of the device obtained from the simulation using three different electron density calculation approaches. The main difference in the results is found in the subthreshold region, where inaccuracies in the interpolation of the electric field play an important role. Fig. 10 also shows a comparison of non self-consistent MC, self-consistent MC and DD simulations of the same device. Current in MC and DD simulations are similar in the subthreshold region, although MC simulations are inaccurate in this region due to excessive numerical noise entering the averaging of a very small quantity, the drain current I_D. However, for high gate voltages, electric field in the semiconductor is too high, the carrier transport becomes highly non-equilibrium and, consequently, the DD mobility model underestimates the current.

8. Conclusion

We have described a parallel 3D device simulator based on the ensemble MC method to model the carrier transport and on the FE method to solve the Poisson equation on unstructured tetrahedral meshes. This device simulator is aimed for modelling of novel, thin-

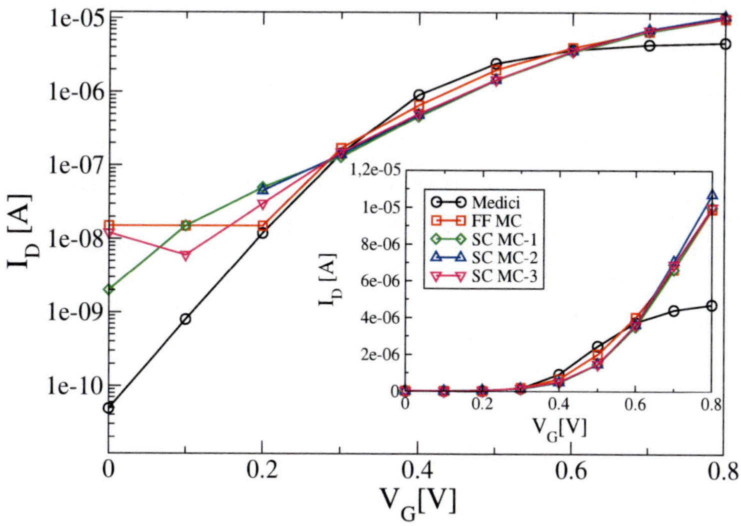

Figure 10. I_D-V_G (at low drain bias, V_D=0.05V) characteristics of the device obtained from different simulation approaches: drift-diffusion (Medici [43]), non self-consistent Monte Carlo Simulation (FF MC) and self-consistent Monte Carlo simulation with three different charge assignment schemes (SC MC-1,2,3).

body transistor architectures with a complex 3D geometry. We have discussed in detail the specifics encountered in the integration of the Monte Carlo engine within the FE solver for Poisson equation which uses a tetrahedral mesh and how to deal with them. The developed 3D MC device simulator has been tested by modelling a 10 nm gate length Si DG MOSFET with a body thickness of 6.1 nm and a gate dielectric stack of 0.5 nm.

The solution of the Poisson equation was parallelised using domain decomposition methods whereas the parallelisation of the MC method was based on particle redistribution. This mixture of parallelisation strategies optimises the parallel performance of the program. The investigations of the parallel performance exhibit superlinear behaviour (overtaking the linear limit) but the maximum optimal number of processors will depend on the size of the mesh. Even the saturation in the parallel performance occurs with increasing of number of processors, the parallel device code is an ideal tool for production runs requiring multiple simulations.

Acknowledgments

This work was supported by Spanish Government (MCYT) under the project TIN2007-67537-C03-01 and by Xunta de Galicia under the project PGIDIT07TIC01CT. We are particularly grateful to CESGA (Galician Supercomputing Centre) for providing access to HP Superdome System. MA thanks Ministerio de Educación y Ciencia de España for the awarded fellowship (FPU). KK would like to thank the UK EPSRC grant (EP/D070236/1) for its support.

References

[1] Jacoboni, C.; Lugli, P. *The Monte Carlo Method for Semiconductor Device Simulation*; Springer–Verlag: Wien–New York, 1989.

[2] Jacoboni, C.; Reggiani, L. *Rev. Mod. Phys.* 1983, 55, 645–705.

[3] Fischetti, M. V.; Laux, S. E. *Phys. Rev. B* 1988, 38, 9721–9745.

[4] Zebarjadi, M.; Bulutay, C.; Esfarjani, K.; Shakouri, A. *Appl. Phys. Lett.* 2007, 90, 092111.

[5] Ramey, S. M.; Ferry, D. K. *J. Comp. Electron.* 2002, 1, 267–271.

[6] Riddet, C.; Brown, A. R.; Alexander, C.; Watling, J. R.; Roy, S.; Asenov, A. *IEEE Trans. Nanotechnol.* 2007, 6, 48–55.

[7] Ranawake, U. A.; Huster, C.; Lenders, P. M.; Goodnick, S. M. *IEEE Trans. Comput-Aided Des. Integr. Circuits Syst.* 1994, 13, 712–724.

[8] Banse, F.; Dekeyser, J.-L.; Fauquenbergue, R.; Dessenne, F. *Lect. Notes Comput. Sc.* 1998, 1401, 193–202.

[9] Vasileska, D.; Ahmed, S. S. *IEEE Tran. Electron Devices* 2005, 52, 227–236.

[10] Kavalieros, J.; Doyle, B.; Datta, S.; Dewey, G.; Doczy, M.; Jin, B.; Lionberger, D.; Metz, M.; Rachmady, W.; Radosavljevic, M.; Shah, U.; Zelick, N.; Chau, R. *2006 Symposium on VLSI Technology, Digest of Technical Papers*; 2006, 50–51.

[11] Cho, K. H.; Suk, S. D.; Yeoh, Y. Y.; Li, M.; Yeo, K. H.; Kim, D.-W.; Park, D.; Lee, W.-S.; Jung, Y. C.; Hong, B. H.; Hwang, S. W. *IEEE Electron Device Lett.* 2007, 28, 1129–1131.

[12] Ghetti, A.; Carnevale, G.; Rideau, D. *IEEE Trans. Nanotechnol.* 2007, 6, 659–666.

[13] Aldegunde, M.; García-Loureiro, A. J.; Kalna, K.; Asenov, A. *J. Comput. Electron.* 2006, 5, 311–314.

[14] García-Loureiro, A. J.; Pena, T. F.; López-González, J. M.; Prat, L. *Int. J. Numer. Methods Eng.* 2000, 49, 639–652.

[15] Pfister, G. *In Search of Clusters*; Prentice Hall PTR: Upper Saddle River, NJ, 1998.

[16] Group, W.; Lusk, E.; Skjellum, A. *Using MPI*; The MIT Press: Boston, 1996.

[17] The message passing interface (MPI) standard.
URL `http://www.mpi-forum.org/`

[18] Selberherr, S. *Analysis and Simulation of Semiconductor Devices*; Springer–Verlag: Wien–New York, 1984.

[19] Ravaioli, U. *Semicond. Sci. Technol.* 1998, 13, 1–10.

[20] Markowich, P. A. *The Stationary Semiconductor Device Equations*; Springer–Verlag: Wien–New York, 1986.

[21] Boltzmann, L. *Wien. Ber.* 1877, 76, 373–435; ibid. 1887, 95, 153–164; *Nature* 1895, 51, 413–415.

[22] Reif, F. *Fundamentals of Statistical and Thermal Physics*; McGraw-Hill: NY, 1965; pp 509.

[23] Kalna, K. *Non-Linear Carrier Dynamics in Semiconductor Microstructures*; Institute of Electrical Engineering, Slovak Academy of Sciences, Bratislava, Slovakia, 1994. [http://userweb.elec.gla.ac.uk/k/kalna/Papers/project_to_thesis.pdf].

[24] Kohn, W.; Luttinger, J. M. *Phys. Rev.* 1957, 108, 590–611.

[25] Schiff, L. I. *Quantum Mechanics* 3rd ed.; McGraw-Hill: NY, 1968; pp 24–30.

[26] Becker, E. B.; Carey, G. F.; Oden, J. T. *Finite Elements*; Prentice-Hall: Englewood Cliffs, NJ, 1981.

[27] Aldegunde, M.; Pombo, J. J.; García-Loureiro, A. J. *Int. J. Numer. Model.-Electron. Netw. Device Fields* 2006, 19, 473–489.

[28] Karypis, G.; Kumar, V. *METIS 4.0: A Software Package for Partitioning Unstructured Graphs, Partitioning Meshes, and Computing Fill–Reducing Orderings of Sparse Matrices*; Tech. rep., Univ. of Minnesota, 1998.

[29] Scharfetter, D. L.; Gummel, H. K. *IEEE Trans. Electron Devices* 1969, 16, 64–77.

[30] Garcia-Loureiro, A. J.; Kalna, K.; Asenov, A. *Int. J. Numer. Model.-Electron. Netw. Device Fields* 2005, 18 327–340.

[31] Barrett, R.; Berry, M.; Chan, T.; Demmel, J.; Donato, J.; Dongarra, J.; Eijkhout, V.; R.Pozo, ; Romine, C.; van der Vorst, H. *Templates for the Solution of Linear Systems: Building Blocks for Iterative Methods*; SIAM: Philadelphia, PA, 1994.

[32] Saad, Y.; Lo, G.-C.; Kuznetsov, S. *PSPARSLIB users manual: A portable library of parallel sparse iterative solvers*; Tech. rep., Univ. of Minnesota, 1997.

[33] Seoane, N.; García-Loureiro, A. J. *Int. J. Numer. Model.-Electron. Netw. Device Fields* 2006, 19, 15–32.

[34] Saad, Y. *Iterative Methods for Sparse Linear Systems*; PWS Publishing Co.: Boston, MA, 1996.

[35] Metropolis, N.; Ulam, S. M. *J. Amer. Stat. Assoc.* 1949, 44, 335–341.

[36] Jensen, G. U.; Lund, B.; Fjeldly, T. A.; Shur, M. *Comput. Phys. Commun.* 1991, 67, 1–61.

[37] Rees, H. D. *J. Phys. Chem. Solids* 1969, 30, 643–655.

[38] Rees, H. D. *Phys. Lett. A* 1968, 26, 416–417.

[39] González, T.; Pardo, D. *Solid–State Electron.* 1996, 39, 555–562.

[40] Hockney, R. W.; Eastwood, J. W. *Computer Simulation Using Particles* ; IOP Publishing: Bristol and Philadelphia, PA, 1988.

[41] Laux, S. E. *IEEE Trans. Comput-Aided Des. Integr. Circuits Syst.* 1996, 15, 1266–1277.

[42] Sota, T.; Suzuki, K. *J. Phys. C: Solid State Physics* 1983, 16, 4347–4364.

[43] Synopsis, Medici User Guide. Version X-2005.10 (2005).

In: MOSFETs: Properties, Preparations and Performance ISBN: 978-1-60456-762-5
Editors: N.T. Andre et al. pp. 85-114

Chapter 3

CRYOGENIC OPERATION OF POWER MOSFETS

Hua Ye and Pradeep Haldar
State University of New York at Albany

I. Introduction

Development of cryogenic power electronic system has attracted much attention recently due to the substantially better performances of semiconductor power devices such as power MOSFETs at cryogenic temperature, e.g., higher efficiency and reduced weight and size[1-4]. For instance, at low temperatures, power MOSFETs demonstrate much lower on-resistance and higher switching speed resulting from increased carrier mobility and saturation velocity [5-7]. At lower temperatures, the thermal conductivities of the device itself and the substrate materials in the packaging improve drastically, which lead to better and simpler thermal management and better reliability. Recent advances in $YBa_2Cu_3O_x$-based coated conductor development are enabling long lengths of Second Generation High Temperature Superconductor (HTS) tapes to operate above liquid nitrogen temperature (>77K) in high magnetic fields. Adapting power electronics control systems for low-temperature operation in combination with HTS devices (such as HTS motors or generators) is a natural step that provides synergistic benefits such as higher power density, further reductions in size, reduced loads on the cryogenic system due to fewer connections to outside world, higher reliability and improved performance. Being able to operate power MOSFETs at cryogenic temperatures is also very beneficial for deep space exploration spacecrafts since radioisotope heating unit (RHU) will no longer be needed to maintain these electronics within the tolerable temperature range (normally near room temperature on earth) [8], thereby reducing system size and weight. In this chapter, the present status of CPE is discussed in order to provide an outlook on emerging device technologies that are gaining in interest. Advanced power electronic packaging/interconnect methods that adapt well to cryogenics and help further improve system performance is discussed. Development of cryogenic integrated power package is discussed in detail. Furthermore, understanding how the structure of the device unit-cell affects the cryogenic performance of a Si power MOSFETis an important step toward optimizing of the device for cryogenic operations. Physically based numerical simulations of the Si power DMOS are performed in order to analyze the effects of device structure on their performances at cryogenic temperatures. The simulation results suggest that the device optimized for room temperature operation can be further optimized at cryogenic temperatures. In the end of the chapter, the possibility of utilizing superjunction power MOSFET for cryogenic operation is also discussed based on experimental testing results.

II. Background and Challenges

Part of the key to the successful operation of electronics at cryogenic temperatures depends on the type of electronic components used. Bipolar Junction Transistors (BJT) rely on thermal excitation to allow for conduction and are subject to carrier freeze out and reduced gain due to bandgap narrowing with decreasing temperature [9]; therefore, bipolar transistors have a built in degradation beneath their specified temperature range. On the other hand, MOSFET is able to operate at cryogenic temperature since the gain of MOSFET actually improves at lower temperatures due to increased carrier mobility, higher transconductances, steeper sub threshold slope, higher threshold voltage, and reduced junction leakage [10-12].

One of the main goals for the design of electronic power modules is to optimize the conversion efficiency. For power MOSFET, its on-resistance determines that efficiency to a large extent in all switch-model applications [13]. In Mueller's experiments [13], he showed that on-resistance of a commercially available high-voltage MOSFETs (500-1000 V) decreases by a factor of 10-20 or more depending on the drain current if cooled down to 77 K (Figure 1). This reduced on-resistance is also due to the increase carrier mobility at lower temperatures. Shenai reported the performance potential of a power MOSFET at cryogenic temperature based on a simple analysis and the measured performance of scaled silicide 30-V power MOSFETs [14]. It was shown that an order of magnitude improvement in on-state resistance can be achieved at liquid nitrogen temperature. This performance improvement results in an order of magnitude improvement in optimum power conversion frequency for a given die size, a factor of 2 reduction in die size at a given conversion frequency, and a factor of 3 reduction in total power loss for switched-mode power converters operated at 77K.

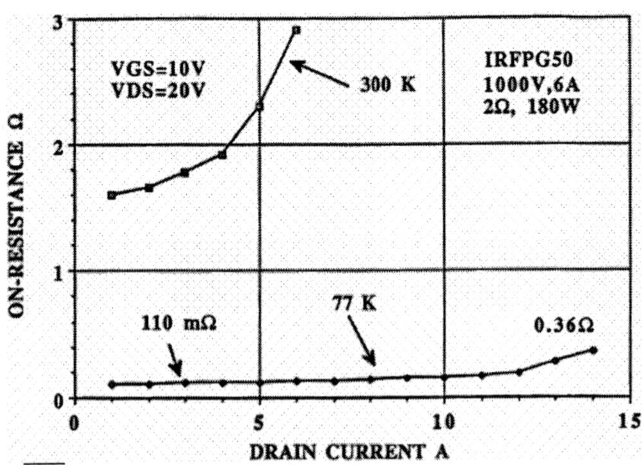

Figure 1. Measured on-resistance of a power MOSFET at 77k and 300K (courtesy of Mueller et al. [13]).

Higher power conversion efficiency also comes from the fact that the switching or dynamic losses of power electronics decreases at cryogenic temperatures. These improvements are due to a moderate decrease of inherent device capacitance at lower temperatures. Furthermore, substantial switching losses are reduced relatively on a per-unit-power basis. For example, the switching loss of a power MOSFET P_s is given as $P_s = k \cdot f \cdot C_0 \cdot V_{ds}^{\,2}$, where k is a constant, f is switching frequency, C_0 is device capacitance, and V_{ds} is drain-source voltage. As aforementioned, at the same drain-source voltage, the device has much higher drain current at cryogenic temperature, which means that the device has the same switching loss at much higher power level.

The switching speed and high frequency performance of power electronic devices also improves dramatically at cryogenic temperatures due to the higher majority carrier mobility and saturation velocity. Lu et al. showed (Figure 2) that the overall high frequency performance of a power MOSFET increased dramatically by exposing the device to the liquid nitrogen temperature [15].

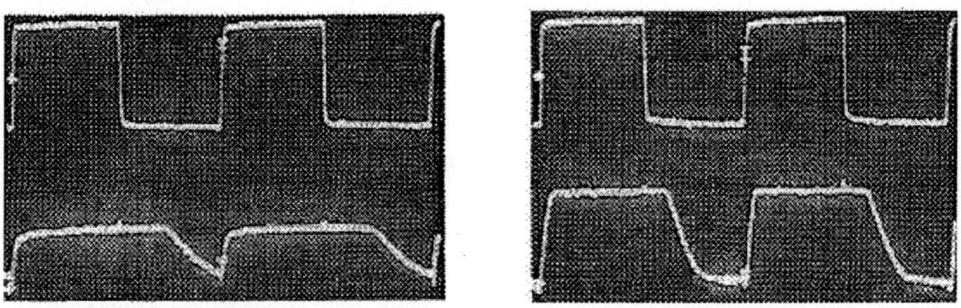

(a) (b)

Figure 2. Switching characteristic of a power MOSFET at 100KHz (a)295K (upper curve: input signal; lower curve: MOSFET output); (b)77K (upper curve: input signal; lower curve: MOSFET output) (courtesy of Lu et al. [15])

The thermal conductivity of silicon and most substrate materials increase dramatically at reduced temperatures and can lead to improved heat sinking capability. Mueller found in his testing of power MOSFETs that, because of this, thermal run-away problems were eliminated at cryogenic temperatures [13]. Coupled with much reduced on-resistance, the power density in cryogenically cooled power MOSFETs can be much higher than their room temperature counterparts.

Using commercially available MOSFETs, Mueller built and tested several full-bridge switch mode circuits at liquid nitrogen temperature [13]. He found that 3-4 times higher current pulse could be obtained at 77 K and extremely high conversion efficiencies of better than 99% might be achievable at cryogenic temperature for switch mode circuits when high temperature superconductors were used for the high current inductor [13;16]. Gardiner et al. demonstrated the operation of a 50 KW MOSFET based inverter at 77 K [17]. They found that at 77 K substrate diode conduction can be practically avoided, allowing for very short soft switching transition times. Switching frequencies for large power converters an order of magnitude higher than those practical with room temperature IGBT technology were found possible. NASA Glenn Research Center has reported substantial research on cryogenic electronics for space applications [6;18-29]. A Pulse Width Modulation (PWM) buck DC/DC power converter using commercially available components, which demonstrated a full-load efficiency of 97% recorded at liquid-nitrogen temperature. This is in contrast to demonstrated full-load efficiency of 95.8% recorded at room temperature [30].

Cryogenic operation of power MOSFET provides promising benefits for power conditioning system. However, there are also challenges that need to be addressed. Jackson et al. argued that it was not likely that existing devices were truly optimized for cryogenic environment although it was fortunate that they were both structurally and electrically compatible [31]. Giesselmann et al. studied the behavior of n-channel power MOSFETs at cryogenic temperature and found that at cryogenic temperatures in addition to expected benefits, the breakdown voltage of the device decreased moderately (20%) when cooling from 297 K to 78 K [32]. This was due to the increased impact ionization efficiency at lower temperatures. Singh and Baliga's found that threshold voltage and transconductance of the device increased with a decrease in operating temperature [33]. This was mainly due to the large reduction in the intrinsic carrier concentration in silicon from 300 to 77K resulting in a higher voltage requirement at the gate for the inversion layer formation in the P-basse region. Device optimizations targeted to specific application and specific temperature range are needed. Optimization is also required at the circuitry and system level to ensure the overall performance of power conditioning system.

Packaging is also critical to the success of cryogenic power electronics based on power MOSFETs. The packaging provides electrical and thermal connections to the device, and protects the device from mechanical and environmental stresses. In order to achieve a significant system level advantage from cryogenic operation of power MOSFETs, equivalent improvements such as interconnect electrical performance in packaging are vital for achieving reduced size/weight and overall performance goals. The thermo-mechanical structural performance of the power packaging is another critical factor for cryogenic modules, which are operated under high thermal residue stress at cryogenic temperature and additional stresses generated by cyclic cryogenic cooling and by cyclic operation while it is cold. Failure mechanisms of the existing package under cryogenic condition need to be studied. Based on this, packaging structure optimization for cryogenic operation is needed to ensure low interconnect parasitics, good thermal management, and high structural resistance to cryogenic stresses. In order to achieve this goal, material selection, based on the thermal and mechanical property characterizations of packaging materials at low temperature, is also important. The use of high temperature superconducting thin films for device interconnection can further improve the performance of cryogenic power electronic system.

III. Optimization of Power MOSFET for Cryogenic Applications

Understanding how the structure of the unit-cell affects the cryogenic performance of a Si MOSFET is an important step toward optimizing of the device for cryogenic operations. In this section, physically based numerical simulations of the Si power Double Diffused MOSFET (DMOS) are performed at both 300K and 77K. The simulation results were compared with experimental results. The effects of the cell structure on the on-resistance and breakdown voltage of the device are analyzed. The simulation results suggest that the device optimized for room temperature operation can be further optimized at cryogenic temperatures.

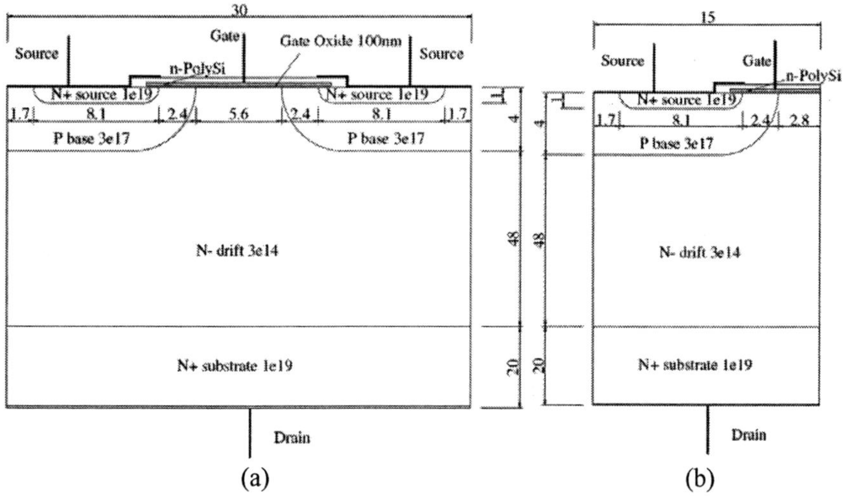

Figure 3. (a) Cross-sectional view of the DMOS unit-cell showing the dimensions and impurity concentrations (b) half unit-cell used in the simulation

Power DMOS' are vertical conduction devices which comprise a repetitive array of unit-cells arranged in a topological layout (such as linear, square array, circular array, atomic lattice layout, hexagonal array, etc.). Therefore, simulation of power DMOS is rather the simulation of the elementary unit-cell. Figure 3 (a) and (b) show the baseline unit-cell of a medium-voltage DMOS and the half unit-cell used in the numerical simulation due to symmetry, respectively. In the baseline unit-cell, the p-base depth was 4 μm and n^+ source depth was 1 μm, yielding a channel length of 2.4 μm (with lateral diffusion as 80% of the vertical depth). Gaussian profiles have been assumed for n^+ source and p-base with the peak concentrations of 3×10^{17} and 1×10^{19} cm^{-3}, respectively. The doping concentration of the epi-layer was 3×10^{14} cm^{-3}. The n^+ substrate had a concentration of 1×10^{19} cm^{-3} with the actual thickness of n^+ substrate was 300 μm. However, due to the very high doping concentration in the n^+ substrate, which behaves like a conductor in the simulation, a thickness of 20 μm was used in order to save the computational time without lose any accuracy. The gate oxide thickness was 100 nm with a gate length of 12 μm in the baseline unit-cell. The topological design was assumed to be circular cell in square array. Cylindrical symmetry was assumed in simulating the half unit-cell, which gave the characteristic current values for a circular unit cell. In order to give the current density (A/cm^2), the unit-cell current values were multiplied by the number of unit cells that can be accommodated in 1 cm^2 area (e.g., for the circular baseline unit-cell with a diameter of 30 μm, a multiplier of 1.11×10^5 will be used).

In this simulation study, the physically-based semiconductor device simulation software ATLAS has been used. The operation of the semiconductor device is modeled by a set of partial differential equations (PDEs) with appropriate models for various temperature dependent semiconductor properties. ATLAS uses Box Integration Method to solve these simultaneous PDEs numerically on a discrete grid of points which is referred to as mesh points. All the simulations in this study have been performed in two dimensions. A variable mesh spacing has been used by refining the mesh in the critical regions (such as in the channel and junctions regions) in order to increase the accuracy without overly lengthening the computational time. Cylindrical symmetry was also employed for this purpose.

The drift-diffusion models for the steady-state are based on the three PDEs:

Poisson's Equation:
$$\vec{\nabla}\cdot(\varepsilon\vec{\nabla}\Psi) = q(n - p - N_D^+ + N_A^-) \quad (1)$$

electron continuity equation:
$$\vec{\nabla}\cdot(\vec{J}_n) = qR \quad (2)$$

hole continuity equation:
$$\vec{\nabla}\cdot(\vec{J}_p) = -qR \quad (3)$$

and
$$\vec{J}_n = q(D_n\vec{\nabla}n - \mu_n n\vec{\nabla}\Psi) \quad (4)$$

$$\vec{J}_p = q(-D_p\vec{\nabla}p - \mu_p p\vec{\nabla}\Psi) \quad (5)$$

where Ψ is the electrical potential, n and p are the concentrations of electrons and holes, respectively, D_n, D_p, μ_n, μ_p are diffusivity and mobility of electrons and holes, respectively, ε is the electric permittivity of the material, q is electron charge, R is the net electron-hole recombination rate, $\vec{J}_n$ and $\vec{J}_p$ are electron and hole current density, respectively, N_D^+ and N_A^- are the concentration of the ionized donor and ionized acceptor.

In order to model the device characteristics at low temperatures, incomplete ionization and Fermi-Dirac statistics are employed in the simulations. At room temperature, impurities are usually assumed to be fully ionized. However, impurity freeze-out effect kicks in at low temperatures. Moderately doped semiconductors are modeled using Fermi-Dirac statistics with appropriate degeneracy factors for conduction and valance bands [34]. For degenerately ($> 3 \times 10^{18}$ cm^{-3}) doped semiconductors, complete ionization is assumed. The effective density of states in silicon are considered temperature dependent. In the simulations, the temperature dependence of the bandgap energy for silicon is considered Sze [35]. Bandgap narrowing was not considered since it has little effect on the performance of power DMOS at low temperatures.

The temperature dependence of carrier mobility is the most important factor that affects the performance of power DMOS' at low temperatures. At lower temperatures, the vibration of the lattice is reduced resulting in less frequent scattering from the lattice. Therefore, carrier mobility increases as temperature decreases. In order to accurately model carrier mobility, various scattering mechanisms need to be considered. For low field bulk mobility, Klaassen's [36;37] unified mobility model is employed, which considers the effects of lattice scattering, impurity scattering, carrier-carrier scattering, and impurity clustering effects at high concentration. The model agrees well with the empirical data for a wide range of donor and acceptor concentrations with temperature dependence over the range of 70 K to 500 K. For the mobility degradation within the inversion layers, Lombardi *et al.*'s model [38] is used to include the effects of surface acoustic phonon scattering and surface roughness scattering. Caughey and Thomas's model [39] is used to calculate the carrier mobility where electrical field is high. The saturation velocities are considered as temperature-dependent [40], where the saturation velocity of silicon increases about 20% when cooled from 300 K to 77 K. Shockley-Read-

Hall (SRH) and Auger recombination were considered in this simulation work. SRH recombination accounts for phonon transitions in the presence of a trap (or defect) within the bandgap of the semiconductor. The SRH carrier lifetime is considered to be concentration and temperature-dependent as proposed by Klaassen [37]. Auger recombination describes a three-particle transition whereby a mobile carrier is either captured or emitted [41]. When the reverse bias is sufficiently large, the electric field in the space charge region will be high enough to accelerate free carriers up to sufficiently high kinetic energy to cause avalanche breakdown. At lower temperatures the mean free path of carriers increases, giving them more energy for a given electric field prior to collisions, thus resulting in a reduced avalanche breakdown voltage. The phenomenological local electric field model proposed by Selberherr [41] was employed. Numerical simulation of breakdown at low temperatures can be difficult because the depletion regions have too few carriers to start an avalanche due to the extremely low intrinsic carrier concentration at 77 K. In order to overcome this numerical difficulty, a small beam of light was artificially shined on the top surface of the DMOS to provide carriers which can start the avalanche process. This technique is routinely practiced even at room temperature where extremely low intrinsic carrier concentration causes numerical convergence problems such as simulation studies in certain SiC based devices. The wavelength of this artificial light was 800 *nm* and the beam intensity was 0.001 $Watt/cm^2$ in the simulations. The effects of different beam intensities on simulation results were analyzed to ensure that this artificial light did not affect the accuracy of the breakdown voltage.

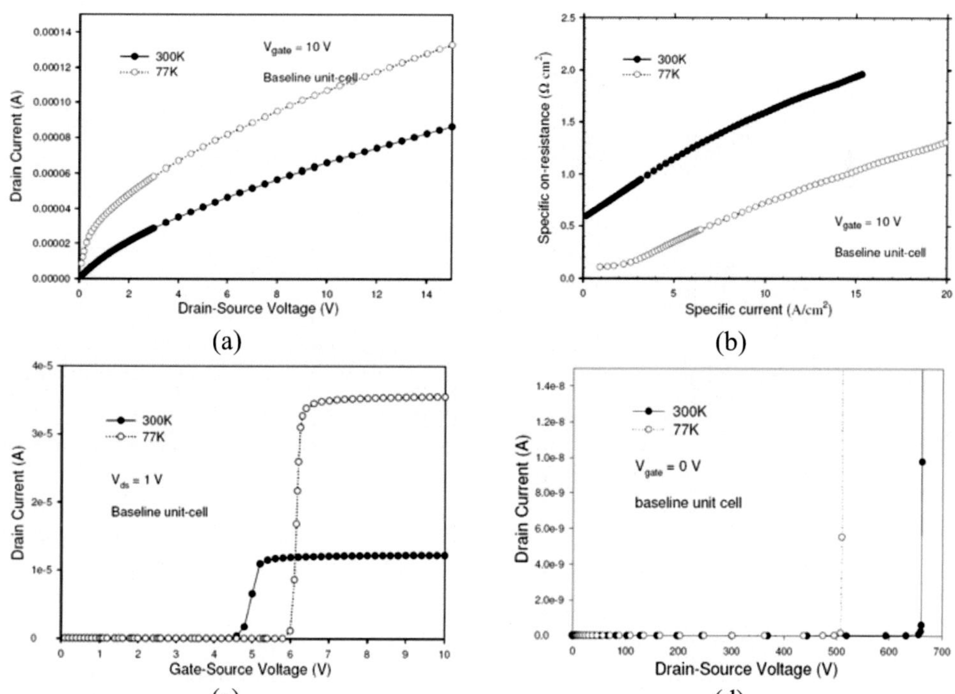

Figure 4. Characteristics of baseline DMOS unit-cell at 300 K and 77 K (a) I-V characteristics at V_{gate} = 10 V (b) specific on-resistance vs. current (c) Threshold characteristics at V_{ds} = 1 V (d) Breakdown characteristics.

Steady-state I-C characteristics, threshold characteristics, and breakdown characteristics of baseline power DMOS unit-cell have been simulated as shown in Figure 4. It is shown in Figure 4(a) and (b) that the on-resistance of the power DMOS unit-cell decreases with temperature. This is due to the higher electron mobility at lower temperatures. In Figure 4(b), the specific current (A/cm^2) is calculated by multiplying the drain current of the unit-cell by the number of unit-cells that can be accommodated in a 1 cm^2 area. The specific on-resistance ($\Omega \cdot cm^2$) is calculated by dividing the drain-source voltage with the corresponding specific current. Figure 4(c) shows that, at 77 K, the DMOS has higher threshold voltage (from 4.75 V at 300 K to 6.0 V at 77 K), transconductance, and current saturation level. The intrinsic carrier concentration decreases 30 orders of magnitude at 77 K resulting in increased band-bending in the p-base region, which needs to be compensated by a higher gate voltage in order to inverse the channel. The higher transconductance at 77 K is due to the increased channel mobility whereas the higher current saturation level is due to the higher electron mobility in all the device regions. Figure 4(d) shows that the blocking voltage of the baseline unit-cell decreases significantly from 660 V at 300 K to 510 V at 77 K, a 23% reduction. At lower temperatures, the mean free path of the carrier increases giving free carrier more energy for a given electric field prior to collision, thus resulting in a reduced avalanche breakdown voltage. Although the on-resistance of the baseline unit-cell shows moderate improvement at 77 K, it is far less than the improvement we expected. As shown in Figure 4(a), the drain-

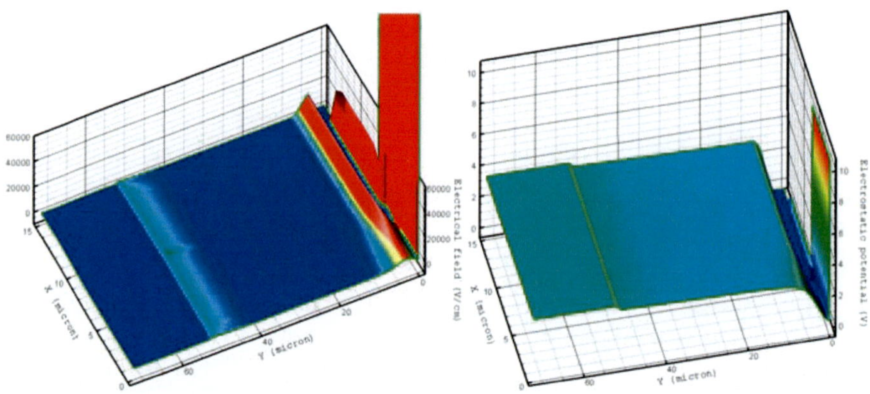

(a) (b)

Figure 5. (a) Electrical field and (b) Electrostatic potential of the baseline unit-cell with Vds=3V and Vgate=10V at 300K.

source current of the baseline unit-cell increased by a factor of 5.6 at 77 K compared to that at 300 K at a very low specific current level and much less at higher current level. Let's now look into the electrical field within the unit-cell.

Figure 5 shows the potential and electrical field of the baseline unit-cell at $V_{ds} = 3$ V with a gate voltage of 10 V at 300 K. The electrical field along the p-base and n-epi junction expands into the region between the p-bases which is sometimes referred to as the JFET region (Figure 5(b)). Along the current flow path, most of the applied drain-source voltage is taken by the JFET region, which contributes to the most of the series resistance of the baseline unit-cell.

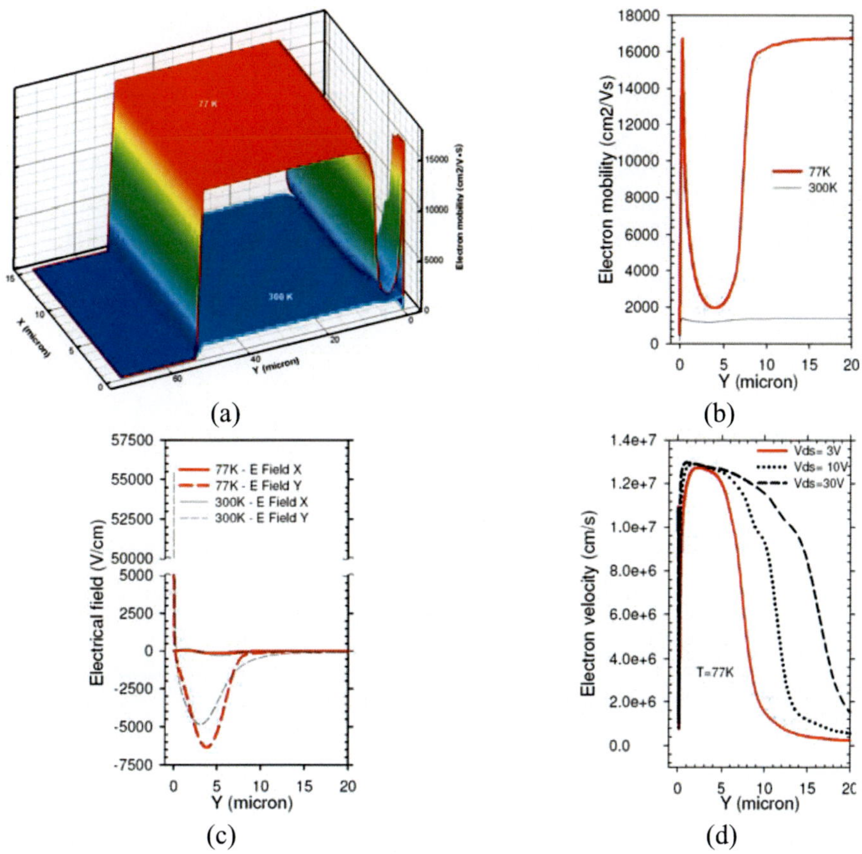

Figure 6. (a) Electron mobility distribution of the baseline unit-cell with Vg=10V and Vds=3V at 300K and 77K (b)Electron mobility along the vertical line through the center of the baseline unit-cell with Vg=10V and Vds=3V at 300K and 77K (c) Electrical field along the vertical line through the center of the baseline unit-cell with Vg=10V and Vds=3V at 300K and 77K (d) Electron velocity along the vertical line through the center of the baseline unit-cell with Vg=10V, Vds=3V, 10V, and 30V at 77K.

As aforementioned, the on-resistance improvement of power DMOS at lower temperature is a direct result of the increased carrier mobility. Figure 6(a) compares the electron mobility distribution within the baseline half unit-cell at 300 K and 77 K under a drain-source voltage of 3 V. Due to the impurity scattering, the electron mobility in the high doping region doesn't increase much at 77 K. The electron mobility in the accumulation and the channel regions has moderate increase at 77 K due to surface scattering. While electron mobility increases by a factor of over 10 in the lightly doped epi-layer at 77 K compared to that at 300 K, it only increases by a small amount in the JFET region for the baseline unit-cell. Figure 6(c) shows that, in the JFET region, while the horizontal electrical fields are low, the vertical electrical fields (parallel to the current flow) are high at both temperatures. This high parallel electrical field in the JFET region corresponds to the limited improvement of electron mobility at 77 K in this region as shown in Figure 6(b). We believe the reason for the limited electron mobility improvement in JFET region is due to electron velocity saturation. Figure 6(d) shows the electron velocities in the JFET region with various drain-source voltages at 77 K. It is clearly shown that the peak electron velocities in the JFET region saturate at about $1.3\times10^{7}\,cm/s$.

Therefore, in order to achieve the highest cryogenic improvement, the structure of the baseline unit-cell needs to be modified in such a way that the electrical field in the JFET region can be reduced. Fortunately, the electrical field in the JFET region in a unit cell optimized for room temperature operation is normally much smaller than that in the baseline cell.

The effect of the gate length on the cryogenic improvement of unit-cell was studied by performing numerical simulations of unit-cells with various gate lengths. It is well known that a larger gate length leads to smaller JFET region resistance but larger accumulation and channel resistance because of the longer path along the oxide-Si surface as well as smaller equivalent channel density [42]. As the gate length increases, the electrical field in the JFET region decreases; therefore, we can also expect on-resistance improvement at cryogenic temperatures.

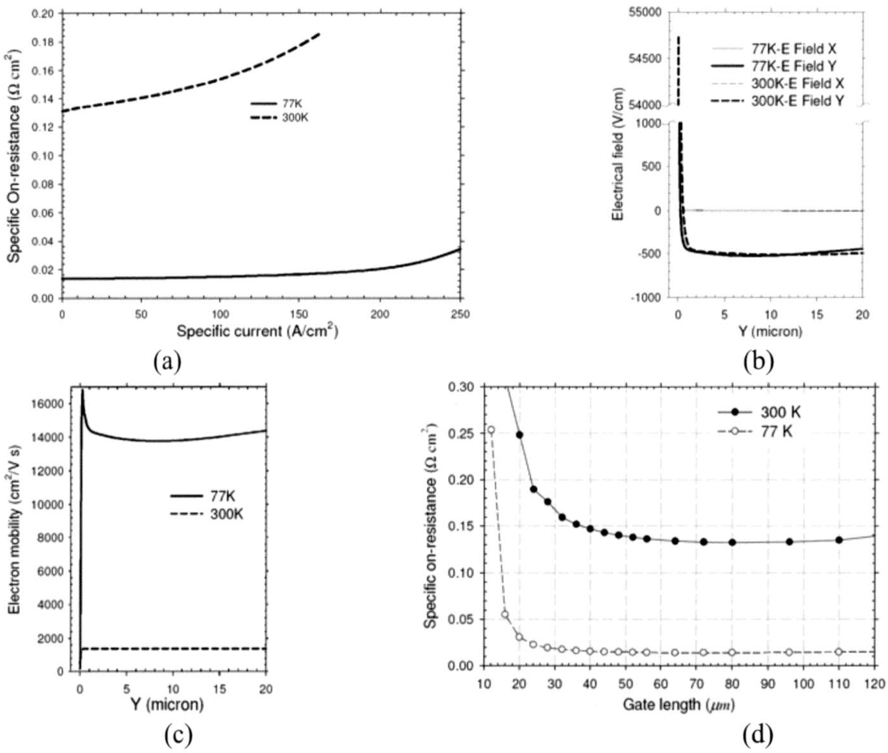

Figure 7. (a) Specific on-resistance vs. specific current and (b) Electrical field along the central vertical line, (c) Electron mobility along the central vertical line of the 72μm-gate-length unit-cell with Vg=10V and Vds=3V at 300K and 77K; (d) Optimization of gate length for the DMOS unit-cell.

Figure 7(a) shows, that as the gate length increases to 72 *μm*, not only the DMOS on-resistance decreases substantially at room temperature, but also the much higher (10x) cryogenic improvement is achieved. Figure 7(b) and Figure 7(c) show the electrical field and electron mobility along the center vertical line of the 72 *μm*-gate-length unit-cell at 300 K and 77 K. Compared with the baseline unit-cell, under a drain-source voltage of 3 V, the vertical electrical field is reduced by a factor of 10. Without the existence of the high parallel electrical field in the JFET region, the electron mobility in the region increases more than 10 times when cooled to 77 K without suffering from the velocity saturation effect. Since the drift region and the JFET region resistance contribute to the majority of the on-resistance of the unit-cell, the on-resistance of the DMOS decreases by a factor of 10 at 77 K. Figure 7(d) shows the plot of the specific on-resistance versus the gate length, where the value of specific on-resistance for a particular gate length is taken as the value at V_{ds} = 0.5V. At 300 K, the optimized value of the gate length is found to be 80 μm while the optimized value of the gate length is 72 μm at 77 K. However, the difference is less than 0.5% between the 80 μm and 72 μm gate-length cells at 77K. This means gate length optimized for 300K is effectively optimized for 77K operation. The gate length optimization of DMOS at cryogenic temperatures has been analyzed before by Singh using an analytical approach [43]. He claimed that the power DMOS with optimized gate-length design for room temperature operation would have more than 15% higher on-resistance than that optimized for 77K operation. We believe the discrepancy might come from his calculation of JFET region resistance, where mobility in the region was assumed to follow the same temperature dependence as that in the epi-layer. It has been shown in our simulations that it does not always hold true for smaller gate lengths where parallel electrical field is high.

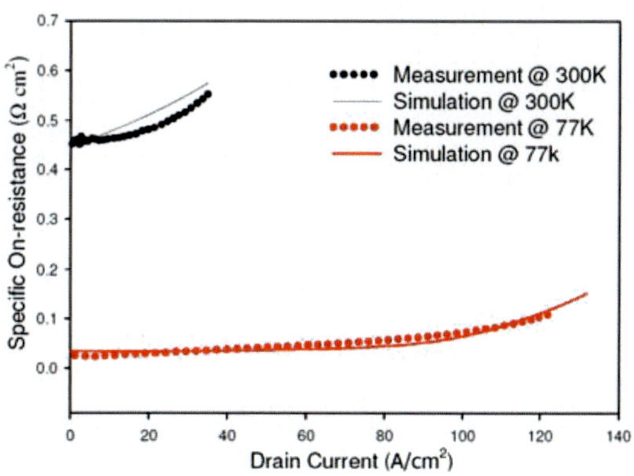

Figure 8. Comparison of Ron vs. drain current between measurements and simulation results.

Experimental characterizations were performed on commercially available power MOSFETs and compared to the simulation results. The tested device is a 1000V rated power MOSFET with a measured breakdown voltage of 1100 V at room temperature. The tested device has a die area of 1.9 cm^2. In Figure 8, both the measured current and on-resistance have been normalized to 1 cm^2 for comparison with simulation results. The power MOSFETs were mounted on the Direct Bonded Copper (DBC) and wire bonded for electrical connections. A Tektronix 371A curve tracer was used to measure the I-V characteristics of the integrated module at both room temperature and liquid nitrogen temperature. Since the curve tracer provides pulsed current with a pulse width of 250μs and repetition rate of 15Hz, little temperature rise in the MOSFET was observed due to minimum heat dissipation during testing. The power DMOS cell structure used in the simulation has an epi-layer doping concentration of $1.25 \times 10^{14}\ cm^{-3}$, epi-layer thickness of 84 μm, and gate length of 70 μm and the simulation found its breakdown voltage to be 1120 V. As shown in Figure 7, the simulation results matched with experimental measurement very well at both room temperature and 77 K even though the exact cell structure of the tested devices is not known. The simulation correctly predicted more than 14x reduction of the on-resistance for this particular device. At 77 K, the breakdown voltage of the tested devices was measured to be 860V, which is also very close to the breakdown voltage of the simulated cell, 884 V.

One of the purposes of this section is to answer the question: “does the commercially available power MOSFETs also give optimized on-resistance although its breakdown voltage is degraded?” The analysis in the gate-length optimization showed that no additional optimization of gate-length is needed for cryogenic operation. However, we found that epi-layer thickness could be re-optimized to gain the most performance improvement at 77 K.

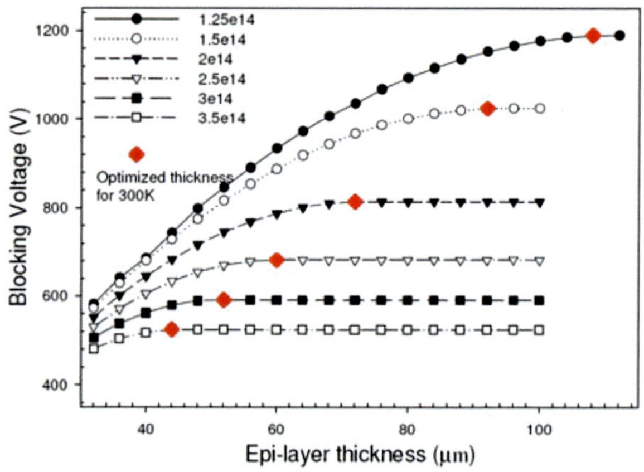

(a)

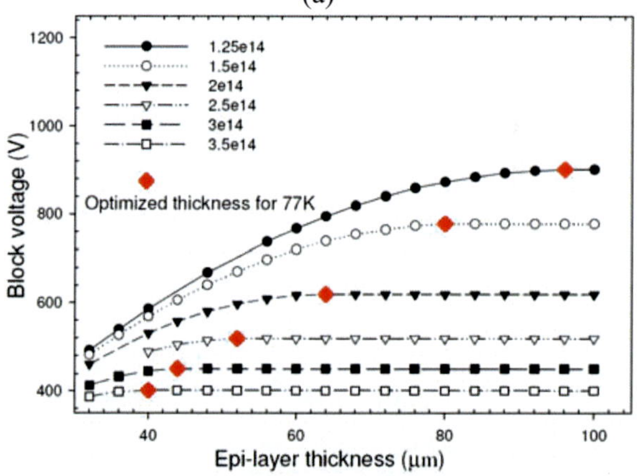

(b)

Figure 9. Blocking voltage vs. Epi-layer thickness (a) 300K and (b) 77K.

As mentioned earlier, the resistance from the epi-layer contributed to the majority of the on-resistance for medium to high voltage power MOSFETs. Higher voltage blocking capability requires thicker epi-layer and lower doping concentration. Both factors increase device on-resistance. Device design therefore requires optimization of a combination of the epi-layer thickness and the doping concentration so that lowest specific on-resistance can be achieved. For certain epi-layer doping levels, there is a corresponding threshold thickness, above which a maximum blocking voltage will be reached no matter how thick the epi-layer is; on the other hand, if the thickness is smaller than the threshold value, the blocking voltage will decrease. The cell with this threshold thickness will have the lowest on-resistance with the highest reachable blocking voltage for a specific epi-layer doping level. The commercially available devices have an optimized combination of epi-layer doping level and thickness for room temperature operation. However, we found that the optimized design for room temperature operation is not optimal at 77 K. In our analysis, we conducted simulations for cells with epi-layer doping concentrations from 1.25×10^{14} cm^{-3} to 3.5×10^{14} cm^{-3}. This covers power MOSFETs with breakdown voltages between 500 V and 1200 V at room temperature. The epi-layer thickness under consideration ranged from 32 μm to 112 μm.

The simulation results of blocking voltage dependence on epi-layer thickness for various epi-layer doping concentrations at both 300 K and 77 K are shown in Figure 1. They clearly demonstrate that there exists a threshold epi-layer thickness for any specific doping level at certain temperature. Increasing the thickness above this threshold value will increase the specific on-resistance without increasing the blocking voltage. It also clearly shows that the threshold epi-thickness for certain doping levels at 77 K is significantly smaller than that at 300 K. It indicates that commercially available devices, which are optimized for room temperature operation, are not optimal for cryogenic operation. The epi-layer thickness of device optimized for room temperature operation can be reduced significantly without reducing its blocking voltage at 77 K. This is because, at 77 K, the mean free path of carriers increases giving free carrier more energy for a given electric field prior to collision, which effectively reduces the critical electrical field for avalanche breakdown. As the critical field decreases, the threshold epi-layer thickness required to reach critical field before punch-through also decreases due to the triangular shape of the electrical field near the p-base and epi-layer junction. The summary of epi-layer thickness optimization for various doping levels at both 300 K and 77 K are listed in Table 1. It indicates while more than 8 to 13 times on-resistance improvement (depending on the voltage rating) can be expected by cooling the commercially available power MOSFETs to 77 K, the device optimized for 77 K by reducing the epi-layer thickness can provide additional 5% to 10% on-resistance reduction depending on its voltage rating.

Table 1. Summary of epi-layer thickness optimization at 300K and 77K

	Optimization for 300K							Optimization for 77K			
Doping (cm^{-3})	**Optimal thickness (μm)**	**Blocking voltage (V)**			**Ron (Ω cm^2)**			**Optimal thickness (μm)**	**Blocking Voltage (V)**	**Ron @ 77K (Ω cm^2)**	**Additional reduction**
		300K	**77K**	**Reduction**	**300K**	**77K**	**Improvement**				
1.25e14	108	1190	901	24%	0.5565	0.0415	13.4 x	96	901	0.0377	9.1%
1.50e14	92	1025	778	24%	0.4058	0.0322	12.6 x	80	778	0.0290	10.0%
2.00e14	72	814	619	24%	0.2514	0.0224	11.2 x	64	619	0.0207	7.8%
2.50e14	60	682	519	24%	0.1767	0.0175	10.1 x	52	519	0.0159	9.0%
3.00e14	52	591	451	24%	0.1344	0.0145	9.3 x	44	451	0.0131	9.4%
3.50e14	44	524	401	23%	0.1042	0.0122	8.5 x	40	401	0.0116	5.0%

In this section, two dimensional numerical simulation study of the Si power DMOS' at cryogenic temperatures is presented. The simulation results are compared with experimental measurements with good agreement. It is suggested that any medium to high voltage power MOSFETs will get huge on-resistance improvement by cooling down to 77 K. The simulation results also suggest that the gate length of a power DMOS cell optimized for room temperature operation is sufficiently optimized for cryogenic operations. However, the epi-layer thickness optimal for room temperature operation can be reduced without reducing its breakdown voltage at 77 K. This leads to another 10% in on-resistance reduction.

IV. Development Example of Integrated Cryogenic Power Modules

As aforementioned, in order to take full advantage of cryogenic power electronics, it is necessary to integrate control systems and to develop compatible packaging technologies that can accommodate low-temperature device operation at very high power density. In liquid nitrogen temperature environments, all components in a power electronics module are subject to thermal stresses due to CTE (Coefficient of Thermal Expansion) mismatches that are considerably more severe than at room temperature. The difficulty in developing cryogenic modules results from the fact that most commercially available electronic devices and materials are not rated for operation in a cryogenic environment. In addition to reliability, good electrical performance of the cryogenic module is also important. In this section, we discuss the development of cryogenic power modules by examples of integrated power modules developed at SUNY at Albany and MTECH Labs. In these examples, both cryogenically qualified power MOSFETs and driver circuit components were integrated on a single substrate in order to achieve high power density.

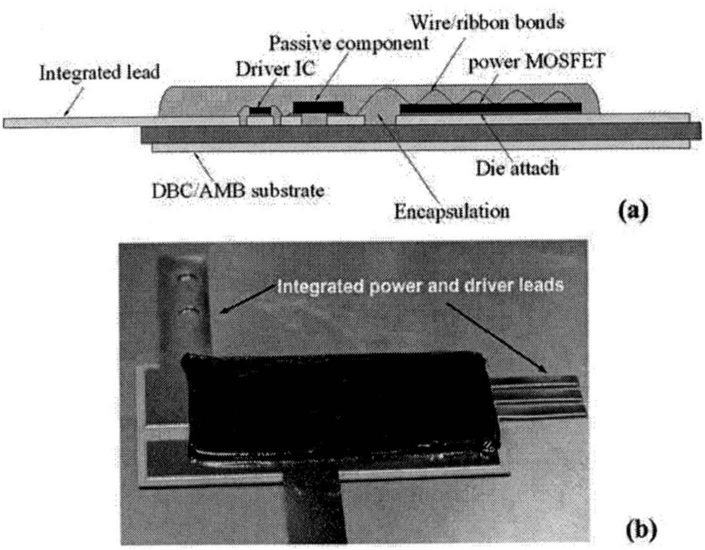

Figure 10. (a) Schematic of the integrated power module structure (b) Prototype of the packaged integrated module.

The first cryogenic power module developed was for use in a 50kW liquid nitrogen cooled full-bridge cryogenic inverter. The module features a single power MOSFET. In order to achieve the highest reliability, compactness, and electrical performance, an integrated power package was chosen over a discrete assembly. Compared with a discrete design, the integrated module (IM) has far fewer soldered interconnection interfaces leading to improved reliability. The size of the IM is a critical consideration in fabrication since a large number of modules are required for the cryo-inverter. Fewer interconnect interfaces and closer distances between driver components and power MOSFETs in the IM leads to better electrical characteristics due to reduced packaging parasitics. Another advantage is the improved thermal dissipation capability of the IM compared to the discrete assembly. The basic structure for the integrated power package for the cryo-inverter is shown in Figure 10. The footprint of the package is 2.5 cm by 5 cm, where both the power MOSFET and driver circuit components are integrated on a common substrate patterned with the circuit and protected by epoxy encapsulation. The module features heavy-copper leads to facilitate easy and reliable interconnection to the power bus of the inverter with high current carrying capability. The integrated leads eliminate additional steps of connecting to the substrate and thus improve module reliability.

Si_3N_4, AlN, BeO, and Al_2O_3 Direct-Bonded Copper (DBC) and Active Metal Bonding (AMB) substrates from various vendors were evaluated as candidate substrates under accelerated thermal cycling between -180°C and 125°C with a measured ramping rate of 22°C/min. DBC and AMB have high current-carrying capacities and low thermal impedance [44] and provide high electrical isolation with low electrical capacitance[7]. The mechanical advantage of these ceramic-based substrates is their stiffness. They require large amount of stresses to initiate any damage [45]. Figure 11 shows the measured thermal cycling profile that was used in a typical test. We observed cracks in the ceramic layer and delamination between the copper and ceramic layers for Al_2O_3 and AlN DBC substrates at less than 100 thermal cycles. Both Si_3N_4 AMB and BeO DBC substrates showed no stress induced cracks with microscopic investigation after 200 thermal cycles and were qualified for cryogenic use. Si_3N_4 AMB was ultimately selected for fabricating the cryogenic power modules based on cost considerations.

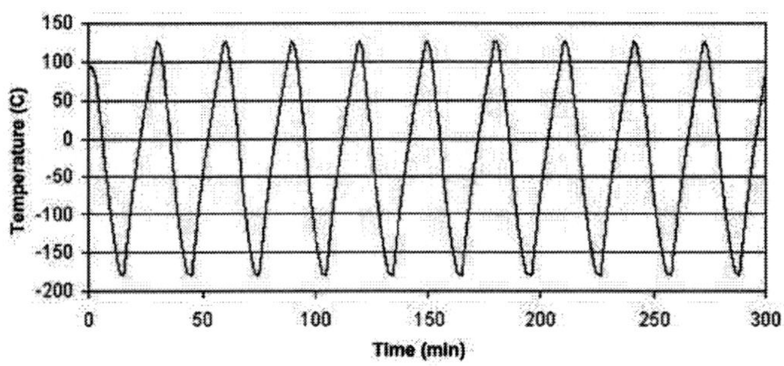

Figure 11. Typical cryogenic thermal cycling profile used for testing the modules and components.

Both the power MOSFETs and driver circuit components need to be reliably attached to the DBC substrates to create the IM. Therefore, die-attach materials were qualified to ensure cryogenic reliability. For powering MOSFETs that have Ni/Ag metallization on the backside, low melting point (157°C) indium solder was selected as the die-attach material due to its excellent ductility at cryogenic temperatures. Various combinations of solder fluxes and substrate surface finishes were tested to determine the optimal performance. A fluxless indium solder preform together with an immersion gold surface finish (on top of electrolessnickel-plated copper) was found to provide the best solderability. Vigorous cryogenic thermal-cycle tests (more than 200 cycles) were performed on the samples where power MOSFETs were attached to the substrate. This material combination was qualified after observing crack-free attachments after thermal cycling. For the driver Integrated Circuit (IC) chips, on the other hand, organic die-attach materials (epoxies) were used because the backside of the driver chip did not have metallization and was not solderable. However, none of the die-attach epoxies on the market were rated for cryogenic temperature application by their vendors. Thermal cycling on test samples was performed, where epoxies from various vendors were used to attach the driver chips to the substrates. Two epoxies (Henkel QMI519 and Emerson & Cumming AbleBond 84-3) were qualified after 200 thermal cycles without observations of any micro-crack formation. We selected QMI519 because it is electrically conductive and can be used to attach other driver circuit components such as capacitors, diodes, and resistors.

Aluminum (Al) wire/ribbon bonds were used to electrically connect the power MOSFETs and driver chips to the circuits on the substrate. Al wedge wire bonds ($50\mu m$ diameter) and 2 *mm* Al ribbon bonds (PowerRibbon™ from Orthodyne Electronics) were also tested for cryogenic reliability by thermal cycling. The ribbon bonds were selected over heavy wire bonds due to their higher current carrying capacity which leads to better performance. No sign of fatigue failure was observed for these wires and ribbons after extensive cryogenic thermal cycling.

The components and interconnecting wires/ribbons were encapsulated from environmental hazards such as moisture, electric static discharge, and corrosion. Dam-and-fill scheme was used to create the encapsulation. Again, no commercial encapsulant is rated for liquid nitrogen temperature operation by their vendors. Candidate encapsulants were cured and subject to cryogenic thermal cycling. The purpose of these cycling tests was to find encapsulants that would survive at least 200 cryogenic cycles without failing or delamination from the substrate. Several encapsulants including Henkel FP4470 epoxy were qualified based on this criterion. However, these encapsulants have a different CTE than the substrate (18ppm vs. 2.8ppm) or other components in the module. The deformation of the encapsulant during curing and thermal cycling causes stresses on the substrate, components, and delicate wire bonds to be protected. The encapsulated modules were periodically taken out of the thermal chamber during cycling for electrical testing. No electrical parameter shifts were found after 100 thermal cycles for the module encapsulated with Henkel FP4470 epoxy, and where Henkel FP4451TD was used as the dam epoxy. Therefore, these epoxies were selected for encapsulating the IMs. After all components and materials were qualified for cryogenic packaging, prototype IMs were electrically tested. A Tektronix 371A curve tracer was used to measure the I-V characteristics of the IM at both room temperature and liquid nitrogen temperature. Since the curve tracer provides pulsed current with a pulse width of $250\mu s$ and repetition rate of 15Hz, little temperature rise in the MOSFET was observed due to the minimum heat dissipation during testing. As shown in Figure 12(a), a factor of more than 14x improvement was measured in the modules on-resistance when cooled down to liquid nitrogen temperature (77K). The testing of switching characteristics of the module is underway. The on-resistance improvement is mainly due to increased electron mobility at lower temperatures due to less phonon scattering as shown by the following equation:

$$\mu_{n,L} = \mu_{n,\max}(300/T)^{\theta_n} \quad (1)$$

where $\mu_{n,L}$ is the bulk electron mobility, $\mu_{n,\max}$ is bulk electron mobility at 300K, T is the absolute temperature, and θ_n is a constant. The resistance of the module is composed of both the resistance from the MOSFET and that from the interconnections. However, wire/ribbon bonds, lead and substrate only contribute to a very small portion of the total resistance. Therefore, most of the resistance and subsequently most of the reduction of resistance at liquid nitrogen temperature comes from the MOSFET. At the same current level, the cryogenic integrated power module generates significantly less resistive heat than at room temperature. As a result, the cryogenic integrated power module provides higher efficiency and increased power density. For example, at a current level of 40A, the cryogenic module generates 28 watts of heat, while a MOSFET module maintained at room temperature generates 436 watts. The current carrying capacity of the power module is determined by the maximum heat that can be effectively dissipated to the surroundings in order to keep the temperature of the power device within the allowable range. The on-resistance of the power module will increase as the operating temperature of the MOSFET increases due to self-heating. This will lead to more heat generation. The advantage of cooling the power module in liquid nitrogen, in addition to the greatly reduced heat generation, is that a large amount of heat can be effectively dissipated by liquid nitrogen boiling cooling without attaching a bulky heat sink to the module. This is especially beneficial since the cryogenic cooling system is already in place in the HTS motor or generation systems where the cryogenic power modules will be incorporated. This would greatly reduce the size and weight of the power module as well as the complexity of system integration.

Experiments were conducted to determine the heat dissipation capability of the prototype cryogenic IM in liquid nitrogen. A power supply capable of providing 100A DC constant current and 10V constant voltage was used to feed the current to the power module. The gate of the power MOSFET was biased at 15V. A thermocouple was embedded in the encapsulation to measure the MOSFET surface temperature. Note that the measured temperature is likely to be lower than the real temperature of the power MOSFET due to the fact that the encapsulation has very poor thermal conductivity. As current was increased, the measured temperature increased. As current was gradually increased to 45A the measured temperature increased only by 4°C at -192°C. However, as current was increased beyond 47A, the measured temperature increased dramatically by more than 75°C to reach -120°C. This is due to the phenomenon called critical heat flux (CHF) in liquid boiling cooling. As the temperature on the surface of the power module reached a critical value the gaseous bubbles from liquid nitrogen boiling cover a large portion of the surface, preventing heat from efficiently transferring to the liquid. A detailed analysis of CHF in cooling a cryogenic power module can be found in [46]. Since we set the maximum output voltage of the power supply at 2.5V, a snapback of current was observed, which is shown as the dotted line in Figure 12(b). The temperature eventually stabilized at -120°C with a current level of 39A after a very short period of time. Since the output voltage was 2.5V, the module was dissipating 98 watts of heat at this point. This is because the on-resistance of the power MOSFET is much higher at -120°C. The shape of the dashed line is dependent on the maximum output voltage. For instance, when the output voltage is set to 10V, the temperature stabilized at -100°C with a current of 18A. In any case, the liquid nitrogen cooled cryogenic module was demonstrated to carry 40A of continuous current without significant increase in temperature and on-resistance. In comparison, at room temperature, the maximum current that the module can carry is dependent on the cooling scheme. It is estimated the module can carry a maximum of 11.7A of current with a junction temperature of 150°C, assuming an active air cooled heat sink is attached to the module with a junction to ambient (25°C) thermal resistance ($R_{\theta ja}$) of 1.5°C/W and a heat dissipation of 83W. At this current level, the liquid nitrogen cooled module only dissipates less than 3W of heat. The estimated current-temperature curve is shown in Figure 12(b) for comparison.

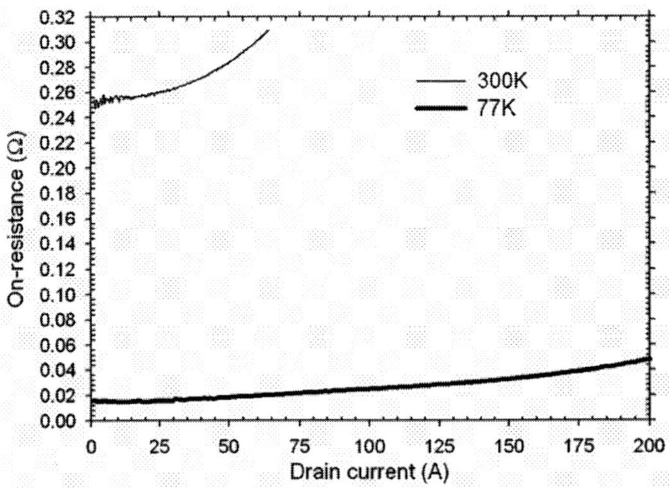

(a)

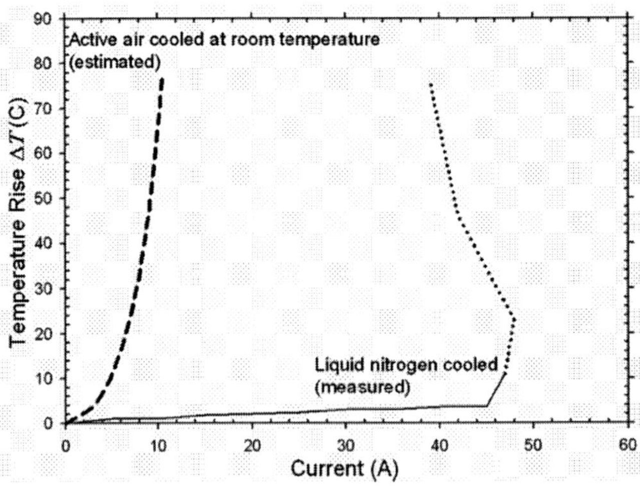

(b)

Figure 12. (a) On-resistance vs. current at both room temperature and 77K for the integrated module showing 14X improvement; (b) Measured temperature vs. current showing ability to carry up to 40A with liquid nitrogen cooling. The current-temperature curve for the module operating at room temperature is estimated by assuming the module is attached to an active air cooled heat sink with a junction to ambient thermal resistance of 1.5 °C/W.

(a)

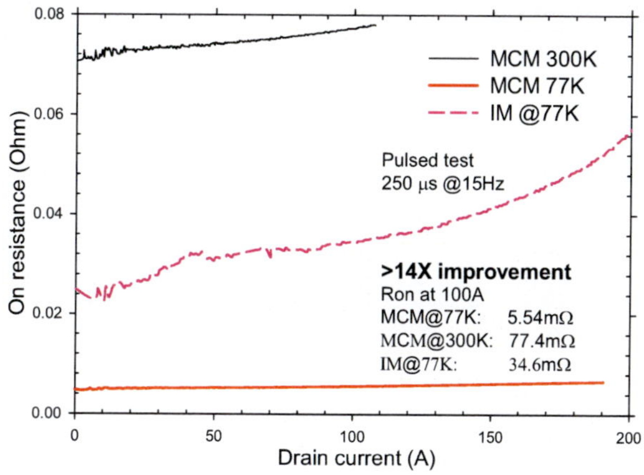

(b)

Figure 13. (a) Comparison of footprint between the single power MOSFET module and the multi-MOSFETs module; (b) On-resistance comparison between the two cryogenic modules

These highly compact 700V rated integrated cryogenic power modules that showed more than 14x improvements in the on-resistance and continuous current carrying capability exceeding 40A were integrated into the 50kW liquid nitrogen cooled full-bridge cryogenic inverter. The whole circuits including the IMs were cooled in liquid nitrogen a cryogenic dewar. The cryo-inverter demonstrated exceptional efficiency and power density. In order to further increase the efficiency and power density of the cryo-inverter, integerated cryogenic power modules features three power MOSFETs per substrate (called multi-chip module or MCM) were developed. Figure 13(a) shows the footprint comparison between the single MOSFET IM and MCM. The footprint of the MCM is compatible with the single MOSFET IM in the cryo-inverter. Inclusion of multiple power MOSFETs on a single substrate is possible because of the high thermal conductivity of Si_3N_4 substrate and the high heat dissipation capability by pool boiling of liquid nitrogen. Since the multi-MOSFETs module has bigger footprint and designed to carry much higher current, both the thickness of the Si_3N_4 ceramic layer and the copper layers in the AMB substrate were increased based on the numerical thermomechanical simulations. Figure 13(b) shows the on-resistance vs. current for both MCM and IM. 14x of cryogenic improvement was found for the MCM. At a current level of 100A, the MCM has an on-resistance of 5.54m , only one sixth of IM. By replacing the IMs in the cryo-inverter with MCMs, the power density of cryo-inverter was easily increased by 3 times without increase the physical envelop.

V. Characterization of Superjunction MOSFET and Its Cryogenic Applications

Superjunction (SJ) power MOSFET, a new generation device utilizing charge-balance principle, has been proposed in the late 1980s [47;48] and become commercially available since the late 1990s. A schematic of the superjunction power MOSFET is shown in Figure 14(a), where the N-drift layer in a conventional DMOS is replaced by alternatively stacked, more heavily doped N and P pillars. During the on-state, the N pillar will be conducting the current. In the reverse blocking state, the N and P pillars are fully depleted before breakdown if the doping concentrations in the N and P pillars ensure the charges in the N and P pillars are exactly balanced. This leads to a flat distribution of electrical field (as shown in Figure 14(b)) in the pillar regions instead of a triangular one across the junction between P-base and N-drift layer in conventional power MOSFETs. This structure enables the doping in the current conducting N pillars to be greatly increased compared to doping levels in the epi-layers in a conventional power MOSFET without sacrificing the breakdown voltage.

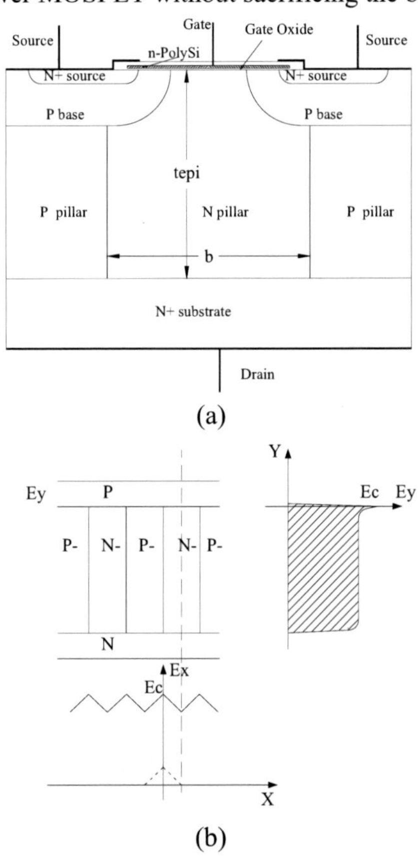

Figure 14. (a) Schematic of SJ MOSFET (b) Electrical field distribution in SJ structure under reverse bias condition.

Furthermore, the highest doping concentration of each pillar for certain breakdown voltage is inversely proportional to the pillar width. As a consequence, at the same voltage rating, SJ power MOSFET can achieve orders of magnitude smaller specific on-resistance than a conventional power MOSFET at room temperature. The achievable specific on-resistance is restricted by the minimum pillar width that can be created by the state-of-the-art device processing technologies. Several companies (e.g. Infineion, Toshiba, etc.) have already producing SJ MOSFETs that break "silicon limit", the ideal specific on-resistance can be achieved by a conventional power MOSFET for a given breakdown voltage.

While the characteristics of SJ MOSFETs at room temperature have been actively investigated, very few reports [49] on the electrical characteristics of SJ MOSFETs at cryogenic temperature has been reported in the literature. It is interesting to know the performance of SJ MOSFET at cryogenic temperatures and assess its viability for cryogenic applications. In this section, a typical SJ power from Infineon (CoolMOS™ SIPC26N80C3) is characterized at both room temperature and liquid nitrogen temperature. The device is rated at 800V and has a typical on-resistance of $25m\Omega$ at room temperature with a die size of $0.57\times0.451cm^2$. The MOSFETs were tested in the bare-chip form in order to exclude the effects of parasitics from the discrete packages. This study includes both the static (i.e., on-resistance and breakdown voltage) and the switching characterizations (i.e., switching energy loss, switching time, etc.) of SJ MOSFET at room and liquid nitrogen temperatures.

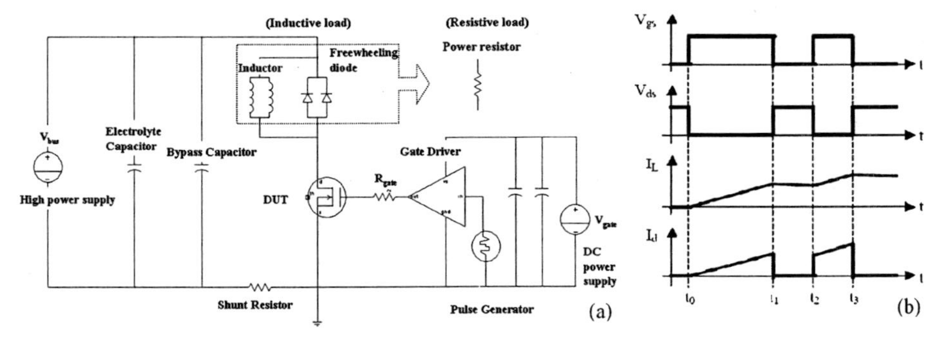

Figure 15. (a) Schematic of switching test circuit for Inductive-load and Resistive-load test and (b) Double-pulse waveform.

In the switching characterization set-up, special attention was paid to minimize errors due to measurement issues such as coupling effects of the measuring systems and operating conditions. The simplified circuit diagram for switching testing under the diode-clamped inductive-loading as well as resistive-loading is shown in Figure 15(a). Silicon Carbide (SiC) diode (Infineon SDT12S60) was used as the freewheeling diode due to its low reverse recovery loss and time. Double-pulse scheme was employed for the tests under inductive-loading. As shown in Figure 15(b), the drain current (I_d) through the SJ MOSFET can be controlled by the width of the first pulse (t_1-t_0) of the gate voltage (V_{gs}). The turn-off waveform is measured at the falling edge of the first pulse (i.e., t_1) and the turn-on waveform is measured at the rising edge of the second pulse (i.e., t_2) so that they have about the same current levels. For the test under resistive-loading, the inductor and the freewheeling diode were replaced by a power resistor and only a single square-wave pulse was used. A BNC

6040 pulse generator was used to provide the square pulses to the signal input of the MOSFET driver through a coaxial cable.

The test circuit is composed of two separate modules: the Device-Under-Test (DUT) module (i.e., module B) and the loading module (i.e., module A) that contains the rest of the test circuits, including the MOSFET driver and the loading circuits as shown in Figure 16. The bare chip SJ MOSFET and the surface mount gate resistor (R_g) were bonded on the gold-plated DUT board using conductive epoxy (Figure 16(a)). The source and gate of the MOSFET are wire-bonded to the corresponding bonding pads on the board. The electrical connections between the two modules are direct soldering in order to minimize the parasitic inductance (Figure 2(b)). To further reduce the circuit parasitics, components such as MOSFET driver, load capacitors, load resistor and shunt resistor were placed physically very close to the DUT. Three electrolytic capacitors and ten ceramic bypass capacitors were used in parallel to minimize their parasitic inductances and resistances. When testing the device at liquid nitrogen temperature, only the SJ MOSFET and its gate resistor were submerged in liquid nitrogen. The upper portion of the DUT module and the all portions of module A were kept in the atmosphere. Tektronix TDS3054B oscilloscope and P6139A high-impedance passive probe were used for voltage measurement. Both the oscilloscope and the voltage probe have a combined system bandwidth of 500MHz. For current measurement, we used T&M shunt resistor (SDN-414-10, 0.1Ω) which has a bandwidth of 2GHz. The shunt resistor was connected to the oscilloscope by a coaxial cable through its BNC connector. The system bandwidth for current viewing is 500MHz. The rise/fall time of a 500MHz measuring system is about 0.7ns ($t_{rise/fall}$=0.35/*bandwidth*), which is good for faithfully measuring signals with rise/fall time around 3.5ns.

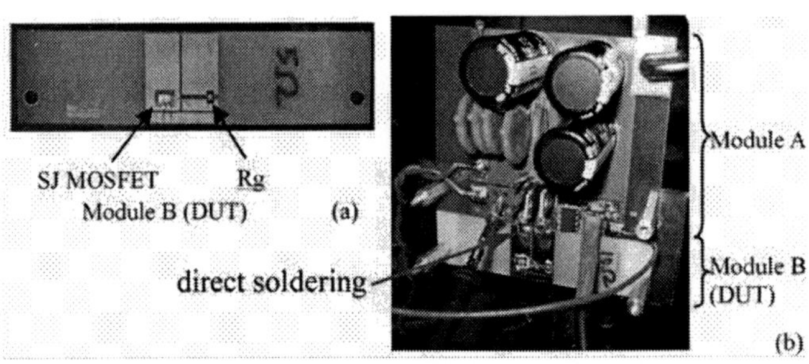

Figure 16. Test circuits (a) DUT (Module B) and (b) assembled test circuits.

The same DUT design for switching testing is also used in the static characterizations of the SJ MOSFET. High power curve tracer Tektronix 371B was used for measuring the I-V and breakdown characteristics. Since the curve tracer provides pulsed current with a pulse width of 250 μs and repetition rate of 15Hz, little temperature rise in the MOSFET was expected due to the minimum heat dissipation during testing.

Figure 17(a) and (b) show the typical curves of specific on-resistance and breakdown voltage for a SJ MOSFET at both 300K and 77K, respectively. The specific on-resistance was calculated from the I-V curve and based on the device area of the SJ MOSFET. More than 5x on-resistance reduction was measured and 20% reduction of breakdown voltage was observed when the device was cooled down from 300K to 77K. The 5x on-resistance reduction is smaller than the over 10x improvement of the conventional power MOSFET with similar breakdown voltage as discussed earlier. This is because the electron mobility, the main reason for on-resistance reduction at lower temperatures, is not only temperature dependent, but also dependent on doping concentration. SJ MOSFETs use higher doping concentration (e.g., on the order of 10^{15}~$10^{16}cm^{-3}$ depending on the width of the alternating p and n pillars) in the voltage-blocking drift layer to achieve lower on-resistance compared to that (e.g., on the order of $10^{14}cm^{-3}$) used by the conventional power MOSFETs at the 800-1000V rating. The major contributions to the electron mobility in the bulk silicon (e.g., voltage-blocking drift layer in power MOSFETs) can be described by Mathiessen's rule [36;37]:

$$\frac{1}{\mu_{e,b}} = \frac{1}{\mu_{e,L}} + \frac{1}{\mu_{e,DAP}} \tag{6}$$

where $\mu_{e,b}$ is the bulk electron mobility, $\mu_{e,L}$ is electron mobility due to phonon (lattice) scattering, $\mu_{e,DAP}$ is electron mobility due to donor, acceptor, and screening and carrier-carrier scattering. At low doping levels (e.g., $10^{14}cm^{-3}$), phonon scattering is dominant; as doping level increases, impurity and carrier-carrier scattering becomes the limiting factor. Both and $\mu_{e,DAP}$ are temperature dependent. $\mu_{e,L}$ increases as temperature decreases due to less lattice scattering at lower temperatures; on the other hand, $\mu_{e,DAP}$ decreases as temperature decreases due to more scattering from charged impurities and carriers. Therefore, electron mobility improvement due to the increasing of $\mu_{e,L}$ at lower temperatures is significantly restricted by $\mu_{e,DAP}$ as doping level increases. As a consequence, at liquid nitrogen temperature, the electron mobility in the drift layers of conventional high voltage MOSFET increases more than that of SJ MOSFET due to its lower doping levels. Nonetheless, it is worth noting that the specific on-resistance of high-voltage SJ MOSFET at room temperature is already an order of magnitude smaller than that of a conventional power MOSFET; despite smaller cryogenic improvement, the specific on-resistance of SJ MOSFET is still several times smaller than that of a conventional power MOSFET at cryogenic temperatures, making cryogenic operation of SJ power MOSFET still more efficient than conventional power MOSFETs. The 20% reduction in breakdown voltage of SJ MOSFET is about the same as that of conventional power MOSFETs. It is due to the lower critical electric field of silicon at cryogenic temperatures, a result of longer mean free path of carriers, and needs to be considered in the cryogenic applications.

The transient characteristics under hard switching inductive-loading conditions were investigated at both room temperature and liquid nitrogen temperature at various drain currents (5, 10, and 15A) and drain voltages (100,200, and 300V). The turn-on and turn-off waveforms for a typical case (200V drain voltage, 15A drain current, 15V gate voltage, and 5Ω gate resistance) are shown in Figure 18(a) and (b), respectively. As shown on the gate voltage waveforms, the charging and discharging time of gate capacitance becomes faster at 77K due to the increased electron mobility.

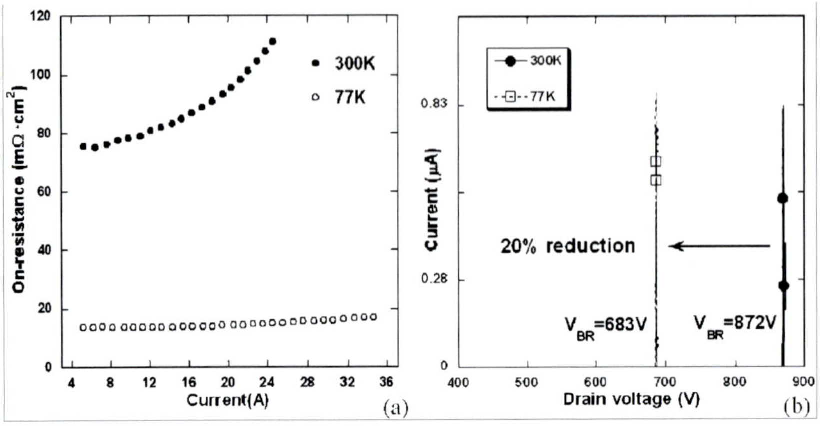

Figure 17. (a) Specific on-resistance vs. current at Vg=10V, and (b) breakdown characteristics of a SJ power MOSFET

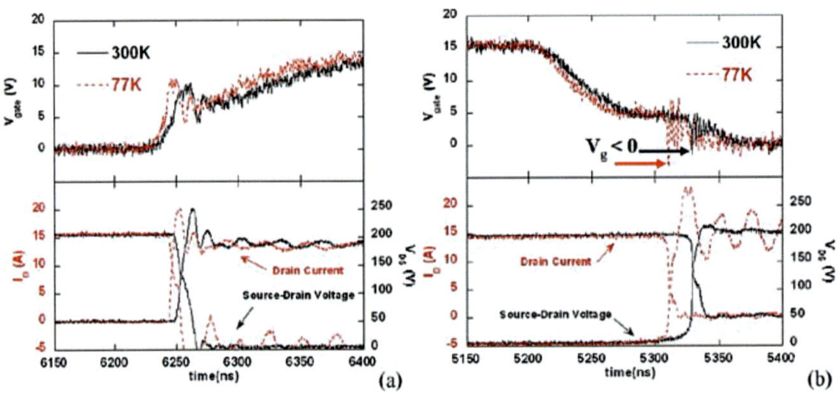

Figure 18. Switching waveforms: (a) turn-on and (b) turn-off for the case: V_{DS}=200V, I_D=15A, V_{gate}=15V, R_{gate}=5Ω.

The switching characteristics derived from the waveform measurement such as rise/fall times, etc., are listed in Table 2. The detailed definition of the quantities in the table can be found in reference [50]. In this particular case, drain voltage fall time ($t_{f,V}$) and drain current rise time ($t_{r,I}$) during turn-on are reduced by 30% and 50% at 77K whereas drain voltage rise time ($t_{r,V}$) and drain current fall time ($t_{f,I}$) during turn-off are reduced by 2% and 15%, respectively, indicating the SJ power MOSFET switches faster at 77K. The total turn-on and turn-off time were reduced by 20% and 17%, respectively. The turn-off was a lot slower than turn-on due to the much longer turn-off delay time. It is worth noting that the 17% reduction of turn-off time is mainly due to the 13% reduction of turn-off delay time. When looking at the switching characteristics at different voltage and current levels, we observed that the current rise time and voltage fall time during turn-on always decreased by 10% to 50% at 77K; the current fall time and voltage rise time during turn-off did not always decrease at 77K although the variation is moderate. The total turn-on/off time were reduced by 10%~30% at 77K. We observed high frequency oscillations (>20MHz) especially for drain voltage during switching transients at both temperatures. These high frequency ringings are due to the non-desirable parasitics from the components and PCB boards.

Table 2. Switching characteristics under inductive-loading (for the case of V_{DS}=200V, I_D=15A, V_{gate}=15V and R_{gate}=5Ω)

	Turn-on (ns)				Turn-off (ns)			
Temperature	$t_{f,v}$	$t_{r,I}$	$t_{d,on}$	t_{on}	$t_{r,V}$	$t_{f,I}$	$t_{d,off}$	t_{off}
300K	14.6	6.2	11.7	27.8	8.9	9.2	110	125
77K	10.2	3.1	10.7	22.2	8.8	7.8	95	104
Reduction (%)	30	51	9	20	2	15	13	17

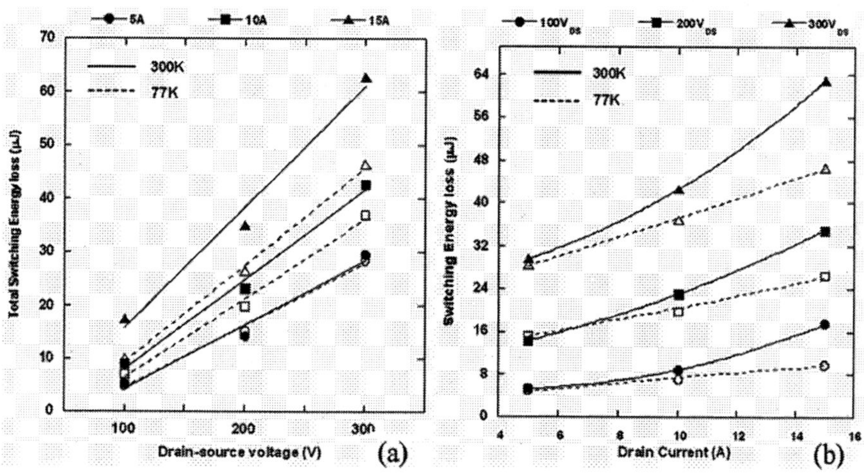

Figure 19. (a) Switching energy loss vs. drain voltage, and (b) Switching energy loss vs. drain current under inductive loading.

Turn-on energy loss is calculated as the integral of the instantaneous product of the drain current and the drain voltage between the time when drain current begins to rise, and the time when drain voltage drops from bus voltage to on-state voltage during the turn-on transient. Turn-off energy loss is calculated in a similar way. The total switching energy loss is the sum of the two. While operating the SJ MOSFET at cryogenic temperatures generally reduces the switching time by 10%~30%, it leads to up to 45% reduction in the switching energy loss. There is a linear dependence of total switching energy loss on drain-source voltage for different current levels at both temperatures as shown in Figure 19(a); higher drain-source voltage level leads to higher switching energy loss. Figure 19(b) shows the dependence of total switching energy loss on drain current. It can be seen that higher drain current leads to higher switching energy loss at both temperatures. However, the switching energy loss is linearly dependent on the current level at cryogenic temperatures, whereas it has a second order dependence at room temperature. This indicates that operating SJ MOSFET at cryogenic temperatures is more suitable for higher current operations, since the switching loss does not increase with current as much as the case of room temperature operation.

Switching characterizations under the resistive-loading were also performed with various bus voltages (100V, 150V and 200V) and gate resistances (2Ω, 5Ω and 10Ω). In the resistive-loading tests, the drain current was controlled by the drain-source voltage and the load resistor

(10.8Ω used) in the current loop. Therefore, in our tests, the drain currents were dependent on and proportional to the drain voltage. The switching characteristics derived from the waveform measurement for a typical case are shown in Table 3. In general, for the various voltage and gate resistance we have tested, operating SJ power MOSFET at cryogenic temperature, turn-on and turn-off times are reduced by 10%~30% and the rise and fall time of both drain voltage and current decrease by 10%~70%. In general, bigger drops in the voltage rise time ($t_{r,V}$) and current fall time ($t_{f,I}$) during turn-off were observed than those under inductive loading. Overall effect of temperature on the switching time of SJ MOSFET is quite similar to the results of the inductive-loading tests. Figure 20 shows the switching energy loss vs. the drain voltages with different gate resistances at both temperatures. The reduction ratio of the switching energy loss at 77K is similar to that of inductive-load test as about 10%~40%.

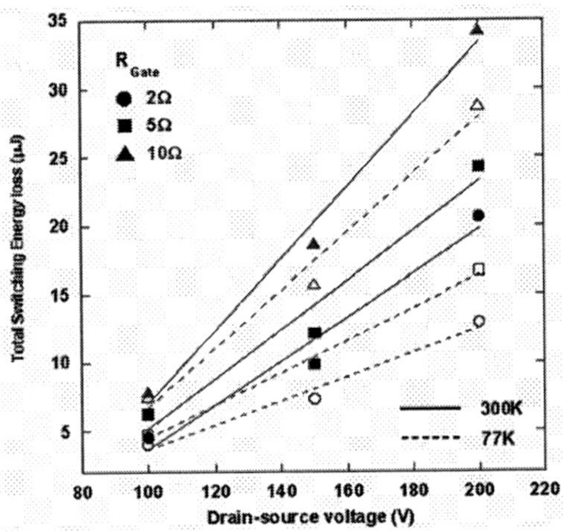

Figure 20. Total switching energy loss vs. drain voltage under resistive loading.

Table 3. Switching characteristics under resistive-loading (for the case of V_{DS}=200V, I_D=18A, V_{gate}=15V and R_{gate}=5Ω)

	Turn-on (ns)				Turn-off (ns)			
Temperature	$t_{f,v}$	$t_{r,I}$	$t_{d,on}$	t_{on}	$t_{r,V}$	$t_{f,I}$	$t_{d,off}$	t_{off}
300K	9.3	11.7	11.3	61.4	12.7	13.5	107.8	122.2
77K	7.4	10.5	11.0	53.9	9.5	3.7	94.9	99.5
Reduction (%)	20	10	3	12	25	73	12	19

The power MOSFETs are normally switched at certain frequency with certain duty ratio in applications such as inverter and converter. Total loss of the power MOSFET during continuous switching operation can be estimated by the sum of the static and the switching losses at various frequencies. For instance, assuming operating SJ MOSFET at 15A and 300V with a duty ratio of 1, at lower frequencies (<100 KHz), the majority of the total energy loss comes from the conduction loss; therefore, operating SJ MOSFETs at 77K leads to about 5x reduction in total energy loss. As the frequency increases, the switching loss becomes more dominant resulting in decrease in the reduction ratio of total energy loss at 77K; as shown in Figure 21, more than 2x reduction of total energy loss is expected at 1MHz.

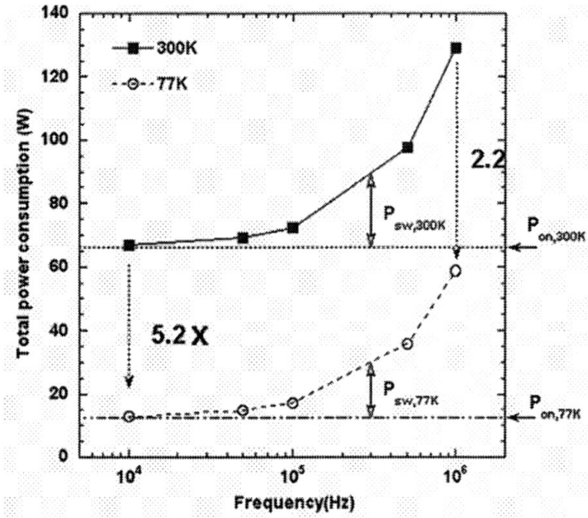

Figure 21. Total energy losses SJ power MOSFET (at 15A, 300V) vs. frequency (P_{on} and P_{sw} are static loss and switching loss, respectively).

In this section, static and switching characteristics of the SJ power MOSFET at both room temperature and liquid nitrogen temperature have been investigated at room temperature and liquid nitrogen temperature. More than 5 times on-resistance reduction was measured when the device is cooled at 77K. It was shown that 10%~30% improvement of switching time and 10%~50% reduction of switching energy loss can be achieved at 77K under both inductive and resistive loadings. Under inductive-loading, the switching energy loss was linearly dependent on the drain current at 77K while it has a non-linearly dependence at room temperature. Operating SJ power MOSFET at cryogenic temperature leads to more than 2~5 times reduction on total energy loss when operating continuously at frequencies up to 1MHz. For the operating frequency below 100 kHz, the total energy loss reduction is very significant at about 4~5x. Although the 5x cryogenic improvement on power less is smaller than that of the conventional MOSFET, cryogenic operation of SJ power MOSFET is still much more efficient than that of conventional power MOSFETs. Therefore, it is worthwhile to consider using SJ MOSFETs in the place of conventional power MOSFETs in cryogenic applications.

VI. Summary

In this chapter, we have summarized the benefit and challenges for the cryogenic applications of power MOSFETs. Physically based numerical simulations of the Si power DMOS are performed and the optimizations of device structure at cryogenic temperatures are discussed. Advanced power electronic packaging/interconnect methods that adapt well to cryogenics and help further improve system performance is discussed. Development of cryogenic integrated power package is discussed in detail. In the end of the chapter, the possibility of utilizing superjunction power MOSFET for cryogenic operation is also discussed based on experimental testing results.

References

[1] R. K. Kirschman, "Low-temperature electronics," *Circuits and Devices Magazine, IEEE*, vol. 6, no. 2, pp. 12-24, 1990.

[2] O. Mueller, "Properties of high-power Cryo-MOSFETs," in *Proc.of IEEE 31st Industry Applications Conference,* 3 ed pp. 1443-1448, 1996.

[3] P. Haldar, H. Ye, H. Efstathiadis, J. Raynolds, M. J. Hennessy, O. M. Mueller, and E. K. Mueller, "Improving Performance of Cryogenic Power Electronics," *IEEE Trans. Appl. Superconductivity*, vol. 15, no. 2, pp. 2370-2375, 2005.

[4] H. Ye, C. Lee, R. Simon, P. Haldar, M. J. Hennessy, and E. K. Mueller, "Liquid-Nitrogen Cooled Integrated Power Electronics Module with High Current Carrying Capability and Lower On-Resistance," *Appl. Phys. Lett.*, vol. 89, no. 19, p. 192107, 2006.

[5] M. Dean, D. Foty, N. Saks, S. Raider, and G. Oleszel, "Low temperature microelectronics: opportunities and challeges," *Proc. Symp. Low Temperature Electronic Device Operation, Electrochemical Society*, vol. 91, no. 14, pp. 25-37, 1991.

[6] B. Ray, S. S. Gerber, R. L. Patterson, and I. T. Myers, "Power control electronics for cryogenic instrumentation," *Advances in Inst. and Control, 50, 1, Int. Soc. for Measurement and Control*, pp. 131-139, 1995.

[7] R. K. Kirschman, "Cold electronics: an overview," *Cryogenics*, vol. 25, no. 3, pp. 115-122, Mar.1985.

[8] M. A. Newell, R. Stern, D. Hykes, G. Bolotin, T. Gregoire, T. McCarthy, C. Buchanan, and S. Cozy, "Extreme temperature (-170C to +125C) electronics for nanorover operation," in *Proceeding of IEEE 2001 Aerospace Conference,* 5 ed pp. 2443-2456, 2001.

[9] G. W. Nuedeck, *Modular Series on Solid State Devices, Volume III: The Bipolar Junction Transistor*, Second ed. Reading Massachusetts: Addison-Wesley Pub Co., 1989.

[10] R. F. Pierret, *Modular Series on Solid State Devices, Volume I: Semiconductor Fundamentals*, Second ed. Reading Massachusetts: Addison-Wesley Pub Co., 1989.

[11] R. J. Baker, H. W. Li, and D. E. Boyce, *COMS Circuit Design, Layout and Simulation* Wiley-IEEE Press, 1998.

[12] P. Ghazavi and D. H. Fat, "A numerical model for MOSFET's from liquid-nitrogen temperature to room temperature," *Electron Devices, IEEE Transactions on*, vol. 42, no. 1, pp. 123-134, 1995.

[13] O. Mueller, "Cryogenic MOSFET power conversion," *Proc. of the Workshop on Low Temperature Semiconductor Electronics, 1989*, pp. 94-98, 1989.

[14] K. Shenai, "Performance potential of low-voltage power MOSFETs in liquid-nitrogen-cooled power systems," *Electron Devices, IEEE Transactions on*, vol. 38, no. 4, pp. 934-936, 1991.

[15] W. Lu, R. J. Mauriello, K. B. Sundaram, and L. C. Chow, "Switching characteristics of a power MOSFET with varying temperature," *Southeastcon '98. Proceedings. IEEE, Orlando, Florida*, pp. 268-270, 1998.

[16] O. M. Mueller and K. G. Herd, "Ultra-high efficiency power conversion using cryogenic MOSFETs and HT-superconductors," *24th Annual Power Electronics Specialists Conference, PESC '93*, pp. 772-778, 1993.

[17] A. I. Gardiner, S. A. Johnson, and E. Schempp, "Operation of power electronic converters at cryogenic temperatures for utility energy conditioning applications," *Proc.*

of the 31st Energy Conversion Engineering Conference, IECEC 96, vol. 4, pp. 2209-2214, 1996.

[18] R. L. Patterson, A. Hammoud, and J. E. Dickman, "Development of Electronics for Low-Temperature Space Missions,"NASA TM-2001-210235, 2001.

[19] S. S. Gerber, R. Patterson, B. Ray, and C. Stell, "Performance of a spacecraft DC-DC converter breadboard modified for low temperature operation," *Energy Conversion Engineering Conference, 1996. IECEC 96. Proceedings of the 31st Intersociety, Washington DC USA*, vol. 1, pp. 592-598, 1996.

[20] R. L. Patterson, A. Hammoud, J. E. Dickman, and S. Gerber, "Electronic components and circuits for extreme temperature environments,"2003.

[21] M. E. Elbuluk, A. Hammoud, S. S. Gerber, and R. L. Patterson, "Cryogenic Evaluation of an Advanced DC/DC Converter Module for DeepSpace Applications," Technical Memorandum NASA TM-2003-212085, 2003.

[22] M. E. Elbuluk, S. S. Gerber, A. Hammoud, R. L. Patterson, and E. Overton, "Performance of High-Speed PWM Control Chips at Cryogenic Temperatures," Technical Memorandum NASA TM-2001-211123, 2001.

[23] M. E. Elbuluk, S. S. Gerber, A. Hammoud, and R. L. Patterson, "Performance of Power Converters at Cryogenic Temperatures," Technical Memorandum NASA TM-2001-211130, 2001.

[24] M. E. Elbuluk, A. Hammoud, S. S. Gerber, R. L. Patterson, and E. Overton, "Low Temperature Performance of High Power Density DC/DC Converter Modules," Technical Memorandum NASA TM-2001-210973, 2001.

[25] M. E. Elbuluk, S. S. Gerber, A. Hammoud, and R. L. Patterson, "Efficiency and Regulation of Commercial Low Power DC/DC Converter Modules at Low Temperatures," Technical Memorandum NASA TM-2000-210251, 2000.

[26] R. L. Patterson, M. E. Elbuluk, and E. Overton, "Electronic Components and Systems for Cryogenic Space Applications," Technical Memorandum NASA TM-2001-211129, 2001.

[27] R. L. Patterson, A. Hammoud, and S. S. Gerber, "Evaluation of Capacitors at Cryogenic Temperatures for Space Applications," Technical Memorandum NASA TM-1998-206983, 1998.

[28] R. L. Patterson, J. E. Dickman, S. S. Gerber, M. E. Elbuluk, and E. Overton, "Electronics for Deep Space Cryogenic Applications," Technical Memorandum NASA TM-2002-211695, 2002.

[29] R. L. Patterson, A. Hammoud, J. E. Dickman, S. S. Gerber, E. Overton, and M. E. Elbuluk, "Electrical Devices and Circuits for Low Temperature Space Applications," Technical Memorandum NASA TM-2003-212600, 2003.

[30] B. Ray, S. S. Gerber, R. L. Patterson, and I. T. Myers, "Low-temperature operation of a buck DC/DC converter," *Proc. of Tenth Annual Applied Power Electronics Conference and Exposition, APEC '95, Dallas, TX USA*, no. 0, pp. 941-946, 1995.

[31] W. D. Jackson, O. S. Mazzoni, and E. Schempp, "Characteristics of semiconductor devices at cryogenic temperatures," *Proceedings of the Intersociety Energy Conversion Engineering Conference*, vol. 31st, pp. 676-681, 1996.

[32] M. Giesselmann, Z. Mahund, and S. Carson, "Investigation of power MOSFET switching at cryogenic temperatures," *22nd International Power Modulator Symposium*, pp. 47-50, 1996.

[33] R. Singh and B. J. Baliga, "Cryogenic operation of asymmetric n-channel IGBTs," *Solid-State Electron.*, vol. 38, no. 3, pp. 561-566, 1995.

[34] R. C. Jaeger and F. H. Gaensslen, "Simulation of impurity freezeout through numerical solution of Poisson's equation with application to MOS device behavior," *Electron Devices, IEEE Transactions on*, vol. 27, no. 5, pp. 914-920, 1980.

[35] S. M. Sze, *Physics of Semiconductor Devices*, 2nd ed. New York: Wiley-Interscience, 1981.

[36] D. B. M. Klaassen, "A unified mobility model for device simulation--I. Model equations and concentration dependence," *Solid-State Electron.*, vol. 35, no. 7, pp. 953-959, July1992.

[37] D. B. M. Klaassen, "A unified mobility model for device simulation--II. Temperature dependence of carrier mobility and lifetime," *Solid-State Electron.*, vol. 35, no. 7, pp. 961-967, July1992.

[38] C. Lombardi, S. Manzini, A. Saporito, and M. Vanzi, "A physically based mobility model for numerical simulation of nonplanar devices," *Computer-Aided Design of Integrated Circuits and Systems, IEEE Transactions on*, vol. 7, no. 11, pp. 1164-1171, 1988.

[39] D. M. Caughey and R. E. Thomas, "Carrier mobilities in silicon empirically related to doping and field," *Proceedings of the IEEE*, vol. 55, no. 12, pp. 2192-2193, 1967.

[40] S. A. Schwarz and S. E. Russek, "Semi-empirical equations for electron velocity in silicon: Part II—MOS inversion layer," *Electron Devices, IEEE Transactions on*, vol. 30, no. 12, pp. 1634-1639, 1983.

[41] S. Selberherr, *Analysis and Simulation of Semiconductor Devices*. New York: Springer-Verlag, 1984.

[42] B. J. Baliga, *Power Semiconductor Devices*. Boston: PWS Publishing Company, 1996.

[43] R. Singh and B. J. Baliga, *Cryogenic operation of silicon power devices*. Boston: Kluwer Academic Publishers, 1998.

[44] E. Savrun and C. Toy, "An aluminum nitride package for 600C and beyond," *High Temperature Electronics Conference HITEC. 1998 Fourth International, Albuquerque, NM USA*, pp. 265-268, 1998.

[45] J. Schulz-Harder, "Advantages and new development of direct bonded copper substrates," *Microelectronics Reliability*, vol. 43, no. 3, pp. 359-365, Mar.2003.

[46] H. Ye, H. Efstathiadis, and P. Haldar, "Numerical Thermal Simulation of Cryogenic Power Modules under Liquid Nitrogen Cooling," *Journal of Electronic Packaging*, vol. 128, no. 3, pp. 267-272, 2006.

[47] D. J. Coe, "High voltage semiconductor device," US Patent 4754310, 1988.

[48] X. Chen, "Semiconductor power devices with alternating conductivity type high-voltage breakdown regions," US5216275, 1993.

[49] A. E. Schlogl, G. Deboy, H. W. Lorenzen, U. Linnert, H. J. Schulze, and J. P. Stengl, "Properties of CoolMOS between 420K and 80K - the ideal device for cryogenic applications," in *Proc.of the 11th International Symposium on Power Semiconductor Devices and ICs,* Toronto, ON, Canada: pp. 91-94, 1999.

[50] J. B. Witcher, "Methodology for switching characteristics of power devices and modules." M.S. Thesis, Virginia Polytechnic Institute and State University, 2002.

In: MOSFETs: Properties, Preparations and Performance ISBN: 978-1-60456-762-5
Editors: N.T. Andre et al. pp. 115-167

Chapter 4

REDESIGN AND OPTIMIZATION OF MOSFET DEVICES

Petru Andrei
Florida State University and Florida A&M University
2525 Pottsdamer St., Tallahassee, FL 32310, USA

Abstract

The problem of redesign and optimization of semiconductor devices is revised from a different perspective from what has been done in the past, namely by using the formalism of doping sensitivity functions. The doping sensitivity functions are introduced in this chapter as the set of operators on the adjoint space of the space of square integrable functions generated by all possible doping configurations. On this functional space we are able to construct analytical and numerical algorithms for the redesign and optimization of semiconductor devices and circuits subject to technological or practical constraints. These constraints are taken into consideration by using the Lagrange multipliers technique. Our redesign technique relies on first-order perturbations, for which reason is computationally very efficient and requires only the solution of linear systems of equations. The technique is applied the design of random doping fluctuation resistant structures (fault-tolerant devices) and to the optimization of semiconductor devices and circuits. The design is performed by starting with an initial device structure and applying an iterative multidimensional maximization procedure to "readjust" the doping configuration and device structure in order to obtain the desired final properties for the device. The presented technique is universal in the sense that it can be applied to any semiconductor device such as field-effect transistors or heterojunctions devices and can be used in the framework of any transport model. It requires only the solution of a few linear sparse systems of equations, for which reason is computationally very efficient. Numerical results related to the suppression of random doping induced fluctuations in a 35-nm channel length transistor and suppression of the effects of the drain-induced barrier lowering (DIBL) on the value of the threshold voltage are presented for a 30 nm channel length MOSFET device. It is found that by carefully designing the doping profiles in these devices the effects of the random doping concentration on threshold voltage can be decreased by almost one order of magnitude, while the effects of the DIBL can be reduced at least 50%.

1. Introduction

Device and circuit optimization is one of the most challenging problems in nanoelectronics because it usually requires a large computational effort. Depending on the type of the optimization problem, parallel computers are usually needed in order to perform many trial-and-error simulations and find the device structures that have the optimum electrical properties [1-3]. For instance, in the case of the optimization of relatively simple circuits such as logic gates, memory cells, or even single field-effect-transistors, the optimization of the doping profiles requires solving the transport equations (i.e. the Poisson, drift-diffusion, Density-Gradient, or Boltzmann transport equation) for many devices with different doping profiles. Although each simulation could take a relatively low computational effort in the case of large devices because of various approximations that might be used, in the case of three-dimensional nanoscale devices the computation of each bias point might take hours or even longer on standard computers. Even if faster or more accurate optimization procedures such as the Levenberg-Marquardt [3] or steepest descent techniques are used to optimize various parameters, the total computational time is still relatively large, which restricts existing optimization methods to very simplified device structures.

In this chapter we introduce perturbation techniques for the optimization of semiconductor devices and circuits and focus on two important problems: (1) the design of random doping fluctuation resistant structures (fault-tolerant devices) and (2) optimization of semiconductor devices and circuits. These techniques should have a broad impact on the computer-aided-design and fabrication procedures for the next generation integrated circuits and devices.

The rest of the chapter is structured as follows. In Section 2 we state the redesign and optimization problems in a concise mathematical form. The doping sensitivity functions that stay at the basis of our techniques are presented in Section 3. We focus on the calculation of doping sensitivity functions of terminal currents, potentials, and static noise margins in static random access memories (SRAM) circuits. These circuits are by far the most affected circuits by random doping fluctuations in the semiconductor industry today. Sections 4 and 5 are dedicated to solving the redesign and optimization problems, respectively. The numerical algorithms are presented in detail. Numerical results related to the suppression of random doping induced fluctuations in a 35-nm channel length transistor and to the suppression of the effects of the drain-induced barrier lowering (DIBL) for a 30 nm channel length MOSFET are also presented in these sections. Conclusions are drawn in Section 6.

2. Statement of the Problem

The redesign and optimization problem can be stated in general as to compute the doping profiles and device structures in an electronic circuit in order to achieve certain nominal parameters for the circuit. Depending the parameter that is optimized we device these problems in two categories: (1) the design of random doping fluctuation resistant structures in which we compute the doping profiles and device structures that minimize the effects of random doping fluctuations and (2) optimization of semiconductor devices and circuits in which we compute the doping profiles and device structures in order to minimize or maximize the values of some parameters. The importance of these two problems and their mathematical formulation are presented in the following subsections.

2.1. Suppressing Random Doping Fluctuations in Semiconductor Devices and Circuits

The demand for high operational frequency and low power consumption of digital and analog electronic circuits has led to the development of ultra-small semiconductor devices with characteristic dimensions in the decananometer regime [4]. Unfortunately, such small devices are very susceptible to random dopant fluctuations [5-18], inhomogeneities, and structural nonuniformities [19,20], which are inevitably present in semiconductor devices. These fluctuations lead to undesirable variations of threshold voltages, terminal currents, and frequency response characteristics of field-effect-transistors, which often make the electronic circuits unoperational right after the construction stage. The magnitude of intrinsic parameter fluctuations between macroscopically identical devices is becoming the main source of device mismatch in the decanano regime and represents a fundamental limitation for conventional device scaling. Mismatch and intrinsic parameter variations begin to impinge on the performance and functionality of digital (e.g. SRAMs, logic gates) and analog circuits (e.g. operational amplifiers, low noise amplifiers), as shown by the increasing number of articles published on this topic recently [5-10,18-32].

In order to show the impact of intrinsic parameter fluctuations on CMOS integrated circuits we have represented the probability distribution function of the threshold voltage of a set of MOSFET devices in Fig. 1. The MOSFETs have microscopically different doping profiles (the location of the dopant atoms was randomly chosen), however, are macroscopically identical in the sense that the local average doping concentrations are identical. The average value of the threshold voltage is denoted by V_{T0}, while $V_{T\min}$ and are the minimum and maximum admissible values of the threshold voltage in order for 1 GB static- random access memory (SRAM) integrated circuit to function properly. If the integrated circuit contains at least one transistor with the threshold voltage in the dark regions then it will not function properly. Fig. 2 shows the simulated static transfer characteristics of a set of 200 SRAM cells fabricated by using the previously generated MOSFET devices. Keeping in mind the thumb rule that is order for an SRAM call to function properly one needs that the standard deviation of its static noise margins (SNM) to be at least 6 times larger that the value of the SNM it is estimated that the number of defectuous transistors in our SRAM is of the order of a few hundreds, which makes it unoperational. This number depends exponentially on the standard deviations of threshold voltage fluctuations (σ_{VT}), as can be inferred from the inset presented in Fig. 1. Due to this exponential behavior, any reduction in the standard deviation of random doping induced fluctuations will reduce the number of defectuous integrated circuits and, in this way, reduce the fabrication cost significantly. It is estimated that by reducing σ_{VT} by 30% the number of defectuous transistors is decreased more then two orders of magnitude, which would improve the static noise margins and make the SRAM device fully operational [24,25,32,33]. Therefore, there is a continuous need to develop models and techniques that reduce the standard deviation of fluctuating parameters and optimize semiconductor devices and circuits. In this chapter we introduce perturbation techniques for the design of random doping fluctuation resistant structures and optimization of semiconductor devices and circuits. Such techniques are expected to have a broad impact on the computer-aided-design and fabrication procedures for the next generation integrated circuits and devices.

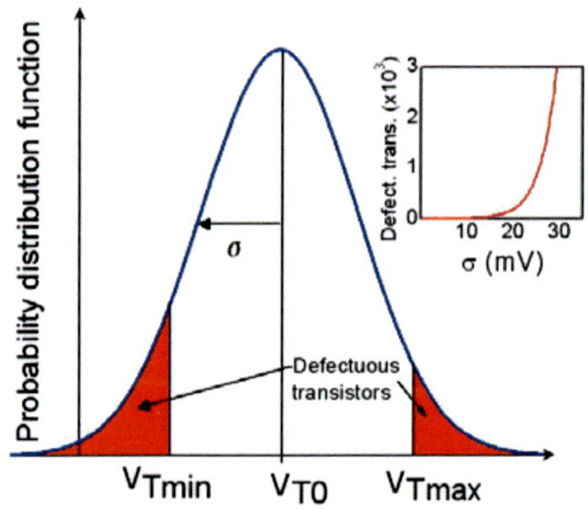

Figure 1. Probability distribution function showing the regions of defectuous transistors. The inset shows the number of defectuous transistors in a 1 GB SRAM circuit as a function of the standard deviation of V_T.

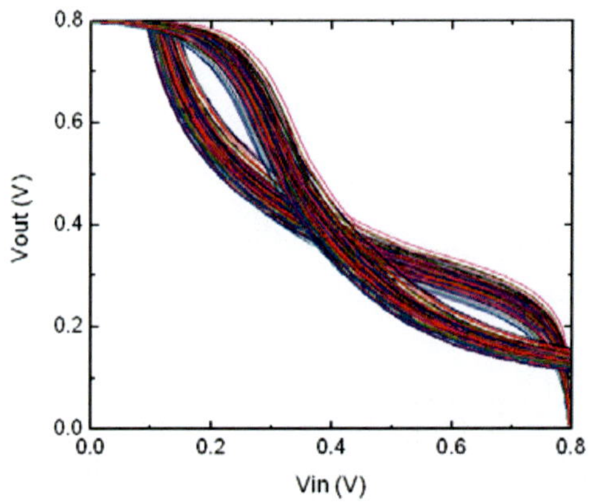

Figure 2. Static transfer characteristics of 200 SRAM circuits fabricated by using 35-nm channel length MOSFETs.

It should be mentioned that there are a few attempts in the literature to redesign semiconductor devices in order to obtain structures that are resistant to random doping fluctuations [13,34-37]. These attempts discover that some particular doping profiles (such as undoped channel or δ-doping profiles) result in relatively low random doping induced fluctuations but do not give any method on how to compute or design those profiles, or whether or not they are the only or the optimum doping profiles that decrease the most the effects of random doping fluctuations. The relatively small number of attempts in the literature is mostly due to the high computational cost required to simulate many trial-and-error devices in order to come up with the best doping profiles.

In order to set the design problem in a mathematical form, let us denote by P some parameter of the semiconductor device whose fluctuations are to be minimized and by A, B, , etc. parameters whose expected values should be kept constant during the minimization process. For instance, to minimize the fluctuations of the threshold voltage in a MOSFET device, we set P to the threshold voltage, while parameters A, B, and C could be the cutoff frequency, gain, and/or even the expected value of the threshold voltage itself. Due to the stochastic nature of ion implantation and diffusion processes doping concentration $D(\boldsymbol{r})$ and parameters A, B, C, and P are random variables. Therefore, the design problem can be stated as the following minimization problem with constraints:

$$\min_{\langle D(\boldsymbol{r})\rangle} \sigma_P^2 \text{ such that } \langle A\rangle, \langle B\rangle, \langle C\rangle, \ldots \text{ are constant,} \tag{1}$$

where σ_P^2 denotes the variance of random doping induced fluctuations on parameter P, $\langle D(\boldsymbol{r})\rangle$ is the average value of the doping concentration at location $\boldsymbol{r}$, and $\langle A\rangle$, $\langle B\rangle$, and $\langle C\rangle$ are the average values of parameters A, B, and C. All these average values are computed by summing over all semiconductor devices and taking the average. Note that in this minimization problem one does not have to compute the exact doping profiles but only the average values of the doping concentration.

2.2. Optimizing Semiconductor Devices and Circuits

The problem of optimization of semiconductor devices and circuits has been in the literature for a long time. As in the case of designing random doping fluctuation structures the existing approaches are based more or less on trial-and-error techniques which are trying to find particular device structures and doping profiles that results in better (optimal) characteristics of the device or circuit. These approaches are computationally very expensive and quite often are not feasible in practical applications.

Using the same notations as before let A, B, C, etc. be the parameters that are supposed to stay constant during the optimization, and P be the parameter that should be optimized. Without loosing any generality we can state the optimization problem as the following maximization:

$$\max_{D(\boldsymbol{r})} P \text{ such that } A, B, C, \ldots \text{ are constant,} \tag{2}$$

where D is the doping concentration in the device or circuit. Note that the minimization problem can be solved by applying the same approach to parameter $-P$.

Although problems (1) and (2) are very similar, there is a fundamental difference between them. As will be discussed in Section 4, the doping concentration function in problem (1) is a random variable which makes parameters A, B, and C to be random as well. For this reason, problem (1) can be regarded as a stochastic problem with minimization. The doping concentration function in problem (2) is a real function of position and all quantities entering this equation are deterministic. Despite this fundamental difference, it turns out that the two problems can be solved approximately in the same way.

3. Doping Sensitivity Functions

The first part of this section introduces the concept of doping sensitivity functions that are at the basis of the redesign and optimization presented in this chapter. The second part of the section presents the technique for the computation of doping sensitivity functions presents a few numerical results related to the analysis of random doping induced fluctuations of static noise margins (SNM) in static random access memories (SRAMs).

3.1. Definition of Doping Sensitivity Functions

Denote by X some parameter of the semiconductor device or circuit, such as the threshold voltage, terminal current, or cut-off frequency. In general, parameter X can be written as a *functional* $X(D)$, where D is the doping concentration, which depends on position $\boldsymbol{r}$, $D(\boldsymbol{r})$. We define functional X as acting on the space of all possible doping profiles, which is assumed to be the space of square integrable functions. Note that we do not make any assumption about the particular transport model used, so the analysis that follows can be applied to the drift-diffusion model, semiclassical (e.g. Monte-Carlo techniques based on the Boltzmann transport equation) or any quantum transport model (e.g. the nonlinear Green's function technique).
Let us compute now how much the value of parameter X changes when the doping concentration is subject to an infinitesimally small variation $\tilde{D}(\boldsymbol{r})$. The new value of parameter X can be written in terms of the Fréchet derivative $X'(\tilde{D})$ as follows [38]:

$$X\left(D_0+\tilde{D}\right)=X\left(D_0\right)+X'\left(\tilde{D}\right), \tag{3}$$

where D_0 the initial doping concentration function in the device. equation (3) is exact when the norm of $\tilde{D}$ goes to zero. Since the Fréchet derivative is a linear operator, according to the Riesz representation theorem [39], there exist a function $\gamma_X(\boldsymbol{r})$ such that:

$$X'\left(\tilde{D}\right)=\int_{\Omega}\gamma_X(\boldsymbol{r})\tilde{D}(\boldsymbol{r})\,d\boldsymbol{r}, \tag{4}$$

where Ω is the region of all semiconductor devices. For instance, in the case of logic CMOS gates, Ω includes the substrate, drain and source junctions, and the polysilicon gates of all MOSFET devices. The last two equations yield:

$$X\left(D_0+\tilde{D}\right)=X\left(D_0\right)+\int_{\Omega}\gamma_X(\boldsymbol{r})\tilde{D}(\boldsymbol{r})\,d\boldsymbol{r}. \tag{5}$$

The physical interpretation of function $\gamma_X^{Dop}(\boldsymbol{r})$ can be found by taking $\tilde{D}(\boldsymbol{r})=\delta(\boldsymbol{r}-\boldsymbol{r}_0)$ in the last equation, where δ is the Dirac distribution function. We obtain:

$$X\left(D_0+\tilde{D}\right)-X\left(D_0\right)=\int_{\Omega}\gamma_X\left(\boldsymbol{r}\right)\delta\left(\boldsymbol{r}-\boldsymbol{r}_0\right)d\boldsymbol{r}=\gamma_X\left(\boldsymbol{r}_0\right), \tag{6}$$

which shows that $\gamma_X\left(\boldsymbol{r}_0\right)$ gives how much parameter X changes when the doping concentration at location $\boldsymbol{r}_0$ is increased by one δ-function, which is equivalent to saying that one additional dopant atom is introduced at location $\boldsymbol{r}_0$. Function $\gamma_X\left(\boldsymbol{r}\right)$ is a measure of how sensitive parameter X is to variations of the doping concentrations at location $\boldsymbol{r}$ inside the device and, for this reason, we call this function the *doping sensitivity function*. Mathematically, the doping sensitivity functions are introduced in equation (5) as functionals on the adjoint space of the space of square integrable functions generated by all possible doping variations $\tilde{D}$.

Equation (5) can be written in a more compact form as:

$$\tilde{X}=\int_{\Omega}\gamma_X\left(\boldsymbol{r}\right)\tilde{D}\left(\boldsymbol{r}\right)d\boldsymbol{r}, \tag{7}$$

where $\tilde{X}=X\left(D_0+\tilde{D}\right)-X\left(D_0\right)$ shows how much parameter X has varied as a result of fluctuations.

The doping sensitivity functions can be computed by using RandFlux [40] which is a software for the redesign and optimization of semiconductor devices and circuits. The user can define various parameters by writing analytical equations for which RandFlux should compute the sensitivity functions. For instance, one can compute the sensitivity functions of threshold voltages in filed-effect-transistors, terminal currents, cutoff frequencies, static noise margins (SNM) of SRAM circuits, small-signal parameters, etc. All simulation results presented in this chapter were performed by using RandFlux.

3.2. Estimation of the Variability Induced by Random Doping Induced Fluctuations

The sensitivity functions defined in (7) show how sensitive parameter X is to fluctuations of doping concentration at different locations. Due to the stochastic nature of ion implantation and diffusion processes doping concentration is a random variable and can be expressed as the sum of its average value $\left\langle D\left(\boldsymbol{r}\right)\right\rangle$ and fluctuations $\tilde{D}\left(\boldsymbol{r}\right)$ [41]:

$$D\left(\boldsymbol{r}\right)=\left\langle D\left(\boldsymbol{r}\right)\right\rangle+\tilde{D}\left(\boldsymbol{r}\right), \tag{8}$$

where by definition, the expected values of $\tilde{D}$ is equal to zero. The variance of parameter X can be computed as:

$$\sigma_X^2=\int_{\Omega}\int_{\Omega}\gamma_X\left(\boldsymbol{r}\right)\gamma_X\left(\boldsymbol{r}'\right)ACF\left[D\left(\boldsymbol{r}\right),D\left(\boldsymbol{r}'\right)\right]d\boldsymbol{r}\,d\boldsymbol{r}', \tag{9}$$

where, $ACF\left[D(\boldsymbol{r}),D(\boldsymbol{r}')\right]$ denote the autocorrelation function of the doping concentrations at locations $\boldsymbol{r}$ and $\boldsymbol{r}'$. Usually this function is evaluated by using that the doping concentrations at different locations are independent random variable, hence:

$$\sigma_X^2\left[Dop\right]\approx\int_\Omega\left[\gamma_X\right]^2\sigma_D^2(\boldsymbol{r})\Delta V\,d\boldsymbol{r}\,, \tag{10}$$

where $\sigma_{Dop}^2(\boldsymbol{r})$ is the variance of doping fluctuations at location $\boldsymbol{r}$, and $\Delta V\equiv d\boldsymbol{r}'$ is the volume over which the averaging is computed. In devices such as modulation doped n-GaAs heterostructures, the silicon donors are deposited at elevated temperature and tend to self-avoid during growth. In these situations approximation (10) does not hold and one should use equations (9) instead.

The variance of doping concentration σ_D^2, can be easily computed if we assume that the total number ΔN of dopant atoms in volume ΔV has Poisson distribution. In the case of Poisson distributions, the expected value and the variance coincide. Hence, $\sigma_{\Delta N}^2=\langle\Delta N(\boldsymbol{r})\rangle=\langle D(\boldsymbol{r})\rangle\Delta V$, where, $\langle\Delta N(\boldsymbol{r})\rangle$ is the average number of dopants in volume ΔV. By using this fact we derive $\sigma_D^2=\sigma_N^2/(\Delta V)^2=\langle D(\boldsymbol{r})\rangle/\Delta V$, which leads the following equation for the variance of parameter A induced by doping fluctuations:

$$\sigma_X^2\left[Dop\right]\approx\int_\Omega\left[\gamma_X(\boldsymbol{r})\right]^2\langle D(\boldsymbol{r})\rangle d\boldsymbol{r}\,. \tag{11}$$

This equation can be applied to compute the variance of random acceptor and donor induced fluctuations in electronic circuit. It is natural to distinguish between donor and acceptor concentrations and uses the following equation for the estimations of random doping induced fluctuations:

$$\begin{aligned}\sigma_X^2\left[Dop\right]&=\sigma_X^2\left[N_A\right]+\sigma_X^2\left[N_D\right]=\\&=\int_\Omega\left[\gamma_X^{N_A}(\boldsymbol{r})\right]^2\langle N_A(\boldsymbol{r})\rangle d\boldsymbol{r}+\int_\Omega\left[\gamma_X^{N_D}(\boldsymbol{r})\right]^2\langle N_D(\boldsymbol{r})\rangle d\boldsymbol{r}\,.\end{aligned} \tag{12}$$

It is important to note for the numerical implementation that the sensitivity functions of N_A and N_D are equal to each other so there is no need to compute both functions. The variance of parameter X induced by random doping concentration becomes:

$$\sigma_X^2\left[Dop\right]=\int_\Omega\left[\gamma_X(\boldsymbol{r})\right]^2\left[\langle N_A(\boldsymbol{r})\rangle+\langle N_D(\boldsymbol{r})\rangle\right]d\boldsymbol{r}\,. \tag{13}$$

3.3. Computation of Doping Sensitivity Functions of Terminal Currents

In this section we present the technique for the computation of doping sensitivity functions of terminal currents. The electron and hole transport model assumed this work is based on the Poisson, current continuity, and Density-Gradient equations:

$$\nabla \cdot (\varepsilon \nabla \varphi) + q\left(p - n + N_D^+ - N_A^-\right) = 0, \tag{14}$$

$$\frac{2\nabla \cdot \left(b_n \nabla \sqrt{n}\right)}{\sqrt{n}} + \varphi - \phi_n - \frac{kT}{q}\ln\frac{n}{n_i} = 0, \tag{15}$$

$$\frac{2\nabla \cdot \left(b_p \nabla \sqrt{p}\right)}{\sqrt{p}} - \varphi + \phi_p + \frac{kT}{q}\ln\frac{n_i}{p} = 0, \tag{16}$$

$$\nabla \cdot \left(\mu_n n \nabla \phi_n\right) + q\left(R_n - G_n\right) = 0, \tag{17}$$

$$\nabla \cdot \left(\mu_p p \nabla \phi_p\right) - q\left(R_p - G_p\right) = 0, \tag{18}$$

where φ is the electrostatic potential, ϕ_n and ϕ_p are the electron and hole quasi-Fermi potentials, n and p are the electron and hole densities, R_n, G_n and R_p and G_p are the electron and hole recombination and generation rates, respectively. N_D^+ and N_A^- are the concentration of ionized donor and acceptor impurities, n_i and ε are the intrinsic concentration and electric permittivity of the semiconductor, and T is the temperature. The quantum mechanical effects are "controlled" by parameters:

$$b_n = \frac{\hbar^2}{12qm_n^*},\; b_p = \frac{\hbar^2}{12qm_p^*}, \tag{19}$$

where m_n^* and m_p^* denote the effective masses of the electrons and holes, which in our simulations were calibrated by using the technique presented in [42]. The system of equations (14)-(18) is subject to appropriate boundary conditions and solved self-consistently for all devices in the circuit.

In order to present how to compute the sensitivity functions of terminal currents, let us denote the discretized transport equations (14)-(18) in vector form [43]:

$$\boldsymbol{F}(\boldsymbol{X}, \boldsymbol{D}) = 0, \tag{20}$$

where $\boldsymbol{F}$ is a nonlinear vector function of the unknown "state" vector $\boldsymbol{X}$ and doping vector $\boldsymbol{D}$. The sate vector consist of the mesh-point values of the electric potential, electron and hole concentrations, and quasi-Fermi potentials. The doping vector consists of the mesh-point values of the doping concentration. If $\tilde{\boldsymbol{X}}$ denotes the fluctuations of the state variable and $\tilde{\boldsymbol{D}}$ the fluctuations of the doping concentration, one can use the first-order approximation to write:

$$\hat{F}_X \tilde{X} + \hat{F}_D \tilde{D} = 0, \tag{21}$$

where $\hat{F}_X$ and $\hat{F}_D$, are the derivatives of F with respect to X and D, respectively. All derivatives are computed at the given dc bias point and by assuming constant (non-fluctuating) values of the doping concentration.

Let us denote now the value of the current through terminal α by I_α and write it as an explicit function of state vector $\boldsymbol{X}$ and doping concentration $\boldsymbol{D}$:

$$I_\alpha = I^\alpha(\boldsymbol{X}, \boldsymbol{D}). \tag{22}$$

The fluctuations of terminal currents can be found by linearizing (22) with respect to the fluctuating quantities:

$$\tilde{I}_\alpha = \left(I_X^\alpha\right)^t \tilde{X} + \left(I_D^\alpha\right)^t \tilde{D}, \tag{23}$$

where I_X^α and I_D^α are the derivatives of I^α with respect to the state variable and doping concentration, while superscript t denotes the transpose of the given vector. Equations (21) and (23) cab be solved for the fluctuations of the terminal current:

$$\tilde{I}_\alpha = -\left[\boldsymbol{g}_\alpha^t \cdot \hat{F}_D - \left(I_D^\alpha\right)^t\right] \cdot \tilde{D}, \tag{24}$$

where $\boldsymbol{g}_\alpha^t$ is the transpose of column vector $\boldsymbol{g}_\alpha$ and can be found by solving the following linear system of equations:

$$\hat{F}_X^t \boldsymbol{g}_\alpha = I_X, \tag{25}$$

where $\hat{F}_X^t$ denotes the transpose of matrix $\hat{F}_X$. By comparing (7) and (24), it can be inferred that the following coefficients:

$$\gamma_{I_\alpha}^i = -\left(\boldsymbol{g}_\alpha^t \cdot \hat{F}_D\right)_i + \left(I_D^\alpha\right)_i^t \tag{26}$$

are the discretized values of the doping sensitivity function. The most expensive computational effort in the calculation of these coefficients is to solve linear system (25). For 2-D simulations we solve this system by using the classical Gauss-Seidel method, while for 3-D simulations we solve it by using the Successive Over-Relaxation (SOR) technique.

3.4. Computation of Doping Sensitivity Functions of Terminal Potentials and Other Parameters

The technique presented above for the computation of sensitivity functions of terminal currents can be extended to the computation of other sensitivity functions, such as the sensitivity functions of terminal potentials, gains, cutoff frequencies, etc. For instance, in order to compute the sensitivity function of terminal potential we only have to replace the expression of the terminal current I_α with the expression of the terminal voltage V_α in equations (22)-(26). Both the terminal current and terminal potential enter as unknowns in the transport system of equations and can be treated in the same manner. For more information related to the computation of doping sensitivity functions we recommend Refs. [44,45] for threshold voltages, [46] for terminal currents, [47,48] for small-signal parameters, [49] for frequency characteristics, and [50] for subthreshold voltage characteristics. In the following we present the technique for the computation of sensitivity functions of the SNM in SRAM cells. This technique is extremely important for circuit design because SRAMs are among the worst affected by random doping fluctuation circuits in the semiconductor industry. A more detailed analysis of variability can also be found in [41,43].

3.5. Computation of Doping Sensitivity Functions in SRAM Circuits

Definition of SNM

Consider the 6-T SRAM bitcell shown in Fig. 3. The SNM of SRAM cell is defined as the amount of voltage noise required at the internal nodes of a cell to flip its state. Geometrically, the SNM can be identified as the length of the side of a square having the longest diagonal that can be fitted between the voltage transfer characteristics of the two inverters of the SRAM cell (see Fig. 1b). Using the notations from Fig. 3 we denote the voltage transfer characteristic of the inverter on the left (L) by $f(x,D)$ and of the inverter on the right (R) by $h(y,D)$, where x and y denote the input voltages of the inverters and D denotes the doping distribution function of the cell. The voltage transfer characteristics f and h are functions of the input voltages and depend on the exact doping configuration of each semiconductor device $D(\boldsymbol{r})$. Here $\boldsymbol{r} \in \Omega$, where Ω is the total region of semiconductors such as channels, junctions, and polygates used for building the logic cell. In the case of the SRAM cell represented in Fig. 1a, Ω is the sum of the semiconductor regions belonging to each of the six cell transistors, i.e. $\Omega = \Omega_{PR} + \Omega_{NR} + \Omega_{AR} + \Omega_{PL} + \Omega_{NL} + \Omega_{AL}$.

Note that by writing the voltage transfer characteristics as a function of the doping concentration, $f(x,D)$ and $h(y,D)$ can be regarded as *functionals* acting on the space of all possible doping profiles, which in this analysis is assumed to be the space of the square integrable functions $\mathcal{L}_2$. By restricting the doping concentration to be an $\mathcal{L}_2$ function we do not impose any constraints on the continuity of $D(\boldsymbol{r})$. Hence, the analysis below can be applied to both continuum and "atomistic" [51,52] simulations.

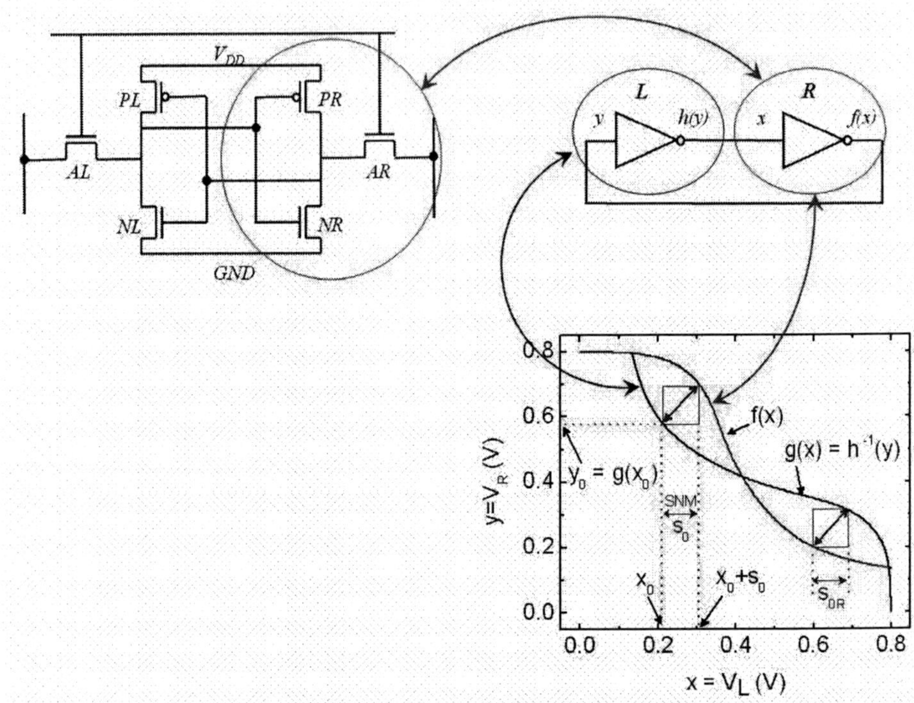

Figure 3. 6T SRAM schematics used in the self-consistent simulations and the definition of the static noies margins. The inverter on the left (L) includes transistors PL, NL, and AL, while the inverter on the right (R) includes transistors PR, NR, and AR. Functions f and h are the voltage transfer characteristics of the R- and L-inverters, respectively.

If is convenient for our next analysis to introduce the inverse function of the voltage transfer characteristic of the R-inverter as $g(x,D)=h^{-1}(y,D)$, i.e. $g(h(y,D),D)=y$ (see also Fig. 3). With this notation the computation of the SNM is reduced to finding the length of the side of a square fitted between functions $f(x,D)$ and $g(x,D)$ and having the longest diagonal. In order to find the SNM for a given doping configuration $D_0(\boldsymbol{r})$, one has to solve the following system of nonlinear equations:

$$\begin{cases} f(x_0+s_0,D_0)=y_0+s_0 \\ \dfrac{\partial f}{\partial x}(x_0+s_0,D_0)=\dfrac{\partial g}{\partial x}(x_0,D_0) \end{cases} \tag{27}$$

where $y_0=g(x_0,D_0)$ and $\dfrac{\partial f}{\partial x}$ and $\dfrac{\partial g}{\partial x}$ denote the derivatives with respect to the first argument. The first equation in (27) expresses the condition that the height and the width of the rectangle fitted between the two voltage characteristics are equal to each other, while the second equation gives the first-order necessary condition to maximize the length of the diagonal of the inside square. If f and g are known, system (27) can be solved for the value of the static noise margin s_0 as well as for the input voltage of the R-inverter x_0 corresponding to the point on the voltage transfer characteristics where the SNM is estimated (see Fig. 2). The solution of (27) can be formally written as $s_0 = s(D_0)$, which shows that the SNM is also a *functional* acting on the same space generated by all possible doping configurations $\mathcal{L}_2$.

It is important to note that, in general, system (27) has 2 solutions corresponding to the two loops of the butterfly curves shown in Fig. 2. The two values of the SNM correspond to the case when the SNM cell is in state "0" and "1", respectively. In the following, we consider only the case when the initial state of the SRAM cell is "1" (the upper-left loop); the other case can be treated similarly. Also note that the cell is most vulnerable to noise during read-access mode. Therefore, in the simulations presented in this section we consider the cell in the read-access mode.

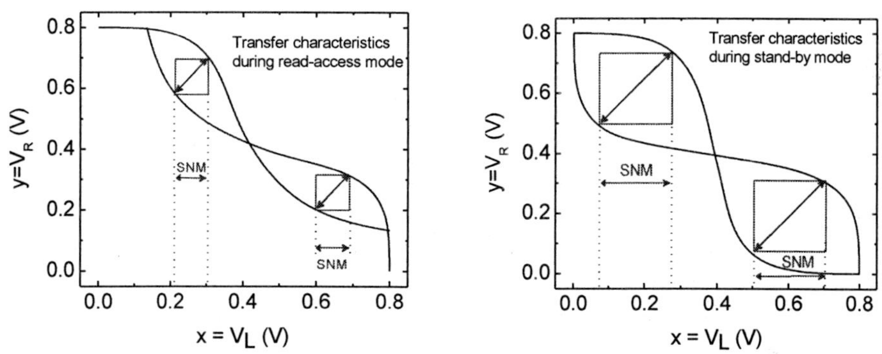

Figure 4. Transfer characteristics of the SRAM cell during read-access mode (left) and stand-by mode (right). The cell is most vulnerable to noise during access mode.

Doping Sensitivity Functions of SNM

Let us suppose now that there is infinitesimally small change in the doping profile $\tilde{D}(\boldsymbol{r})$, and the new doping concentration is $D(\boldsymbol{r}) = D_0(\boldsymbol{r}) + \tilde{D}(\boldsymbol{r})$. The new value of the static noise margin $s(D)$ can be computed by solving:

$$\begin{cases} f\left(x+s, D_0+\tilde{D}\right) = g\left(x, D_0+\tilde{D}\right) + s \\ \dfrac{\partial f}{\partial x}\left(x+s, D_0+\tilde{D}\right) = \dfrac{\partial g}{\partial x}\left(x, D_0+\tilde{D}\right) \end{cases} \tag{28}$$

This system of equations will be solved analytically in the next section. For now let us observe that the solution s corresponding to doping $D_0 + \tilde{D}$ can be written as:

$$s\left(D_0+\tilde{D}\right)=s\left(D_0\right)+s'\left(\tilde{D}\right)=s\left(D_0\right)+\int_{\Omega}\gamma_s\left(\boldsymbol{r}\right)\tilde{D}\left(\boldsymbol{r}\right)d\boldsymbol{r}\ , \tag{29}$$

where $s'\left(\tilde{D}\right)$ the Fréchet derivative of SNM and $\gamma_s\left(\boldsymbol{r}\right)$ is the *doping sensitivity function* of SNM. The last equation was obtained by using the Riesz representation theorem [53], and can be regarded as the formal solution of system (28). Function $\gamma_s\left(\boldsymbol{r}\right)$ is a measure of how sensitive the SNM is to variations of the doping concentrations at different locations inside the cell. Solving system (28) is equivalent to finding the doping sensitivity function $\gamma_s\left(\boldsymbol{r}\right)$ that satisfies equation (29).

Computation of Doping Sensitivity Functions of SNM

The main equations for the evaluation of the doping sensitivity functions of SNM are deduced in this section. To this goal we suppose again that the doping profile is changed from D_0 to $D_0+\tilde{D}$ and write the transfer characteristics of the two inverters by using equations similar to (29):

$$f\left(x,D_0+\tilde{D}\right)=f\left(x,D_0\right)+\int_{\Omega_R}\gamma_f\left(x,\boldsymbol{r}\right)\tilde{D}\left(\boldsymbol{r}\right)d\boldsymbol{r}\ , \tag{30}$$

$$g\left(x,D_0+\tilde{D}\right)=g\left(x,D_0\right)+\int_{\Omega_L}\gamma_g\left(x,\boldsymbol{r}\right)\tilde{D}\left(\boldsymbol{r}\right)d\boldsymbol{r}\ , \tag{31}$$

where Ω_R denotes all the semiconductor regions belonging to the R-inverter and containing transistors PR, NR, and AR, while Ω_L denotes the semiconductor regions belonging to the L- inverter, which contains transistors PL, NL, and AL. $\gamma_f\left(x,\boldsymbol{r}\right)$ is the doping sensitivity function of the voltage transfer characteristics of the R-inverter, while $\gamma_g\left(x,\boldsymbol{r}\right)$ is the doping sensitivity function of the *inverse* of the voltage transfer characteristics of the R-inverter. Introducing equations (30) and (31) in the first equation in (28), we obtain:

$$f\left(x+s,D_0\right)+\int_{\Omega_R}\gamma_f\left(x+s,\boldsymbol{r}\right)\tilde{D}\left(\boldsymbol{r}\right)d\boldsymbol{r}=g\left(x,D_0\right)+\int_{\Omega_L}\gamma_g\left(x,\boldsymbol{r}\right)\tilde{D}\left(\boldsymbol{r}\right)d\boldsymbol{r}+s\ . \tag{32}$$

It is convenient now to introduce the following variables:

$$\varepsilon_x=x-x_0\ ,\ \varepsilon_s=s-s_0\ , \tag{33}$$

which denote how much the bias point and the SNM have varied when the doping concentration was changed by $\tilde{D}\left(\boldsymbol{r}\right)$. Equations (32) and (33) yield:

$$\begin{aligned} &f\left(x_0+s_0+\varepsilon_x+\varepsilon_s,D_0\right)+\int_{\Omega_R}\gamma_f\left(x_0+s_0+\varepsilon_x+\varepsilon_s,\boldsymbol{r}\right)\tilde{D}\left(\boldsymbol{r}\right)d\boldsymbol{r} \\ &\qquad=g\left(x_0+\varepsilon_x,D_0\right)+\int_{\Omega_L}\gamma_g\left(x_0+\varepsilon_x,\boldsymbol{r}\right)\tilde{D}\left(\boldsymbol{r}\right)d\boldsymbol{r}+s_0+\varepsilon_s\ , \end{aligned} \tag{34}$$

which can be expended in Taylor series around the initial values of the bias point x_0 and static noise margin s_0. Restricting only to first-order terms, we obtain:

$$
\begin{aligned}
& f\left(x_0+s_0,D_0\right)+\frac{\partial f}{\partial x}\left(x_0+s_0,D_0\right)\left(\varepsilon_x+\varepsilon_s\right) \\
& +\int_{\Omega_R}\gamma_f\left(x_0+s_0,\boldsymbol{r}\right)\tilde{D}(\boldsymbol{r})\,d\boldsymbol{r}+\int_{\Omega_R}\frac{\partial\gamma_f}{\partial x}\left(x_0+s_0,\boldsymbol{r}\right)\left(\varepsilon_x+\varepsilon_s\right)\tilde{D}(\boldsymbol{r})\,d\boldsymbol{r} \\
& =g\left(x_0,D_0\right)+\frac{\partial g}{\partial x}\left(x_0,D_0\right)\varepsilon_x+\int_{\Omega_L}\gamma_g\left(x_0,\boldsymbol{r}\right)\tilde{D}(\boldsymbol{r})\,d\boldsymbol{r} \\
& +\int_{\Omega_L}\frac{\partial\gamma_g}{\partial x}\left(x_0,\boldsymbol{r}\right)\varepsilon_x\tilde{D}(\boldsymbol{r})\,d\boldsymbol{r}+\varepsilon_s
\end{aligned}
\quad . \tag{35}
$$

Using now equation (27), after a few rearrangements equation (35) can be written as:

$$
\begin{aligned}
& \varepsilon_s\left[1-\frac{\partial f}{\partial x}\left(x_0+s_0,D_0\right)-\int_{\Omega_R}\frac{\partial\gamma_f}{\partial x}\left(x_0+s_0,\boldsymbol{r}\right)\tilde{D}(\boldsymbol{r})\,d\boldsymbol{r}\right]= \\
& \int_{\Omega^l}\gamma_f\left(x_0+s_0,\boldsymbol{r}\right)\tilde{D}(\boldsymbol{r})\,d\boldsymbol{r}-\int_{\Omega_L}\gamma_g\left(x_0,\boldsymbol{r}\right)\tilde{D}(\boldsymbol{r})\,d\boldsymbol{r} \\
& +\varepsilon_x\left[\int_{\Omega_R}\frac{\partial\gamma_f}{\partial x}\left(x_0+s_0,\boldsymbol{r}\right)\tilde{D}(\boldsymbol{r})\,d\boldsymbol{r}-\int_{\Omega_L}\frac{\partial\gamma_g}{\partial x}\left(x_0,\boldsymbol{r}\right)\tilde{D}(\boldsymbol{r})\,d\boldsymbol{r}\right].
\end{aligned}
\tag{36}
$$

This equation should be solved for ε_s. The variation of the state variable ε_x can in principle be computed by using (30), (31), and (33) and expending the second equation of system (28) in Taylor series in the same way we derived equation (35). However, it can be shown that the derivatives of the sensitivity functions with respect to the input voltage are relatively small and, for this reason, the integral in the left-hand side of equation (36) and the last two integrals in the right-hand side of the same equation can be neglected. This approximation leads to the following formula for the variation of the SNM:

$$
\varepsilon_s=\omega_R\left[\int_{\Omega_R}\gamma_f\left(x_0+s_0,\boldsymbol{r}\right)\tilde{D}(\boldsymbol{r})\,d\boldsymbol{r}-\int_{\Omega_L}\gamma_g\left(x_0,\boldsymbol{r}\right)\tilde{D}(\boldsymbol{r})\,d\boldsymbol{r}\right], \tag{37}
$$

where we have introduced notation:

$$
\omega_R=\frac{1}{1-\dfrac{\partial f}{\partial x}\left(x_0+s_0,D_0\right)}. \tag{38}
$$

Since equation (37) is valid for any doping variation $\tilde{D}(\boldsymbol{r})$ and:

$$
\varepsilon_S=s-s_0=s\left(D_0+\tilde{D}\right)-s\left(D_0\right)=\int_{\Omega_R+\Omega_L}\gamma_s(\boldsymbol{r})\tilde{D}(\boldsymbol{r})\,d\boldsymbol{r}, \tag{39}
$$

by comparing equations (29) and (37) it follows that the doping sensitivity function of SNM is given by:

$$\gamma_s(\boldsymbol{r}) = \begin{cases} \omega_R \gamma_f(x_0 + s_0, \boldsymbol{r}), \; if \; \boldsymbol{r} \in \Omega_R, \\ -\omega_R \gamma_g(x_0, \boldsymbol{r}), \; if \; \boldsymbol{r} \in \Omega_L. \end{cases} \tag{40}$$

The doping sensitivity functions of the SNM for six SRAM transistors are represented in Fig. 3 when the SRAM cell is in the "1" state, i.e. x and $h(y)$ are in the "0" state, and $f(x)$ is in the "1" state. When the SRAM cell is in state "0" the doping sensitivity functions of the L- and R-inverters are interchanged because of the symmetry of the cell. Note that, in the case of the presented SRAM cell, the most sensitive regions to doping fluctuations are the conduction channels of the n-channel MOSFETs and access transistors. The doping sensitivity functions shown in this figure allows the estimation of the width of the retrograde doping profile that should be used for each MOSFET in order to suppress the random dopant induced fluctuations [54,55]. It is apparent that an undoped layer of approximately 10 nm under the oxide can reduce by and large the effects of the random dopant fluctuations on SNM.

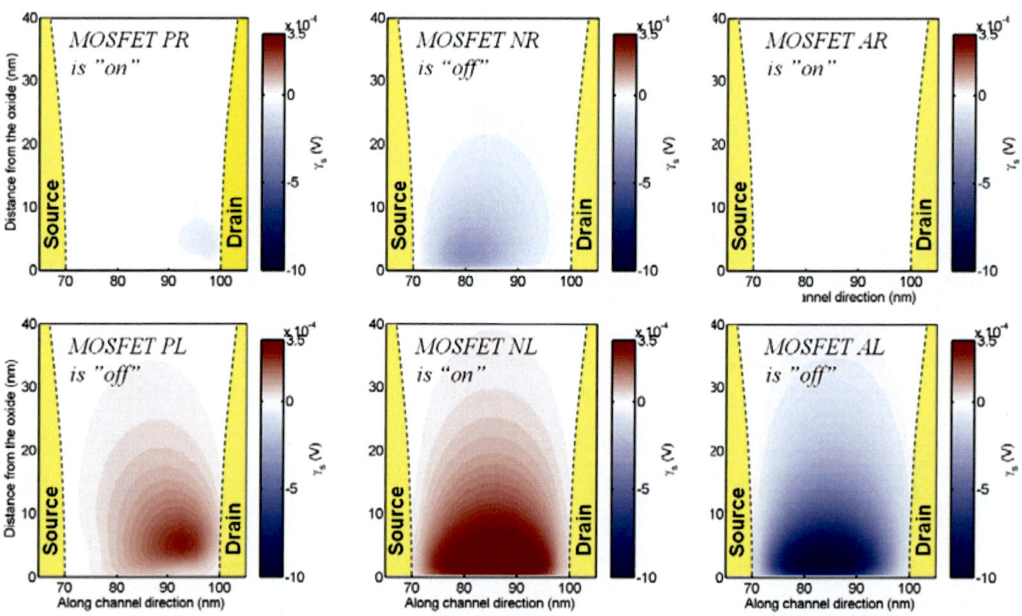

Figure 5. Doping sensitivity functions of the SNM for the six transistors of the SRAM. The cell is the read-state, i.e. x and $h(y)$ are the "0" state, $f(x)$ is in the "1" state.

All n-channel and p-channel transistors were assumed to have identical macroscopic characteristics. The doping in the channel was assumed uniformly distributed with an average doping concentration of 3×10^{18} cm^{-3}. Unless otherwise specified the oxide thickness of both types of transistors is 1 nm, and the oxide semiconductor interfaces contain a fixed-charge density of 5×10^{11} cm^{-2} in the case of p-channel and -5×10^{11} cm^{-2} in the case of n-channel devices. The doping concentration at the polygates is 5×10^{19} cm^{-3} (acceptor type in the case of p-channel and donor type in the case of n-channel devices), the doping profiles of the source an drain junctions are Gaussian with a maximum value of the 10^{20} cm^{-3} at the metal contacts and junctions depths of 20 nm. The lateral standard deviation of the doping profiles induces gate-to-source and gate-to-drain overlaps of approximately 1.5 nm. The channel length of all transistors is 27 nm, while the widths are 100 nm for the pull-up and pull-down transistors (PL, PR, NL, and NR) and 50 nm for the access transistors (AL and AR). The threshold voltages of the n-channel and p-channel transistors are 210 mV 190 mV respectively. These values were computed for an oxide thickness of 1 nm.

Variance of SNM

The variance of SNM can now be computed using equation (11). The contribution of random dopant fluctuations in the L-inverter (containing transistors PL, NL, and AL) can be separated from the contribution of the R-inverter (containing transistors PR, NR, and AR) and the variance of the SNM can be written as:

$$\sigma_{SNM}^2 = \omega_R^2\left[\int_{\Omega_R}\gamma_f^2(x_0+s_0,\boldsymbol{r})\langle D(\boldsymbol{r})\rangle d\boldsymbol{r}+\int_{\Omega_L}\gamma_g^2(x_0,\boldsymbol{r})\langle D(\boldsymbol{r})\rangle d\boldsymbol{r}\right], \tag{41}$$

where $\bar{D}(\boldsymbol{r})=\langle D(\boldsymbol{r})\rangle$ is the average value of the doping concentration at location $\boldsymbol{r}$. The last equation has a relatively simple physical interpretation: the fluctuations of the SNM can be expressed as the sum of the fluctuations of the positions of the two end points of the diagonal in Fig. 2. The first integral in the right-hand side of equation (41) denotes variations of the upper-right point of the diagonal (which corresponds to random dopant fluctuations in the R-inverter), while the second integral denotes variations of the lower-left point of this diagonal (which corresponds to fluctuations in the L-inverter). The random dopant induced fluctuations of the L- and R-inverters are independent random variables and, hence, the variance of the SNM can be computed by adding the variances induced by the two inverters. It is important to note that γ_g denotes the doping sensitivity function of the inverse of the voltage transfer characteristic of the left inverter. However, in most device simulators, one can evaluate the doping sensitivity function of the direct voltage transfer characteristic but not of the inverse characteristic $g=h^{-1}$. Therefore, it is essential now to evaluate the doping sensitivity function of g as a function of γ_h. To this purpose we write:

$$\begin{aligned}x_0 = h(g(x_0,D),D) &= h\left[g(x_0,D_0)+\int_{\Omega_L}\gamma_g(x_0,\boldsymbol{r})\tilde{D}(\boldsymbol{r})\,d\boldsymbol{r},D\right]\\ &= h\left[y_0+\int_{\Omega_L}\gamma_g(x_0,\boldsymbol{r})\tilde{D}(\boldsymbol{r})\,d\boldsymbol{r},D_0\right]\\ &+\int_{\Omega_L}\gamma_h\left[y_0+\int_{\Omega_L}\gamma_g(x_0,\boldsymbol{r}')\tilde{D}(\boldsymbol{r}')d\boldsymbol{r}',\boldsymbol{r}\right]\tilde{D}(\boldsymbol{r})d\boldsymbol{r}.\end{aligned} \tag{42}$$

The first term in the right-hand side of the last equation can be expanded in Taylor series:

$$h\left[y_0 + \int_{\Omega_L} \gamma_g(x_0,\boldsymbol{r})\tilde{D}(\boldsymbol{r})\,d\boldsymbol{r}, D_0\right]$$
$$= h(y_0,D_0) + \frac{\partial h}{\partial x}(y_0,D_0)\int_{\Omega_L}\gamma_g(x_0,\boldsymbol{r})\tilde{D}(\boldsymbol{r})\,d\boldsymbol{r}\,. \quad (43)$$

The last term in the right-hand side of equation (42) can be approximated by using the same line of reasoning as in the derivation of equation (37):

$$\int_{\Omega_L}\gamma_h\left[y_0 + \int_{\Omega_L}\gamma_g(x_0,\boldsymbol{r}')\tilde{D}(\boldsymbol{r}')d\boldsymbol{r}',\boldsymbol{r}\right]\tilde{D}(\boldsymbol{r})d\boldsymbol{r}$$
$$= \int_{\Omega_L}\gamma_h(y_0,\boldsymbol{r})\tilde{D}(\boldsymbol{r})d\boldsymbol{r} + \int_{\Omega_L}\int_{\Omega_L}\frac{\partial\gamma_h}{\partial x}(y_0,\boldsymbol{r})\gamma_g(x_0,\boldsymbol{r}')\tilde{D}(\boldsymbol{r})\tilde{D}(\boldsymbol{r}')d\boldsymbol{r}d\boldsymbol{r}'$$
$$\approx \int_{\Omega_L}\gamma_h(y_0,\boldsymbol{r})\tilde{D}(\boldsymbol{r})d\boldsymbol{r}\,. \quad (44)$$

Using $h(y_0,D_0)=x_0$ and introducing (43) and (44) into (42), we obtain the following expression for the doping sensitivity function of g:

$$\gamma_g(x_0,\boldsymbol{r}) = -\frac{\gamma_h(y_0,\boldsymbol{r})}{\frac{\partial h}{\partial x}(y_0,D_0)}, \quad (45)$$

which, together with (40) implies:

$$\gamma_s(\boldsymbol{r}) = \begin{cases}\omega_R\gamma_f(x_0+s_0,\boldsymbol{r}), & \text{if } \boldsymbol{r}\in\Omega_R\\ \omega_L\gamma_h(y_0,\boldsymbol{r}), & \text{if } \boldsymbol{r}\in\Omega_L\end{cases} \quad (46)$$

where we have introduced the notation:

$$\omega_L = \frac{\omega_R}{\frac{\partial h}{\partial x}(y_0,D_0)}\,. \quad (47)$$

Equation (46) can be used to compute the doping sensitivity function of SNM when the voltage transfer characteristics of the two inverters of the SRAM cell are known. Using (41) and (46) we obtain that the variance of SNM is:

$$, \quad (48)$$

A final simplification to this equation can be made if we assume what is usually done in the literature that the two inverters have similar macroscopic (average) characteristics. In this case the doping sensitivity functions of the two inverters are equal to each other and:

$$\frac{\partial h}{\partial x}(y_0, D_0) = \frac{\partial f}{\partial x}(y_0, D_0) = \frac{1}{\frac{\partial g}{\partial x}(x_0, D_0)} = \frac{1}{\frac{\partial f}{\partial x}(x_0 + s_0, D_0)}, \tag{49}$$

where we have used that, for macroscopically identical inverters, $f\left(g(x, D_0), D_0\right) = x$. Thus:

$$\omega_L = \omega_R \frac{\partial f}{\partial x}(x_0 + s_0, D_0) = \frac{\frac{\partial f}{\partial x}(x_0 + s_0, D_0)}{1 + \frac{\partial f}{\partial x}(x_0 + s_0, D_0)} = \omega_R - 1 \tag{50}$$

and the variance of the SNM can be computed using:

$$\sigma_{SNM}^2 = \int_{\Omega_R} \left[\omega_R^2 \gamma_f^2 (x_0 + s_0, \boldsymbol{r}) + \omega_L^2 \gamma_f^2 (y_0, \boldsymbol{r})\right] \langle D(\boldsymbol{r}) \rangle d\boldsymbol{r} \,. \tag{51}$$

Note that if the two invertors have macroscopically identical distributions the integrals in (48) can be performed on the semiconductor region of only one inverter.
In the case of two-dimensional simulations, the integral along the z-direction in equation (51) can be performed independently on the other variables. Also, since the doping sensitivity functions are inversely proportional to the width of the devices (W), we obtain that:

$$\sigma_{SNM}^2 \sim \frac{1}{W}. \tag{52}$$

This equation implies that the variance fluctuations of the SNM in doubled when the width of the device gets twice smaller and all other dimensions are kept constant. Similar scaling equations have been obtained for the fluctuations of most device parameters, such as threshold voltages, saturation currents, cutoff frequencies, etc. [41,56].
The doping sensitivity function of the voltage transfer characteristic, γ_f, can be evaluated by using point-by-point evaluations of γ_f or the perturbation (linearization) technique developed in the previous section. In the first case the doping concentration at various locations inside the SRAM cell is modified point-by-point and the changes in the output voltage of the two inverters are computed by solving the transport equations each time. This method is computationally very expensive because it requires solving the transport equations many times. For instance, if we want to evaluate the doping sensitivity function for a 6-T SRAM cell in 1000 points for each transistor, we have to compute the bias point for 6000 different doping profiles. The computation of each bias point involves solving the transport equations for all transistors self-consistently, which requires a relatively large computational time, particularly for two- and three-dimensional simulations.

Numerical Results

The dependence of the standard deviation of the SNM on the oxide thickness t_{ox} of the cell is represented in Fig. . In this simulations V_{DD}=0.8 V and the oxide thicknesses of all MOSFETs are varied in the same time from 0.5 nm to 1.4 nm. The inset of the figure shows the value of the SNM (left axis) and the ratio between the SNM and the standard deviation of the SNM, s/σ_{SNM}^2 (right axis) as a function of the oxide thickness. While the value of the SNM is more or less independent on the oxide thickness, the standard deviation of the SNM is increasing linearly in t_{ox}. That means one should use smaller oxide thicknesses in order to suppress the fluctuations of the SNM in the SRAM cells. The increase of σ_{SNM}^2 with the oxide thickness can be explained because, when t_{ox} is increased the threshold voltage of the transistors increase and the cell is starting to operate in the subthreshold regime. This fact is also supported by the results presented in Fig. 5 that show the dependence of the standard deviation of the SNM as a function of V_{DD}. The standard deviation of the SNM depends as $1/V_{DD}$, which shows that SRAM cells working in the subthreshold regime are more vulnerable to random dopant fluctuations than in the saturation regime. Hence, there is tradeoff between the random dopant induced fluctuations of the SNM and the biasing voltage (that is proportional to the power dissipated in the SRAM cell). The relatively high values obtained for s/σ_s are due to the fact that the SRAM cell is biased in saturation, i.e. V_{DD} is larger than the sum of the two threshold voltages of the pull-up and pull-down transistors.

The results for the standard deviation of SNM presented in the last two figures can be summarized by using approximation:

$$\sigma_{SNM} = \frac{\sigma_0}{\sqrt{W}} \frac{t_{ox}}{V_{DD}}, \tag{53}$$

where W is the width of the transistors and σ_0 is a proportionality constant that depends on all other properties of the cell, such as channel lengths, doping concentrations, β-ratio, etc. The dependence on W has been introduced in (53) using of equation (52).

Finally, note that in most simulations the ratio between the average value of the SNM and standard deviation of the fluctuations is larger than six, which is the minimum required value for SRAM design. Although these results might be optimistic for SRAM design, they might not be so good if we bias the cell in the subthreshold regime or if we also consider the effects of other fluctuations such as fluctuations induced by oxide roughness, line edge roughness, or geometrical dimensions. When added to random dopant fluctuations [22], the combined effects of these fluctuations might impede further scaling of SRAM cells.

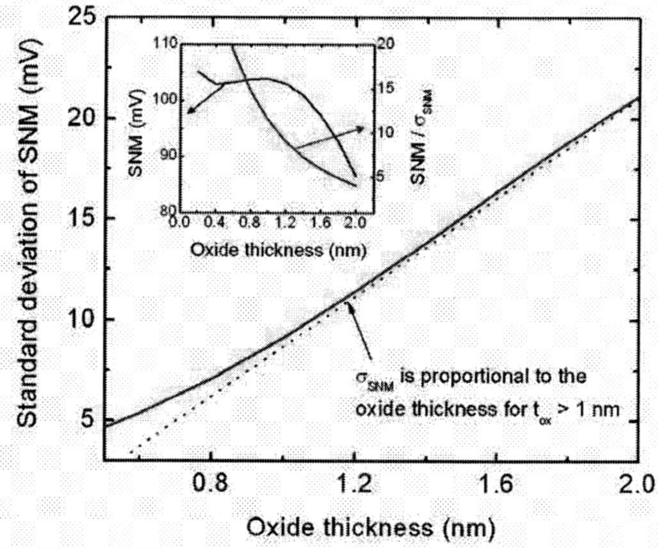

Figure 6. Doping sensitivity functions of the SNM for the six transistors of the SRAM in the read-state: x and $h(y)$ are the "0" state, $f(x)$ is in the "1" state.

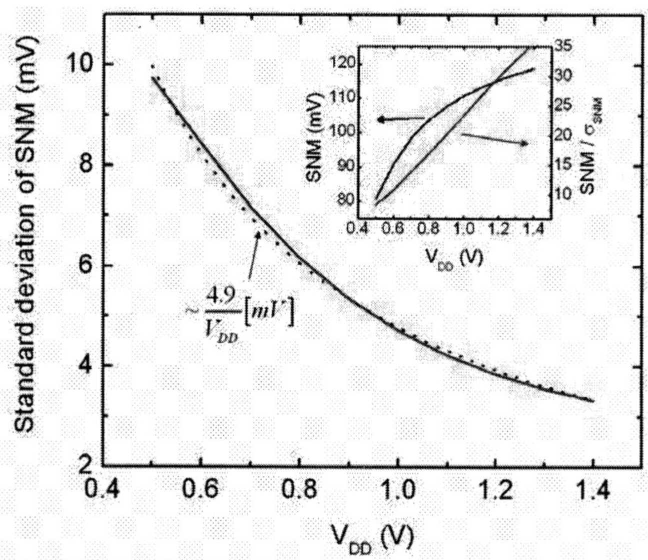

Figure 7. Standard deviation of the SNM as a function of the oxide thickness (a) and bias voltage (b). The insets show the values of the SNM as function of the oxide thickness and bias voltage.

3.6. Is the Perturbation Technique an Accurate Approach?

The reliability of the technique for the evaluation of random doping induced fluctuations depends on how accurate the first-order perturbation technique is for the evaluation of the magnitude of fluctuations in semiconductor devices. Therefore, it is natural to ask if the perturbation (linearization) of transport equations with respect to the fluctuating quantities provides a good approximation. In order to check the accuracy of the perturbation technique we have computed the standard deviation of the threshold voltage for a MOS capacitor by using the perturbation and the standard Monte-Carlo technique. In the case of the perturbation technique, the standard deviation of the threshold voltage is computed by solving self-consistently the Poisson and effective-mass Schrödinger equation and using the first-order (quantum) perturbation technique to estimate the fluctuations. In the case of the Monte-Carlo technique we have accumulated statistics for 100, 200, 500, and 1000 randomly generated capacitors, which were modeled by using the same transport model; the standard deviations were computed by using $\sigma_{V_T} = \sqrt{\sum_{i=1}^{N}\left(V_{T_i} - \langle V_T\rangle\right)^2 \big/ (N-1)}$, where $\langle V_T\rangle$ is the average value of the threshold voltage. The results of these simulations show that the perturbation approach provides results which are more than 90% accurate in a time which is three orders of magnitude faster.

A second test that we have performed is in which we consider various semiconductor devices and carry out simulation experiments in which the doping concentration at different points is varied and the intrinsic parameters of these device (e.g. terminal currents, threshold voltage, small-signal parameters, etc.) are computed as a function of the doping concentrations at these points. We consider the following three devices with simplified architectures: (a) a fully depleted, double-gate SOI device with the thickness of the semiconductor 10 nm, the oxide thickness 2 nm, and doping concentration of 10^{18} cm^{-3} in the channel; (b) a MOSFET device with improved architecture (retrograde well and halo implantation) with oxide thickness of 3 nm, doping concentration in the channel between 10^{16} cm^{-3} and 10^{18} cm^{-3}; (c) a p-n junction diode with the doping concentration at the cathode 10^{17} cm^{-3} built on a p-type substrate doped with 10^{16} cm^{-3}. Fig. 8 shows the structures of the double-gate SOI and MOSFET devices.

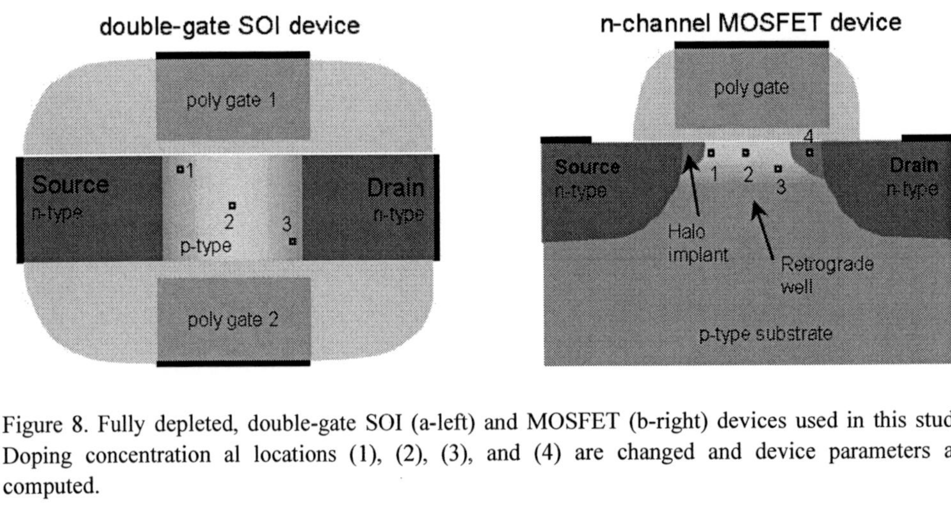

Figure 8. Fully depleted, double-gate SOI (a-left) and MOSFET (b-right) devices used in this study. Doping concentration al locations (1), (2), (3), and (4) are changed and device parameters are computed.

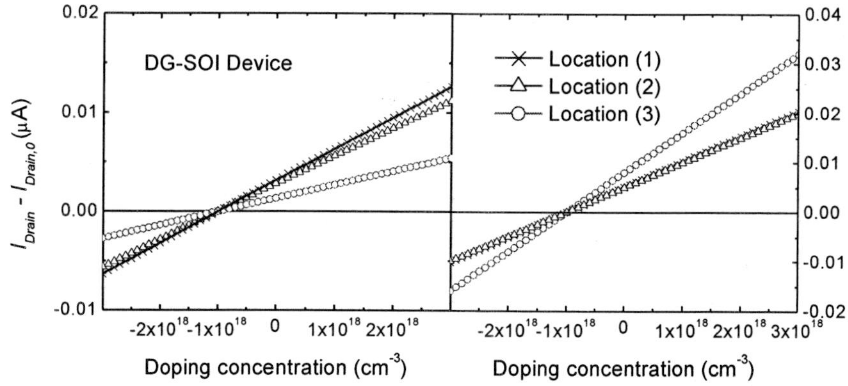

Figure 9. Dependence of the drain current and transconductance on the doping concentration at mesh points denoted by (1), (2), and (3) in the SOI device presented in Fig. 8.

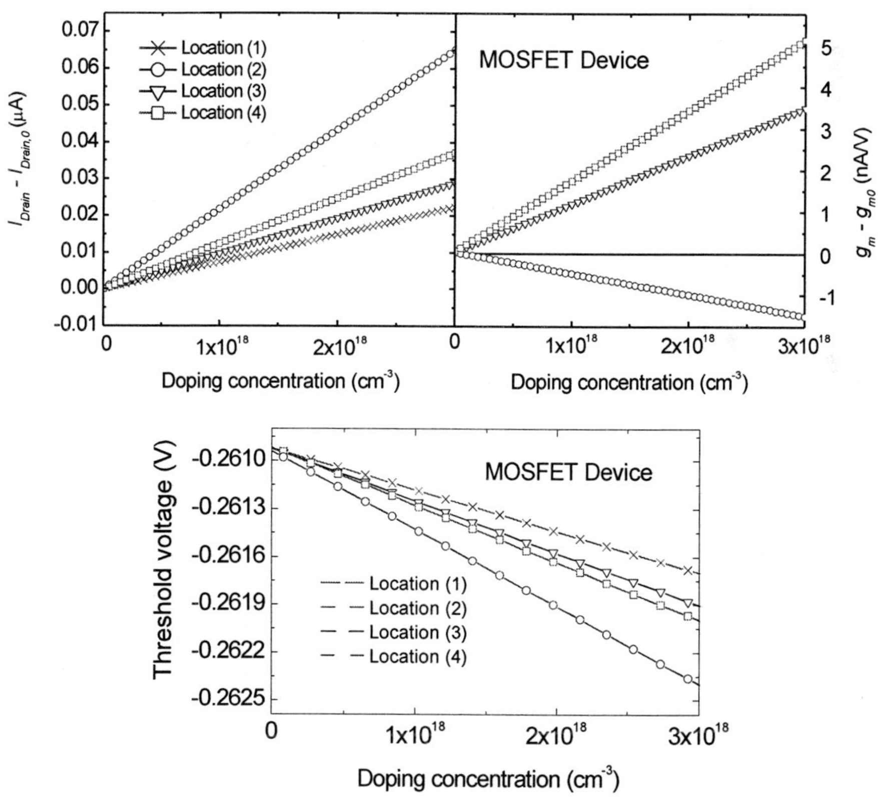

Figure 10. Dependence of the drain current, transconductance, and threshold voltage on the doping concentration at mesh points denoted by (1), (2), (3), and (4) in the MOSFET device presented in Fig. 8.

The locations where the doping concentrations were varied are represented by numbers from (1) to (4) in Fig. 8. The results of the simulations are represented in Figs. 9-11 for the double gate SOI device, MOSFET, and a p-n junction diode. There is a remarkable linear dependence of the parameters of the SOI and MOSFET on the doping concentrations, which strongly supports the basic assumptions of the perturbation technique. In the case of the p-n junction diode the current characteristics are not linear for a large range of variation of the doping concentration. However, if the fluctuations of the doping concentration are small, the dependence of the terminal current on the doping concentrations remains linear. To test the accuracy of the linearization technique in the case of p-n junctions we have generated 200 diodes with different doping concentrations but the same average values of the doping and computed the terminal currents for given forward potentials across the diode. The statistical (Monte Carlo) techniques give a standard deviation of the terminal currents of 0.26 μA, while the linearization technique gives 0.31 μA; the average value of the drain current is 1 μA in both cases. The relatively good agreement between the two techniques suggests that the linearization of transport equations is still a fairly good approximation even for the p-n junction diode whose doping characteristic is represented in Fig. 11.

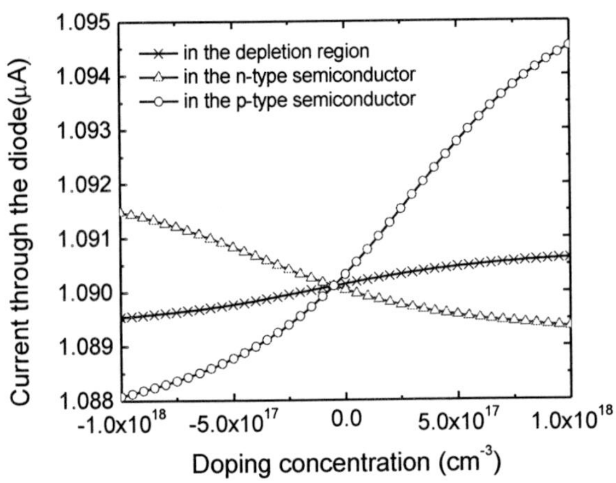

Figure 11. Current through a p-n junction diode as a function of the doping concentration at different locations.

It should not be surprising that the perturbation technique gives such a good accuracy for the computation of the random doping induced fluctuations in semiconductor devices. In fact, many other well-established theories in nanoelectronics are based on a first-order perturbation approach. For instance, the *Fermi's golden rule* which provides a way to calculate the transition rate (probability of transition per unit time) from one energy eigenstate of a quantum system into a set of final states, due to a perturbation, is a first-order perturbation theory that stays at the basis of all semiclassical (Monte-Carlo) simulations. Using the Fermi's golden rule we usually compute the scattering rates in the Boltzmann transport equation, from which drift-diffusion equations are derived. Impurity atoms (dopants) are assumed as perturbations on the background, lattice potential in the same way we assume them as perturbations in our random dopant fluctuation analysis. For this reason the perturbation technique developed in this project is as accurate as the transport model that is used (drift-diffusion, density-gradient, Boltzmann equation, etc.). Another similar example is the Born approximation in scattering theory, which is again a first-order perturbation theory applied to scattering by an extended body. Although the *Born approximation* is valid only when the scattered field is small, this approximation has been used in a very large range of applications and it stays at the basis of most quantum or semiclassical transport models in nanoelectronics. Our perturbation technique starts from first principles and relies on the same assumptions as any of the above mentioned theories and the validity and accuracy of our technique are beyond doubt.

4. Design of Random Doping Fluctuation Resistant Structures

4.1. Mathematical Formalism

The automated design of random dopant fluctuation resistant structures of semiconductor devices that is presented in this section is based on the perturbation technique developed at the beginning of the previous section.. Again, we do not make assumption about the particular transport model used and the analysis below can be applied to the drift-diffusion model, semiclassical (e.g. Monte-Carlo techniques based on the Boltzmann transport equation [57]) or any quantum transport model (e.g. such as the nonlinear Green's function technique [58-60]).

Mathematically, the problem of design of random doping fluctuation resistant structures is reduced to the computation of the doping concentration profiles than minimize the variance of parameter P in (1). This problem can be rewritten as:

$$\min_{\langle D(r)\rangle} \sigma_P^2(D) \text{ such that } \langle A(D)\rangle = A(D_0), \langle B(D)\rangle = B(D_0), \langle C(D)\rangle = C(D_0)... \quad (54)$$

where $\langle D\rangle$ denotes the average value of the doping concentration and D_0 is the initial doping configuration. It is important to note that, in general, equation (54) might not have unique solution and more doping concentration profiles might satisfy this problem. We will not consider this situation here and neither the case of more than one local minimum.

An approximate solution of (54) can be obtained by computing the change in the average doping concentration $\langle \tilde{D}\rangle = \langle D\rangle - D_0$ that solves:

$$\min_{\langle \tilde{D}(r)\rangle} \sigma_P^2(D) \text{ such that}$$

$$A(D_0 + \langle \tilde{D}\rangle) = A(D_0),\ B(D_0 + \langle \tilde{D}\rangle) = B(D_0),\ C(D_0 + \langle \tilde{D}\rangle) = C(D_0),... \quad (55)$$

Note that going from (54) to (55) we have assumed that $A(\langle D\rangle) = \langle A(D)\rangle$,... which is not true if functionals A, B, and C are nonlinear. The accuracy of this approximation will be checked numerically later and, for now, we will try to solve (55).

The variance of parameter P induced by random dopant fluctuations can be expressed as a function of the doping sensitivity function γ_P by using equation (10). Problem (54) is in general difficult to solve analytically because of the large number of variables and nonlinearity of functionals involved. However, if we try to find solutions of (54) which are infinitesimally closed to the initial doping configuration $D_0(\boldsymbol{r})$, (54) can be solved analytically by using the formalism of Lagrange multipliers. Indeed, if we impose $\|\tilde{D}\|^2 = \|\langle D\rangle - D_0\|^2 \le \varepsilon^2$, where ε is a positive constant that goes to zero, the previous problem can be written as:

$$\min_{\langle\tilde{D}(\boldsymbol{r})\rangle} \int_\Omega \gamma_P^2(\boldsymbol{r})\langle D(\boldsymbol{r})\rangle d\boldsymbol{r} \text{ such that } \int_\Omega \left|\langle\tilde{D}(\boldsymbol{r})\rangle\right|^2 d\boldsymbol{r} \le \varepsilon^2,$$

$$\int_\Omega \gamma_A(\boldsymbol{r})\langle\tilde{D}(\boldsymbol{r})\rangle d\boldsymbol{r} = \int_\Omega \gamma_B(\boldsymbol{r})\langle\tilde{D}(\boldsymbol{r})\rangle d\boldsymbol{r} = \int_\Omega \gamma_C(\boldsymbol{r})\langle\tilde{D}(\boldsymbol{r})\rangle d\boldsymbol{r} = \ldots = 0 \tag{56}$$

The first constraint expresses the condition that the variation of the doping profile should be infinitesimally small, while the other constraints were obtained by using equations (5) and (54) in which we replaced $\tilde{D}$ by $\langle\tilde{D}\rangle$. Defining a proper Lagrangian for (56) and taking the first-order necessary conditions we get the following system of equations:

$$\int_\Omega \gamma_A(\boldsymbol{r})\langle\tilde{D}(\boldsymbol{r})\rangle d\boldsymbol{r} = \int_\Omega \gamma_B(\boldsymbol{r})\langle\tilde{D}(\boldsymbol{r})\rangle d\boldsymbol{r} = \int_\Omega \gamma_C(\boldsymbol{r})\langle\tilde{D}(\boldsymbol{r})\rangle d\boldsymbol{r} = \ldots = 0, \tag{57}$$

$$\int_\Omega \left|\langle\tilde{D}(\boldsymbol{r})\rangle\right|^2 d\boldsymbol{r} = \varepsilon^2, \tag{58}$$

$$2\lambda_\varepsilon \tilde{D}(\boldsymbol{r}) = \gamma_P^2(\boldsymbol{r}) - \lambda_A\gamma_A(\boldsymbol{r}) - \lambda_B\gamma_B(\boldsymbol{r}) - \lambda_C\gamma_C(\boldsymbol{r}) - \ldots, \tag{59}$$

where λ_ε, λ_A, λ_B, λ_C,… are the Lagrange multipliers. The last two equations can be solved for $\langle\tilde{D}(\boldsymbol{r})\rangle$ and λ_ε. The solution for $\langle\tilde{D}(\boldsymbol{r})\rangle$ is:

$$\langle\tilde{D}(\boldsymbol{r})\rangle = -\varepsilon \frac{\gamma_P^2(\boldsymbol{r}) - \lambda_A\gamma_A(\boldsymbol{r}) - \lambda_B\gamma_B(\boldsymbol{r}) - \lambda_C\gamma_C(\boldsymbol{r}) - \ldots}{\int_\Omega \left[\gamma_P^2(\boldsymbol{r}) - \lambda_A\gamma_A(\boldsymbol{r}) - \lambda_B\gamma_B(\boldsymbol{r}) - \lambda_C\gamma_C(\boldsymbol{r}) - \ldots\right]^2 d\boldsymbol{r}}. \tag{60}$$

To compute the Lagrange multipliers, we multiply (60) by $\gamma_A(\boldsymbol{r})$ and integrate over Ω. Using the first equation in (57), we obtain:

$$0 = \int_\Omega \left[\gamma_P^2(\boldsymbol{r})\gamma_A(\boldsymbol{r}) - \gamma_A(\boldsymbol{r})\gamma_A(\boldsymbol{r}) - \gamma_B(\boldsymbol{r})\gamma_A(\boldsymbol{r}) - \gamma_C(\boldsymbol{r})\gamma_A(\boldsymbol{r}) - \ldots\right] d\boldsymbol{r}. \tag{61}$$

By using the same procedure for all doping sensitivity functions it can be shown that the Lagrange multipliers satisfy the following system of linear equations:

$$\begin{cases} \lambda_A \gamma_{A,A} + \lambda_B \gamma_{A,B} + \lambda_C \gamma_{A,C} + \dots = \gamma_{P^2,A} \\ \lambda_A \gamma_{B,A} + \lambda_B \gamma_{B,B} + \lambda_C \gamma_{B,C} + \dots = \gamma_{P^2,B} \\ \lambda_A \gamma_{C,A} + \lambda_B \gamma_{C,B} + \lambda_C \gamma_{C,C} + \dots = \gamma_{P^2,C} \\ \quad\dots \end{cases} \tag{62}$$

where:

$$\gamma_{X,Y} = \int_\Omega \gamma_X(\boldsymbol{r})\gamma_Y(\boldsymbol{r})d\boldsymbol{r} \tag{63}$$

is the dot product between the doping sensitivity functions of parameters X and Y.
Equations (60) and (62) represent the closed-form solution of minimization problem (56). These equations should be solved numerically in order to compute the optimum doping profiles that minimize the standard deviation of random dopant-induced fluctuations of parameter P. It should be noted that these equations were obtained by assuming that the change in the doping profile is small. In order to accurately optimize the device one should use these equations for as many times (iterations) as needed, in order to achieve an acceptable final standard deviation of parameter P. The conditions under which equations (60) and (62) are valid will be discussed in the next section when we present the optimization technique. A few special cases are presented below.

Minimization without constraints
If no constraints are imposed in the minimization problem (56) equation (60) becomes:

$$\langle \tilde{D}(\boldsymbol{r}) \rangle = -\frac{\varepsilon \gamma_P^2(\boldsymbol{r})}{\int_\Omega \gamma_P^4(\boldsymbol{r})d\boldsymbol{r}}, \tag{64}$$

which shows that the only information necessary to compute the change in the doping profile is the doping sensitivity function $\gamma_P(\boldsymbol{r})$. For instance, equation (64) can be used to compute the doping concentration profile in a MOSFET that result in minimum fluctuations of the threshold voltage ($P = V_T$). Note that according to this equation, when no constraints are imposed, the doping concentration inside the semiconductor device should always be reduced in order to suppress random dopant fluctuations. The doping concentration should be decreased more (ideally the semiconductor device should be undoped) in the regions where the doping sensitivity functions have high values.

Minimization with one constraint
If only one constraint is imposed during the minimization problem, equations (60) and (62) can be solved easily for the change in the doping concentration:

$$\langle \tilde{D}(\boldsymbol{r}) \rangle = -\varepsilon \frac{\gamma_P^2(\boldsymbol{r}) - \lambda_A \gamma_A(\boldsymbol{r})}{\int_\Omega \left[\gamma_P^2(\boldsymbol{r}) - \lambda_A \gamma_A(\boldsymbol{r})\right]^2 d\boldsymbol{r}}, \text{ where } \lambda_A = \frac{\int_\Omega \gamma_P^2(\boldsymbol{r})\gamma_A(\boldsymbol{r})d\boldsymbol{r}}{\int_\Omega \gamma_A^2(\boldsymbol{r})\boldsymbol{r}}, \tag{65}$$

and A is the parameter that is supposed to remain constant during the minimization process. Note that no constraint is imposed on the nature of parameters A and P and they could even denote the same quantity. For instance, one could use equation (65) to minimize the fluctuations of the threshold voltage when the average value of the threshold voltage is fixed ($P = A = V_T$). In this special case, equation (65) reads:

$$\left\langle \tilde{D}(\boldsymbol{r}) \right\rangle = \varepsilon \frac{\gamma_P(\boldsymbol{r})\left[\gamma_P(\boldsymbol{r}) - \lambda_P\right]}{\int_\Omega \gamma_P^2(\boldsymbol{r})\left[\gamma_P(\boldsymbol{r}) - \lambda_P\right]^2 d\boldsymbol{r}}, \quad \lambda_P = \frac{\int_\Omega \gamma_P^3(\boldsymbol{r}) d\boldsymbol{r}}{\int_\Omega \gamma_P^2(\boldsymbol{r}) \boldsymbol{r}}. \tag{66}$$

In the case of minimization with one constraint the doping concentration should be increased in the regions of the device where $\gamma_P(\boldsymbol{r}) < \lambda_P$ and decreased in regions where $\gamma_P(\boldsymbol{r}) > \lambda_P$. Line $\gamma_P(\boldsymbol{r}) = \lambda_P$ has particular importance in the design of the device because it separates the regions that should be highly doped from the regions that should be undoped.

Minimization with Two or More Constraints

If two or more constraints are imposed in the minimization problem, equations (60) and (62) should be solved self-consistently. A particular case is when the determinant associated with the right-hand side of system (62) is equal to zero. In fact, due to the special structure of this determinant, it can be shown that it vanishes only if the doping sensitivity functions are linearly dependent and it is positive in all other cases. When the doping sensitivity functions are linearly dependent one can remove the dependent constraints in (56) and reduce the total number of constraints to a smaller number.

Doping profiles for long-channel FET devices

The doping profiles that minimize random dopant-induced fluctuations in long-channel (>100 nm) MOSFET devices can be computed analytically by using equation (64) in the case of unconstrained minimization and (65) in the case of constrained minimization. Indeed, as shown in [61] the doping sensitivity function of the threshold voltage is:

$$\gamma_{V_T}(x) = \frac{q t_{ox}}{\varepsilon_{ox} L W}\left(1 - \frac{x}{W_d}\right), \tag{67}$$

where t_{ox} is the oxide thickness, ε_{ox} is the oxide permittivity, L and W are the length and width of the conduction channel, W_d is the width of the depletion region, and x is the distance from the oxide/semiconductor interface inside the semiconductor device. It should be mentioned that the last expression is quite general and is applicable to nonuniformly doped channels as well. The doping profiles that minimize σ_{V_T} can be obtained from (64) for the unconstraint case:

$$\left\langle D(x) \right\rangle = D_0 - \varepsilon \frac{5\varepsilon_{ox}^2 L^2 W^2}{q^2 t_{ox}^2 W_d}\left(1 - \frac{x}{W_d}\right)^2 = D_0 - \varepsilon'\left(1 - \frac{x}{W_d}\right)^2, \tag{68}$$

where D_0 is the initial doping concentration in the channel and $\varepsilon' = \varepsilon \frac{5\varepsilon_{ox}^2 L^2 W^2}{q^2 t_{ox}^2 W_d}$ is a parameter proportional to how much the doping concentration should be modified in the channel. In general, the larger ε' the smaller are the random dopant-induced fluctuations of the threshold voltage. The doping profiles obtained by using equation (68) are represented in Fig. 12 with continuous lines for different values of ε'. In the case of the constrained minimization, for instance when we minimize the fluctuations of the threshold voltage but keep the value of the drain-to-source current constant, the Lagrange multiplier is $\lambda_{I_{DS}} = \frac{3qt_{ox}}{4\varepsilon_{ox} LW}$ and equation (65) gives:

$$\langle D(x) = \rangle D_0 - \varepsilon'' \left(1 - \frac{x}{W_d}\right)\left(\frac{1}{4} - \frac{x}{W_d}\right), \tag{69}$$

where ε'' is another proportionality parameter. The doping profiles obtained by using the last equation for different values of ε'' are represented with dash lines in Fig. 12. As can be seen from this figure we obtain an undoped channel for the unconstrained minimization and a δ-doping profile for the constrained minimization. In the case of the δ-doping profile the separation line between the lower and higher doped regions is at $1/4$ of the width of the depletion region. It is remarkable that we were able to obtain analytical equations for the doping concentrations as well as for the depth of the undoped regions and the width of the δ-doping barrier, which can be directly used in the design of random dopant fluctuation resistant structures. The standard deviation of the threshold voltage can also be computed analytically by using equation (10).

Numerical implementation

The numerical algorithm for the minimization of random dopant-induced fluctuations is summarized below:

1. Start with an initial doping concentration profile in the device and compute the doping sensitivity functions $\gamma_X(\boldsymbol{r})$ for all parameters under consideration (i.e. parameter P whose fluctuations are to be minimized and parameters A, B, C,... that are supposed to stay constant during the minimization. Only the doping sensitivity functions that are linearly independent are used at the next steps.
2. Solve linear system (62) for Lagrange multipliers λ_A, λ_B, λ_C, ...
3. Select a relatively small value for ε and compute the change in the doping concentration by using equation (60). In our simulations we have adjusted the value of ε such that the doping concentration is reduced by approximately 15% at each iteration step.
4. Compute the new doping profile $\langle D(\boldsymbol{r})\rangle = D_0(\boldsymbol{r}) + \langle \tilde{D}(\boldsymbol{r})\rangle$ and the new value of $\sigma_P^2(D)$ by using (13).
5. Repeat steps 2-4 until an acceptable final variance of parameter P is obtained.

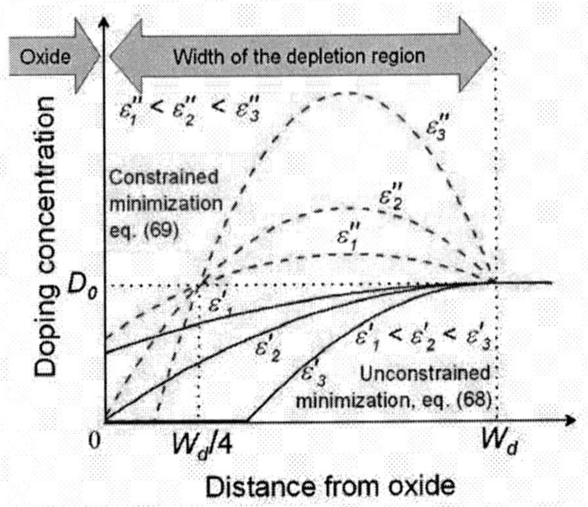

Figure 12. Doping profiles that minimize the random dopant induced fluctuations of threshold voltage for long-channel MOSFET devices. Continuous lines were obtained by using the unconstrained minimization, while the dash lines were obtained by using the constrained minimization. In the first case we obtained an undoped doping profile while in the second case a δ-doping profile.

These steps should be performed independently for donors ($D = N_d$) and acceptors ($D = N_a$). In the case where there are more doping species one should perform steps 1-5 for each individual specie separately.

Related to the computational efficiency of the minimization algorithm it should be noted that the computation of the doping sensitivity functions (step 1) usually takes a relatively small time because it requires the solution of a few linear sparse systems of equations [41,56]. For a two-dimensional device discretized in 10,000 mesh points the computation of the doping sensitivity functions usually takes a few seconds on a single processor personal computer (PC) working at 3 GHz. The rest of the iteration steps can be performed much faster. The integration in (60) can be done be using quadratures and it takes a few milliseconds on the same PC. System (62) usually involves at most 2 or 3 linear equations and can be solved easily.

A few modifications that can be done to simplify the numerical implementation are discussed below. First, it is usually not necessary to compute the integral in the denominator of (60) because it can be included in ε. Denoting the new coefficient by ε', equation (60) can be written as:

$$\left\langle \tilde{D}(\boldsymbol{r}) \right\rangle = -\varepsilon' \left[\gamma_P^2(\boldsymbol{r}) - \lambda_A \gamma_A(\boldsymbol{r}) - \lambda_B \gamma_B(\boldsymbol{r}) - \lambda_C \gamma_C(\boldsymbol{r}) - \ldots \right], \tag{70}$$

where ε' is of the order of $10^{18} \ldots 10^{22}$, depending of the structure of the device and the doping concentration.

Second, as mentioned in [54], the doping sensitivity functions of many parameters do not change much when the doping concentration is varied. For instance, in the case of the threshold voltage of MOSFET devices, the doping sensitivity function usually changes by less than 10% when the doping concentration in the channel is decreased by one order of magnitude. Hence, the values of the Lagrange multipliers do not change significantly from one iteration to another and the algorithm presented at the beginning of this section can be much simplified by using large values of ε and performing only one iteration step. In this way, the only computational intensive step is the one-time evaluation of the doping sensitivity functions, which anyway requires a relatively short computational time. The independence of the doping sensitivity functions of threshold voltage on doping profiles has been also observed by Taur and Wang [61-63] for long-channel devices.

Applying the one-step technique presented above to the case of minimization without constraints [see equation (64)] the doping concentration in the optimized device becomes:

$$\langle D(\boldsymbol{r})\rangle = D_0(\boldsymbol{r}) - \varepsilon'\gamma_P^2(\boldsymbol{r}). \tag{71}$$

This equation gives another useful interpretation of the doping sensitivity functions. Namely, the square of the sensitivity functions of some parameter are directly proportional to the amount of dopants that should be removed in order to minimize the random dopant-induced fluctuations of that parameter.

Finally, let us discuss how much the fluctuations of same parameter can be reduced by using our optimization technique. In principle steps 1-4 can be applied for as many iterations as needed in order to obtain a desirable standard deviation of parameter P. However, if too many iterations are used the doping concentration might get values which are either negative or too large to be able to manufacture the semiconductor. In both cases the iteration should be stopped. An estimate of how much the variance can be reduced can be obtained from formula (71):

$$\sigma_P^2(D_0) - \sigma_P^2(\langle D\rangle) = \int_\Omega \gamma_P^2(\boldsymbol{r})\left[D_0(\boldsymbol{r}) - \langle D(\boldsymbol{r})\rangle\right]d\boldsymbol{r} = \varepsilon'\int_\Omega \gamma_P^4(\boldsymbol{r})\,d\boldsymbol{r}$$

$$= \min_{\boldsymbol{r}_0} \frac{D_o(\boldsymbol{r}_0)}{\gamma_P^2(\boldsymbol{r}_0)} \times \int_\Omega \gamma_P^4(\boldsymbol{r})\,d\boldsymbol{r}, \tag{72}$$

The last equation gives that the variance of random dopant-induced fluctuating can be decreased at least one order of magnitude in the case of long-channel devices and between 30%-70% in devices with channel length less than 30 nm.

4.2. Design of MOSFET Devices

Next we present a few simulation results obtained in RandFlux for the 35 nm n-channel length MOSFET device represented in Fig. 13. The initial doping concentration in the channel was uniform and equal to 3×10^{18} cm^{-3}. The doping profile of the drain and source junctions is Gaussian, the doping concentration of the polysilicon is equal to 5×10^{19} cm^{-3}, and the oxide thickness is 1.2 nm. The drain and source were also doped uniformly with 5×10^{19} cm^{-3}, and the concentration of the polysilicon is 5×10^{19} cm^{-3}. A uniform sheet charge density of -5×10^{16} C/cm^2 was assumed at the semiconductor/oxide interface. The width of the transistor is 0.1 μm. The threshold voltage was computed as the potential on the gate for which the drain-to-source current is equal to $10^{-8} \times W$ at $V_{DS} = 0.1$ V, where W is the width of the device expressed in μm. First, the fluctuations of the threshold voltage were minimized by using the unconstrained [the fluctuations of the threshold voltage are minimized with no constrains in (54)] and then the constrained optimization algorithm. During the constrained optimization, the standard deviation of the threshold voltage was minimized such that the value of drain-to-source current was kept constant and equal to $10^{-8}W$ (A). Note that this constraint is equivalent to imposing the value of the threshold voltage to be constant.

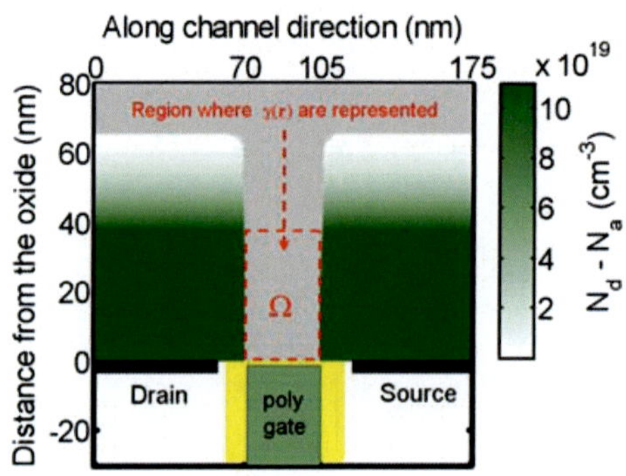

Figure 13. A 35-nm channel length MOSFET structure showing the region where the optimization was performed. The channel extends from 70 to 105 nm in the along channel direction.

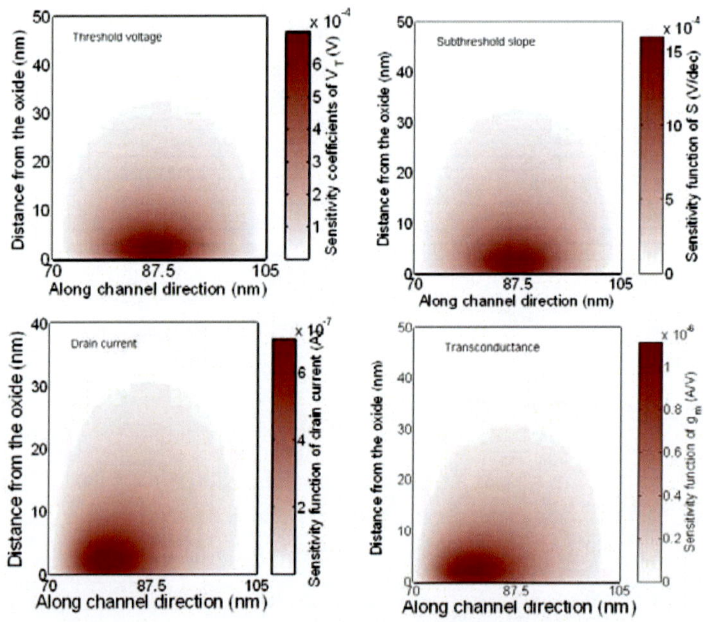

Figure 14. Doping sensitivity functions of the threshold voltage, subthreshold slope, drain current, and transconductance for the 35-nm channel length MOSFET presented in Fig. 13. Note that the most sensitive region to random doping concentration is the middle of the conduction channel.

The doping sensitivity functions of the threshold voltage, subthreshold slope, drain current, and transconductance and drain current were computed by solving self-consistently the Poisson, electron and hole continuity, and the Density-Gradient equations (see Fig. 14). As shown these plots of the doping sensitivity functions the channel region is the most sensitive region to doping fluctuations. Therefore, even though we solve the transport equations and optimize the whole semiconductor device (including the polysilicon gate and the drain and source junctions), in the results presented below we show only the doping concentration of acceptor impurities in the channel region. Initially (before starting the iterations) the doping concentration in the channel was uniform and equal to the values given above.
The average doping concentration profiles obtained during the unconstraint minimization after each iteration are represented in Fig. 15. In Fig. 16 we present the standard deviation of the threshold voltage (σ_{V_T}) and the expected value of the threshold voltage (see the inset) as a function of the number of iterations performed. Note that the more iterations we perform, the lower the value of σ_{V_T} becomes; for example, after 8 iterations σ_{V_T} has decreased more by than 35%. Also, the value of the threshold voltage decreases as the number of iterations is increased, and it becomes negative after a relatively large number of iterations.

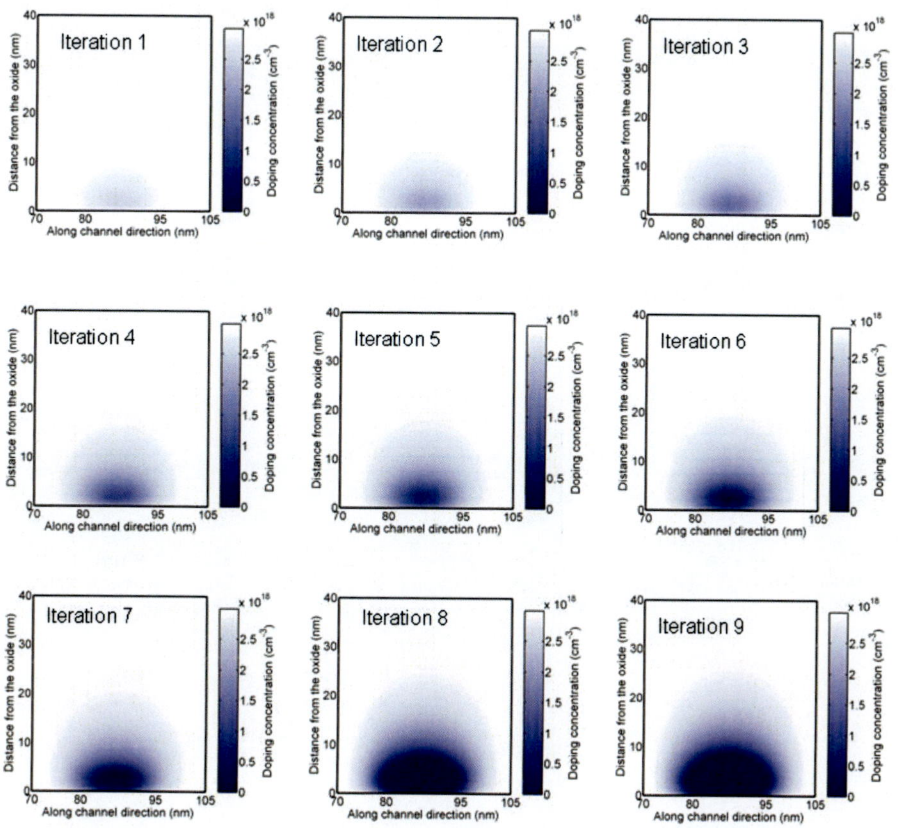

Figure 15. Average doping profiles in the channel of the 35 nm channel length MOSFET obtained after each iteration. The doping profiles were obtained by minimizing the standard deviation of threshold voltage fluctuations without any constraint.

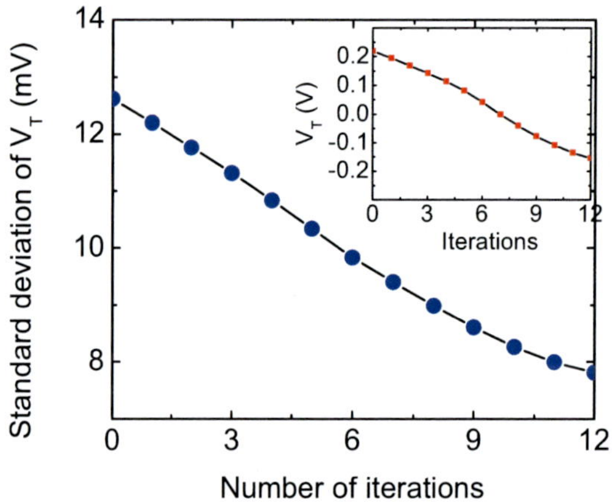

Figure 16. Standard deviation of the threshold voltage and the values of the threshold voltage (inset) as a function of the number of iterations during the *unconstrained minimization* for the 35 nm channel length MOSFET represented in Fig. 13. The corresponding doping profiles are represented in Fig. 15.

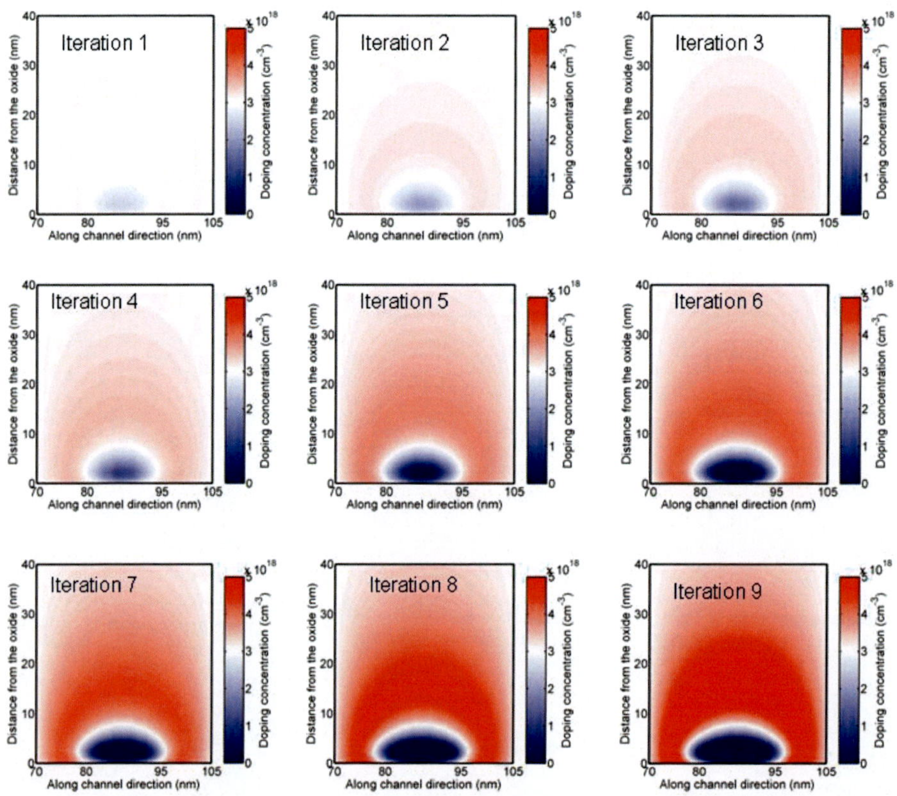

Figure 17. Average doping profiles in the channel of the 35 nm channel length MOSFET obtained after each iteration. The doping profiles were obtained by minimizing the fluctuations of σ_{VT} and imposing the constraint that the saturation current be constant (which is equivalent to saying that the average value of the threshold voltage is constant).

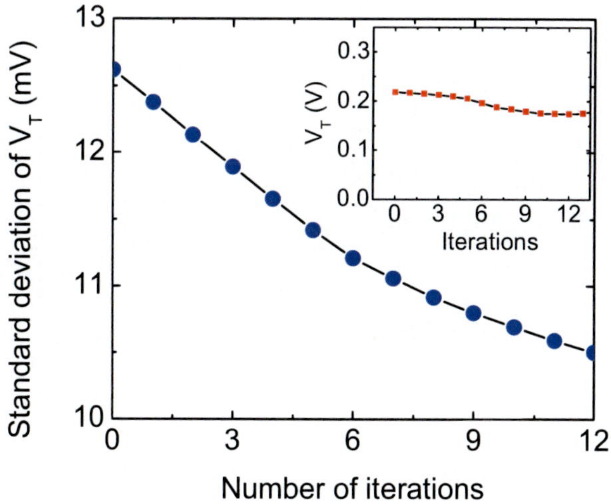

Figure 18. Standard deviation of the threshold voltage and the values of the threshold voltage (inset) as a function of the number of iterations during the *constrained minimization* for the MOSFET device. The corresponding doping profiles are represented in Fig. 17.

In order to stabilize the value of the threshold voltage we use the constrained minimization. The average doping concentration profiles obtained during the constrained minimization after each iteration are represented in Fig. 17. Note that we obtain an approximately semicircular δ-doping. The semicircular shape is mainly due to short-channel and quantum effects in the structures. It is important to observe that the ¼ ration between the location where the δ-doping starts and the width of the depletion region that was computed for long-channel devices (see Section 3.D) seems to remain the same for short-channel devices. Indeed, the line of separation between the undoped region and the δ-doping represented in Fig. 17 is at about 7 nm from the middle point of the channel, which is very close to ¼ of the width of the depletion region, which is approximately 27 nm for our device.

4.3. Design of SOI Devices

Now we present simulation results obtained for the fully depleted 25 nm n-channel length double-gate SOI device represented in Fig. 19. The initial doping concentration in the channel device was uniform and equal to 10^{19} cm^{-3}. The drain and source were also doped uniformly with 5×10^{19} cm^{-3}, and the concentration of the polysilicon is 5×10^{19} cm^{-3}. The thicknesses of the top and bottom oxides are equal to 1 nm and 5 nm, respectively. A uniform sheet charge density of -5×10^{16} C/cm^2 was assumed again at all semiconductor/oxide interfaces. The width of the SOI device is 0.1 μm and the threshold voltage was computed by using the same definition like in the case of the MOSFET device. The same types of simulations were performed for the SOI device. First, the fluctuations of the threshold voltage were minimized by using the unconstrained [the fluctuations of the threshold voltage are minimized with no constrains in (54)] and then the constrained optimization algorithm. During the constrained optimization, the standard deviation of the threshold voltage was minimized such that the value of drain-to-source current stays constant and equal to $10^{-8}W$ (A).

The doping sensitivity function of the drain-to-source current, threshold voltage, transconductance, and drain conductance are represented in Fig. 20. The fact that the doping sensitivity function is confined towards the middle of the channel in the device is a pure quantum effect and is due to the electron confinement induced by the top and bottom oxides.

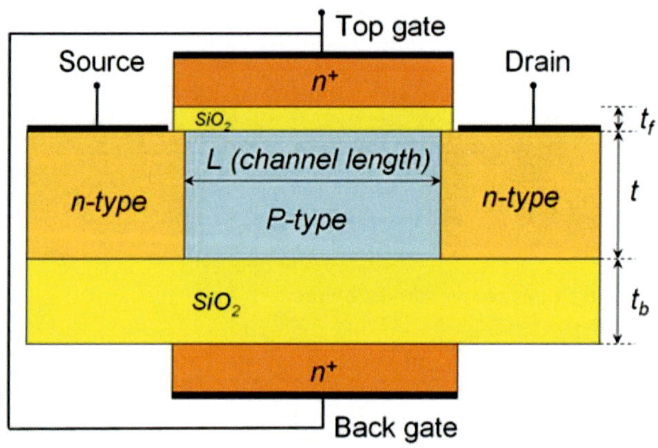

Figure 19. A 35-nm channel length MOSFET structure showing the region where the optimization was performed.

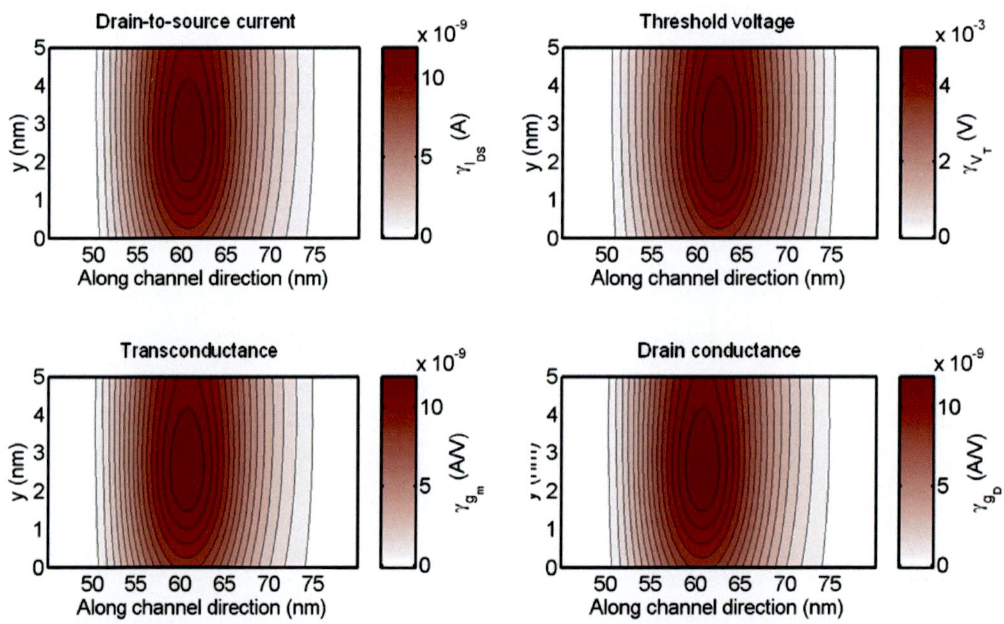

Figure 20. Doping sensitivity functions of the drain-to-source current, threshold voltage, transconductance, and drain conductance for the 25-nm channel length double-gate SOI device presented in Fig. 19. Note that the most sensitive region to random doping concentration is the middle of the conduction channel.

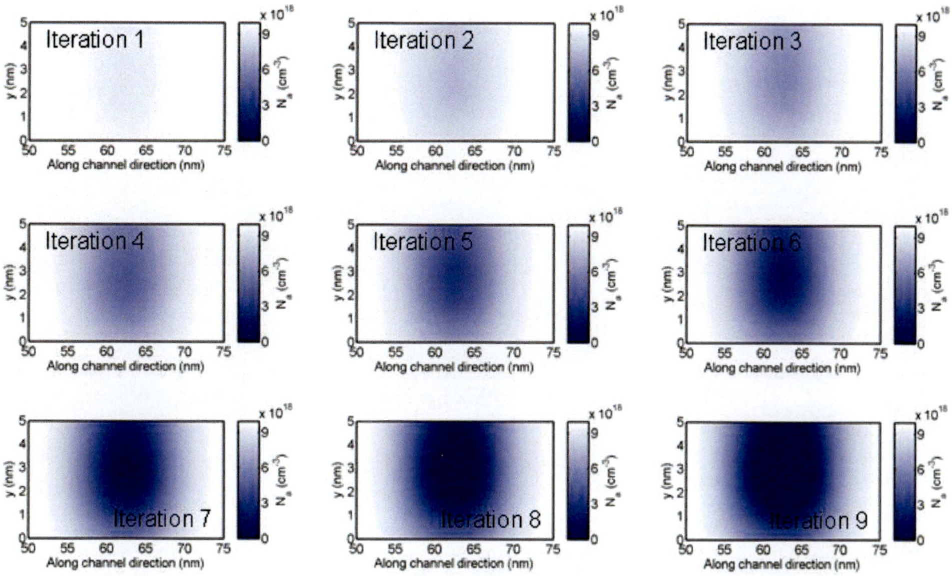

Figure 21. Average doping profiles in the channel of the 25 nm channel length double-gate SOI device obtained after each iteration. The doping profiles were obtained by minimizing the standard deviation of threshold voltage fluctuations without any constraint.

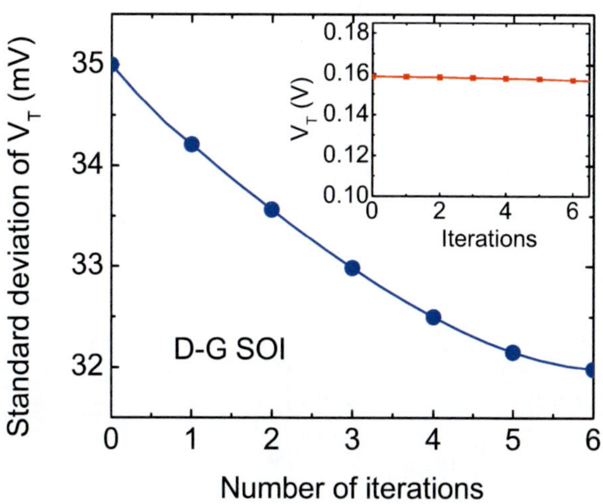

Figure 22. Standard deviation of the threshold voltage and the values of the threshold voltage (inset) as a function of the number of iterations during the *unconstrained minimization* for the double-gate SOI device represented in Fig. 19. The corresponding doping profiles are represented in Fig. 21.

The doping concentration profiles obtained during the unconstraint minimization after each iteration are represented in Fig. 21. In Fig. 22 we present the standard deviation of the threshold voltage and the expected value of the threshold voltage (in the inset) as a function of the number of iterations performed. Note that the more iterations we perform, the lower the value of σ_{V_T} becomes; for example, after 10 iterations σ_{V_T} has decreased more than 50%.

Also, the value of the threshold voltage decreases as the number of iterations is increased, and it becomes negative relatively fast.

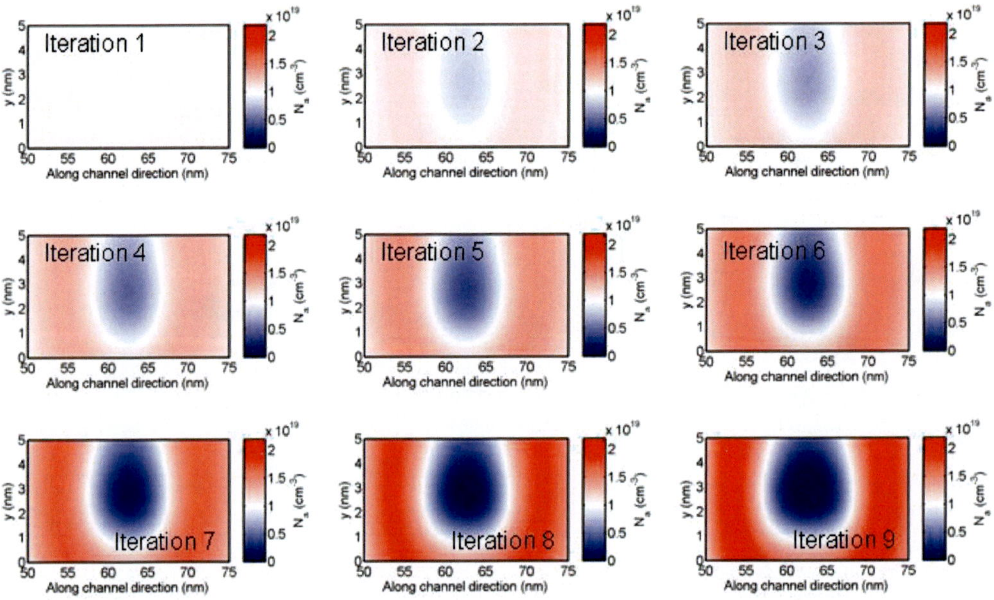

Figure 23. Average doping profiles in the channel of the 25 nm channel length double-gate SOI device obtained after each iteration. The doping profiles were obtained by minimizing the standard deviation of threshold voltage fluctuations an imposing the constraint that the saturation current be constant (which is equivalent to saying the average value of the threshold voltage is constant).

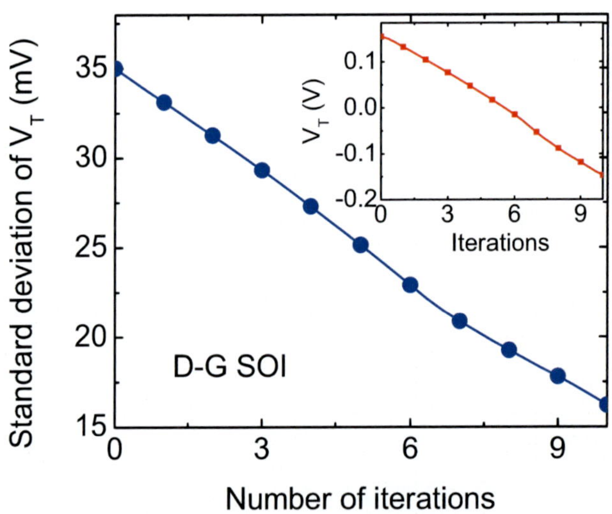

Figure 24. Standard deviation of the threshold voltage and the values of the threshold voltage (inset) as a function of the number of iterations during the *unconstrained minimization* for the double-gate SOI device. The corresponding doping profiles are represented in Fig. 23.

The doping concentration profiles obtained during the constrained minimization after each iteration are represented in Fig. 23. The standard deviations and the expected values of V_T as a function of the number of iterations are represented in Fig. 24. Although σ_{V_T} is decreased for only approximately 20%, the value of the threshold voltage and the drain-to-source current are more or less constant and equal to their initial values. The asymmetry in the doping profiles obtained by optimizing the device is due to the asymmetric geometry of the transistor (the back gate is 5 times thicker than the top gate). In the case of symmetric SOI devices the optimum doping profiles are symmetric and have a minimum in the middle of the channel and a circular barrier of high doping concentration around this minimum.

5. Optimization of Semiconductor Devices and Circuits

In this section we consider problem (2) of optimization devices and circuits. As mentioned before, the doping concentration in this problem can be assumed deterministic and we do not have to work with random variables. Next we will consider separately the unconstrained and constrained problems.

5.1. Unconstrained Optimization

Let us denote by P some parameter of the semiconductor device that should be optimized. The optimization problem can be stated as:

$$\max_{D(\boldsymbol{r})} P(D), \tag{73}$$

in which we have considered only the maximization case. This problem is somewhat similar to the inverse modeling problem [64-69] that seeks to extract a set of parameters of the semiconductor structure that results in the best fit between the simulated and experimental data. However, the difference is that we try to compute the doping profile that results in the optimum values of some parameters, without making any assumption about the experimental data. The existing inverse modeling techniques are usually developed for particular device structures, require a well defined set of experimental data to extract the parameters, and are often computationally very demanding for three-dimensional simulations for three-dimensional simulations. The technique presented below is computationally very efficient since it is based on the estimation of the doping sensitivity functions, which can be computed relatively fast and can be applied to a large range of device structures and be used in the framework of any transport model.

Problem (73) is in general difficult to solve analytically because of the large number of variables and the nonlinearity of functionals involved. However, if we try to find solutions of (73) that are infinitesimally closed to some given doping configuration $D_0(\boldsymbol{r})$, (73) can be solved analytically by using the technique of Lagrange multipliers. Indeed, if we impose that the change in the doping concentration $\tilde{D}(\boldsymbol{r})$ is infinitesimally small, the previous problem can be written as the following optimization problem with one constraint:

$$\max_{\tilde{D}(\boldsymbol{r})} P\left(D_0+\tilde{D}\right) \text{ such that } \int_\Omega \left|\tilde{D}(\boldsymbol{r})\right|^2 d\boldsymbol{r} \le \varepsilon^2, \tag{74}$$

where ε is a positive constant that goes to zero. Defining Lagrangian:

$$L = P\left(D_0+\tilde{D}\right) - \lambda\left(\int_\Omega \left|\tilde{D}(\boldsymbol{r})\right|^2 d\boldsymbol{r} - \varepsilon^2\right) \tag{75}$$

and taking the first-order necessary conditions we get the following system of equations:

$$\begin{cases} \int_\Omega \left|\tilde{D}(\boldsymbol{r})\right|^2 d\boldsymbol{r} = \varepsilon^2 \\ 2\lambda\tilde{D}(\boldsymbol{r}) = \gamma_P(\boldsymbol{r}) \end{cases} \tag{76}$$

in which we have used definition of the sensitivity functions. This system of equations can be solved for the change in the doping profile:

$$\tilde{D}(\boldsymbol{r}) = \varepsilon \frac{\gamma_P(\boldsymbol{r})}{\int_\Omega \gamma_P^2(\boldsymbol{r})d\boldsymbol{r}} = \varepsilon'\gamma_P(\boldsymbol{r}), \tag{77}$$

where we have introduced notation $\varepsilon' = \varepsilon \Big/ \int_\Omega \gamma_P^2(\boldsymbol{r})d\boldsymbol{r}$. Equation (60) is the closed-form solution of maximization problem (74) and can be applied to optimize the doping profile that maximizes the value of parameter P. It should be noted that (77) was obtained by assuming that the change in the doping profile is small. In order to accurately optimize the device one should apply this equations for as many times (iterations) as needed, in order to achieve an acceptable value for parameter P.

5.2. Constrained Optimization

Let us consider now that the optimization is subject to constraints and denote the parameters whose values should be kept constant by A, B, C, etc. The optimization (maximization) problem can be stated as follows:

$$\max_{D(\boldsymbol{r})} P(D) \text{ such that } A(D) = A(D_0),\, B(D) = B(D_0), C(D) = C(D_0), \ldots \tag{78}$$

By following the same line of reasoning as in the previous section we look for solutions that are infinitesimally closed to the initial doping configuration $D_0(\boldsymbol{r})$. Hence, the maximization problem can be written as:

$$\max_{\tilde{D}(\boldsymbol{r})} P\left(D_0+\tilde{D}\right) \text{ such that}$$

$$\int_{\Omega}\left|\tilde{D}(\boldsymbol{r})\right|^2 d\boldsymbol{r} \le \varepsilon^2, A(D)=A(D_0), B(D)=B(D_0), C(D)=C(D_0),... \tag{79}$$

Defining Lagrangian:

$$L=\max_{\tilde{D}(\boldsymbol{r})} P\left(D_0+\tilde{D}\right)-\lambda\left(\int_{\Omega}\left|\tilde{D}(\boldsymbol{r})\right|^2 d\boldsymbol{r}-\varepsilon^2\right)-\lambda_A\left[A\left(D_0+\tilde{D}\right)-A\left(D_0\right)\right]-$$
$$\lambda_B\left[B\left(D_0+\tilde{D}\right)-B\left(D_0\right)\right]-\lambda_C\left[C\left(D_0+\tilde{D}\right)-C\left(D_0\right)\right]-... \tag{80}$$

where λ, λ_A, λ_B, λ_C,... are the corresponding Lagrange multipliers and taking the first-order necessary conditions we get the following system of equations:

$$\begin{cases}\int_{\Omega}\left|\tilde{D}(\boldsymbol{r})\right|^2 d\boldsymbol{r}=\varepsilon^2 \\ 2\lambda\tilde{D}(\boldsymbol{r})+\lambda_A\gamma_A(\boldsymbol{r})+\lambda_B\gamma_B(\boldsymbol{r})+...=\gamma_P(\boldsymbol{r}) \\ \int_{\Omega}\gamma_A(\boldsymbol{r})\tilde{D}(\boldsymbol{r})d\boldsymbol{r}=0 \\ \int_{\Omega}\gamma_B(\boldsymbol{r})\tilde{D}(\boldsymbol{r})d\boldsymbol{r}=0 \\ ... \end{cases} \tag{81}$$

which can be solved for $\tilde{D}(\boldsymbol{r})$, λ, λ_A, λ_B,... Indeed, the first two equations in (81) can be solved for $\tilde{D}(\boldsymbol{r})$:

$$\tilde{D}(\boldsymbol{r})=\varepsilon\frac{\gamma_P(\boldsymbol{r})-\lambda_A\gamma_A(\boldsymbol{r})-\lambda_B\gamma_B(\boldsymbol{r})-\lambda_C\gamma_C(\boldsymbol{r})-...}{\int_{\Omega}\left[\gamma_P(\boldsymbol{r})-\lambda_A\gamma_A(\boldsymbol{r})-\lambda_B\gamma_B(\boldsymbol{r})-\lambda_C\gamma_C(\boldsymbol{r})-...\right]^2 d\boldsymbol{r}}. \tag{82}$$

Multiplying (82) by $\gamma_A(\boldsymbol{r})$ and integrating over Ω yields:

$$\int_{\Omega}\gamma_A(\boldsymbol{r})\tilde{D}(\boldsymbol{r})d\boldsymbol{r}=\frac{\varepsilon}{\int_{\Omega}\left[\gamma_P(\boldsymbol{r})-\lambda_A\gamma_A(\boldsymbol{r})-\lambda_B\gamma_B(\boldsymbol{r})-\lambda_C\gamma_C(\boldsymbol{r})-...\right]^2 d\boldsymbol{r}}\times$$
$$\int_{\Omega}\gamma_A(\boldsymbol{r})\left[\gamma_P(\boldsymbol{r})-\lambda_A\gamma_A(\boldsymbol{r})-\lambda_B\gamma_B(\boldsymbol{r})-\lambda_C\gamma_C(\boldsymbol{r})-..\right]d\boldsymbol{r} \tag{83}$$

From (83) and the third equation in (81) we obtain the following equation for the Lagrange multipliers:

$$\lambda_1\gamma_{A,A}+\lambda_2\gamma_{A,B}+\lambda_3\gamma_{A,C}+...=\gamma_{P,A} \tag{84}$$

where: $\gamma_{X,Y} = \langle \gamma_X, \gamma_Y \rangle$ is the do product between the doping sensitivity functions of parameters X and Y ([see equation (63)]. Note that in order to obtain (84) we have also supposed that the denominator of (83) is not equal to zero.

By using a similar procedure we can derive analogous expressions for $\gamma_{P,B}$, $\gamma_{P,C}$, etc. Finally, we arrive at the following linear system of equations for the Lagrange multipliers:

$$\begin{cases} \lambda_A \gamma_{A,A} + \lambda_B \gamma_{A,B} + \lambda_C \gamma_{A,C} + \ldots = \gamma_{P,A} \\ \lambda_A \gamma_{B,A} + \lambda_B \gamma_{B,B} + \lambda_C \gamma_{B,C} + \ldots = \gamma_{P,B} \\ \lambda_A \gamma_{C,A} + \lambda_B \gamma_{C,B} + \lambda_C \gamma_{C,C} + \ldots = \gamma_{P,C} \\ \quad \ldots. \end{cases} \tag{85}$$

which has a very similar structure to system (62).

Equations (82) and (85) represent the closed-form solution of optimization problem (79). These equations should be solved numerically in order to compute the optimum doping profiles that maximize parameter P.

Note again that equations (82) and (85) were obtained by assuming that the change in the doping profile is small. In order to accurately optimize the device one should use these equations iteratively for as many times as needed, in order to obtain an acceptable maximum value for parameter P. More details about the numerical implementation will be presented later in this Section.

5.3. A Few Special Cases

In the case of the maximization without constraints we recover equation (77). When one constraint is imposed during the maximization problem, equations (82) and (85) can be solved easily for the change in the doping concentration:

$$\tilde{D}(\mathbf{r}) = -\varepsilon \frac{\gamma_P(\mathbf{r}) - \lambda_A \gamma_A(\mathbf{r})}{\int_\Omega [\gamma_P(\mathbf{r}) - \lambda_A \gamma_A(\mathbf{r})]^2 \, d\mathbf{r}}, \text{ where } \lambda_A = \frac{\int_\Omega \gamma_P(\mathbf{r}) \gamma_A(\mathbf{r}) d\mathbf{r}}{\int_\Omega \gamma_A^2(\mathbf{r}) d\mathbf{r}} = \frac{\lambda_{P,A}}{\gamma_{A,A}}, \tag{86}$$

where A is the parameter that is supposed to remain constant during the maximization.

In the case of maximization with two or more constraints, equations (82) and (85) should be solved self-consistently. The additional computational overhead introduced by solving system (62) is negligible because the number of constraints is usually small.

5.4. Necessary and Sufficient Conditions for the Existence of the Solution

In the following we discuss necessary and sufficient conditions under which the solution (82) of the optimization problem exists. First, we have to check the existence of the solution of system (62). It is important to note that the determinant associated with the left-hand side of this system is always larger or equal to zero. For instance, in the case of the two constraints the determinant is:

$$\gamma_{A,A}\gamma_{B,B} - \gamma_{A,B}\gamma_{B,A} = \int_\Omega \gamma_A^2(\boldsymbol{r})d\boldsymbol{r} \int_\Omega \gamma_B^2(\boldsymbol{r})d\boldsymbol{r} - \left[\int_\Omega \gamma_A(\boldsymbol{r})\gamma_B(\boldsymbol{r})d\boldsymbol{r}\right]^2 \geq 0, \qquad (87)$$

which is nothing else but the Cauchy(-Schwartz-Buniakowski) inequality [70]. The determinant is equal to zero *if and only if* the doping sensitivity functions are linearly dependent. In this case one can remove the dependent constraints in (79) and reduce the total number of constraints to a smaller number. If the maximization problem has n constraints among which m are linearly dependent, one can remove $m-1$ of the linearly dependent constraints and reduced the initial optimization problem to one with $n-m+1$ constraints that are linearly independent. Therefore, the case when the determinant of (85) is equal to zero does not imply that the solution of the optimization problem does not exist but only means that too many equivalent constraints have been imposed during the optimization and some of them should be dropped out without affecting the final result.
Next we suppose that the linearly dependent constraints have been dropped out and the determinant of system (62) is not null. The necessary and sufficient condition under which the solution of the optimization problem exists is that the denominator of (82) does not vanish:

$$\int_\Omega \left[\gamma_P(\boldsymbol{r}) - \lambda_A\gamma_A(\boldsymbol{r}) - \lambda_B\gamma_B(\boldsymbol{r}) - \lambda_C\gamma_C(\boldsymbol{r}) - \ldots\right]^2 d\boldsymbol{r} \neq 0, \qquad (88)$$

which is equivalent to

$$\gamma_P(\boldsymbol{r}) \neq \lambda_A\gamma_A(\boldsymbol{r}) + \lambda_B\gamma_B(\boldsymbol{r}) + \lambda_C\gamma_C(\boldsymbol{r}) + \ldots \qquad (89)$$

for at least one point $\boldsymbol{r} \in \Omega$. The last inequality is true if and only if the doping sensitivity function of P cannot be written as a linear combination of the doping sensitivity functions of the parameters under constraints. This statement is more complicated to prove in general, but it can be easily probed for a small number of constraints. For instance, in the case of only one constraint, the denominator in equation (65) is equal to zero if

$$\gamma_P(\boldsymbol{r}) - \lambda_A\gamma_A(\boldsymbol{r}) = 0, \text{ for all } \boldsymbol{r} \in \Omega. \qquad (90)$$

Using the expression of λ_A from equation (65) we obtain

$$\gamma_P(\boldsymbol{r}) \int_\Omega \gamma_A^2(\boldsymbol{r})d\boldsymbol{r} = \gamma_A(\boldsymbol{r}) \int_\Omega \gamma_P(\boldsymbol{r})\gamma_A(\boldsymbol{r})d\boldsymbol{r}, \text{ for all } \boldsymbol{r} \in \Omega, \qquad (91)$$

which is true if and only if $\gamma_P(\boldsymbol{r}) = const. \times \gamma_A(\boldsymbol{r})$, i.e. functions $\gamma_P(\boldsymbol{r})$ and $\gamma_A(\boldsymbol{r})$ are linearly dependent. In other words, it is impossible to maximize (or minimize) a parameter of the device that has the same sensitivity function as of another parameter that is set to remain constant during the optimization. For instance, an example of an incorrectly set problem is to maximize the value of the drain-to-source current when the gate voltage is equal to the threshold voltage, and in the same time to keep the value of the threshold voltage constant; the doping sensitivity function of the threshold voltage are proportional to the doping sensitivity function of the drain-to-source current and the optimization problem has no solution.

5.5. Optimization of More Than One Parameter

Sometimes it is useful to optimize more than one parameter of the device. Such an example is to maximize the value of the saturation current in a device and, in the same time, to minimize the value of the subthreshold voltage swing. This maximization can be performed by slightly modifying the previous method in order to take into account more optimization parameters. Let us denote the parameters that should be optimized by R, S, T etc. In order to optimize the values of all parameters we introduce the new quantity

$$P = w_R R^2 + w_S S^2 + w_T T^2 + \ldots, \tag{92}$$

where w_R, w_S, w_T, etc. are some coefficients that give how much weight is given to the optimization of each parameter. These weighting coefficients are also introduced in order for $w_R R^2$, $w_S S^2$, $w_T T^2$,... to have no units. They should have positive values if the parameters that they denote are to be maximized and negative values if the parameters are to be minimized. The doping sensitivity function of P can be computed by using the algebra of sensitivity functions presented in [49,50]:

$$\gamma_P(\boldsymbol{r}) = 2\left[w_R R \gamma_R(\boldsymbol{r}) + w_S S \gamma_S(\boldsymbol{r}) + w_R T \gamma_T(\boldsymbol{r}) + \ldots\right], \tag{93}$$

where $\gamma_R(\boldsymbol{r})$, $\gamma_S(\boldsymbol{r})$, and $\gamma_T(\boldsymbol{r})$ are the doping sensitivity functions of parameters R, S, and T, respectively. Now, the analysis presented at the beginning of this section can be applied in order to optimize parameter P. It can be shown that if the weighting coefficients are chosen properly all terms in the summation in (92) are maximized simultaneously, which was exactly our goal. The necessary and sufficient conditions for the existence of the solution of this optimization problem are more complex then in the case of only one optimization parameter and are not presented here.

5.6. Suppressing Random Dopant Induced Fluctuations

A special case of the above optimization procedure is the minimization of the random dopant induced fluctuations of some parameter (call it R). As shown in the previous section the variance of parameter R can be written as a function of the doping sensitivity function $\gamma_R(\boldsymbol{r})$. Therefore, if we want to minimize the variance of random dopant induced fluctuations we set parameter P to $-\sigma_R^2$. The doping sensitivity function of $P=-\sigma_R^2$ is $\gamma_P(\boldsymbol{r})=\gamma_R^2(\boldsymbol{r})$, so the solution of the optimization problem can be formally written as:

$$\tilde{D}(\boldsymbol{r})=-\varepsilon\frac{\gamma_R^2(\boldsymbol{r})-\lambda_A\gamma_A(\boldsymbol{r})-\lambda_B\gamma_B(\boldsymbol{r})-\lambda_C\gamma_C(\boldsymbol{r})-\dots}{\int_\Omega\left[\gamma_R^2(\boldsymbol{r})-\lambda_A\gamma_A(\boldsymbol{r})-\lambda_B\gamma_B(\boldsymbol{r})-\lambda_C\gamma_C(\boldsymbol{r})-\dots\right]^2 d\boldsymbol{r}}, \tag{94}$$

where the Lagrange multipliers satisfy the following system of linear equations:

$$\begin{cases} \lambda_A\gamma_{A,A}+\lambda_B\gamma_{A,B}+\lambda_C\gamma_{A,C}+\dots=\gamma_{R^2,A} \\ \lambda_A\gamma_{B,A}+\lambda_B\gamma_{B,B}+\lambda_C\gamma_{B,C}+\dots=\gamma_{R^2,B} \\ \lambda_A\gamma_{C,A}+\lambda_B\gamma_{C,B}+\lambda_C\gamma_{C,C}+\dots=\gamma_{R^2,C} \\ \quad\quad\quad\dots\dots \end{cases} \tag{95}$$

Note that we recover the results obtained in the previous section for the design of random fluctuation resistant structures of semiconductor devices [see equations (60) and (62)]. This formal derivation of the optimal doping profiles hides the fact that the doping concentration is a random variable and, at least in principle is not correct.

5.7. Numerical Algorithm

The numerical algorithm of the optimization problem is summarized below (see also Fig. 25).

1. Start with an initial doping concentration profile throughout the devices and compute the doping sensitivity functions $\gamma_X(\boldsymbol{r})$ for all parameters under consideration (i.e. P, A, B, C,...). Only a set of linearly independent doping sensitivity functions are used at the next steps.
2. Solve linear system (85) for the Lagrange multipliers λ_A, λ_B, λ_C, ...
3. Select a relatively small value for ε and compute the change in the doping concentration by using equation (82). In our simulations we have adjusted the value of ε such that the doping concentration is reduced by approximately 15% at each iteration step.
4. Compute the new doping profile $D(\boldsymbol{r})=D_0(\boldsymbol{r})+\tilde{D}(\boldsymbol{r})$ and the new value of P.
5. Repeat steps 2-4 until an acceptable final value of parameter P is obtained.

These steps should be performed independently for donors ($D=N_d$) and acceptors ($D=N_a$). In the case where there are more doping species one should perform steps 1-5 for each individual specie separately.

As it concerns to the computational efficiency of the optimization algorithm it should be noted that the computation of the doping sensitivity functions (step 1) usually takes a relatively small time because it requires solving only a few linear sparse systems of equations [41,56]. For a two-dimensional device discretized in 10,000 mesh points the computation of the doping sensitivity functions usually takes a few seconds on a single processor personal computer (PC) running at 3 GHz. The rest of the iteration steps can be performed much faster. The integration in (60) can be done be using quadratures and it takes a few milliseconds on the same PC. System (62) usually involves at most 2 or 3 linear equations and can be solved easily.

An important simplification that can be made to this algorithm in order to decrease the computational cost is to compute the doping sensitivity functions once, at the beginning of the simulation, and use them at each iteration loop. Indeed, as shown in [56] the doping sensitivity functions are practically independent on the exact doping profile in the channel so they can be computed only once.

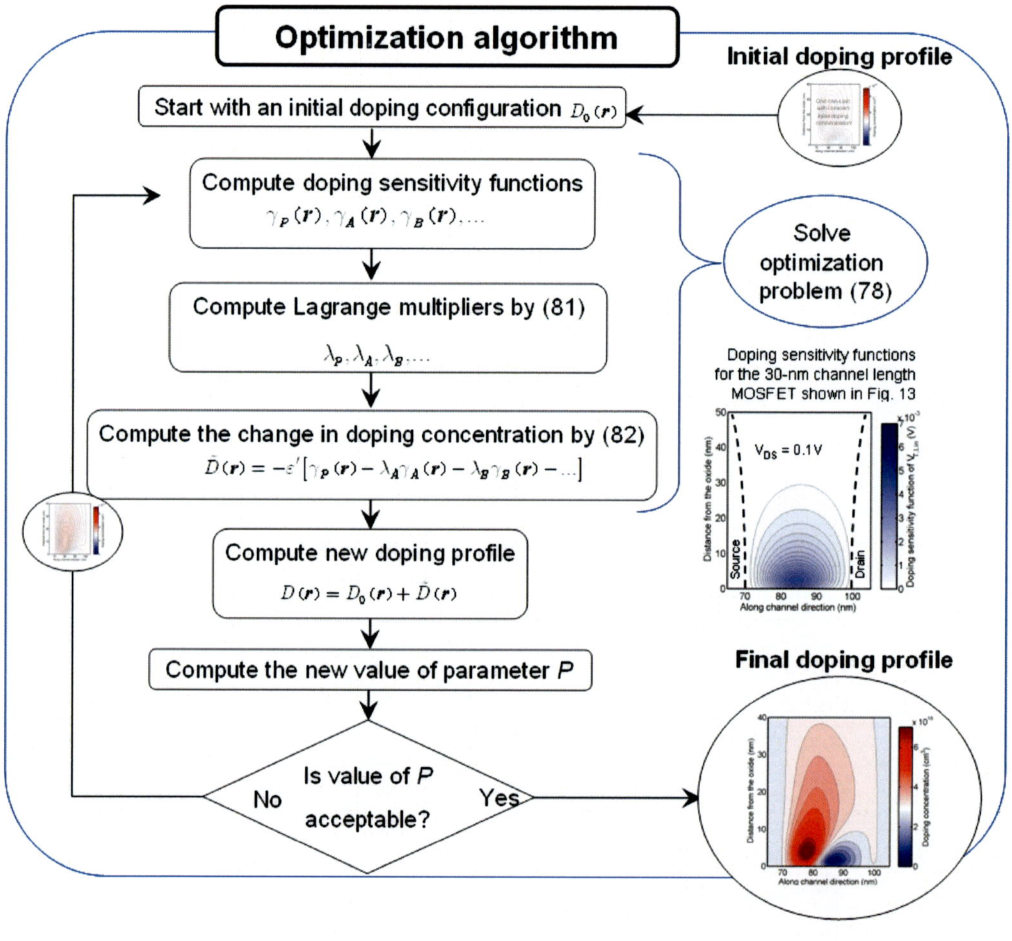

Figure 25. Numerical algorithm for the optimization of semiconductor devices.

5.8. Numerical Results

The optimization procedure presented in the previous sections has been numerically implemented in our device and circuit simulator RandFlux© [40]. RandFlux© allows the computation of the doping sensitivity functions of most parameters (e.g. terminal currents, V_T, subthreshold characteristics, small-signal parameters, cutoff frequencies, etc.) by using perturbation techniques applied to various 2-D and 3D transport models. The transport equations are solved self-consistently with the Poisson equation and quantum corrections are taken into account using various approximate models, the effective-mass Schrödinger equation, or the Density-Gradient model.

One phenomenon that negatively affects the performance of digital and mixed-signal circuits and, for this reason, has received much attention in the literature in the last decade is the DIBL in MOSFET devices. The effects induced by DIBL are detrimental for the power dissipation of digital circuits and shift of logical levels in ultrasmall semiconductor devices. Therefore, as a direct application of the optimization technique presented in the previous sections, we try to optimize the doping concentration in a 30-nm channel length MOSFET device in order to suppress the effects of DIBL on the value of the threshold voltage. The structure of the n-channel MOSFET is similar to the one represented in Fig. 13 but the channel length is 30 nm instead of 35 nm. The initial doping concentration in the channel was uniform and equal to 3×10^{18} cm^{-3}. The doping profiles of the drain and source junctions are Gaussian, the doping concentration of the polysilicon is equal to 5×10^{19} cm^{-3}, and the oxide thickness is 1 nm. A uniform sheet charge density of -5×10^{11} C/cm^2 was assumed at the semiconductor/oxide interface. The width of the device is 100 nm.

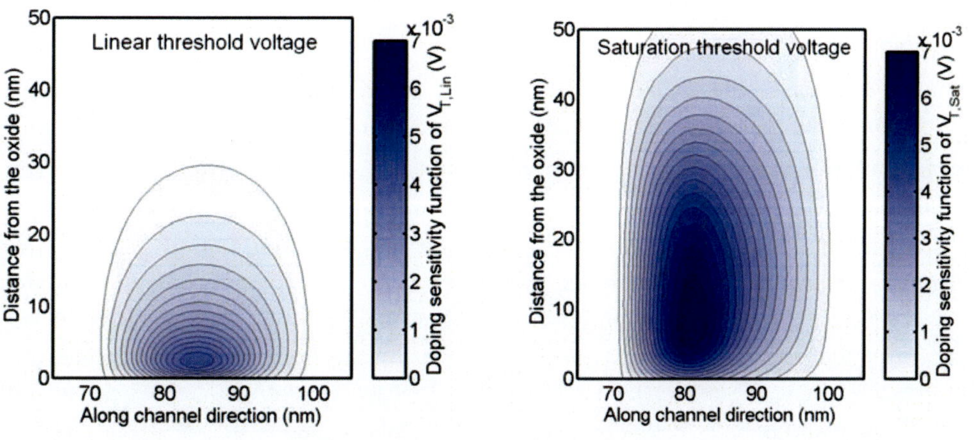

Figure 26. Doping sensitivity functions of the linear (left) and saturation (right) threshold voltages, computed by using the perturbation technique [45]. The linear threshold voltage is computed at $V_{DS} = 0.1\,V$ while the saturation threshold voltage is computed as $V_{DS} = 1\,V$.

The doping sensitivity functions of the linear and saturation threshold voltages are represented in figure 26. These functions were computed for a uniform doping profile by solving self-consistently the Poisson, electron and hole continuity, and the Density-Gradient [7] equations. It is apparent from this figure that the threshold voltages are highly sensitive to variations of the doping concentration in the channel of the MOSFET device. Therefore, even though we solve the transport equations and optimize the whole semiconductor device (including the polysilicon gate and the drain and source junctions), in the results presented in this section we show the doping concentration of acceptor impurities only in the channel region.

In order to suppress the effects of the drain-induced-barrier-lowering (DIBL) we define parameter P as the difference between the saturation and linear threshold voltages:

$$P = V_{T,Sat} - V_{T,Lin} . \quad (96)$$

We compute the optimum doping profile in two case: first using the unconstrained and, second, the constrained optimization of P. During the unconstrained optimization, parameter P is minimized without imposing any constraints on the values of the other device parameters. During the constrained optimization, P is minimized by setting the value of $V_{T,Lin}$ constant and equal to its value when the doping concentration throughout the channel was uniform and equal to 3×10^{18} cm^{-3}. It should be noted that the difference between the two threshold voltages in (96) is proportional to the difference between the drain voltages, which in our case is fixed and equal to $\Delta V_D = 1\text{ V} - 0.1\text{V} = 0.9\text{ V}$:

$$P = \Delta V_T \approx \eta \Delta V_D , \quad (97)$$

where η is a proportionality constant. Hence the optimization problem is equivalent to minimizing proportionality constant η.

The values of $V_{T,Lin}$ and $V_{T,Sat}$ are presented in Figs. 27 as a function of the number of iteration steps during the unconstrained and constrained minimizations, respectively. During each iteration, the proportionality constant ε in equation (81) is computed in such a way that the maximum variation of the doping concentration is 5×10^{17} cm^{-3}. Hence, after 6 iterations the doping concentration might be zero at some points in the channel (because the initial doping concentration was 3×10^{18} cm^{-3}) and the optimizer will set to zero any further decrease of the doping concentration at those points. The doping profiles obtained after 1, 3, 5, 7, 9, and 10 iterations are represented in Figs. 28 and 29 for the unconstrained and constrained minimizations, respectively. It is interesting to observe that the doping profiles that minimize the most efficiently the effects of DIBL on V_T are obtained when dopants are removed from the middle of the conduction channel (low doping concentration) but a high dopant concentration of acceptors is introduced towards the source side and extends between 5 to 10 nm inside the channel. This doping profile results in a decrease of more than 50% of the threshold voltage variation, which represents a significant improvement for device and circuit design. Surprisingly, one can observe that there is not a fundamental difference between the doping profiles obtained during the constrained and unconstrained optimization. Both optimizations result in similar doping profiles, the only difference being that in the case of the unconstrained optimization the peak of the doping concentration in the channel is higher. The increase of the doping concentration can be explained because it becomes harder for the drain voltage to lower the drain-to-source potential barrier at high doping concentrations.

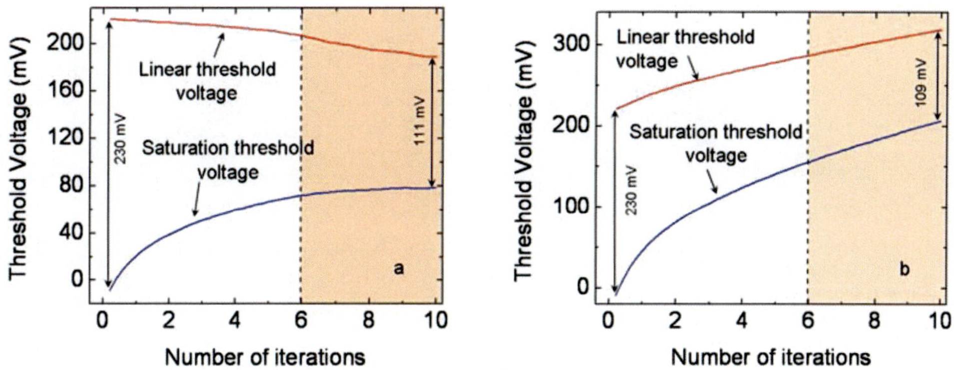

Figure 27. Linear (a) and saturation (b) threshold voltages of the MOSFET device as a function of the number of iterations computed by using the unconstraint and constrained optimizations. Note that the white areas in the channel represents the region with doping concentration is unchanged and equal to its initial value of 3×10^{18} cm^{-3}. During the constrained optimization the value of the linear threshold voltage was set to be constant.

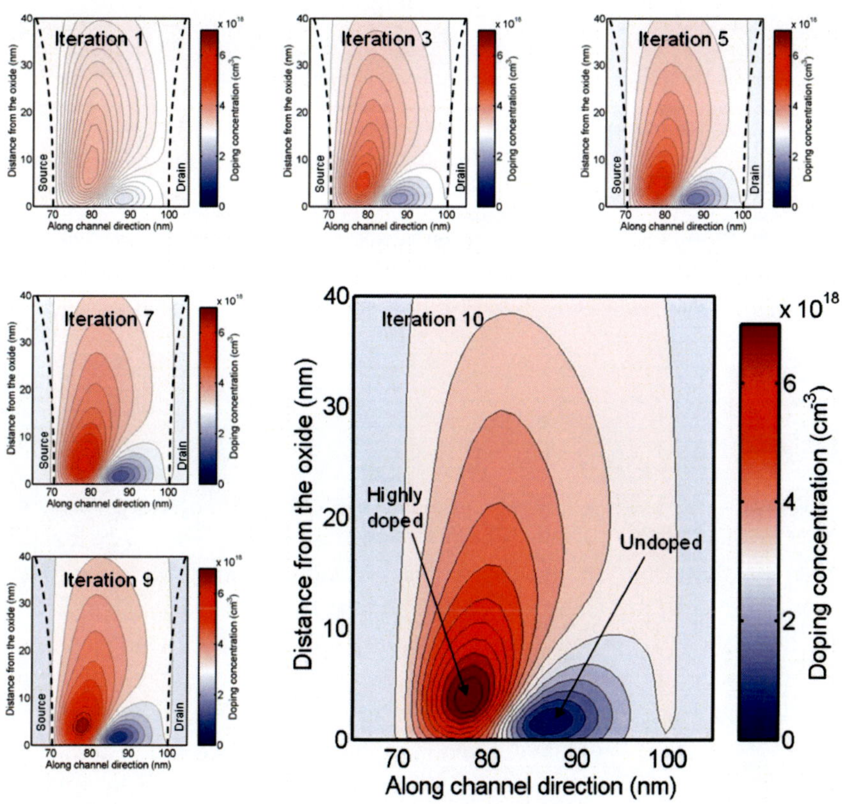

Figure 28. Doping profiles in the channel region of the MOSFET device computed by using the unconstraint optimization after 1, 3, 5, 7, 9, and 10 iterations. The red areas in the channel represent the regions with high doping concentration, while the blue areas denote regions with low doping concentration.

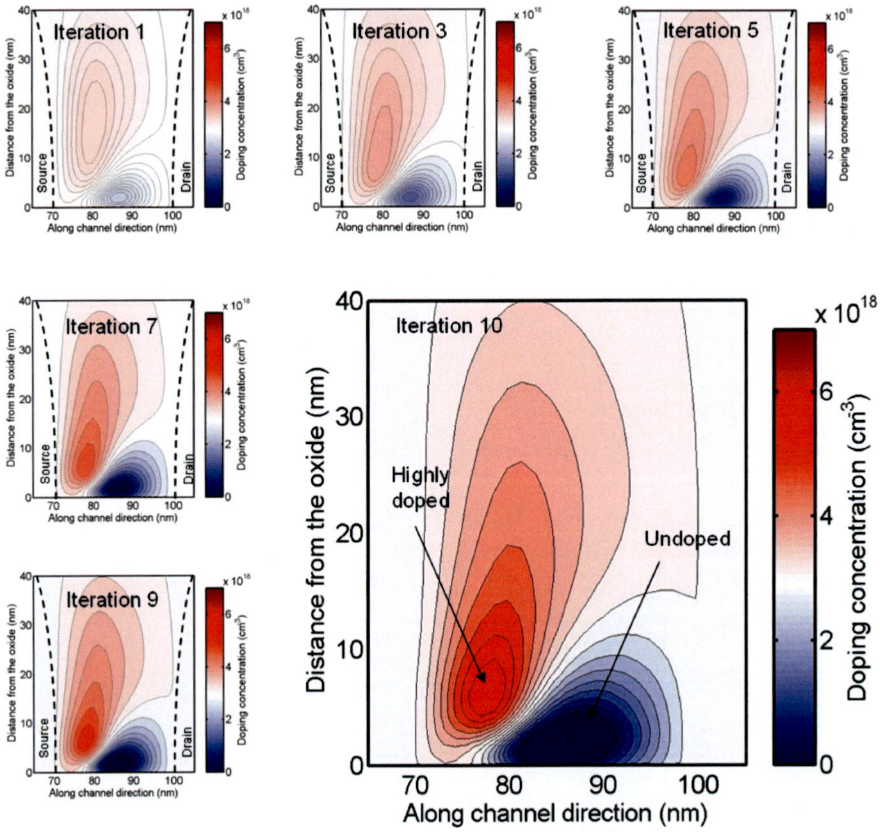

Figure 29. Doping profiles in the channel region of the MOSFET device computed by using the constrained optimization after 1, 3, 5, 7, 9, and 10 iterations. The red areas in the channel represent the regions with high doping concentration, while the blue areas denote regions with low doping concentration.

6. Conclusion

A robust solution is provided to the redesign and optimization of semiconductor devices based on the formalism of doping sensitivity functions. The doping sensitivity functions show how sensitive the parameters of the device are to random doping variations and can be computed very efficiently by using first-order perturbations. An automated technique has been developed for the design of random dopant fluctuation resistant semiconductor devices, i.e. structures in which intrinsic random doing fluctuations have minimum effects on device parameters. It is shown that the standard deviation of the threshold voltage can be decreased significantly by "readjusting" the doping configuration of the device. The design technique has also been applied to the computation of random dopant induced fluctuations in SRAM cells and used to evaluate the effects of these fluctuations on the SNMs. Numerical results for an SRAM cell made with 30 nm channel-length MOSFETs show that the ratio between the value of the SNM and its standard deviation is approaching the minimum allowed limits for SRAM design, which represents a fundamental limitation for further scaling. The relatively high values obtained in our simulations for this ratio are due to the rather low doping concentration in the channel and to the high value of the biasing voltage. It is found that the most sensitive regions to doping fluctuations extend for approximately 10 nm below the oxide/semiconductor interface and are located in the middle of the conduction channels. The most sensitive transistors in the SRAM cell analyzed in this chapter seem to be the access and pull-down transistors.

It is also apparent that it is more difficult to suppress random dopant-induced fluctuations in short-channel than in long-channel devices. This is partly due to the fact that quantum mechanical effects start playing a much more important role in short-channel devices but also because the dopant fluctuations in other regions such as the drain and source junctions become significant. Undoped channels and δ-doping profiles seem to be good alternatives to for constructing fluctuation resistant structures. Analytical computations for long-channel devices show that the δ-doping in bulk MOSFET devices should start at approximately ¼ of the width of the depletion region in order to suppress most efficiently random dopant fluctuations of the threshold voltages. Approximately the same ratio remains valid for short-channel devices, however the shape of the δ-doping becomes almost semicircular because of the short-channel and quantum effects.

References

[1] R. Strasser, R. Plasun, and S. Selberherr, in *12th European Simulation Multiconference - Simulation: Past, Present and Future*, edited by R. Zobel and D. Moeller (1998).

[2] G. Khanduri and B. Panwar, *Solid-State Electronics* **51,** 961-964 (2007).

[3] R. Strasser and S. Selberherr, Ieice *Transactions on Electronics* **E83c,** 1303-1310 (2000).

[4] International roadmap for semiconductors (ITRS), http://public.itrs.net, 2006 edition,

[5] K. A. Bowman, S. G. Duvall, and J. D. Meindl, *Ieee Journal of Solid-State Circuits* **37,** 183-190 (2002).

[6] A. J. Bhavnagarwala, X. H. Tang, and J. D. Meindl, *Ieee Journal of Solid-State Circuits* **36,** 658-665 (2001).

[7] K. A. Bowman, X. H. Tang, J. C. Eble, and J. D. Meindl, *Ieee Journal of Solid-State Circuits* **35,** 1186-1193 (2000).

[8] X. H. Tang, V. K. De, and J. D. Meindl, *Ieee Transactions on Very Large Scale Integration (Vlsi) Systems* **5,** 369-376 (1997).

[9] K. Nishinohara, N. Shigyo, and T. Wada, *IEEE Transactions on Electron Devices* **39,** 634-639 (1992).

[10] H. S. Wong and Y. Taur, *IEDM Tech. Dig.*, 705–708 (1993).

[11] T. Mizuno, J. Okamura, and A. Toriumi, IEEE Transactions on Electron Devices **41,** 2216-2221 (1994).

[12] J. R. Zhou and D. K. Ferry, *Proc. 3rd Int. Workshop on Comput. Electron.*, 74–77 (1994).

[13] P. A. Stolk, F. P. Widdershoven, and D. B. M. Klaassen, *IEEE Transactions on Electron Devices* **45,** 1960-1971 (1998).

[14] Vasileska, W. J. Gross, and D. K. Ferry, *Proc. of the International Workshop of Computational Electronics (IWCE)*, 259–262 (1998).

[15] V. Y. A. Thean, M. Sadd, and B. E. White, *IEEE Transactions on Nanotechnology* **2,** 97-101 (2003).
[16] A. Asenov, *IEEE Transactions on Electron Devices* **45,** 2505-2513 (1998).
[17] A. Asenov, G. Slavcheva, A. R. Brown, J. H. Davies, and S. Saini, *IEEE Transactions on Electron Devices* **48,** 722-729 (2001).
[18] S. Roy and A. Asenov, Science **309,** 388-390 (2005).
[19] A. Brown, J. Watling, and A. Asenov, *Proc. 11rd Int. Workshop on Comput. Electron.*, 27-30 (2006).
[20] A. R. Brown, J. R. Watling, A. Asenov, G. Bersuker, and P. Zeitzoff, *Proc. of SISPAD,* 27-30 (2005).
[21] N. Seoane, A. J. Garcia-Loureiro, K. Kalna, and A. Asenov, *Solid-State Electronics* **51,** 481-488 (2007).
[22] K. Samsudin, F. Adamu-Lerna, A. R. Brown, S. Roy, and A. Asenov, *Solid-State Electronics* **51,** 611-616 (2007).
[23] A. Martinez, J. R. Barker, A. Svizhenko, M. P. Anantram, and A. Asenov, *Ieee Transactions on Nanotechnology* **6,** 438-445 (2007).
[24] K. Samsudin, B. Cheng, A. R. Brown, S. Roy, and A. Asenov, *Solid-State Electronics* **50,** 86-93 (2006).
[25] B. Cheng, S. Roy, G. Roy, F. Adamu-Lema, and A. Asenov, *Solid-State Electronics* **49,** 740-746 (2005).
[26] A. R. Brown, F. Adamu-Lema, and A. Asenov, *Superlattices and Microstructures* **34,** 283-291 (2003).
[27] A. Asenov, S. Kaya, and A. R. Brown, *IEEE Transactions on Electron Devices* **50,** 1254-1260 (2003).
[28] A. R. Brown, A. Asenov, and J. R. Watling, *IEEE Transactions on Nanotechnology* **1,** 195-200 (2002).
[29] A. Asenov, S. Kaya, and J. H. Davies, *IEEE Transactions on Electron Devices* **49,** 112-119 (2002).
[30] K. Kalna, R. Droopad, M. Passlack, and A. Asenov, *Microelectronic Engineering* **84,** 2150-2153 (2007).
[31] B. K. Ridley, *Semiconductor Science and Technology* **3,** 286-291 (1988).
[32] K. Samsudin, B. Cheng, A. R. Brown, S. Roy, and A. Asenov, *Solid-State Electronics* **50,** 660-667 (2006).
[33] B. Cheng, S. Roy, and A. Asenov, *Solid-State Electronics* **51,** 565-571 (2007).
[34] A. Asenov and S. Saini, *IEEE Transactions on Electron Devices* **46,** 1718-1724 (1999).
[35] K. Noda, T. Tatsumi, T. Uchida, K. Nakajima, H. Miyamoto, and C. M. Hu, *Ieee Transactions on Electron Devices* **45,** 809-814 (1998).
[36] K. Takeuchi, T. Tatsumi, and A. Furukawa, *Proc. IEDM'97 Dig. Tech. Papers* (1997).
[37] P. A. W. Stolk, F. P. and D. B. M. Klaassen, *IEEE Trans. Electron Devices* **45,** , 1960–1971 (1998.).
[38] J. R. Munkres, *Analysis on Manifolds* (Addison-Wesley Publishing Company, 1990).
[39] W. Rudin, *Real and Complex Analysis* (McGraw-Hill, 1966).
[40] Florida State University and Florida A&M University, http://www.eng.fsu.edu/ms/RandFlux,
[41] I. D. Mayergoyz and P. Andrei, in *Handbook of Semiconductor Nanostructures and Nanodevices* (2006), p. 257-324.

[42] P. Andrei, *J. of Comput. Electronics* **5,** 315-318 (2006).
[43] P. Andrei, Ph.D. Thesis, University of Maryland (2004).
[44] P. Andrei and I. Mayergoyz, *Journal of Applied Physics* **94,** 7163-7172 (2003).
[45] P. Andrei and I. Mayergoyz, *Journal of Applied Physics* **96,** 2071-2079 (2004).
[46] L. Oniciuc and P. Andrei, *WSEAS Trans. On Circ. and Syst.* **3,** 1842-1848 (2005).
[47] P. Andrei and I. Mayergoyz, *Journal of Applied Physics* **93,** 4646-4652 (2003).
[48] P. Andrei and I. D. Mayergoyz, *J. of Comput. Electronics* **3,** 211-214 (2004).
[49] P. Andrei and I. Mayergoyz, *Solid-State Electronics* **48,** 133-141 (2004).
[50] P. Andrei and I. Mayergoyz, *Solid-State Electronics* **47,** 2055-2061 (2003).
[51] C. Alexander, A. R. Brown, J. R. Watling, and A. Asenov, *Solid-State Electronics* **49,** 733-739 (2005).
[52] A. Asenov, A. R. Brown, J. H. Davies, and S. Saini, *IEEE Transactions on Computer-Aided Design of Integrated Circuits and Systems* **18,** 1558-1565 (1999).
[53] P. Andrei and L. Oniciuc, *Solid-State Electronics* (submitted).
[54] Y. Taur and T. H. Ning, *Fundamental of Modern VLSI Devices* (Cambridge University Press, 1998).
[55] P. Andrei and L. Oniciuc, *IEEE Trans. Electron Devices* submitted (2007).
[56] I. D. Mayergoyz and P. Andrei, *Journal of Applied Physics* **90,** 3019-3029 (2001).
[57] M. Lundstrom, *Fundamentals of carrier transport*, second ed. (Cambridge University Press, 2000).
[58] S. O. Koswatta, S. Hasan, M. S. Lundstrom, M. P. Anantram, and D. E. Nikonov, *Ieee Transactions on Electron Devices* **54,** 2339-2351 (2007).
[59] A. Martinez, M. Bescond, J. R. Barker, A. Svizhenko, M. P. Anantram, C. Millar, and A. Asenov, *Ieee Transactions on Electron Devices* **54,** 2213-2222 (2007).
[60] G. Klimeck, S. S. Ahmed, H. Bae, N. Kharche, R. Rahman, S. Clark, B. Haley, S. H. Lee, M. Naumov, H. Ryu, F. Saied, M. Prada, M. Korkusinski, and T. B. Boykin, *Ieee Transactions on Electron Devices* **54,** 2079-2089 (2007).
[61] Y. Taur and C. Y. Ting, *Fundamentals of Modern VLSI Devices* (Cambridge University Press, pp. 200-2002, 1998).
[62] H. S. P. Wong, Y. Taur, and D. J. Frank, *Microelectronics Reliability* **38,** 1447-1456 (1998).
[63] D. J. Frank, Y. Taur, M. Ieong, and H. S. P. Wong, *Digest of Symp. on VLSI Technol.* (1999).
[64] H. Goto, S. Yamaguchi, and C. Jungemann, *Japanese Journal of Applied Physics Part 1-Regular Papers Short Notes & Review Papers* **37,** 5437-5443 (1998).
[65] N. Khalil, J. Faricelli, C. L. Huang, and S. Selberherr, *Journal of Vacuum Science & Technology B* **14,** 224-230 (1996).
[66] Z. K. Lee, M. B. McIlrath, and D. A. Antoniadis, *Ieee Transactions on Electron Devices* **46,** 1640-1649 (1999).
[67] N. Kikuchi, M. Ogawa, and T. Miyoshi, *Solid-State Electronics* **44,** 1663-1668 (2000).
[68] A. Leitao, P. A. Markowich, and J. P. Zubelli, *Inverse Problems* **22,** 1071-1088 (2006).
[69] M. Burger, H. W. Engl, P. A. Markowich, and P. Pietra, *Inverse Problems* **17,** 1765-1795 (2001).
[70] V. I. Bityutskov, *Bunyakovskii inequality* (Kluwer Academic Publishers, 2001).

In: MOSFETs: Properties, Preparations and Performance ISBN: 978-1-60456-762-5
Editors: N.T. Andre et al, pp. 169-215

Chapter 5

THIN AND ULTRA-THIN SiO_2 GATE OXIDE IN METAL-OXIDE-SEMICONDUCTOR STRUCTURES UNDER ELECTRICAL STRESSES : RELIABILITY PREDICTION AND DEGRADATION MECHANISM MODELS

C. Petit and D. Zander

Centre de Recherche en Sciences et Technologies de l'Information et de la Communication, CReSTIC, Université de Reims, BP 1039, 51687 Reims Cedex 2, France

Abstract

With the increasing of integrated circuit density, the transistor size (channel length and width, gate oxide thickness) must decrease drastically. The reduction of the MOS structure size implies a good knowledge of the physical degradation mechanism and the reliability of the devices when the gate oxide thicknesses are lower than 3 nm. The ultra-thin gate oxide reliability is then an essential factor in CMOS technologies. As actually, no alternative and operative solution can replace the Si/SiO_2 technology, that reaches its physical limits, those studies are important so that the CMOS device development is even interesting. It is then necessary to improve our knowledge on the degradation mechanisms which occur at operating voltage. When the oxide thickness becomes smaller than 3 nm, the gate current becomes measurable when the substrate is in depletion. This current increases with the decrease of the oxide thickness. After electrical stresses, this gate current, called LVSILC (Low Voltage Stress Induced Leakage Current) increases. It does not affect the device and circuit functionality because it is observed in depletion or weak inversion regimes and not at usual operating voltages. The low voltage gate current in ultra-thin oxide metal–oxide–semiconductor devices is very sensitive to electrical stresses, it can be used to monitor the oxide degradation, to predict the charge to breakdown and then to predict the lifetime of the device. For various oxide thicknesses ranging from 1.2 to 4 nm and for P and N-MOSFETs, we have realized an experimental database on the LVSILC evolution, as a function of different stress conditions (voltages, temperatures, etc…). This large database provides a coherent understanding for describing the reliability of ultra-thin SiO_2 gate oxide under a constant voltage tunneling stress for positive and negative bias voltages. We have shown that the voltage dependence of the time to breakdown follows a power law behavior rather than an exponential law. From the LVSILC measurements, we have shown that it can be possible to deduce the time to breakdown as a function of the stress voltage, to predict the device lifetime and to project reliability limits.

Key words: P and NMOSFET, Ultra-thin gate oxide, Tunnel Current, Gate current degradation (SILC and LVSILC), Interface states, Oxide traps, Stress, Time to breakdown (t_{BD}), Reliability.

I. Introduction

Over the past thirty years, the development of the microelectronics industry, in both analogue and digital domain owes its success to the SiO_2. The Internet, the telecommunication revolutions and the mobile phone developments require large integrated circuits. CMOS devices have been scaled down aggressively, which has allowed a higher integration density and performance. All dimensions of the transistors are reduced. Metal-oxide-semiconductor field effect transistor (MOSFET) technology requires gate oxide thicknesses smaller and smaller (2.5 to 1 nm). International Technology Roadmap for Semiconductors (ITRS) forecasts for the Low Standby Power (LSTP) logic, an equivalent oxide thickness (EOT) of 1.6 nm in 2008 with gate length channel of MOSFET of 38 nm for a 65 nm technology [1]. Fig. 1 shows the trend forecasted by ITRS-2006 for the fifteen years for technology, gate length and oxide thickness. As it is seen in this figure, SiO_2 thickness continues to decrease. Nevertheless, for high performance and low operating power (LOP) logic, the projected implementation of high-k gate dielectric and metal gate electrode is delayed from 2008 (as forecasted in the 2005 ITRS) until 2010. The reason for these delays is that it now seems unlikely that the integrated circuit industry will find it feasible to deploy these innovations as early as 2008. However, the sizeable decrease of the dimensions of the transistors has as a consequence a drastic increase of the leakage current. Fig. 2 shows the gate current density characteristics (J_G) as a function of the gate voltage (V_G) for different oxide thicknesses (t_{ox}) ranging from 2.5 to 1.2 nm. In Fig. 3, we have plotted the current density measured at $V_G = 1$ V, corresponding to the operating voltage of the devices with these oxide thicknesses, as a function of the oxide thickness, extracted from Fig. 2. We can observe that current density increases exponentially when the oxide thickness decreases. The leakage current increases one order of magnitude for each 0.2 nm thickness decrease in SiO_2 thickness. Therefore, from these observations, two immediate consequences can be raised: fistly, the power consumption, which may affect the device, requires a maximum tolerable gate leakage current density which limits t_{ox}. Secondly, this gate leakage current involves reliability problems and can lead to oxide breakdown. If some articles [2, 3] have indicated that intrinsic gate oxide reliability has probably not been a real issue for thicker oxides ($t_{ox} > 3$ nm), at the operating voltages employed, on the other hand the gate oxide lifetime is definitively a crucial issue for thinner oxides ($t_{ox} < 2$ nm) [4]. The oxide reliability is characterized by the charge or the time to breakdown at its operating voltage, and these quantities are reached by extrapolation from oxide lifetime data obtained under accelerated stress conditions (high voltage and/or high temperature).

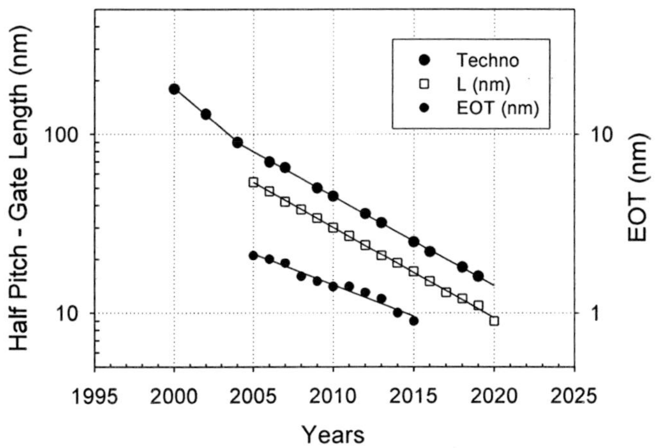

Figure 1. Decrease in gate length, gate equivalent oxide thickness as a function of the expected year of implementation (adapted from ITRS [1]).

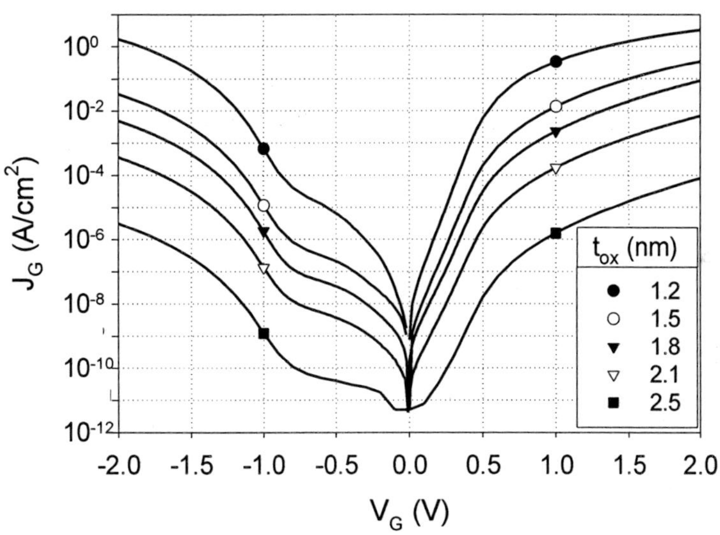

Figure 2. Gate current density characteristics (J_G) as a function of the gate voltage (V_G) for different oxide thicknesses (t_{ox}) ranging from 2.5 to 1.2 nm.

In thin oxides (3–7 nm in thickness), when a high field electrical stress is applied to a MOS capacitor, an increase of the low field leakage current is measured, as it is seen in Fig. 4. This current referred to as SILC (Stress Induced Leakage Current) was first observed by Maserjian and Zamani [5] in 1982 and its characteristics have been studied extensively since 1988 [6-9]. In these thin oxides, it is well known that electrical stress creates bulk oxide neutral electron traps and/or interface traps and some positive charge [10-13]. In MOSFETs, stress-created interface traps cause the degradation of the transconductance g_m and of the

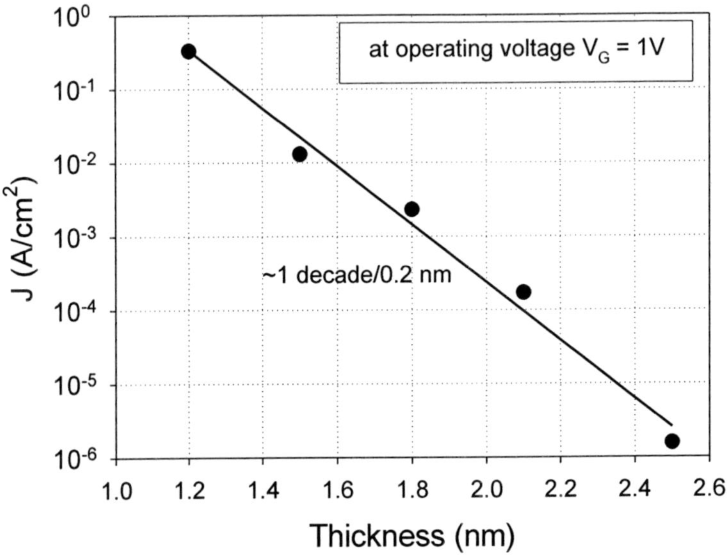

Figure 3. Gate leakage current density measured at $V_G = 1$ V, corresponding to the operating voltage, as a function of the oxide thickness. The leakage current increases one order of magnitude for every 0.2 nm decrease in SiO_2 thickness.

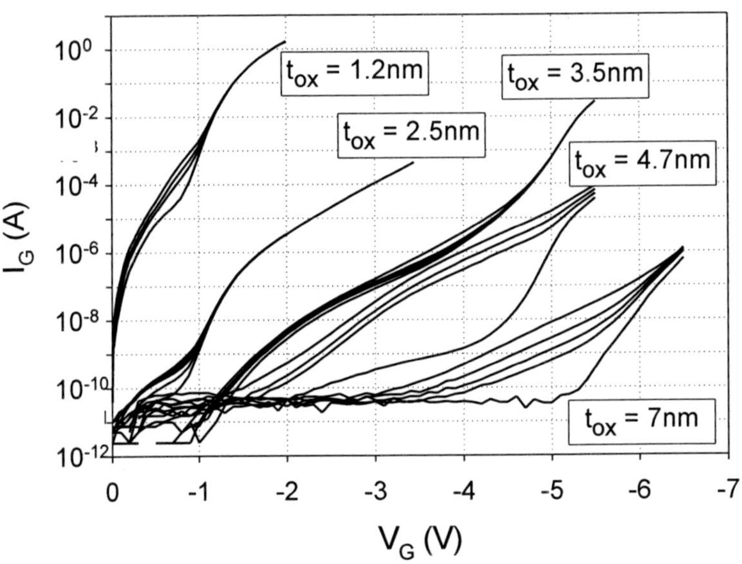

Figure 4. Typical I_G-V_G curves of NMOSFETs in depletion ($-1 < V_G < 0$) and accumulation ($V_G < V_{FB}$) regime, before and after negative stress voltages for various oxide thicknesses. For gate-oxide thicknesses above to 3 nm, no gate current can be measured in depletion regime before and after stress. For these gate-oxide thicknesses, SILC is the current observed after stress below V_{FB}. For gate-oxide thicknesses below 3 nm, a gate current can be measured in depletion regime before stress and increases with the stress (LVSILC).

threshold voltage (V_{TH}), and neutral electron traps induce a gate current increase called stress induced leakage current (SILC). One or both of these defects are at the origin of the soft or hard oxide breakdown. More recently, however, it has been shown that breakdown occurs, in thin oxides, when the number of bulk neutral traps reaches a critical value [14-17]. Since the density of neutral traps reaches a critical value at breakdown, it is obvious that the SILC will also reach a critical value at breakdown and can therefore be used as a degradation monitor [8]. To characterize the oxide degradation, a SILC generation probability P_g has been introduced by DiMaria et al [8]. This probability is defined by $P_G = \Delta J / J_0 \Delta Q_{inj}$, where ΔJ is the SILC and is equal to $(J - J_0)$; J_0 and J are the gate current before and after stress, respectively, and ΔQ_{inj} is the injected charge during Fowler-Nordheim or direct tunnelling (FN/DT) injection regime.

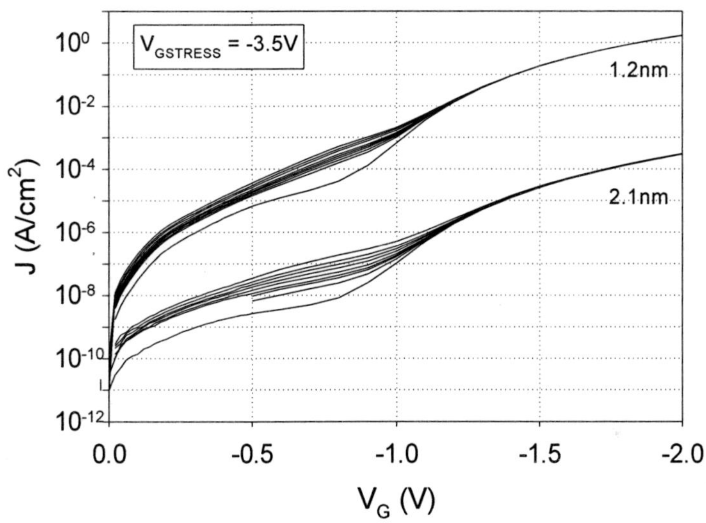

Figure 5. I_G-V_G curves of ultra-thin gate oxide NMOSFETs before and after a stress at $V_{GSTRESS} = -3.5$ V for two oxide thicknesses 1.2 and 2.1 nm.

In ultra-thin oxides (t_{ox} = 2.5 to 1 nm), the gate current is mainly a direct tunneling current which increases drastically as the oxide thickness decreases. Therefore, as it is seen in Fig. 4, contrary to what we have observed in thicker oxides, the traditional SILC is hardly detectable and ceases to be a sensitive tool to monitor the oxide degradation, and to predict the oxides behavior under low stress voltage. As we have seen in Fig.2 and 3, when t_{ox} decreases, J_0 increases exponentially and $\Delta J/J_0$ decreases. As an example, for 2.5 nm gate-oxide, $\Delta J/J_0$ is approximately equal to 4% just before soft or hard oxide breakdown. Nevertheless, when the oxide thickness becomes smaller than 2.5 nm, in the range voltage corresponding to the depletion regime, a gate current becomes measurable before stress and a large increase of this current is observed after stress, as it is seen in Fig. 5 [18]. When the substrate is heavily doped, as is required in ultrathin oxide MOS devices, the low voltage gate current (voltages ranging from 0 V to the flat band voltage (V_{FB})) cannot be explained, at room temperature, by a direct tunneling current from the cathode conduction band to the anode conduction band. To explain this current, two models have been proposed. In the first

[19], a physical interpretation of the LVSILC is based on a trap-assisted tunneling through positively-charged oxide traps (recombination- and trap-assisted tunneling (RTAT) model). In the second [20], the trap assisted tunneling (TAT) current is attributed to gate electron tunneling from the cathode into the anode interface traps. Usually, oxide degradation is quantified by the normalized SILC ($\Delta J/J_0$), because it is proportional to the trap density. In the first model, ($\Delta J/J_0$) is proportional to the relative increase of the traps N ($\Delta N/N_0$), in the second model, ($\Delta J/J_0$) is proportional to the relative increase of the interface states density D_{it}.

The increased current, which arises at low voltages ranging from 0 V to the flat band voltage (V_{FB}), called low voltage stress induced leakage current (LVSILC), is an important reliability problem. Previous studies on NMOSFETs, with oxide thickness ranging from 1.2 nm to 2.5nm, have shown that a large LVSILC increase is always observed before the oxide breakdown [21-23]. Therefore LVSILC can be used to monitor the oxide reliability.

Oxide reliability is characterized by the charge or the time to breakdown at the operating voltage. These quantities are reached by extrapolation from oxide lifetime data obtained under accelerated stress conditions (high voltage and/or high temperature). However, at operating voltage, the direct determination of oxide lifetime requires too long time. This is the reason that the oxide lifetime is deduced from the statistical breakdown studies at large voltages. Can we extrapolate oxide lifetime data from accelerated stress conditions (high voltage, high temperature) under which data can be collected within a reasonable time, to lower operating voltage, at which the failure rate should be very low ? In order to improve extrapolation accuracy, it is important to investigate LVSILC with low stress voltages (to the operating voltage). For different stress conditions, previous studies on NMOSFETs with oxide thicknesses ranging from 2.5 nm to 1.2 nm, have shown two behaviors :

- on the one hand, stress created interface defects (measured by the charge pumping method after Carrier Hot Chanel (CHC) injections) induce the transistor parameter degradation but not the oxide breakdown [22].
- on the other hand, a large LVSILC increase is always observed before the oxide breakdown after CVS injections [21, 22].

This last behavior shows that a relation seems to exist between LVSILC and oxide breakdown.

The aim of this article is to present a thorough study of the SILC and LVSILC variations for different stress conditions, when t_{ox} ranges from 1.2 to 5 nm. We have realized an experimental database on the SILC evolution, as a function of different stress conditions (voltages, temperatures, etc…). We will see the effects on LVSILC at very low stress voltage. From only these results and without statistical oxide breakdown study (SOBD), we introduce a method to deduce the oxide lifetime (t_{BD}).

Oxide traps and defects are known to play a key role in device degradation [24, 25]. Finally, in the latter part of this work, we will present a physical technique of characterization, the inelastic electron tunneling spectroscopy (IETS). Besides the fact that IETS is sensitive tool to detect the interaction between the tunneling electrons and the energy loss modes caused by lattice vibrations (phonons), [26, 27] IETS can also detect and identify impurities as well as defects in the gate dielectrics [28-30] and so to allow the detection and identification of contaminants in the crucial gate oxide elaborations. On the other hand, some authors have shown that it is possible to detect traps [31]. From experimental results, we will demonstrate that the IETS technique, well-used in MIM devices, may be an interesting diagnostic method for the analysis of today's MOS devices with ultra-thin oxides (< 3nm).

II. Device Characteristics and Experimental Procedures

In this study, for the gate oxide thicknesses of 4.6 nm, we have used conventional n^+ polysilicon-gate MOS capacitors, made by CEA-LETI[1], with areas of $3.82x10^{-4}$ cm^2. They were fabricated on P-type (100) Si substrate with a doping concentration of about $10^{17}cm^{-3}$. The sample were characterized by the presence of a n^+ zone in the substrate that behaved as an electron reservoir, when it was connected to the substrate in the inversion regime, in order to felicitate the formation of the inversion layer.

For the ultra-thin gate oxide thicknesses of 1.2 to 3.5 nm, we have used device made by CEA-LETI. The n^+ poly-Si gate NMOSFETs (or capacitors) and p^+ poly-Si gate PMOSFETs (or capacitors) were formed on 7 to 10 Ωcm P-type Si substrate, with N-well implant for PMOS and P-well implant for NMOS using a process compatible with most advanced CMOS technology: Localized Oxidized Silicon (LOCOS) isolation, B-implants to define the active single well and to adjust the threshold voltage, V_{TH}, rapid thermal oxide-type SiO_2 gate oxide, Lightly Doped Drain (LDD) process with As-implanted (NMOS) and BF_2-implanted (PMOS) source and drain, pocket implant, standard Rapid Thermal Annealing (RTA), post metallization annealing and passivation steps. This process was designed to be compatible with a 50 nm CMOS technology. Ultra-thin gate-oxides were grown using Rapid Thermal Oxidation (RTO). The doping of the polycrystalline silicon gate is close to $5x10^{19}$ cm^{-3}, and the doping of the substrate varies from $1.7x10^{17}$ to 10^{18} cm^{-3}.

Two modes of stressing were used in this study: the samples were either stressed at or above room temperature by DT or FN electron injections, or by CHC injections at maximum substrate current. In small area devices, the gate current is too small to be measurable at such low voltage, so LVSILC measurements have been done on relatively large devices.

Large structures, capacitors or P and NMOSFETs with drain and source grounded, with areas of $4x10^{-4}$ or $1x10^{-4}$ cm^2 (W/L = 250x160µm and 100x100µm) were used for uniform injections (DT or FN). For CHC measurements, the channel lengths were equal to 1 µm for a constant channel width of 50 µm. The maximum substrate current for a given source–drain voltage (V_{DS}) was measured on each sample before stressing as a function of the gate-source voltage (V_{GS}). Constant Voltage Stresses (CVS) were performed at various temperatures, but LVSILC measurements are always carried out at room temperature.

A micromanipulator probe station, an Agilent 5270B semiconductor parameter analyser, a picoampermeter HP 4140B, and a HP 4284A LCR meter are used to perform sense measurements and to stress the device under different conditions.

The stress gate voltage and stress temperature are noted, respectively, $V_{GSTRESS}$ and T_{STRESS}, the sense gate voltage is V_G^{sense} and the injected charge is defined by Q_{inj}.

[1] CEA-LETI : Laboratoire d'Electronique et de Technologie de l'Information, Grenoble FRANCE

III. Experimental Results

When a voltage is applied across the gate oxide, a measurable tunneling current will flow if the gate voltage is high enough and/or the oxide is thin enough. For thick oxides at fields above about 7 MV/cm the current is controlled by Fowler-Nordheim tunneling through the triangular barrier, while for thin oxides (t_{ox} < 3 nm) at voltages below about 3 V the current is due to direct mechanical tunneling. The injection mechanisms, the characteristics of the wear out show changes with stress voltage/oxide thickness. For these reasons, it seemed more appropriate to deal separately the experimental results and the reliability of thick and thin oxides. In this paper we will see, from experimental results, the evolution of the gate leakage current degradation in the thick and thin oxide of MOSFETs under constant voltage stresses.

In Section A, we will show the low voltage effects on SILC created in thick oxides (superior to 2.5 nm) by positive and negative stress voltage. In Section B, we will show the effects of electrical stress voltages at low and very low voltage measurements in ultra-thin oxides (inferior to 2.5 nm) and how the LVSILC can be used to predict the reliability of N or PMOSFETs. We will show also, the effects of the temperature annealing on defects creation. In section C, we will see what can make IETS technique in the study and characterization of ultra-thin oxides.

A. Degradation in Thick Oxides (t_{ox} > 2.5 nm)

A typical example of the observed behavior during the stress is shown in Fig. 6, where we can distinguish the initial pre-stress curve exhibiting quantum oscillations which are studied by different authors [32, 33], the progressive increase with the stress in the pre-tunneling gate current I_G, called SILC, a sudden increase (between two and five orders of magnitude) over

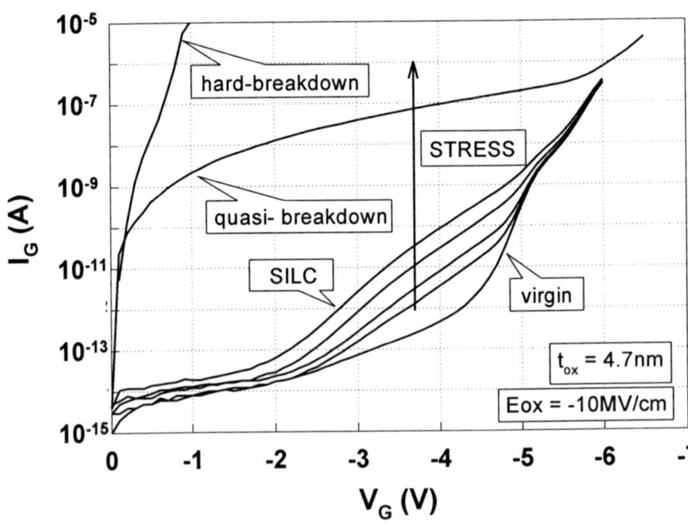

Figure 6. Typical behavior of the gate current as a function of a negative sense voltage showing the successive SILC, quasi-breakdown and breakdown regimes during an uniform Fowler-Nordheim stress.

the whole gate voltage measurement range (quasi-breakdown) and the final classical breakdown failure mechanism after which the oxide is in an ohmic regime, definitely different from the quasi-breakdown regime.

1. SILC Variations as a Function of Stress Conditions and Temperature Annealing in 4.7 nm Gate Oxide

As seen in Fig. 7, as a function of the voltage measurement, the I_G-V_G curve before stress, corresponds to three conduction mechanisms (direct F.N. current, F.N. assisted current and direct tunnel assisted current). Consequently, the relative SILC variation $\Delta J/J_0$, will not have the same meaning in the different parts of the voltage measurement.

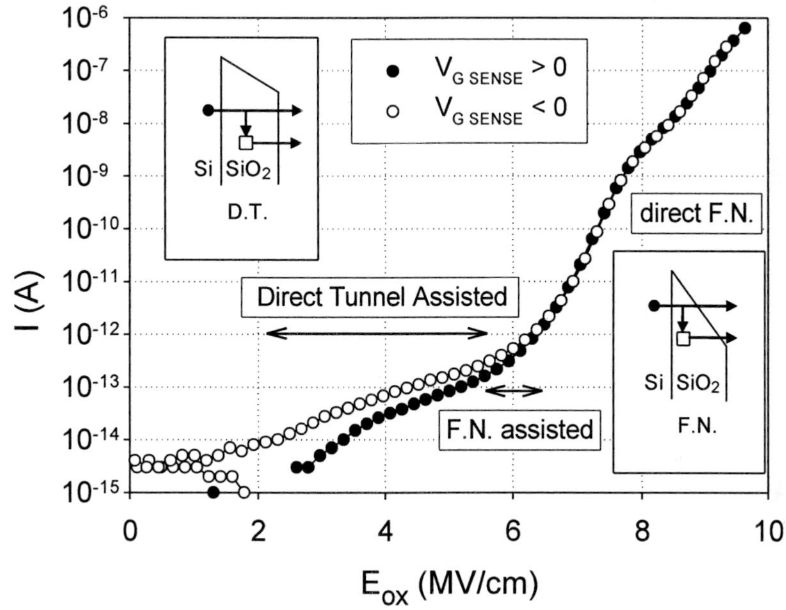

Figure 7. I_G-V_G curve before stress for positive and negative gate sense voltage showing the different electrical conduction mechanisms through the oxide .

We performed stresses in the FN regime, at positive and negative constant voltage of 5.2 V and -6.4 V, corresponding to an oxide electrical field E_{ox} of +10 and -10 MV/cm, respectively. Figs. 8(a) and 9(a) show the typical gate-current I_G versus V_G characteristics, in a fresh device and after stress. The degradation of the leakage current is characterized by the strong SILC increase at low field (for $|E_{ox}|$ between 4 and 8 MV/cm) on I_G-V_G curves after FN stress (corresponding to -3.1 and -4.3 V for negative sense voltages and to 1.9 and 3.1 V for positive sense voltage). We can also observe on these figures no positive charge trapping (no voltage shift at larger fields). Figs. 8(b) and 9(b) show the typical normalized SILC variations $\Delta J/J_0$ versus V_G for the same stress voltages. We observe in these figures that for an electrical field of ± 10 MV, this increase can excess more than two decades. For voltage measurement below -4 V (for negative sense voltages) and greater than +3 V (for positive sense voltages), the $\Delta J/J_0$ decrease is due to the fast gate current increase before stress (I_0) and not to ΔI

decrease. The significantly part of the curve $\Delta J/J_0$-V_G is whose inside -4 to -2 V range, for negative sense measurement and inside 1.5 to 3 V range, for positive sense measurement. As we have seen previously, the signification of $\Delta J/J_0$ depends on the conduction mode before stress. For a direct F.N. conduction, $\Delta J/J_0$ does not allow to measure the number of created traps. For an F.N. assisted or direct tunnel assisted conduction, $\Delta J/J_0$ is proportional to $\Delta N/N_0$.

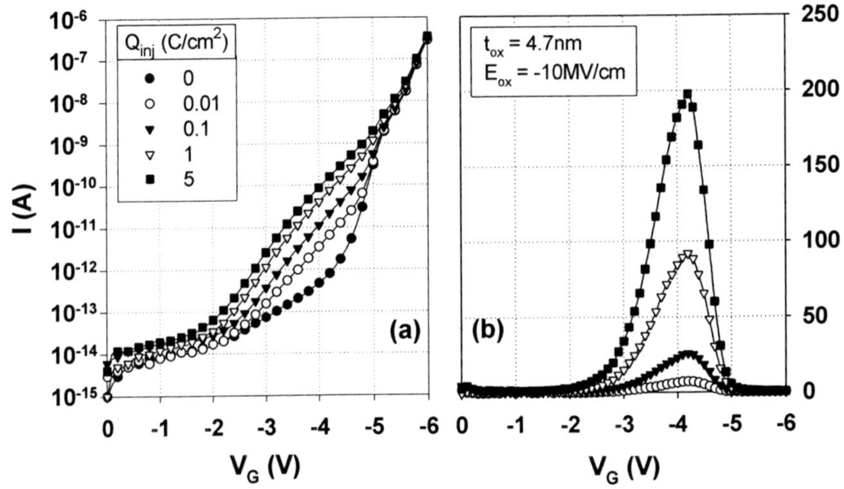

Figure 8. Gate leakage current (a) and normalized SILC variations (b) as a function of the gate voltage before and after a negative constant stress voltage at $V_{GSTRESS}$ = -6.4 V (-10 MV/cm) for various injected charge.

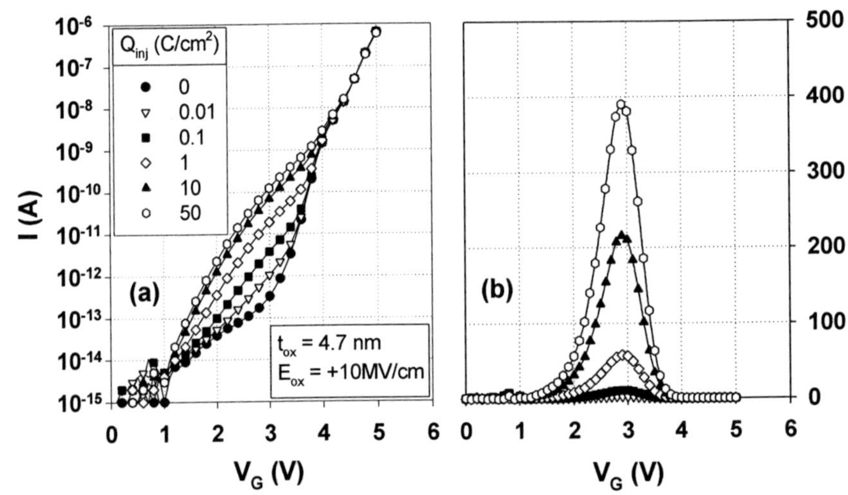

Figure 9. Gate leakage current (a) and normalized SILC variations (b) as a function of the gate voltage before and after a positive constant stress voltage at $V_{GSTRESS}$ = +5.2 V (+10 MV/cm) for various injected charge.

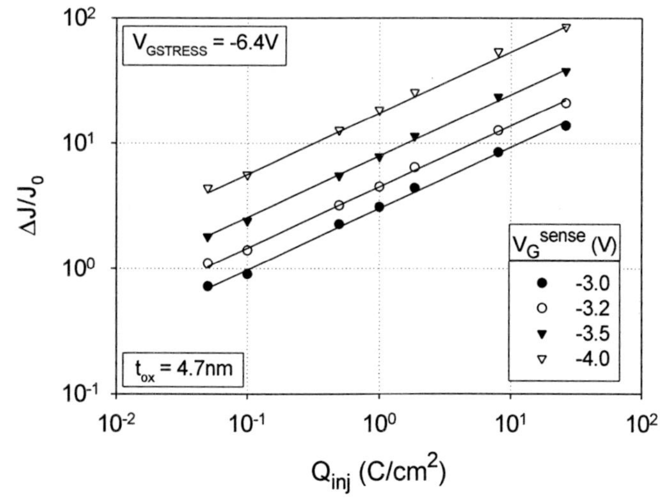

Figure 10. Normalized gate current variations ($\Delta J/J_0$) as a function of the injected charge under $V_{GSTRESS}$ = -6.4 V for various gate sense voltages.

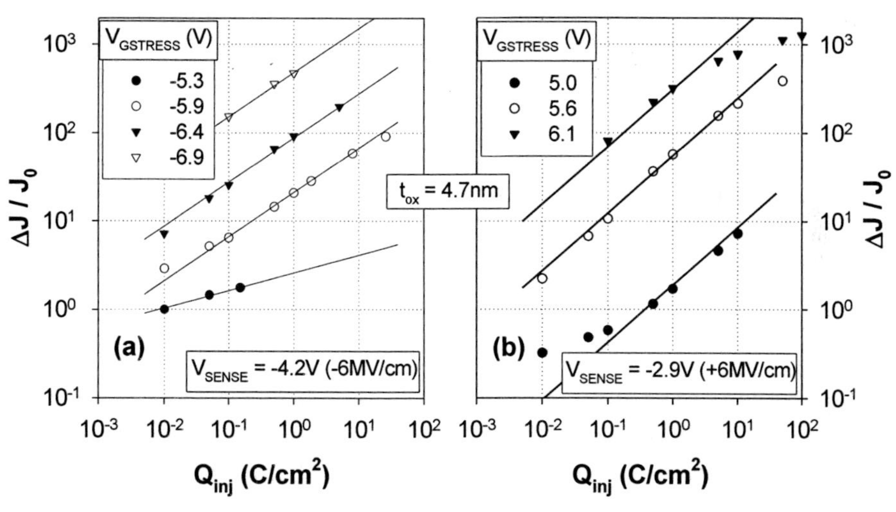

Figure 11. Normalized gate current density variations ($\Delta J/J_0$) as a function of the injected charge, for various stress voltages. (a) For negative stress voltages. (b) For positive gate voltages. V_G^{sense} = -4.2V and the oxide thickness is 4.7 nm.

Fig. 10 shows the SILC creation kinetics as a function of the injected charge at a stress voltage of -6.4 V and for various gate sense voltages. The result does not depend on the measurement gate voltage and $\Delta J/J_0$ increases with the injected charge and follows a same power law. Figs. 11 (a) and (b) present the creation kinetics of FN-induced SILC for different stress voltages. These results were obtained for a measurement gate voltage corresponding to the maximum of the $\Delta J/J_0$ ratio. We can notice a saturation of this ratio at very strong injected charges. Before saturation, degradation mechanisms follow a classical power law $\Delta J/J_0 = A.Q_{inj}^{\alpha}$) with degradation rates α equal to 0.5 for negative stress voltages and 0.65 for positive stress voltages. Theses results are close to those found by Goguenheim et al [34, 35]. We check that α does not depend on the stress gate voltage, excepted for the low stress voltages (due to, perhaps, to the imprecision of measurements). Consequently, it is possible to draw the $(\Delta J/J_0)/Q_{inj}^{\alpha}$ variations as a function of the stress voltage $V_{GSTRESS}$, for both bias stress. We can notice on Fig. 12 showing $(\Delta J/J_0)/Q_{inj}^{\alpha}$ as a function of $V_{GSTRESS}$ that the voltage acceleration is lower for a negative stress than for a positive one. The curve slope is approximately two decades/V when $V_{GSTRESS}$ is positive, and one decade/V when $V_{GSTRESS}$ is negative. The point of the intersection of the two curves is at around |5 V|. At high stress field, the oxide degradation is more important for a positive stress than a negative one, but the opposite must be observed at low stress field. As the operating voltage of the NMOSs with an oxide thickness of 5 nm is around to +2.5 V, this result is very interesting. But, can we extrapolate theses results toward the very low voltage ? This demonstrates the need to carry out measurement at low and very low voltages, but the important thing is to put them into practice with enough precision within a reasonable time. We see on Fig. 12 that, whatever the polarity of the stress voltage, below 5 V, $\Delta J/J_0$ will be very low, so, difficult to detect with precision.

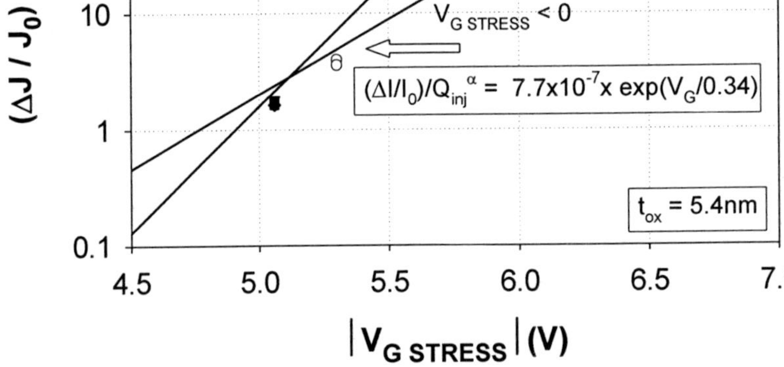

Figure 12. Variations of $(\Delta J/J_0)/Q_{inj}^{\alpha}$ as a function of the stress voltage for various injected charges. Full symbols correspond to positive stresses and open symbols to negative stresses.

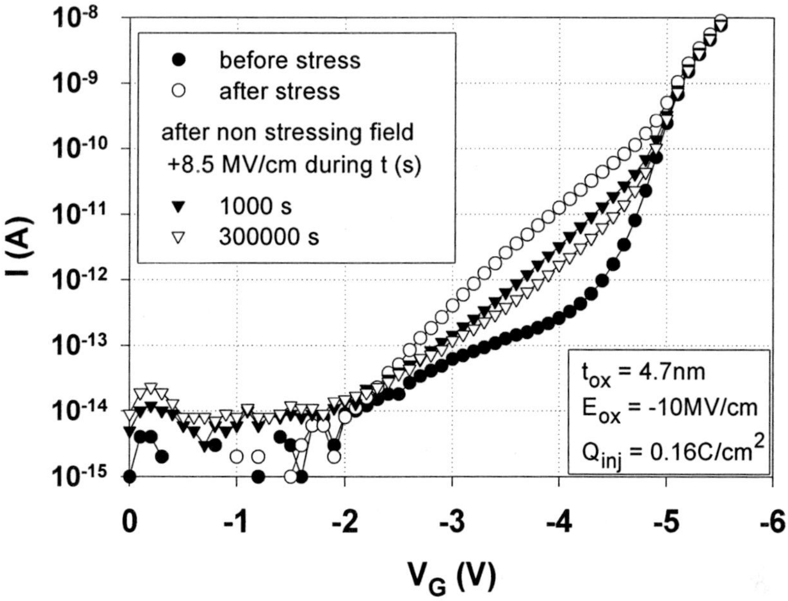

Figure 13. Gate current variation as a function of the gate sense voltage before and after stress and after having biased the gate at $V_G = 4.4$ V (8.5 MV/cm) during 10^3 s and $3x10^5$ s. The stress conditions are $V_{GSTRESS} = -6.4$ V (-10 MV/cm) and $Q_{inj} = 0.16$ C/cm^2.

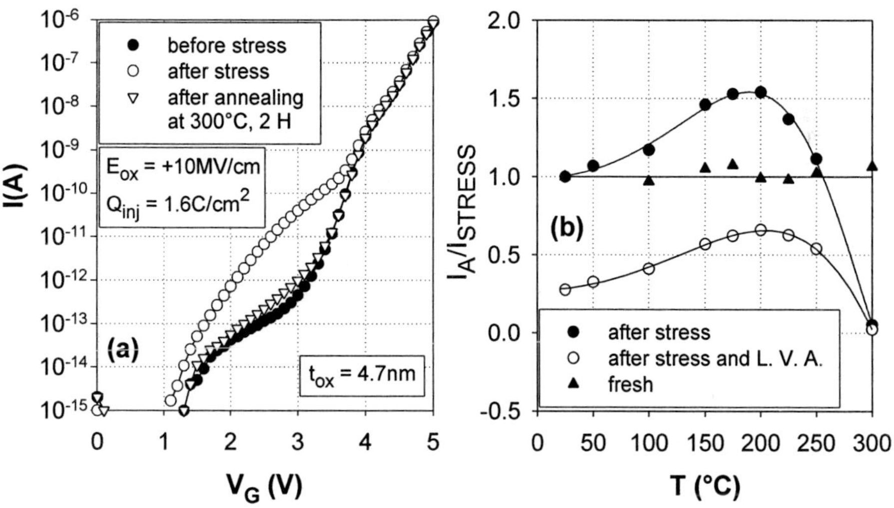

Figure 14. (a) Gate current variation as a function of the gate sense voltage before and after stress and after an annealing temperature at 300°C during 2 hours. The stress conditions are $V_{GSTRESS} = +5.6$ V (+10 MV/cm) and $Q_{inj} = 1.6$ C/cm^2. (b) The SILC variations, before a low voltage bias (2 hours), as a function of the annealing temperature. The stress conditions are $V_{GSTRESS} = +5.6$ V (+10 MV/cm) and $Q_{inj} = 1.6$ C/cm^2. I_{STRESS} is the current after stress, and I_A is the current after temperature annealing.

Nevertheless, the increase of the stress leakage current in the tunnel oxide can caused a read disturb degradation in a Flash EEPROM. Endoh et al. [36] discovered that the degradation of the read disturb life time in a Flash EEPROM memory can be drastically reduced if a low positive pulse is applied after the completion of each write-erase operation. In previous works, we have shown that if a non-stressing field is applied after a stress, the SILC created by a positive or a negative stress decays, and this decay is irreversible as long as the sample undergoes no further stressing [37, 38]. The decay magnitude depends on magnitude and on the duration of application of a non-stressing field. For example, Fig. 13 illustrates the effect of a low positive voltage on I_G-V_G curves after a negative stress.

SILC decay can be also observed after a temperature annealing. Fig. 14 (a) shows the I_G-V_G characteristics before and after stress (E_{ox} = 10 MV/cm, Q_{inj} = 1.6 C/cm^2) and then after a temperature annealing at 300°C during 2 hours. The SILC has strongly decreased. In fact, $\Delta I/I_0$ decrease from 100 (after stress) to 1.3 (after temperature annealing). In order to investigate the temperature effect on the SILC decay, successive annealings were realized on two samples, each anneal cycle lasting 2 hours. On one of the samples, we have realized a low voltage annealing (LVA) (-4.3 V during 2 hours) followed by a temperature annealing, and on the second, we have realized only a temperature annealing. All measurements are performed at room temperature. Fig. 14(b) presents the I_A/I_{STRESS} variations as a function of the annealing temperature, with I_{STRESS}, the current after stress, and I_A, the current after temperature annealing. The gate current of a virgin sample before and after annealing is also indicated in the figure. Whether the sample undergoes a low voltage annealing or not, the I_A/I_{STRESS} variations increase as far as 200°C. There is not temperature effect on the SILC decay before 200°C. The temperature effect on the SILC decay starts at around T = 200°C.

The different experiments related above can give information about the nature of defects responsible of the SILC which is a phenomenon of major importance for the reliability of MOS structures. It is agreed that the SILC conduction mechanism is a tunneling process via stress-induced neutral electron traps (for MOS structures with oxide thickness around 5 nm). Otherwise, we checked that, for stresses at high field, it doesn't have much positive or negative charge, and rules out the possibility of the positive charge as responsible of SILC, in agreement with many authors. For MOS structures with oxide thickness of 4.7 nm, defects are created by positive and negative stresses. These defects are annealed at 300°C, and deactivated when a positive or negative low voltage is applied after stress. We have seen in Fig. 14(b) that I_A/I_{STRESS} increase, both after stress, that after stress and low voltage annealing, with temperature up to 200°C. Therefore, temperature related increase of I_A/I_{STRESS} is not due to the reactivation of deactivated traps, but to the creation of more neutral traps as the temperature increases. After a stress, some traps are not active, but can be activated by thermal effect below 200°C. Above this temperature, the annealing of traps begins.

2. Stress Bias voltage and Temperature Effects on SILC Variations in 3.5 nm Gate Oxide

Before presenting the results obtained on the devices with an oxide thickness of 3.5 nm, we recall an important observation made by Ricco et al. [39] stating that the differences in I_G-V_G characteristics found, before stress, for different values of the oxide thickness, deal with different regions of the same general curve. Ricco et al. showed that experimentally this curve could only be explored within a window limited by the noise level of the measurement apparatus towards the lower currents and by oxide breakdown towards the higher currents. Depending on SILC level and therefore on oxide thickness, different regions fall within such windows, producing the typical experimental results. We have seen in Fig. 8 and 9 on I_G-V_G characteristics in 4.7 nm thick oxides that the curves corresponding to different values of injected charges (or stress times) merge above -2 V for negative sense voltages and below 1 V for positive sense voltages. In this case, it is not possible to see the stress effects at low sense voltages.

Figs. 15 and 16 show the gate current density as a function of the gate voltage characteristics (J_G-V_G) before and after positive and negative stress voltages (normalized SILC variations $\Delta J/J_0$ versus V_G is showed in the insert of theses figures). First, we can remark that, contrary to what we have observed in 4.7 nm thick oxides, a gate leakage current is observed above -2 V for negative sense voltages and below 1 V for positive sense voltages. Secondly, we observe, at low sense voltage, a noticeable difference in the SILC evolution between the two stress polarities. For both stress voltage polarities, at low negative sense voltages, the gate leakage current increase with the stress. On the other hand, at low positive sense voltages, negative stress voltages involve a more important increase of the leakage current than positive stress voltages. This seems to indicate that two kinds of defects, D_1 and D_2, can be envisaged. D_1 defects, created by positive or negative stresses, involve an increase of gate leakage current in the range -2 V to -4.5 V for the negative sense voltages and in range 1.5 V to 3.5 V for the positive sense voltages. D_2 defects, mostly created by negative stresses,

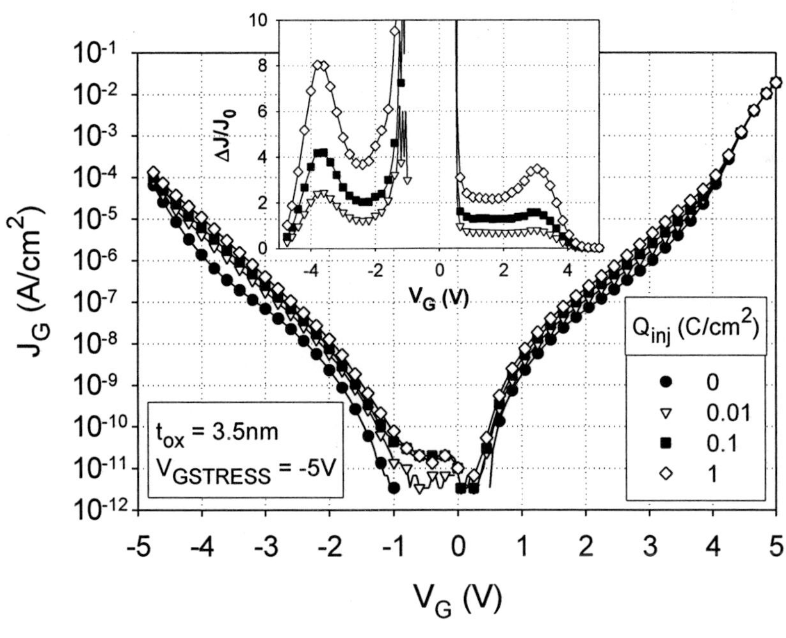

Figure 15. Gate current density variation as a function of the gate voltage (J_G-V_G) before and after a negative constant stress voltage at $V_{GSTRESS}$ = -5 V and for various injected charge. The oxide thickness is 3.5 nm. In the insert is represented the variations of ($\Delta J/J_0$) as a function of the injected charge.

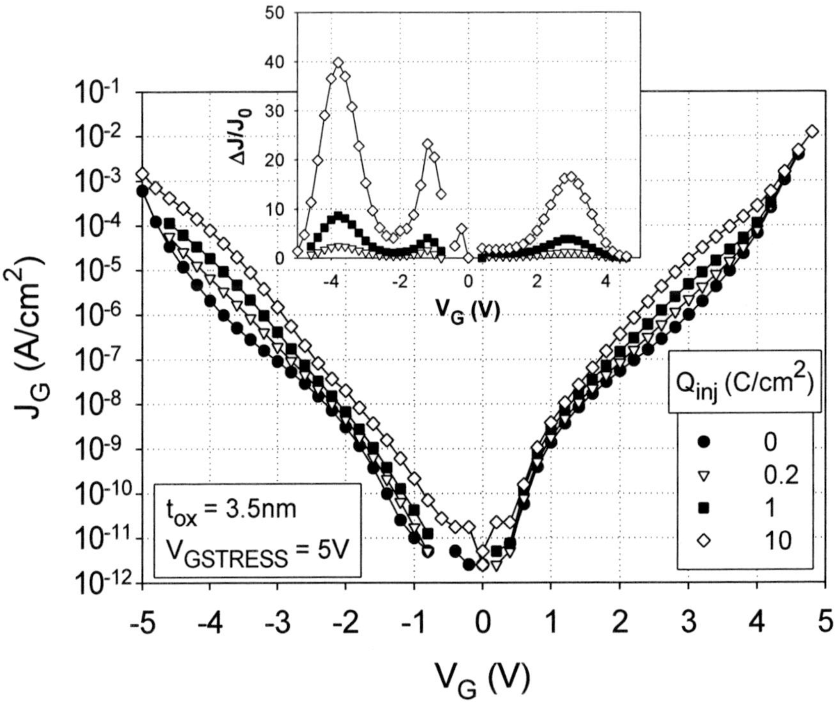

Figure 16. Gate current density variation as a function of the gate voltage (J_G-V_G) before and after a positive constant stress voltage at $V_{GSTRESS} = 5$ V and for various injected charge. The oxide thickness is 3.5 nm. In the insert is represented the variations of ($\Delta J/J_0$) as a function of the injected charge.

involve an increase of the gate leakage current at low voltage. However, after positive stresses, we can observe at low voltage, a slightly increase of the gate leakage current, which is probably due to D_2 defects. At last, at high positive or negative sense voltage, the normalized SILC variations $\Delta J/J_0$ increases with stress and follows a power law with the injected charge.

Temperature annealing experiments were made after stress in ambient atmosphere. All measurements were performed at room temperature. Fig. 17 shows the J_G-V_G characteristics, before and after a negative stress and after 6 h annealing at 300°C. The SILC decay is greater above 3 V than below. Below 2 V, the SILC ceases to decrease. Fig. 18 shows the J_G-V_G characteristics, before and after a positive stress and after 6 h annealing at 300°C. It is seen that when the sensing voltage is superior to 2.5 V, the SILC is fully annealed and that below this value, the annealing effect decreases slightly and ceases to be observed at 1.5 V. This confirms the creation of two kinds of defects after negative stress. D_1 defects are created by positive or negative stresses and are annealed at 300°C. These defects, which we saw previously on the 4.7 nm oxide thicknesses, exist for all oxide thicknesses below 12 nm [40, 41]. D_2 defects are created mainly by negative stresses and are not annealed at 300°C. This component is a steady-state component.

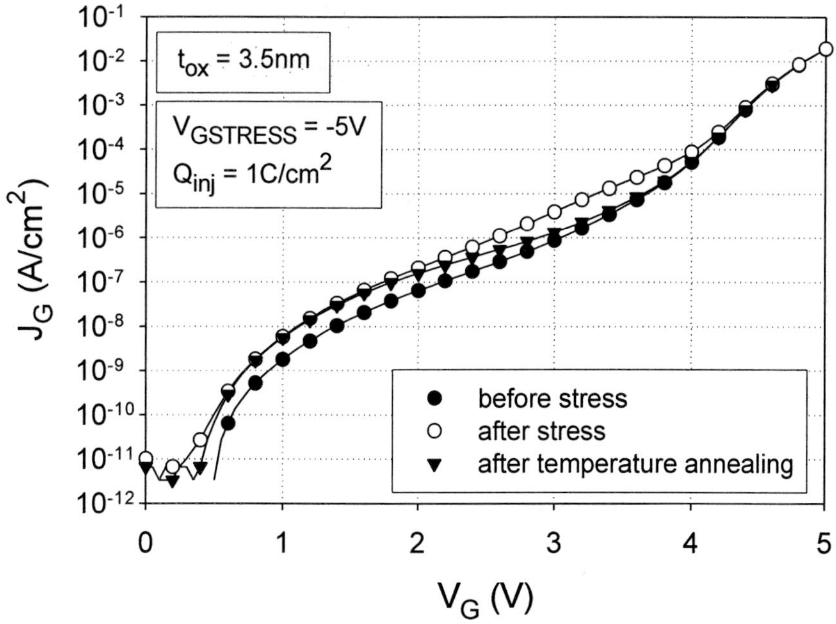

Figure 17. J_G-V_G characteristics before and after a negative stress and after a temperature annealing at 300°C during 4 h. The stress conditions are $V_{GSTRESS}$ = -5 V and Q_{inj} = 1 C/cm^2. The oxide thickness is 3.5 nm.

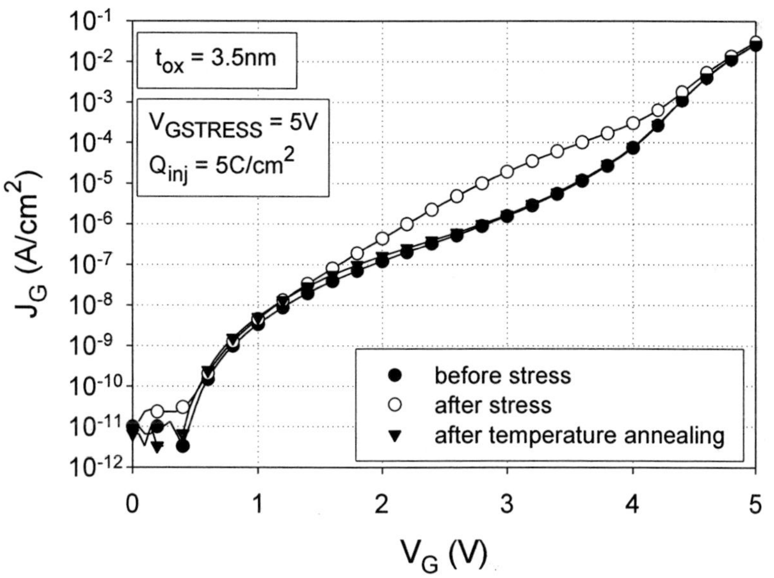

Figure 18. J_G-V_G characteristics before and after a positive stress and after a temperature annealing at 300°C during 4 h. The stress conditions are $V_{GSTRESS}$ = 5 V and Q_{inj} = 5 C/cm^2. The oxide thickness is 3.5 nm.

3. Impact on Device Reliability in 3.5 nm Gate Oxide

Fig. 19(a) shows that the normalized SILC variations $\Delta J/J_0$ increases with stress and follows a power law in Q_{inj}^{α} with an α value equal to 0.7. It is possible to plot the expression defined by $(\Delta J/J_0)/Q_{inj}^{\alpha}$ as a function of the stress voltage $V_{GSTRESS}$ (Fig. 19(b)). From this plot and from a critical value reached by SILC at breakdown, time-to-breakdown at low voltage can be calculated by extrapolation [42, 43]. From this extrapolation, it is possible to obtain an evaluation of the oxide reliability. But, it is know that for the oxides with a thickness of 3-4 nm, the intrinsic gate-oxide reliability is not a real issue at the operating voltages employed. On the other hand, contrary to the 4.7 nm oxide thickness, we saw that a gate leakage current can be measured at low voltages on the 3.5 nm oxide thickness, and after a negative stress voltage, this current at low voltage (so, at operating voltage) increases with the stress time and is not annealed with the temperature. If this gate leakage current (at low voltage) cannot induce the oxide breakdown, it can degrade durably the characteristics and the performances of the devices.

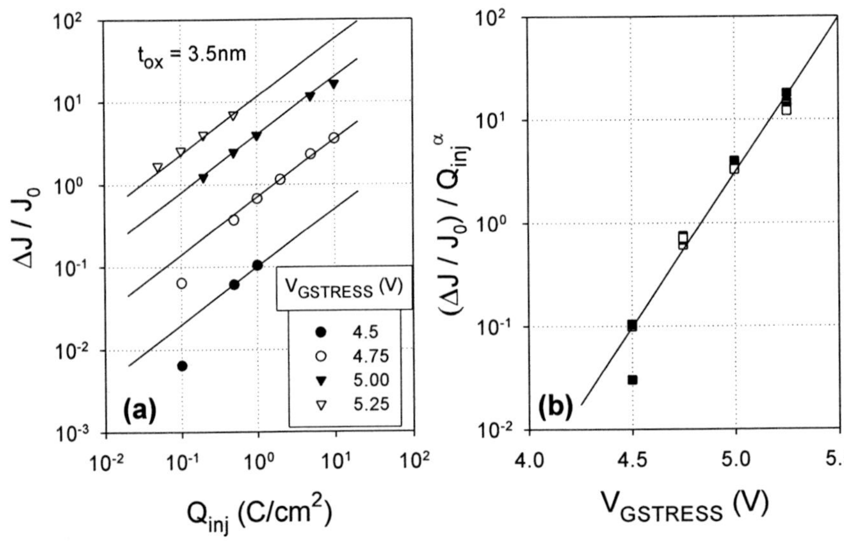

Figure 19. (a) Variations of $(\Delta J/J_0)$ as a function of the injected charge for positive stress voltages. (b) Variations of $(\Delta J/J_0)/Q_{inj}^{\alpha}$ as a function of the stress voltage for various injected charges.

B. Degradation in Ultra-thin Oxides ($t_{ox} < 2.5$ nm)

In the previous presented results, the samples were stressed in FN injection regime. Now, in the thickness oxides ranging from 1.2 to 2.3 nm, Direct Tunnel (DT) mechanism is observed. The electrical stresses are performed in FN regime at high voltage and in DT regime at low voltage.

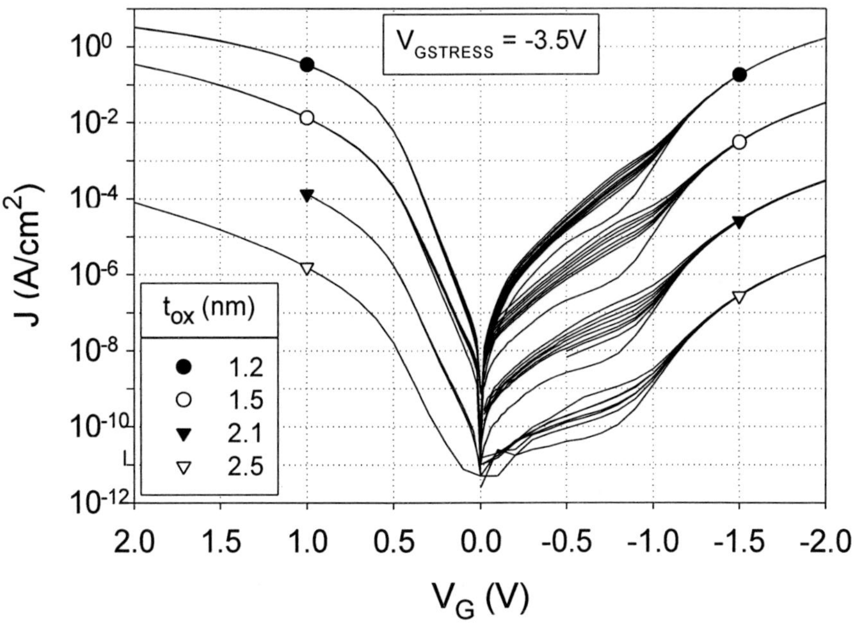

Figure 20. J_G-V_G characteristics of NMOSFETs before and after a constant stress voltage at $V_{GSTRESS}$ = -3.5 V, for various oxide thicknesses.

After stress, for oxide thicknesses smaller than 2.5 nm, a large increase of the gate current is observed in the range voltage corresponding to the depletion regime (0.5 to -1.2 V, for NMOSFETs). In this range voltage, we have seen previously that the gate current is not measurable for sample with an oxide thickness higher than 3.5 nm (the gate current is below the sensibility of the picoamperemeter). Fig. 20 shows the typical J_G-V_G characteristics, for a constant stress voltage (CVS) ($V_{GSTRESS}$ = -3.5 V), for various injected charges and for different gate oxide thicknesses ranging from 1.2 to 2.5 nm.

1. LVSILC Variations as a Function of the Stress Voltage

Figs. 21 show the typical relative LVSILC variations $\Delta J/J_0$ as a function of the gate voltage, after a constant voltage stress for P and NMOSFETs of different oxide thicknesses. Similar curves have always been observed for oxide thicknesses ranging from 1.2 to 2.5 nm in N and PMOSFETs [21, 43, 44]. It is seen that the curves have two maxima, one at gate voltage $V_G^{sense} \approx 0$ V and a second at gate voltage around $V_G^{sense} = 0.8$ V in PMOSFETs and $V_G^{sense} =$ -0.8 V in NMOFETs. The figure shows also that SILC is negligible in the voltage range usually used for the calculation of P_G (> 1.5V and <-1.5 V). If we compare the LVSILC variations, for the NMOSFET of Fig. 21 (at $V_G^{sense} = -0.8$ V, $\Delta J/J_0 \sim 5.4$) and the "conventional" SILC variations (at $V_G^{sense} = -2$ V, $\Delta J/J_0 \sim 2.10^{-3}$), the increase current is a

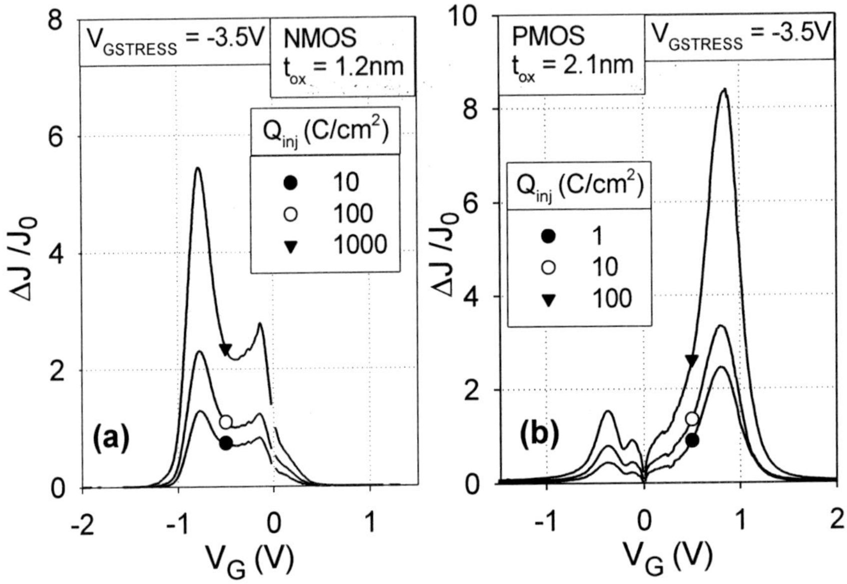

Figure 21. Normalized LVSILC variation ($\Delta J/J_0$) as a function of the gate voltage. (a) for a NMOSFET with an oxide thickness of 1.2 nm and (b) for a PMOSFET with an oxide thickness of 2.1 nm.

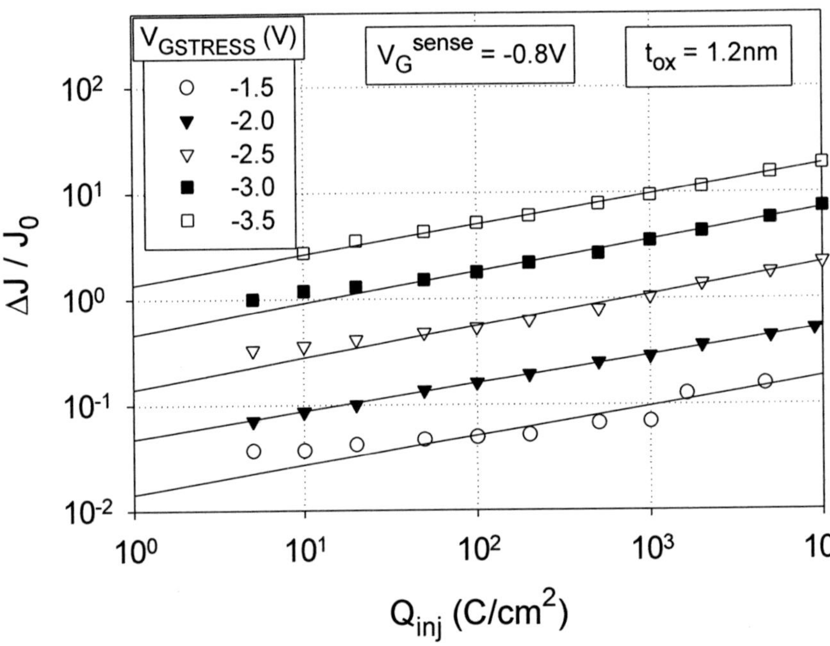

Figure 22. Variations of ($\Delta J/J_0$) as a function of the injected charge, for various stress voltages. $V_G^{sense} =$ -0.8V and the oxide thickness is 1.2 nm.

factor 2700 larger for the LVSILC than that for the conventional SILC. In a previous work [46], we have showed that the magnitude of these two peaks, one compared to the other, depends strongly on the stress voltage magnitude. The first maximum (at $V_G^{sense} = 0$ V) is larger than the second (at $V_G^{sense} = -0.8$ V in NMOFETs) at low stress voltage and smaller at higher stress voltage. For all oxide thicknesses and for both stress polarities, LVSILC increases with Q_{inj}^{α}, with α equal to ~0.3 for the negative stresses and ~0.45 for the positive ones. As it is seen in Fig. 22, α varies little with $V_{GSTRESS}$ and it is possible to represent the variations of $(\Delta I/I_0)/Q_{inj}^{\alpha}$ with $|V_{GSTRESS}|$ for both stress polarities. In the Figs. 23(a) and (b) these curves have been plotted for t_{ox} equals to 1.2 and to 2.3 nm. The measurements were realized at the second LVSILC maximum. α, for each curve, is the average value obtained from our experimental data for either $V_{GSTRESS} > 0$ or for $V_{GSTRESS} < 0$. We can observe that $(\Delta I/I_0)/Q_{inj}^{\alpha}$ expression increases exponentially with $V_{GSTRESS}$:

$$(\Delta J/J_0)/Q_{inj}^{\alpha} = A\exp(B.V_{GSTRESS}) \tag{1}$$

where A and B are parameters deduce from the curve. It should be noted, first, that the value of this last expression depends little on oxide thickness; second, that the stressing voltage threshold is smaller for a negative stress than for a positive one; and third, that after a positive stress, the curve slope is steeper than after a negative one (approximately two decades/V when $V_{GSTRESS}$ is positive and one decade/V when $V_{GSTRESS}$ is negative).

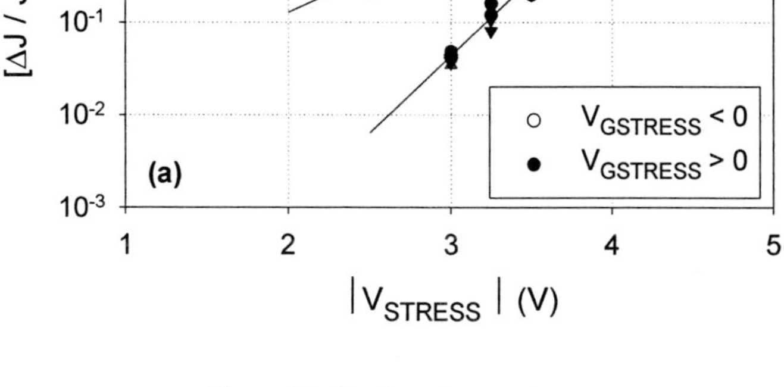

Figure 23. Continued on next page.

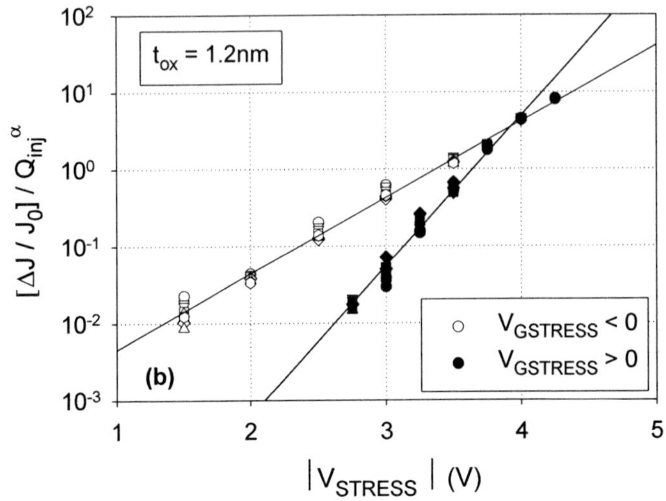

Figure 23. Variations of $(\Delta I/I_0)/Q_{inj}^{\alpha}$ as a function of the stress voltage for various injected charges measured at V_G^{sense} = -0.8 V. Full symbols correspond to positive stresses and open symbols to negative stresses. The oxide thicknesses are 1.2 nm for the upper curve (a) and 2.3 nm for the lower curve (b).

The oxide reliability is characterized by the charge or the time to breakdown at its operating voltage, and these quantities are reached by extrapolation from oxide lifetime data obtained under accelerated stress conditions (high voltage and/or high temperature). At operating voltage, the direct determination of oxide lifetime requires too long time. As we have seen previously, at room temperature, for devices with a gate oxide thickness of 2.3 nm, it is very difficult to observe degradations for stress voltages below 2.5 V. For the purpose of accelerating the degradation process, we have performed constant stress voltages at different temperature ranging from 50 to 250 °C (323 to 523 K). A high operating temperature can have two effects, either increase the stress effects or have an annealing effect. Each J_G-V_G characteristic is performed at room temperature.

2. LVSILC Variations as a Function of Stress Temperature

The first effect of the stress temperature on a device is the increase of the leakage current, as it is seen on Fig. 24, where we have plotted the normalized LVSILC versus V_G, after a CVS at different stress temperatures, for an injected charge and a stress voltage equal to 1000 C/cm^2 and -3 V, respectively. Electrical stresses, at high temperature induce a larger LVSILC than at room temperature. Whichever the stress temperature, the two maxima are at the same sense voltage. In Fig. 25, we have represented the LVSILC variations $\Delta J/J_0$ as a function of the injected charge, for two stress voltage (-1.5 V and -3.5 V) at different stress temperatures T_{STRESS} and for a gate oxide thickness of 2.3 nm. We can observe that $\Delta J/J_0$ follows always a

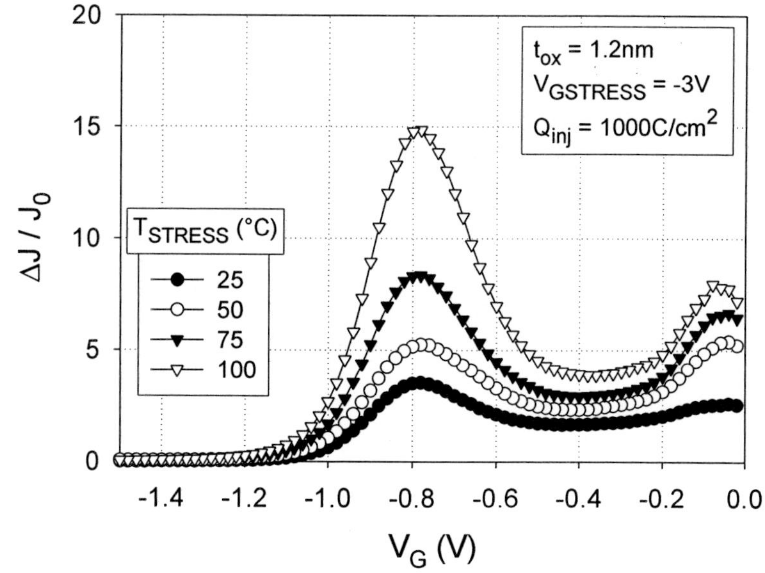

Figure 24. Normalized LVSILC variations ($\Delta J/J_0$) as a function of the gate voltage V_G after a negative stress for $V_{GSTRESS}$ = -3 V and Q_{inj} = 1000 C/cm^2 and different stress temperatures.

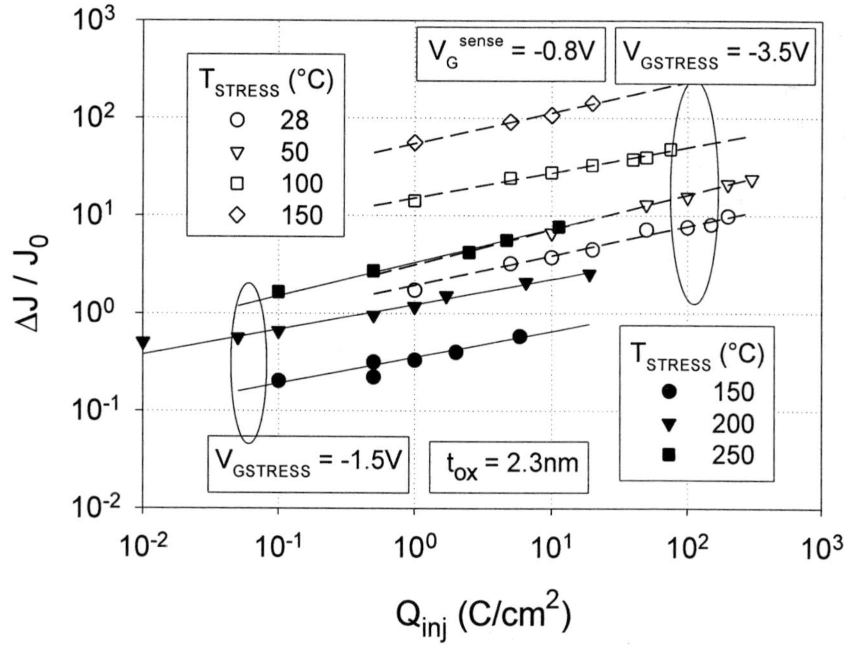

Figure 25. Variation of $\Delta J/J_0$ as a function of the injected charge, for two stress voltage (-1.5V and -3.5V) at different stress temperatures and for a gate oxide thickness of 2.3 nm.

power low as a function of the injected charge Q_{inj}^{α}, , whichever the stress temperature, with a factor α (equals to 0.30 ±0.05) varying little with the stress temperature. We can note that a stress voltage as low as -1.5 V creates defects. So, as it has seen in the lower curve in Fig. 26, it is possible to normalize these curves as a function of the injected charge and the stress temperature, by drawing the expression $\left[(\Delta J/J_0)/Q_{inj}^{\alpha}\right]/\exp(T_{STRESS}/T_0)$ as a function of the stress voltage, where T_0 is a constant. On this curve, drawn with all experimental results that we have obtained, we find the same slope of one decade/V, which we have seen previously. From these results displayed in Fig. 26, for stress voltage at -1.5 V at room temperature, degradations are absolutely conceivable. It is possible to extrapolate these results to operating voltages and so to determine the time-to-breakdown and consequently the oxide reliability. The upper curve of Fig. 26 represents the same expression as a function the stress voltage, but for a gate oxide thickness of 1.2 nm. These two curves show that whichever t_{ox}, after negative stresses at different stress temperatures, the LVSILC increases exponentially by approximately one decade/V.

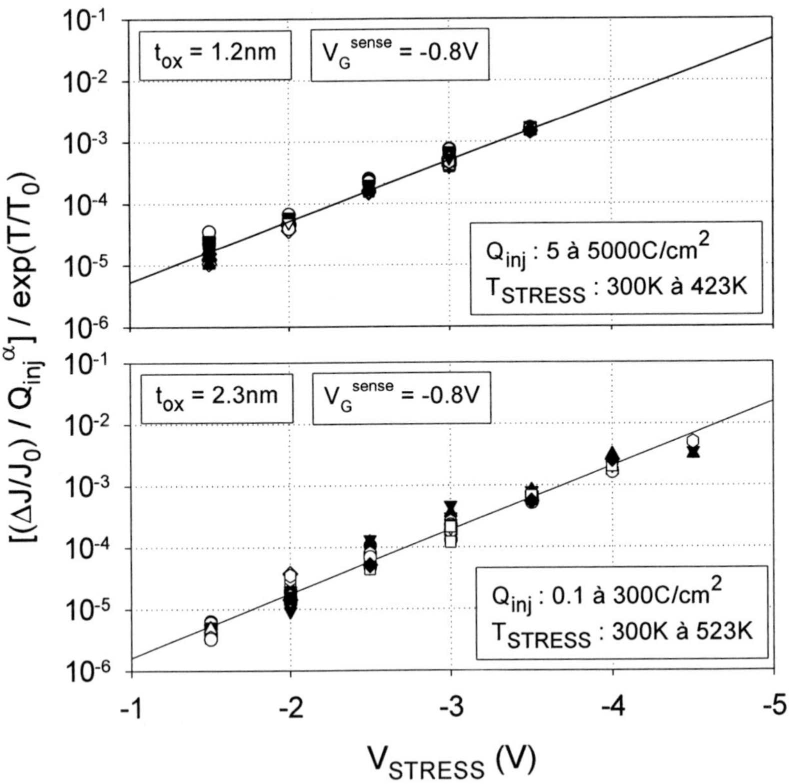

Figure 26. Variation of $\left[(\Delta J/J_0)/Q_{inj}^{\alpha}\right]/\exp(T_{STRESS}/T_0)$ as a function the stress voltage, for different stress temperatures, the upper curve for t_{ox} = 1.2 nm and the lower curve for t_{ox} 2.3 nm.

In the same way, we have drawn in the lower curve in Fig. 27 the evolution of the leakage current degradation, $\left[(\Delta J/J_0)/Q_{inj}^{\alpha}\right]/\exp(V_{GSTRESS}/V_0)$ as a function of the stress temperature, where V_0 is a constant determined in Fig. 25 (Q_{inj} varies from 0.1 to 300 C/cm^2, $V_{GSTRESS}$ from -4.5 to -1.5 V, and t_{ox} =2.3 nm). In the upper curve of Fig. 27, the same expression is represented as a function the stress temperature for a gate oxide thickness of 1.2 nm. The comparison of these two curves shows that after negative stresses the LVSILC increases exponentially by approximately one decade when T_{STRESS} increases by 100 K, independently of $V_{GSTRESS}$, Q_{inj} and t_{ox}. This result is very important, the devices being lead to run at operating temperature increasingly high and it is necessary to integrate them in the determination of the time-life.

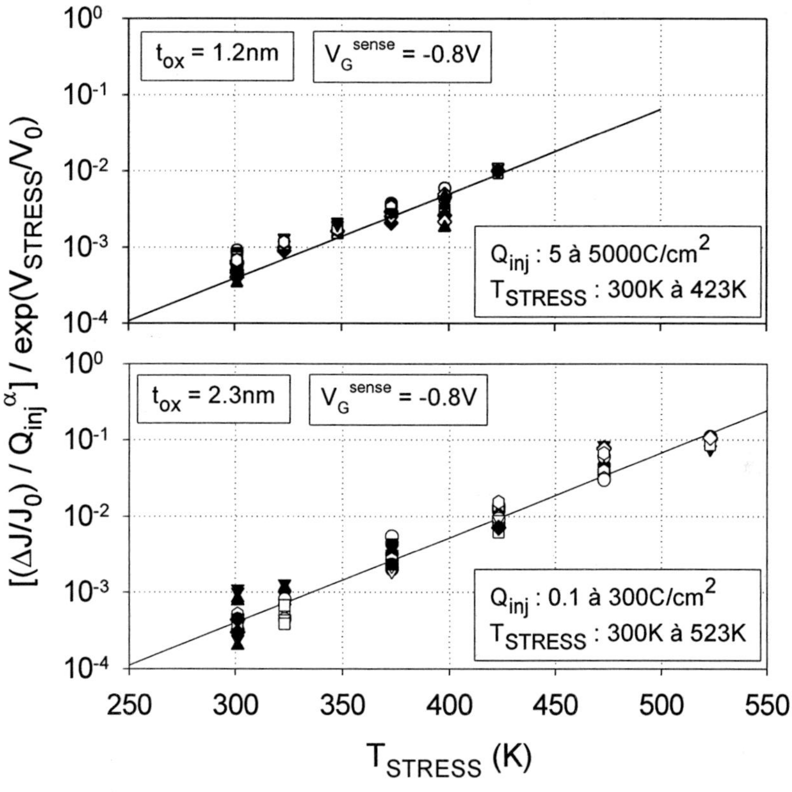

Figure 27. Variation of $\left[(\Delta J/J_0)/Q_{inj}^{\alpha}\right]/\exp(V_{STRESS}/V_0)$ as a function of the stress temperature, for different stress condition, the upper curve for t_{ox} = 1.2 nm and the lower curve for t_{ox} 2.3 nm.

3. Temperature Annealing Effects

The second effect of the temperature is an annealing effect. After a CVS at $V_{GSTRESS}$ = -4 V and an injected charge of 5 C/cm^2 realized at room temperature on a NMOS with a gate oxide thickness of 2.3 nm, we have performed a succession of temperature annealing during 10

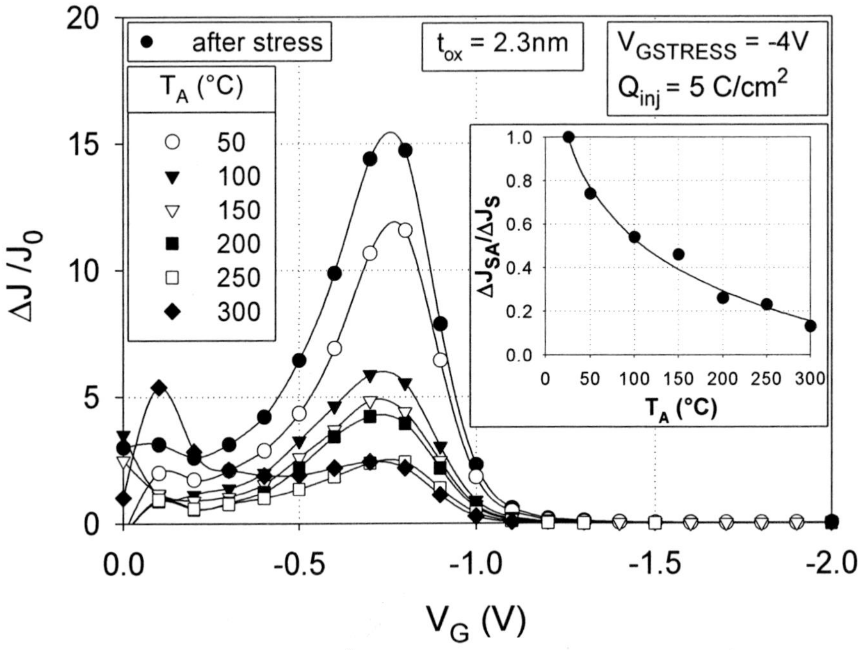

Figure 28. Normalized LVSILC variation as a function of the gate voltage after a negative constant stress voltage and after a succession of temperature annealing. The insert represents the non-annealed LVSILC evolution after successive temperature annealing.

minutes, as we can see in Fig. 28. The normalized LVSILC decreases as a function of temperature annealing. The insert of the Fig. 28 presents the evolution of $\Delta J_A/\Delta J$ (or the non-annealed LVSILC) after successive temperature annealing (ΔJ = J-J_0 and ΔJ_A= J_A-J_0, where J is the current density after stress and J_A is current density after stress and temperature annealing). While after a temperature annealing at 50°C, the LVSILC has decreased of 20% of its initial value, it decreases close to 85% after an annealing at 300°C. On the decreasing of the LVSILC as a function the temperature annealing, we have two comments to be made :

- First, we have followed the evolution of $\Delta J_A/\Delta J$ as a function of the time of annealing for different temperature annealing. We have observed that virtually, almost all of LVSILC which can be annealed at a given temperature is before 10 minutes (Fig. 29(a)). A defect is all the more stable when its annealing temperature is high.

- Secondly, for two different samples, one sample was stressed at room temperature and the other at 100 °C with same conditions of stress voltage and injected charges. The percentage of LVSILC left in the samples after annealing increased from 25% to 50% when the stress temperature varied from 25 °C to 100 °C (Fig. 29(b)). Electrical stresses, at high temperature, not only induce a larger LVSILC than at room temperature, but also increase the proportion of the most stable defects.

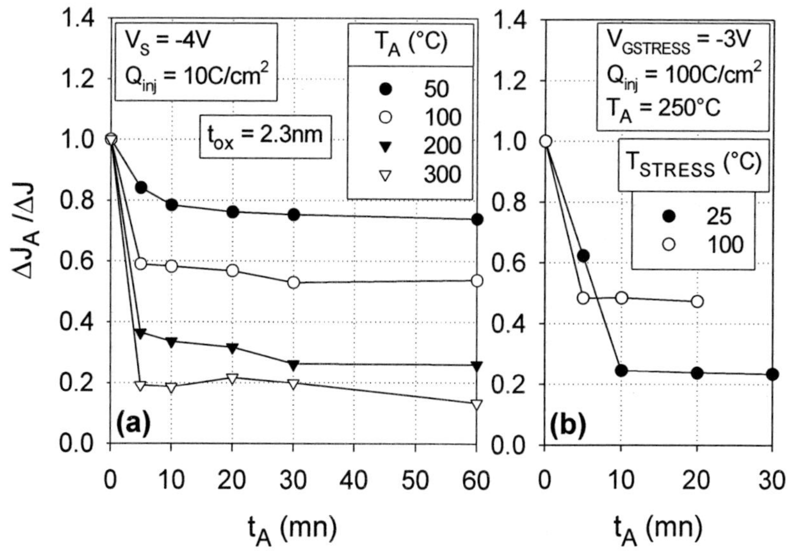

Figure 29. (a) Percentage of LVSILC left in the sample after annealing at different temperature annealing as a function of the annealing time. (b) Percentage of LVSILC left in the sample after annealing at 250°C as a function of the annealing time. One sample has been stressed at 25°C and the other at 100°C.

4. Time to Breakdown Prediction, Oxide Reliability

We have not made a statistical breakdown study, but nevertheless we have noticed that most of the oxide breakdown (soft or hard) happened when $\Delta J/J_0$, sensed at the second peak maximum, reached a critical value which was approximately equal to 80 for negative stress and around 45 for positive stress. This value seems independent of stress temperature, oxide thickness and, as much as it was possible to check, stress voltage. Above these values the oxide breakdown probability is very high. If we suppose that this condition continues to be valid at operating voltage, it is possible to deduce from these curves the time to breakdown (t_{BD}) as a function the stress voltage. In a previous work [23], we have described the methodology to extrapolate the t_{BD} value from the LVSILC measurements. Fig. 30 presents the median time-to-breakdown, t_{BD}, as function of $V_{GSTRESS}$, calculated from our experimental results after positive stress voltage in N-MOSFET for different oxide thicknesses (1.2 to 2.3 nm) (full dots), where the full line is calculated with expression $A\exp(B.V_{GSTRESS})$ equalling to $(\Delta J/J_0)_{BD}/Q_{inj}^{\alpha}$, where $(\Delta J/J_0)_{BD}$, is the value just before breakdown, ($(\Delta J/J_0)_{BD}$ is equal to ~ 45 and ~ 80 for positive and negative stresses, respectively). The t_{BD} values extracted from the analysis of LVSILC can be compared to the intrinsic t_{BD} values obtained from direct time-dependent dielectric measurements by several authors [47-54]. We find an exponential behavior of the time to breakdown values with the gate voltage stress in agreement with the results found by these methods. Their results were obtained from

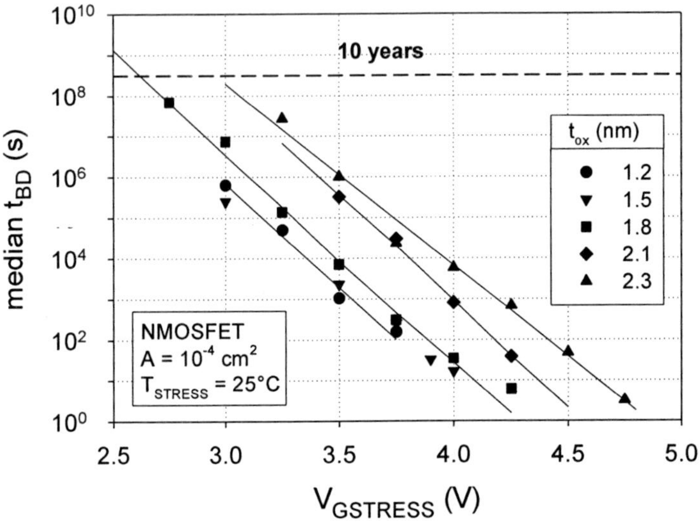

Figure 30. Extrapolated median time to breakdown as a function of a positive gate voltage stress V_S, at room temperature, for N-MOSFET. This result is obtained from the analysis of the LVSILC, for several gate-oxide thicknesses. The full dots are extrapolated from the $I_G - V_G$ characteristic before stress and the experimental data of Figs. 23 and with a value of $(\Delta J/J_0)$ just before breakdown equal to around 45. The full line is calculated from the relation (1).

statistical breakdown measurements, with area devices similar to ours [47, 48, 51, 54] and with smaller area devices than ours [49, 50, 52, 53]. If we compare our results with those obtained from statistical breakdown measurements, which is with area structures similar to ours or with normalized area structures to the same values than ours, we can observe that the slope of the t_{BD}–V_G curve, corresponding to the voltage acceleration, is very similar to those deduced by previous authors. On the other hand, if T_{STRESS} increases by 100 K, the time to breakdown is divided by 10.

5. Stress Induced Gate-Drain Leakage Current in Ultra-thin Gate Oxide

When drain, source and substrate are grounded, direct tunneling currents not only flow between the channel and the gate but also between the shallow source/drain extension (SDE) regions and the gate. In a previous work [55], we have described the different components of the gate leakage currents in a scaled MOSFET device and the methodology to measure these different currents : the gate-to-channel current (I_{GC}) corresponding to the current due to electron tunneling through the oxide between the gate and the substrate and the edge direct tunneling (EDT) component which corresponds to the current between the gate and the SDE region. In the large area devices, the gate current is mainly a gate-to-bulk current, gate-to-drain and gate-to-source currents are negligible even if the substrate is in depletion. However, in the ultra-thin gate-oxide, gate tunneling current between gate and source-drain extension overlap region becomes important. When the channel length is scaled down, the gate-drain current may be of similar magnitude as the gate-to-substrate current when the substrate is depleted or accumulated, even if the overlap gate area (approximately 0.02 μm) is smaller than L.

In the presented results below, gate, drain and source currents are measured simultaneously when the substrate (or bulk), drain and source are grounded. In this case, the gate is biased, the drain and the source are connected to ground and we measure the current on the gate, drain and source electrodes The measured currents on the drain and source terminals correspond to the gate-drain and gate-source currents (I_{GD} and I_{GS}), respectively. From the measured gate current I_G and the measured drain/source current $I_{GD/S}$ as a function of V_G, we have plotted, in the following figures, the gate-to-bulk current defined by $I_{GB} = I_G - (I_{GD} + I_{GS})$.

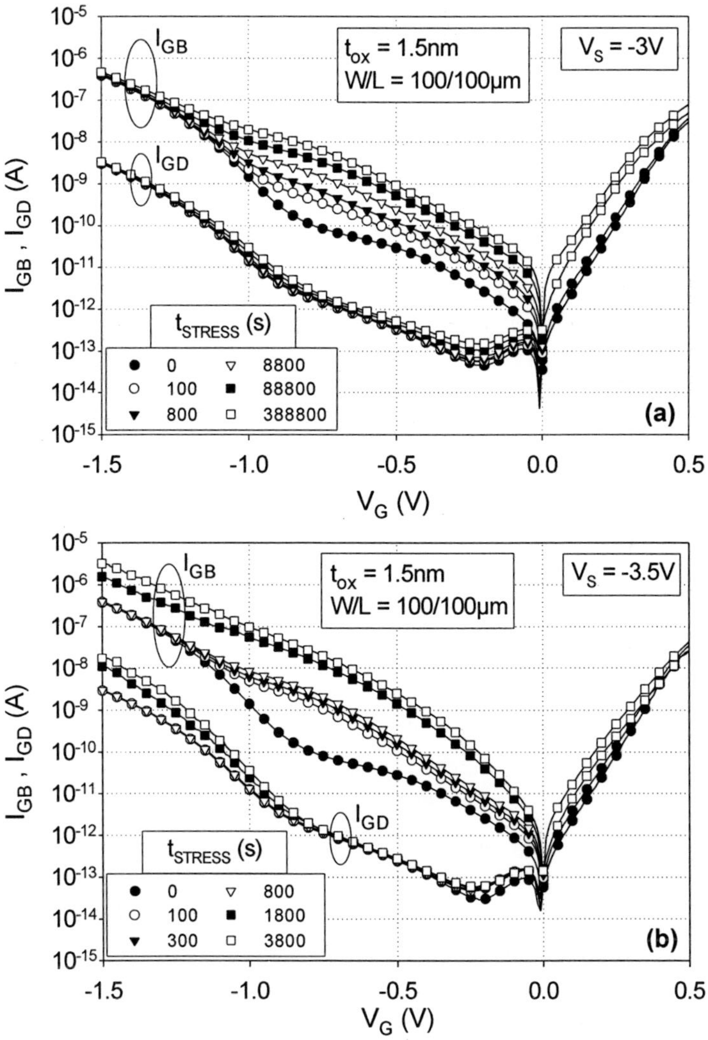

Figure 31. $I_{GB}(V_G)$ and $I_{GD}(V_G)$ curves before and after CVS, for different stress times in a large NMOSFET (W/L = 100/100 µm). (a) at $V_{GSTRESS}$ = -3 V; (b) at $V_{GSTRESS}$ = -3.5 V. The gate-oxide thickness is equal to 1.5 nm.

For W/L = 100/100 μm and 1.5 nm gate-oxide thickness, typical $I_{GB}(V_G)$ and $I_{GD}(V_G)$ curves are shown in Fig.31(a), before and after constant voltage stress (CVS) at $V_{GSTRESS} = -3V$, and various stress times t_{STRESS}. A similar degradation of the gate-source current (I_{GS}) is observed. These curves show a degradation of I_{GB}, as we showed in a previous paragraph, but also a degradation of the I_{GD} (and I_{GS}) currents, particularly at low voltage (-1 to 0 V). The gate-to-drain (or gate-to-source) currents can be decomposed in two components, one current (I_{GD1}), between 0 and -1 V, which flows between the gate and the SDE region, and an other (I_{GD2}), for $V_G < -1$ V, which flows between the gate and the drain (or source) through the channel [55]. The increase of gate-to-drain current between 0 and -1 V reveals a degradation of the SDE region. Although the gate-to-drain (or to–source) current is much lower than the gate-to-bulk current, an increase of the gate-to-drain (or source) current due to a degradation of the gate-drain/source overlap region can cause a breakdown (soft or hard) as we saw in Fig. 31(b). This figure shows the $I_{GD}(V_G)$ curves, before and after constant voltage stress at $V_{GSTRESS} = -3.5$ V and various stress times. We can see an increased of the gate current after a stress time of 1800 s, corresponding to the soft-breakdown. We can also observe a jump on the I_{GD2} component of I_{GD} current (for $V_G < -1$ V) but not on the I_{GD1} component (for $-1V < V_G < 0V$). As the result, the SDE regions are not much affected by a soft-breakdown. The gate-to-bulk leakage current increase will result in an increase of the I_{GD2} (or I_{GS2}) component. This explains the simultaneous degradation of gate-to-bulk current and the part of gate-to-channel current which flows to the drain/source via the channel area.

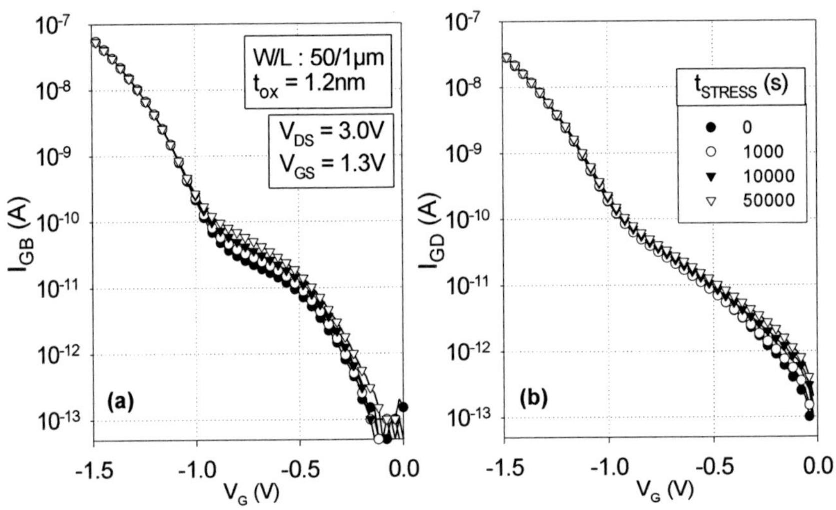

Figure 32. (a) $I_{GB}(V_G)$ and (b) $I_{GD}(V_G)$ curves before and CHC stress at maximum substrate current for $V_{DS} = 3$ V and $V_{GS} = 1.3$ V, for various stress times, in a NMOSFET (W/L = 50/1 μm). The gate-oxide thickness is equal to 1.2 nm.

It is known that stress by Channel Hot Carrier (CHC) at maximum substrate current induces degradation, near or in the drain. Figs. 32(a) and 32(b) represent I_{GB} (V_G) and I_{GD} (V_G) before and after a CHC stress. For a MOSFET with W/L = 50/1μm, we have realized a stress by CHC at maximum substrate current ($V_{DS} = 3V$ $V_{GS} = 1.3V$) for various stress times. We can observe that the difference between I_{GB} and I_{GD} is much weaker than the preceding cases where the active area was larger. Firstly, we can notice that the gate-to-bulk and gate-to-drain or source currents are the same order of magnitude. Consequently, the study of LVSILC curves obtained after stresses on short-channel devices should not be limited to I_G (V_G) curves study, but also to take into account of the gate-to-drain/source current degradation study. I_{GD} and I_{GB} increase mainly in V_G range -1 to 0V. We have seen that in this voltage range, I_{GD} corresponds to I_{GD1}. As a CHC injection degrades the Si/SiO_2 interface mainly near the drain, we checked that the gate-source current was not affected by the stress. We can also observe that for $-1\ V < V_G < 0\ V$, I_{GD} is of similar magnitude as I_{GB}, showing that the increase of I_G (the entire gate leakage current measured on the gate terminal) is due partly to the increase of I_{GD1}.

Only bulk oxide traps in the gate-overlap region seem to be able to explain this increase in current, as it is shown in Fig. 33. However, the fact that the induced defects are localized around the drain region does not favor larger gate leakage current. Compared to the CVS case, the magnitude of the LVSILC increase is smaller in the maximum substrate current case, and the oxide does not break down (soft or/and hard). On the other hand, in a previous study [21] we have shown that after a CHC stress at maximum substrate current, the I_{DSAT} of I_{DS}-V_{DS} MOSFET characteristics has strongly decreased. In spite of MOSFET parameter degradation, no breakdown event was detected during a CHC stress, thus, it is not the oxide lifetime which is critical, but the MOSFET characteristic degradation.

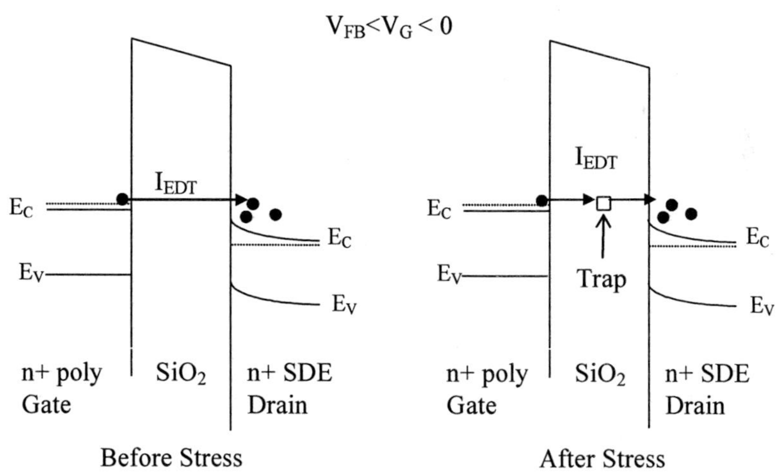

Figure 33. Energy band diagram located at the SDE region describing the EDT current possible conduction mechanism before and after stress, when drain and substrate are grounded.

Therefore, it is important, before explaining the LVSILC curves obtained after stress on short-channel devices, to verify which is the dominate current at low voltage. The study of the LVSILC should thus take into consideration the EDT currents.

6. Interface States and Reliability

Several authors explain the LVSILC in ultrathin SiO_2 films as an assisted tunneling via interfacial traps N_{it} [39, 56, 57]. It is then interesting to compare the interface state density with the LVSILC. The interface states were characterized by classical charge pumping method [58]. The rise and fall times are fixed and equal to 200 ns. Gate pulse frequency and magnitude are equal to 100 kHz and 2.5 V, respectively. In Figs. 34 (a) and (b), the normalized variations of the interface states density ($\Delta D_{it}/D_{it0}$) as a function of Q_{inj} are reported after a constant voltage stress, for different positive and negative stress gate voltages ($V_{GSTRESS}$),. D_{it0} and D_{it} are the interface state densities before and after stress, respectively. Although ΔD_{it} varies as LVSILC with the injected charge (i.e. follows a power law in Q_{inj}^{α}), the dispersion on the α value is greater after a positive stresses ($\alpha = 0.56 \pm 0.02$ for $V_{GSTRESS} >$ 4 V and $\alpha = 0.31 \pm 0.02$ for $V_{GSTRESS}$ below 4 V) than after a negative stresses (α = 0.37 ± 0.05).

In Fig. 34 (a), it is shown that few interface states are created when the positive stress voltage is lower than 4 V. The same behavior is observed, in Fig. 34 (b), after a negative stress when $V_{GSTRESS}$ is larger than -4 V. The Fig. 35 represents the interface density variations normalized by Q_{inj}^{α} as a function of the stress voltage, with α equals to 0.37 and 0.56 for the negative and positive stresses, respectively. After a positive stress the curve slope is steeper than after a negative one (approximately two decades/V when $V_{GSTRESS}$ is positive and 0.6 decade/V when $V_{GSTRESS}$ is negative). The comparison of the results of the Fig. 35 with those of the Fig. 23 (a), where are represented the LVSILC variations normalized by Q_{inj}^{α}, shows

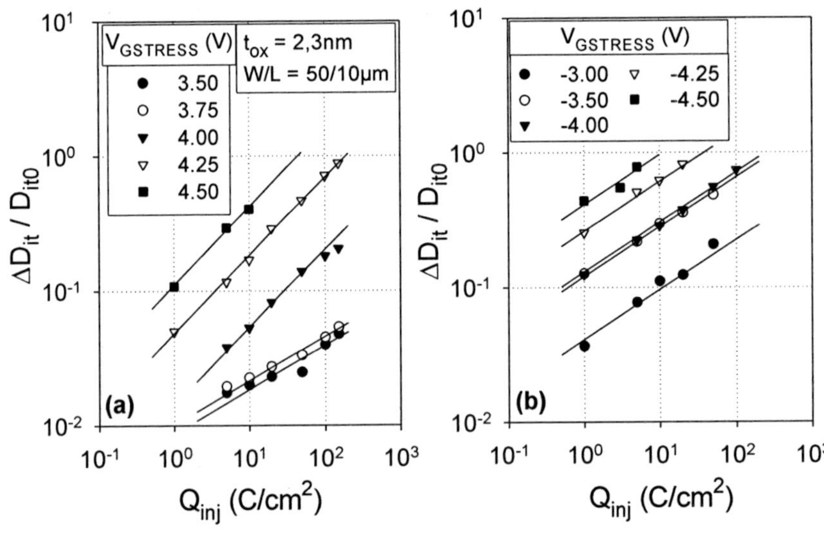

Figure 34. Normalized of interface states density variations ($\Delta D_{it}/D_{it0}$) as a function of the injected charge, for various stress voltages after CVS injections. (a) for positive stress voltages. (b) for negative stress voltages. The gate-oxide thickness is equal to 2.3 nm.

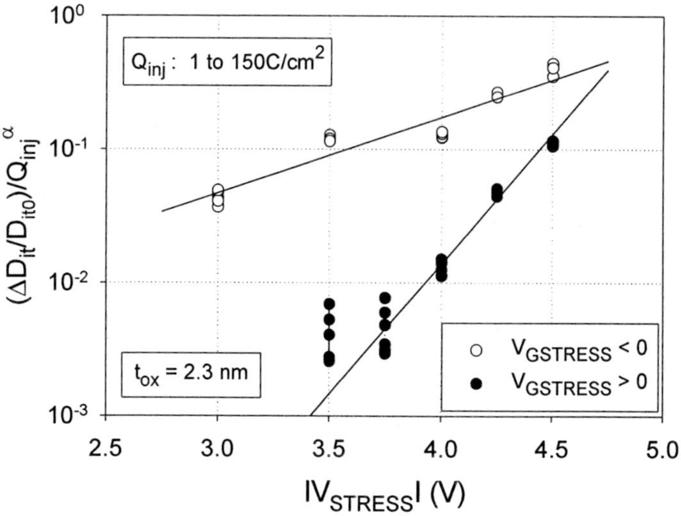

Figure 35. Variations of $\left(\Delta D/D_{it0}\right)/Q_{inj}^{\alpha}$ as a function of the stress voltage for various injected charges. Full symbols correspond to positive stresses and open symbols to negative stresses.

that the $\left(\Delta J/J_0\right)/Q_{inj}^{\alpha}$ and $\left(\Delta D/D_{it0}\right)/Q_{inj}^{\alpha}$ ratios follow the same law with the same slope, for both negative and positive stresses. Nevertheless, there is a gap of around two decades between these two ratios, for both negative and positive stress voltages. CVS injections, for a same injected charge and a same stress voltage, create much more (100 times more) LVSILC than interface states. But we have also noticed that the study of the interface states creation is limited by the device breakdown. Just before breakdown, for both polarities, $\Delta D_{it}/D_{it0}$ does not exceed 1. On the other hand, we have verified that after a CVS injection at $V_{STRESS} = +4.25$ V and for an injected charge of 150 C/cm^2, the threshold voltage does not change which can be explained by the low number of interface states created. After CVS injections, no noticeable characteristic degradation is observed until breakdown.

Fig. 36 shows the normalized variations of the interface states density ($\Delta D_{it}/D_{it0}$) as a function of the stress time (t_{STRESS}) after CHC injections on three MOSFETs, with a channel wide of 50 µm and a channel length of 0.1, 0.15 and 0.2 µm, respectively. For the three NMOSFETs, ΔD_{it} varies in t_{STRESS}^{α}, with a α value equals to 0.40±0.02. We can remark an increase of the interface states when the channel length decreases. We can check, also, that CHC injections create many more interface states ($(\Delta D_{it}/D_{it0})_{max} \approx 80$) than CVS injections ($(\Delta D_{it}/D_{it0})_{max} \approx 1$). In Fig. 37, we present the relative variations of the threshold voltage as a function of the relative variations of interface states density. CHC injections induce many interface states and involve an important degradation of the threshold voltage (V_{th}). V_{th} increases as a function of the stress time and the V_{th} variation reaches around 40% after a CHC at $V_{DS} = 3.3$ V and $V_{GS} = 1.075$ V, for a stress time of 24 000 s. It is important to note that oxide breakdown is not observed during CHC injections. Contrary to CVS injections, after CHC injections, significant characteristic degradation is observed without oxide breakdown. We have also observed that after CHC injections, in the characteristics I_{DS}-V_{DS} a strong decreasing of I_{DSAT} as a function of the stress time.

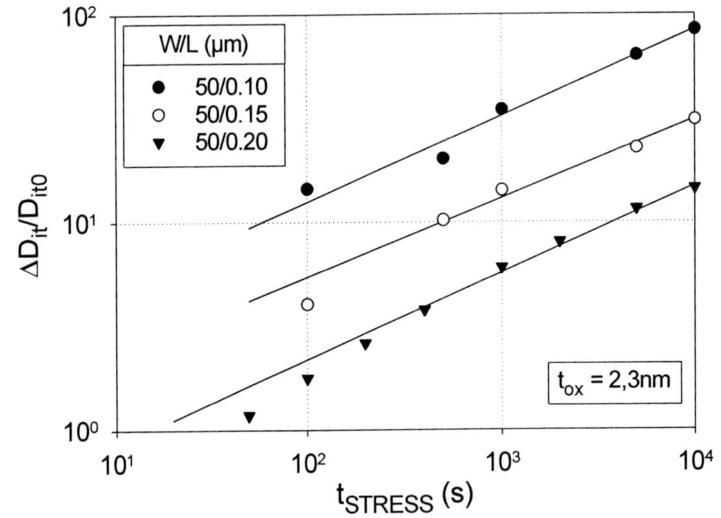

Figure 36. Variations of ($\Delta D_{it}/D_{it0}$) as a function of the stress time, in a NMOSFET with W = 50 µm and various L after CHC injections. The gate-oxide thickness is equal to 2.3 nm.

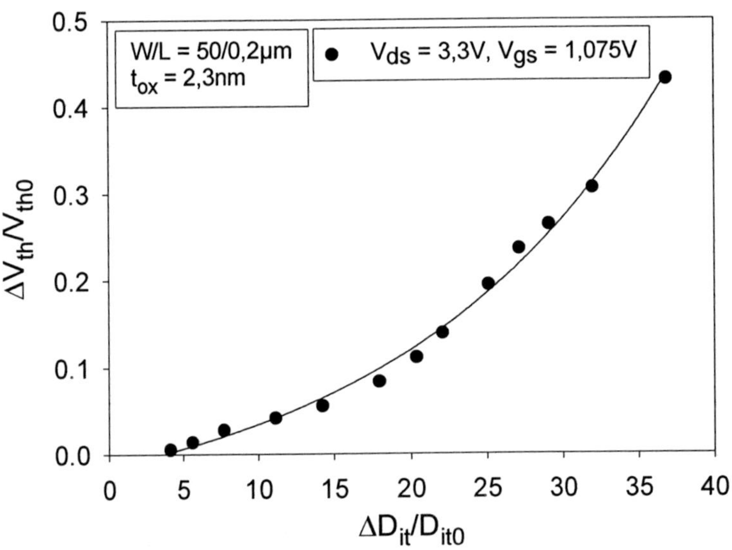

Figure 37. Relative variations of ($\Delta V_{th}/V_{th0}$) as a function of relative variations of ($\Delta D_{it}/D_{it0}$) in a NMOSFET with W/L = 50/0.2 µm. The gate-oxide thickness is equal to 2.3 nm.

Figs. 38 (a) and (b) show that a LVSILC is observable after CHC injections, when t_{ox} equals to 1.2 nm and 1.8 nm, respectively. The first peak at around 0 V which can be observed in Fig. 38 (a) (with a little of noise, the measurements lacking precisions) for the NMOSFET with an oxide thickness of 1.8 nm is not a measurement artefact. This peak is confirmed by measurements on a NMOSFET with an oxide thickness of 1.2 nm as we can see on Fig. 38(b). At approximately 0 V, when t_{ox} = 1.8 nm, the measurement lacks precision, but nevertheless it is seen in both cases, the first maximum is higher that the second one. Fig. 38 (b) shows LVSILC after CHC and DT injections. For the CHC stress conditions indicated in Fig. 38 (b), the surface state density is larger than those obtained after DT injections. On the other hand, at -0.1 V, the LVSILC obtained after CHC injections has the same value than those obtained after DT injections, whereas the CHC stressed sample has much interfaces than the DT stressed sample. Therefore, the breakdown does not seem to be related to the number of stress created interface states.

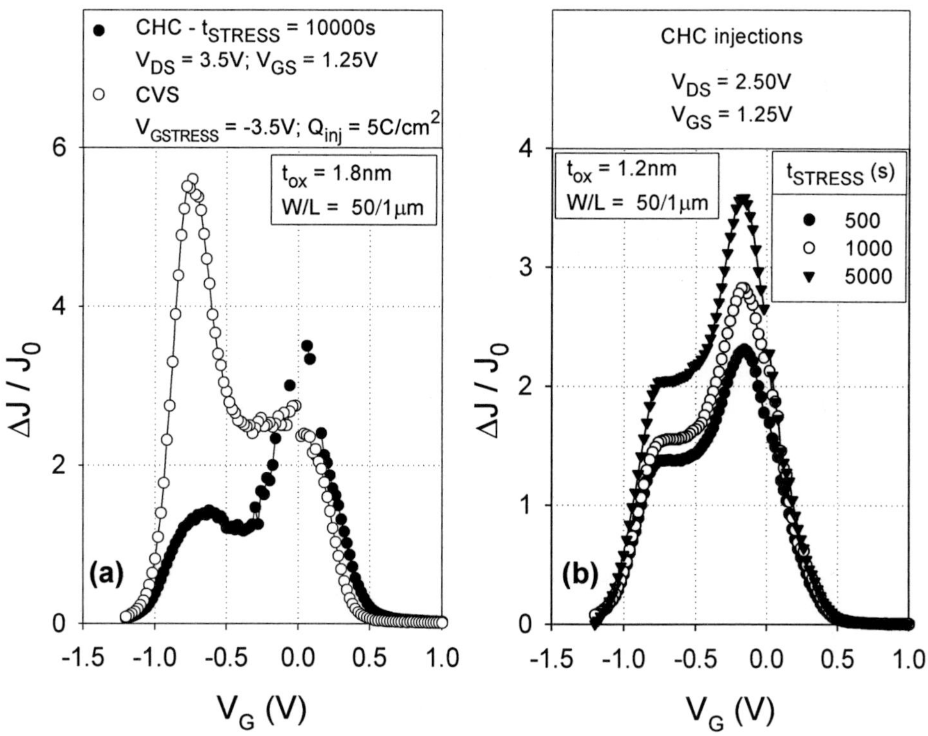

Figure 38. $\Delta J/J_0$ variations as a function of V_G (a) for different CHC stress times in a NMOSFETs with t_{ox} = 1.2 nm and (b) after CHC and DT injections in a NMOSFET with t_{ox} = 1.8 nm.

CVS injections create many LVSILC, but few interface states, on the other hand CHC injections, at maximum current substrate, create many interface states, but few LVSILC. This comparison seems to show that there is not a straightforward relation between LVSILC and interface states, as has been proposed [56]. The high interface state densities, created by CHC injections in MOSFETs, induce the diminution of I_{DSAT}, a shift of V_{th} but not the oxide breakdown, whereas DT/FN injections induce the oxide breakdown but create few interface states and therefore no I_{DSAT} diminution and no shift of V_{th}.

7. LVSILC and Possible Conduction Mechanisms

In a previous work [46], we have shown that the gate leakage current (after stress) decays as a function a waiting time, when no voltage is applied on the sample. The peak in the normalized LVSILC spectrum at V_G^{sense} = -0.8 V decreases continually while the peak at V_G^{sense} = -0.1 V, after a small decrease, seems to reach a limiting value. This seems to indicate that the peak situated at the sense voltage equal to -0.8 V cannot be explained by bulk neutral TAT. On the other hand, for the peak at V_G^{sense} equal to -0.1 V, these results seem to indicate that the two following behaviors are superimposed, tunneling through interface states and neutral trap-assisted-tunneling. The small decrease can explain that the peak at -0.1 V is due to electron tunneling through interface states from the conduction band. Nevertheless, the fact that the current remains constant after a small decrease seems to indicate that for the first peak at -0.1 V, neutral TAT is the main conduction mechanism. We have seen, also, that there is not a straightforward relation between LVSILC and interface states variations.

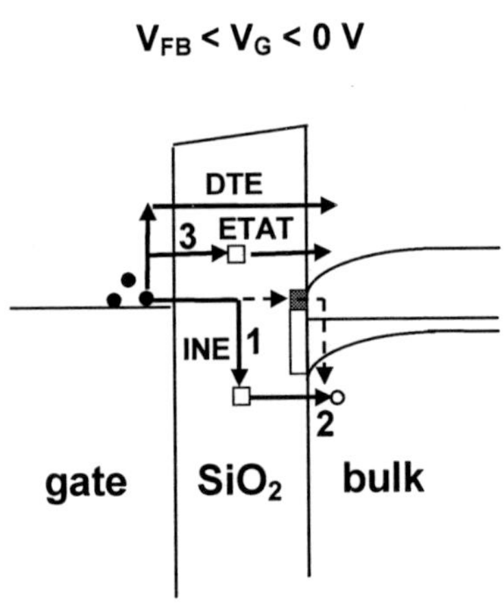

Figure 39. Energy band diagrams and the possible schematic conduction mechanisms: direct tunnel gate electrons (DTE), trap assisted components for electrons (ETAT), and inelastic tunneling of gate electrons (INE) when $V_{FB} < V_G < 0$ V.

Ghetti et al [20] have investigated low voltage tunneling in ultrathin oxides as a function of the substrate doping. For a device with lower substrate doping ($N_A \sim 10^{18}$ cm^{-3}), Ghetti et al have shown that direct tunneling of gate electrons (DTE) and, even better, inelastic tunneling of gate electrons (INE) simulations are in good agreement with measurements, at room temperature and above. These mechanisms are depicted in Fig. 39. In our case, we suggest that the peak behavior of the LVSILC can be due to bulk oxide traps either by elastic or inelastic tunneling mechanisms. First, for V_G^{sense} close to -0.8 V, electrons can tunnel from the cathode conduction band to an oxide trap at the same energy level (ETAT, mechanism 3 in Fig. 39). As trap energy levels are quantified, the trapped electrons cannot escape from the traps at room temperature. However, locally, these trapped electrons can change the oxide field and reduce the potential barrier between the traps and the Si conduction band. Consequently, the electrons can tunnel out of the traps to the Si conduction band without thermal activation. Secondly, the TAT inelastic current is given by the equilibrium of gate electrons tunneling into oxide bulk traps (mechanism 1 in Fig. 39) and the recombination of these electrons with substrate holes (mechanism 2 in Fig. 39). The number of substrate holes, which is temperature dependent, controls the INE mechanism, and we have shown, in a previous work [46], that the gate current is temperature dependent in the voltage range 0 to -1 V. This interpretation is in agreement with the model originally proposed by Ielmini et al [11, 59, 60]

C. Inelastic Electron Tunneling Spectroscopy (IETS) : Study and Characterization of Properties of Thin Insulators and of Their Interface between the Semiconductor and Electrodes

Nowadays, because of device scaling, the insulator thickness has been drastically reduced below 5 nm, and in real devices under development for next-generation MOS technologies, the tunnel leakage current is now a main drawback, affecting device function and reliability. Many spectroscopy analysis methods are used, i.e., infrared, Raman, and x-ray photoelectron spectroscopy, to study these properties. But these methods are not compatible with test structures. Moreover, usual standard techniques as capacitance voltage, admittance spectroscopy, and deep level transient spectroscopy begin to fail or lose their sensitivity to detect interface and oxide defects when oxide films reach thicknesses below about 3 nm. Inelastic electron tunneling spectroscopy (IETS) was discovered by Jaklevic and Lambe [61, 62] in 1966 when they studied the current–voltage (I –V) characteristics of metal– insulator–metal (MIM) junctions. They observed, at bias polarization and at a low temperature, a jump of conductance and a peak by measuring the second-derivative I –V characteristic. The method was used to study the absorption of biologic and organic molecules in polymers and other materials. It is a sensitive spectroscopy technique for the study of various devices [63-65] and complexes of adsorbates [66-68] at the interfaces of metal–insulator–metal (MIM) tunnel junctions. Later, this powerful technique was used by a few authors to investigate the critical insulating barrier of metal–oxide–semiconductor (MOS) devices [28, 70-73].

In the elastic tunneling effect, the electrons flow through the potential barrier without energy loss, which is the case for most electrons. However, another mechanism may occur, for instance, the electrons may excite a vibrational mode $\hbar\omega_0$ of an impurity located in the barrier and lose its specific energy. The condition $eV \geq \hbar\omega_0$ must be satisfied, where $\hbar\omega_0$ is the energy associated with the vibrational mode of the impurity. The total current is the sum of the elastic and inelastic current. The I–V curve exhibits a very small slope change at the voltage $V = \hbar\omega_0 / e$, for a distinctive vibrational mode. The slope change can be observed as a step on the first-derivative spectrum and as a peak in the second-derivative spectrum at this characteristic voltage. An inelastic electron tunneling (IET) spectrum is a series of peaks, each at a typical energy corresponding to a particular vibrational mode.

The methodology and the experimental set-up can be found in a previous work [74]. Experimentally, the second derivative of the I–V curve is obtained by a modulation technique. The voltage applied across the tunnel junction is an AC small alternative signal [V_ω cos(ωt)] superimposed onto a DC bias signal (V_G). The current through the junction can be written as a Taylor expansion :

$$\begin{aligned} I(V) &= I\left(V_0 + V_\omega \cos\omega t\right) \\ &= I(V_0) + \left(\frac{dI}{dV}\right)_{V_0} V_\omega \cos\omega t + \frac{1}{2}\left(\frac{d^2I}{dV^2}\right)_{V_0} V_\omega{}^2 \cos^2\omega t + \ldots\ldots\ldots \\ &= I(V_0) + \left(\frac{dI}{dV}\right)_{V_0} V_\omega \cos\omega t + \frac{1}{4}\left(\frac{d^2I}{dV^2}\right)_{V_0} V_\omega{}^2 + \frac{1}{4}\left(\frac{d^2I}{dV^2}\right)_{V_0} V_\omega{}^2 \cos 2\omega t + \ldots\ldots \end{aligned} \quad (2)$$

Therefore, the amplitude of the first harmonic is proportional to the first-derivative term, and the second harmonic is proportional to the second-derivative term.

Two factors affect the resolution of tunnel spectra : the temperature at which the sample is measured and the modulation voltage used. These parameters affect the measured peak width according the relation :

$$\delta V = \sqrt{(5.4kT)^2 + (1.22eV_\omega)^2} \quad (3)$$

where k is the Boltzmann constant, T the temperature and e the electron charge.

The first MOS devices named "A" is an Al–SiO_2– Si(n+) MOS capacitors, fabricated on highly doped, n-type, Si samples (phosphorous doping, $N_D = 4.7 \times 10^{19}$ cm^{-3}) from Siltronix. The aluminum gate has been preferred rather than the usual n+-polysilicon gate because there is no additional voltage drop due to quantum effects in a metal gate. Ultrathin silicon oxide films were grown by thermal oxidation in dry O_2 at 600 °C for 4 min (oxide thickness 2.1 nm) followed by a postoxidation anneal in dry N_2 at 600 °C for 30 min in the same oven. After thermal oxidation, aluminum gates were deposited (thickness 300 nm) by electron-beam evaporation (using an electron gun in a 10^{-8} Torr vacuum chamber) through shadow masks to define square electrodes with areas of 1×10^{-2} and 3.6×10^{-3} cm^2.

The second MOS devices named "B" is an Al–SiO_2–Si(n+) MOS capacitors, fabricated on highly doped, n-type, Si samples (arsenic doping, $N_D = 7.3 \times 10^{19}$ cm^{-3}). Silicon oxide films, of thickness 1.8 nm, were grown by a rapid thermal oxidation (RTO) process, followed by deposition of the 500 nm aluminum gate.

The inelastic electron tunneling spectra were performed at liquid helium temperature. For the tunneling measurements at 4.2 K, the MOS tunnel junctions were mounted in a sample holder and dipped into a liquid-helium dewar. The temperature may be decreased to 2 K by pumping above the liquid-helium bath.

A typical IET spectrum is shown in Fig. 40, this spectrum is the average of six scans. We can distinguish three regions with specific infrared vibration modes :

- Strong structures that correspond to phonons in the Si substrate in the energy range 30 to 80 meV.

- a small band structure corresponding to the phonons of the SiO_2 barrier in the range 130 to 180 meV.

- small structures sometimes attributed to molecular vibrations present in the tunnel barrier.

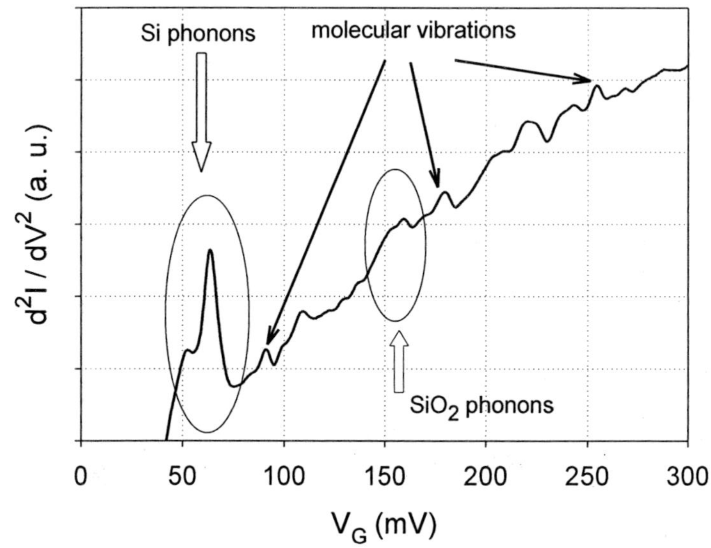

Figure 40. Typical IET spectrum showing the different structures : Si phonons between 45 and 65 mV, SiO_2 phonons between 140 and 170 mV and molecular vibrations.

We go to interest to the last structures which can allow to identify and to characterize defects and impurities in the oxide. What is the capability of IETS to identify specific molecular vibrations located at the critical Si-SiO_2 interface ? The answer of this question is very interesting, insofar as the classical techniques of the characterization of interface states or defects in the oxide, such as capacity or admittance measurements or deep level transient spectroscopy (DLTS), are not very sensitive for the oxide thickness lower than 2.5 nm. In fact, the principal raison is, as we saw in the previous parts, the increase of the leakage current. On the other hand, the IETS technique is very well appropriated to the junctions with a low resistance (some ohms to some kΩ). The principal interest of this spectroscopy is the possibility to detect at high energy, the infra-red or Raman molecular vibrations inside the tunnel barrier, i.e. in the gate oxide.

Fig. 41 presents an IET spectrum of the devices "A" after background removal. We focus here at the peaks that have been observed repeatedly over several IET spectra. We have interpreted the 93.3 meV (753 cm^{-1}) peak as a energy excitation of Si-H wagging vibration, the peak at 110 meV (887 cm^{-1}) as the excitation of the Si_3≡Si-H (or O_3≡Si-H) bending vibration, according to literature [30]. Balk [30] has observed that this peak disappears after

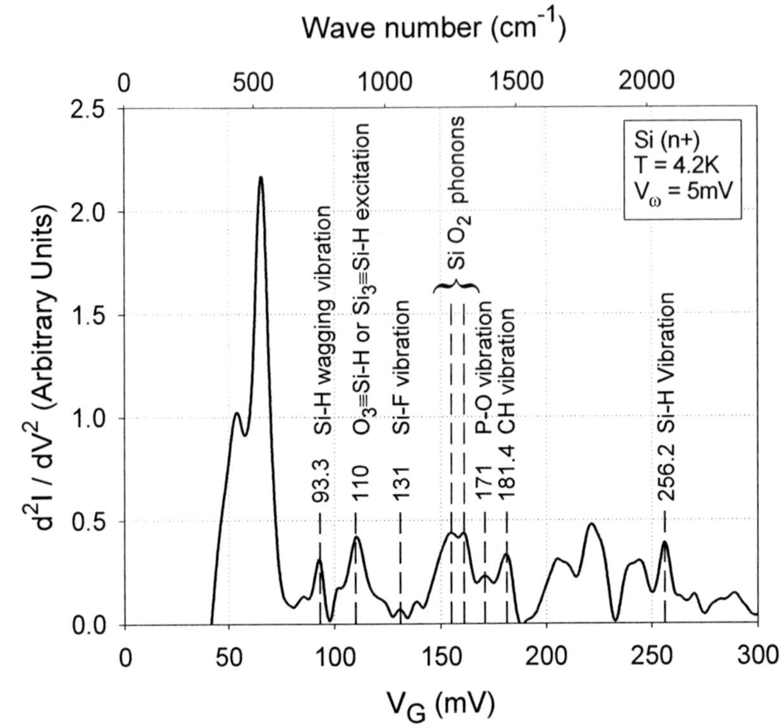

Figure 41. IET spectrum of an Al/SiO_2/Si(n+) junction (sample "A"), obtained after background removal, with the energies of different peak due to molecular vibrations.

subsequent high field stressing, and indicate a correlation between the generation of interface states and the removal of Si-H configurations. This seems to indicate that interfacial defects are being generated by breaking Si-H bonds. The weak structure at 131 meV (1056 cm^{-1}) can be interpreted as an energy excitation of Si-F vibration [30]. The vibration reveals the cleaning process of the wafer with hydrogen fluoride (HF) before oxidation. Moreover, the inelastic peak at 170.9 meV (1378 cm-^{1}) is clearly consistent with a P–O vibration resulting from the diffusion of phosphorous doping impurities from the substrate into the thin oxide during thermal oxidation [75, 76]. The peak at 181.4 meV (1463 cm^{-1}) has been interpreted by Kovchavtsev [77] as an excitation of CH vibration. The CH groups are possible impurities arising from contamination during the evaporation onto the oxidation surface. In the IET spectrum shown in Fig. 40, there is another signature of the presence of Si–H bonding from the IET peak situated at 256.2 meV (2066 cm^{-1}). This inelastic peak may correspond to the localized surface vibration of atomic hydrogen chemisorbed on a cleaved clean (111) silicon surface, as was observed, by Froitzheim, Ibach, and Lehwald [78]. In their high-resolution electron energy-loss spectroscopy, the peak was at the energy of 257 meV (2073 cm^{-1}). In our oxidized sample, this peak is the signature of the hydrogen-passivated Pb center (≡Si-H).

Fig. 42 presents an IET spectrum of the devices "B". The peak at 78.8 meV (635 cm^{-1}) can be attributed to Si-C vibration probably due to a contamination by the carbon. Such vibrations have been identified by infra-red spectroscopy [79]. We interpret the IET peak at

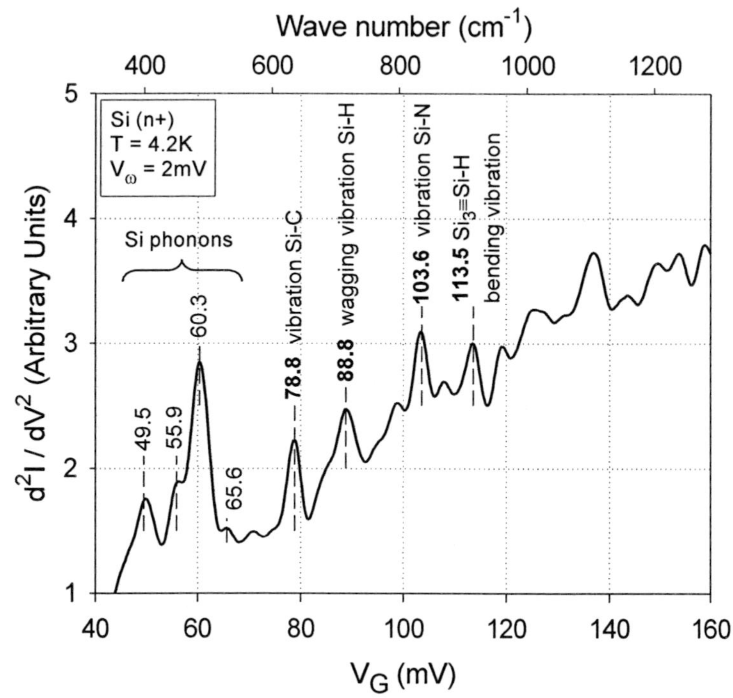

Figure 42. IET spectrum of an $Al/SiO_2/Si(n+)$ junction (sample "B"), with the energies of different peak due to molecular vibrations.

88.8 meV (716 cm^{-1}) as an excitation of a Si–H wagging vibration, already observed the devices "A" with light shift in energy. According to infra-red data [80], we purpose to assign the peak at 103.6 meV (835 cm^{-1}) to a Si-N vibration, reflecting the presence of nitrogen in the oxide. According to the previous works of Bell [81], the peak at 113.5 meV (915 cm^{-1}) can be interpreted as $Si_3{\equiv}Si\text{-}H$ (or $O_3{\equiv}Si\text{-}H$) bending vibration. This last vibration has been also observed in the devices "A".

The interest of this IETS technique is to compare the IET spectra of samples elaborated according several technologies. Our results, extending those of other authors, demonstrate that it is possible to obtain additional information as impurities due to contaminants accidentally incorporated during the fabrication process of the MOS devices.

In the past, Skarlatos [82] has detected traps by IETS in $Al\text{-}Al_2O_3\text{-}Pb$ junctions. More recently, Wei He and T.P. Ma [31] have reported the use of IETS as an effective tool in studying traps in high-k gate dielectrics, in metal-oxide-semiconductor structures. Sometimes, structures as those presented in Fig. 43, are observed in the IET spectra. Traps, in the gate dielectric, can involve different kinds of effects in a MOSFET. Trap-assisted conduction, with as consequence an increased leakage current and charge trapping, with as consequence a shift in threshold voltage, are among these different effects. The differences between these two kinds of traps are illustrated in Fig. 43. Fig. 43 (b) shows the structures at different bias voltages, for a tunnel junction, which could be observed in tunneling spectra (d^2I/dV^2 plot

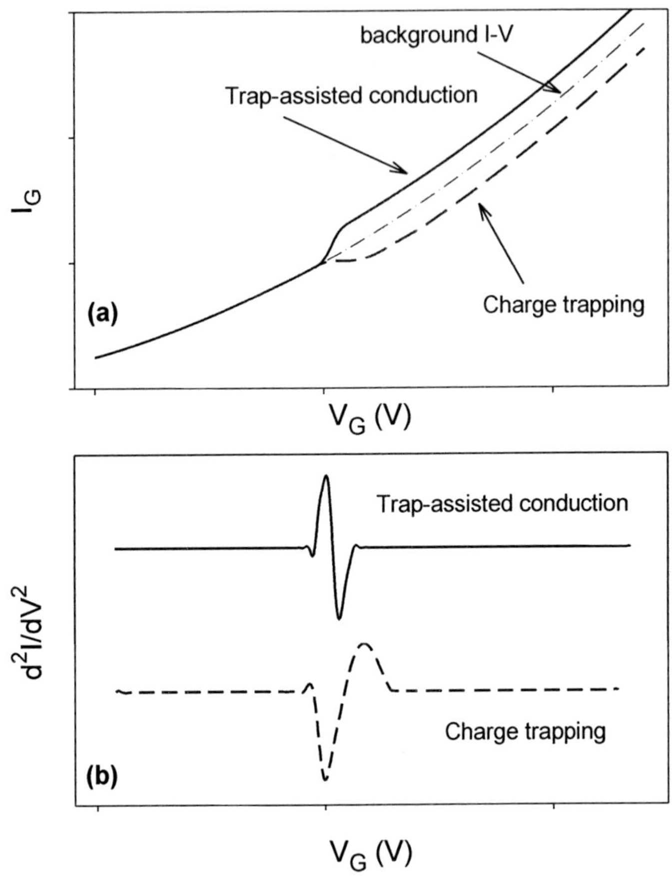

Figure 43. Illustration of two kinds of trap effects on (a) I-V and (b) d^2I/dV^2 –V characteristics.

versus bias) unrelated to inelastic tunneling, due to trap-assisted conduction and to charge trapping. Fig. 43(a) shows the same structures in I-V characteristics. In the I-V characteristics, the structures of Fig. 43 (b) are not visible, but an exaggerated form is shown if Fig 43 (a). A trap-assisted conduction mechanism involves an increase of the current in a small range voltage (where the voltage corresponds to the energy trap). Charge trapping involve in small horizontal plateau in the I-V curve and a shift of the I-V characteristics towards higher voltages.

In Fig. 44, we show four spectra of a same sample, before and after a constant voltage stress (CVS). Curve (a) and (b) were taken before stress; curve (b) was taken 24 h after curve (a). We can observe a good reproducibility of these two tunneling spectra. The most prominent peak at 60 mV corresponds to Si phonon vibrations. Other features appear at 174.3 and 253.2 mV. It seems that one kind of trap effects, as depicted in Fig. 43, is observed and may be attributed to trap-assisted conduction. Curve (c) was taken after a CVS at +2 V during 600 s. Curve (d) was taken after a CVS at 2.5 V during 600s. In these two stresses, we can observe that the structure at 174.3 mV is affected by the stress while the structure at 253 mV

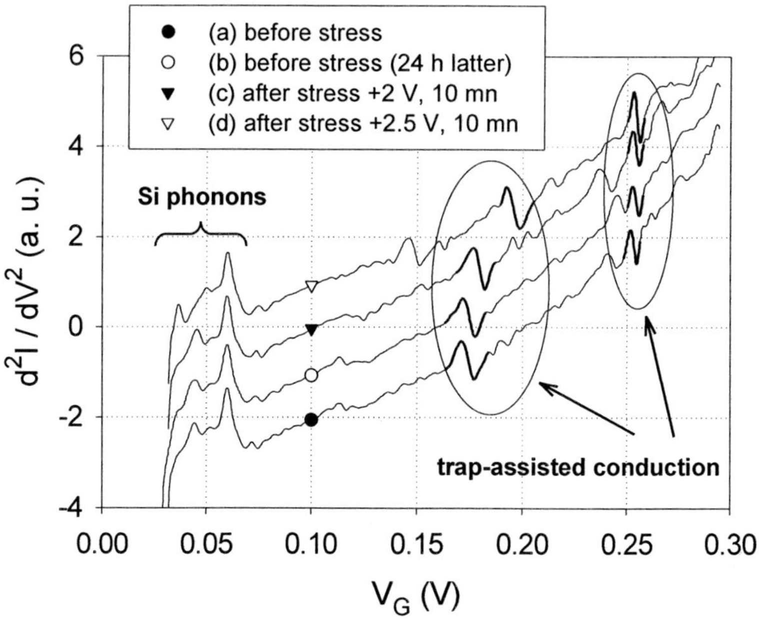

Figure 44. IETS spectra of an $Al/SiO_2/Si(n+)$ junction. The structures, unrelated to inelastic tunneling, but due to trap-assisted conduction are highlighted on the curve.

is not affected. The traps are generated throughout the oxide and at different energy levels [83]. It seems there exist two kinds of electrical-stress-induced traps, at low energy, whose level is dependent on the stress and at high energy whose level is independent of the stress (native traps pre-exist to the stress). It was found that the SILC phenomenon was not correlated with the oxide trap charge build-up and interface state generation, but rather closely correlated with neutral electron trap generation. This result suggested that the SILC originated not from water-related chemical reactions, but from the distortion of the thermally grown SiO_2 bond structure during electrical stressing [84]. These distortions can explain the first kind of electrical-stress-induced traps. These last results show that IETS technique can be used to identify the traps effects and allows the determination of their energy.

IV. Conclusion

In this paper, we have presented a study of the gate oxide reliability regarding the degradation of the gate leakage current of a device under electrical stresses. We have shown that the differences in I_G-V_G characteristics found, before stress, for different values of the oxide thickness (4.7 to 1.2 nm), deal with different regions of the same general curve. We have shown that this curve could only be explored within a window limited by the noise level of the measurement apparatus towards the lower currents and by oxide breakdown towards the higher currents. Depending on SILC level and therefore on oxide thickness, different regions fall within such windows, producing the typical experimental results.

For the gate oxide thickness of 4.7 nm, we have seen that the SILC increases with the injected charge (or the stress time) and that this SILC decays when a non-stressing field is applied on the device after the stress. This SILC decay depends on the magnitude and the duration of application of the non-stressing field and is irreversible as long as the sample is not stressed a second time. However, if the field magnitude increases and stresses the device, there is a competition between a deactivation effect, de-trapping effect and stressing effect. We have shown that the temperature annealing of activated or deactivated traps starts at around 200°C. For a temperature annealing at T = 300°C the SILC has strongly decreased and has practically disappeared.

For the gate oxide thickness of 3.5 nm, we have observed that the SILC have two components, one which can be annealed at a temperature of 300°C. This component is created for a negative and positive stress voltage, increases with the injected charge and follows a power law. The other component cannot be annealed with the temperature and is created mainly by a negative stress voltage.

For the gate oxide thicknesses in the range from 1.2 to 2.5 nm, the SILC become negligible and is hardly detectable, but the LVSILC which arises at low voltages ranging from 0 V to the flat band voltage (V_{FB}), is significant and is an important reliability problem. The normalized LVSILC variations $\Delta J/J_0$, whatever the stress polarity or the temperature polarity, follow a power law with Q_{inj}^{α}. After a positive constant voltage stress, the voltage acceleration is approximately 2 decades/V and around 1 decade/V after a negative one.

We have shown that from the SILC and LVSILC measurement (at high voltage), it was possible to extrapolate the time-to-breakdown towards the operating voltages and so to predict the oxide reliability. Nevertheless, in large area devices used to study the LVSILC, the gate current is mainly a gate-substrate current, the gate-drain and gate-source current are negligible. We have shown that for the short channel devices, in the ultra-thin gate-oxide, gate tunneling current between gate and source-drain extension overlap region becomes important. However, we are going to see that the oxide reliability in the gate-drain overlap region can be more critical than the oxide reliability above the channel. In CMOS circuit, e.g. an inverter, the stress conditions on a transistor gate are not the same as the typical experimental conditions used to study reliability [85].

At last, we have shown that inelastic electron tunnel spectroscopy can be an interesting tool to detect, identify and characterize molecular vibrations located at the critical Si-SiO_2 or distributed in the thin dioxide barrier. Our results demonstrate that it is possible to obtain additional information as impurities due to contaminants accidentally incorporated during the fabrication process of the MOS devices. The ability to probe the good quality of the tunnel oxides in as-fabricated devices, makes IETS a powerful aid in microelectronics for studying these devices.

References

[1] International Technology Roadmap for Semiconductors ITRS 2006 [Online] Available http://www.itrs.net/Links/2006Update/FinalToPost/04_PIDS2006Update.pdf

[2] Stathis, J.H.; DiMaria, D.J. *IEDM tech. Digest* 1998; pp 167-170.

[3] Kaczer, B ; Degraeve, R.; Pangon N.; Groseneken, G. *IEEE Trans. Electron Devices* 2000, 47, 1514-1521.

[4] Stathis, J.H. *IBM J Res & Dev* 2002, 46, 265-285.

[5] Maserjian, J.; Zamani, J. *J. Appl. Phys.* 1982, 53, 559-567.
[6] Olivo, P.; Nguyen, T.; Ricco, B. *IEEE Trans. Electron Devices* 1988, 35, 2259-2267.
[7] Dumin, D.J.; Maddux J.R. *IEEE Trans. Electron Devices* 1993, 40, 986-993.
[8] DiMaria, D.J.; Cartier, E. *J. Appl. Phys.* 1995, 78, 3883-3894.
[9] Rosenbaum, E. *IEEE Trans. Electron Devices* 1997, 44, 317-323.
[10] Takagi, S.; Yasuda N.; Toriumi, A. *IEEE Trans. Electron Devices*, 1999, 46, 335-341.
[11] Ielmini, D.; Spinelli, A.S.; Rigamonti M.A.; Lacaita, A.L. *IEEE Trans. Electron Devices* 2000, 47, 1266-1272.
[12] Meinertzhagen, A.; Zander, D.; Petit, C.; Jourdain M.; Goguenheim, D. *Microelectron. Reliab.* 2000, 40, 711-714.
[13] DiMaria, D.J. J. Appl. Phys. 2000, 87, 8707-8715.
[14] Degraeve, R.; Groeseneken, G.; Bellens, R.; Depas, M.; Maes, H. E. *IEDM 95 Technical Digest* 1995; pp. 863.
[15] Degraeve, R.; Groeseneken, G.; Bellens, R.; Ogier, J. L.; Depas, M.; Roussel, Ph.; Maes, H. E. *IEEE Trans. Electron Devices* 1998, 45, 904-911.
[16] Stathis, J. H. *Microelectron. Eng.* 1997, 36, 325-328.
[17] Degraeve, R.; Kaczer, B.; Groeseneken, G. *Microelectron. Reliab.* 1999, 39, 1445-1460.
[18] Nicollian;, P. E.; Rodder, M.; Grider, D. T.; Chen, P.; Wallace, R. M.; Hattangady, S. V. in *International Reliability Physics Symposium* 1999; pp 400-404.
[19] Ielmini, D., Spinelli, A., Rigamonti M. A., and. Lacaita, A. L *IEEE Trans Electron Devices* 2000, 47, 1258-1272.
[20] Ghetti, A. Sangiorgi, E. Bude, J. Sorsch T.W. and Weber, G. *IEEE Trans Electron Devices* 2000, 47, 2358-2365.
[21] Meinertzhagen, A.; Petit, C.; Zander, D.; Simonetti, O.; Maurel, T.; Jourdain, M. *J. Appl. Phys.* 2002, 91, 2123-2132.
[22] Goguenheim, D.; Trapes C.; Bravaix, A. *J. Non-Crystalline Solids* 2003, 322, 183-190.
[23] Petit, C.; Zander, D. *Microelectron Reliab.* 2007, 47, 401-408.
[24] DiMaria D. J.; Stasiak, J. W. *J. of Appl. Phys.* 1989, 65, 2342-2356.
[25] DiMaria, D.J.; Cartier E.; Arnold, D. *J. of Appl. Phys.* 1993, 73, 3367-3386.
[26] Lye, W.K.; Hasegawa, E.; Ma, T.P.; Barker, R.C.; Hu, Y.;Kuehun, J.; Frystak, D. *Appl. Phys. Lett.* 1997, 71, 2523-2525.
[27] Salace, G.; Petit C.; Vuillaume, D. *J. Appl. Phys.* 2002, 91, 5896-5901.
[28] Balk, P.; Do-Thanh, L.; Ewert, S.; Kuball, M.; Schmitz S. *Appl. Surf. Sci.* 1987, 30, 304-310.
[29] Petit, C.; Salace, G.; Vuillaume D.; *Solid-State Electronics* 2003, 47, 1663–1668.
[30] Balk, P.; Ewert, S.; Schmitz, S.; Steffen, A. *J. Appl. Phys.* 1991, 69, 6510-6516.
[31] He;W. Ma; T. P. *Appl. Phys. Lett.* 2003, 83, 5461-5463.
[32] Maserjian, J. *J. Vac. Sci. Technol.* 1974, 11, 996-1003.
[33] Salace, G.; Hadjadj, A.; Petit, C. J. *Appl. Phys.* 1999, 85, 7768-7773.
[34] Goguenheim, D.; Bravaix, A.; Vuillaume, D.; Mondon, F.; Jourdain, M.; Meinertzhagen, A. *J. Non-Crystalline Solids* 1999, 245, 41-47.
[35] Goguenheim, D.; Bravaix, A.; Ananou, B.; Trapes, C.; Mondon, F.; Reimbold, G. *Non-Crystalline Solids* 1999, 280, 78-85.
[36] Endoh, T.; Shimizu, K.; Iizuka, H.; Masuoka, F. *IEEE Trans. Electron. Devices* 1998, 45, 98-104.

[37] Meinertzhagen, A.; Petit, C.; Jourdain, M.; Mondon, F. *Solid-State Electronics* 2000, 44, 623–630.
[38] Meinertzhagen, A.; Zander, D.; Petit, C.; Jourdain, M.; Goguenheim, D. *Solid-State Electronics* 2001, 45, 1371–1381.
[39] Ricco, B.; Gozzi, G.; Lanzoni, M. *IEEE Trans. Electron. Dev* 1998, 45, 1554-1560.
[40] Meinertzhagen, A.; Petit, C.; Jourdain, M.; Mondon F.; Gogenheim D. *Microelectronic Reliability* 1999, 39, 191-196.
[41] Riess, P.; Ghibaudo, G.; Pananakakis, G.; Brini, *J. Microelectronic Reliability* 1998, 38, 1057-1061.
[42] Petit, C.; Zander, D. *Microelectron Reliab.* 2007, 47, 401-408.
[43] Nigam, T.; Degreave, R.; Groeseneken G.; Heyns, M.M. *Processing IRPS* 1999, 381-388.
[44] Petit, C.; Meinertzhagen, A.; Zander, D.; Simonetti, O.; Fadlallah, M.; Maurel, T. *Microelectron Reliab.* 2005, 45, 479-485.
[45] Fadlallah, M.; Petit, C.; Meinertzhagen, A.; Ghibaudho, G.; Bidaud, M.; Simonetti, O. *Microelectron Reliab.* 20035, 43, 1433-1438.
[46] Petit, C.; Zander, D. *Semicond. Sci. Technol.* 2007, 22, 1165-1173.
[47] Wu, E. Y.; Sune, J. Microelectronic Reliability 2005; 45, 1809-1834.
[48] Degreave, R.; Kaczer, B.; Groeseneken G. *Microelectronic Reliability* 1999, 39, 1445-1460.
[49] Degreave, R.; Kaczer, B.; Groeseneken G. *Microelectronic Reliability* 2000, 40, 697-701.
[50] Weir, B. E.; Alam, M. A.; Bude, J. D.;. Silverman, P. J; Ghetti, A.; Baumann, F.; Diodato, P.; Monreo, D.; Sorsch, T.; Timp, G. L.; Ma, Y.; Brown, M. M.; Hamad, A.; Hwang D.; Mason, P. *Semicond. Sci. Technol.* 2000, 15, 455-461.
[51] Wu, E. Y.; Vayshenker, A.; Nowak E., Sune, J.; Vollertsen, R. P.; Lai W.; Harmon D. *IEEE Trans Electron Devices* 2002; 49, 2244-2253.
[52] Wu, E. Y.; Lai, W.; Khare, M.; Sune, J.; Han, L. K.; McKenna, J.; Bolan, R.; Harmon D.; Strongin A. *International Reliability Physics Symposium.* 2002: pp 60-72.
[53] Vogel, E. M.; Suehle, J. S.; Edelstein, M. D.; Wang, B.; Chen Y.; Berstein J. B. *IEEE Trans Electron Devices* 2000, 47, 1183-1191.
[54] Stathis J. H. *IEEE Trans Devices and Materials Reliability* 2001, 1, 43-59.
[55] Petit, C.; Zander, D. *Microelectron Reliab.* 2007, 47, 2070-2081.
[56] Houssa, M.; Mertens, P.W.; Heys, M.M. *Semicond. Sci. Technol.* **14**, 892-896.
[57] Ghetti, A.; Sangiorgi, E.; Bude, J.; Sorsch, T.W.; Weber, G. *IEEE Trans. Electron. Dev* 2000, 47, 2358-2365.
[58] Groseneken, G; Maes, H.E.; Beltran, N.; DeKerersmaeker, R. F. *IEEE Trans. Electron. Dev* 1984, 31, 42-53.
[59] Ielmini, D.; Spinelli, A.S.; Lacaita, A.L.; Martinelli, A.; Ghidini, G. *Solid State Electronics* 2001, 45, 1361-1369.
[60] Ielmini, D.; Spinelli, A.S.; Lacaita, A.L.; Ghidini G. *Processing IRPS Int. Reliabilty Physics Symp.* 2000, 55-64.
[61] Jaklevic, R. C.; Lambe, *J. Phys. Rev. Lett.* 1966, 17, 1139-1140.
[62] Lambe, J.; Jaklevic, R. C. *Phys. Rev.* 1968, 165, 821-832.
[63] Hayashi, S.; and Aoki, R. *J. Phys. Soc. Jpn.* 1981, 50, 2613 -2618.
[64] Wada, M.; Kubota, T.; Iwamato, M. *Jpn. J. Appl. Phys.* Part 1 1993, 32, 3868-3871.

[65] Kudo, K.; Okazaki, C.; Kuniyoshi, S.; Tanaka, K. *Jpn. J. Appl. Phys*. Part 1 1991, 30, 1452-1456.
[66] Klein, J.; Leger, A. *Phys. Lett*. 1969, 30A, 96 -97.
[67] Bayman, A.; Hansma, P. K. *Nature* 1980, 285, 97-99.
[68] Hansma, P. K Tunneling Spectroscopy, edited by Plenum, New York, 1982.
[69] Jacklevic, R.C.; Gaerttner M.R. *Appl. Phys. Lett.* 1977, 30, 646-648.
[70] Cullen, D.E.; Wolf, E.L.; Dale Compton, W. *Phys. Rev B* 1970 2, 3157-3169.
[71] Bencuya, I.; Thesis, 1984, Yale University.
[72] Lye, W.-K.; Hasegawa, E.; Ma, T.-P.; Barker, R. C.; Hu, Y.; Kuehne, J.; Frystak, D. *Appl. Phys. Lett.* 1997, 71, 2523-2525.
[73] Lye, W.-K. Thesis, 1998 Yale University.
[74] Petit, C; Salace, G. *Rev. Sci. Instrument*. 2003, 74, 4462-4467.
[75] Pliskin W. A., in *Semiconductor Silicon,* edited by H. R. Huff and R. Burgess Electrochemical Society, Princeton, NJ, 1973; pp 506.
[76] Pliskin, W. A. *Appl. Phys. Lett*. 1965, 7, 158-159.
[77] Kovchavtsev, P.; Kurishev, G. L.; Postnikov, K. O.; Sokolov, R. A.; Subbotin, I. M. *Phys. Stat. Sol. a*, 1988, 106, 669-674.
[78] Froitzheim, H.; Ibach, H.; Lehwald, S. *Phys. Lett.* 1975, 55A, 247-248.
[79] Newmann, R. C.; Willis, J. B. *J. Phys. Chem. Solids*, 1965, 26, 373-379.
[80] Knolle W .R .; Osenbach, J. W. *J. Appl. Phys*. 1986, 58, 1248-1254.
[81] Bell L. D.; Coleman, R. V. *Phys. Rev. B,* 1984, 30, 4120-4129.
[82] Skarlatos, Y.; Barker, R. C. J. *Appl. Phys.* 1976, 47, 10, 4593-4597.
[83] Scott, R.S.; Dumin, N.A.; Hugues, T.W.; Dumin, D.J. *International Reliability Physics Symposium* 1995, 131-141.
[84] Kimura, M.; Ohmi, T. *J. Appl. Phys*. 1996, 80 6360-6369.
[85] Petit, C.; Meinertzhagen, A.; Zander, D.; Simonetti, O.; Fadlallah, M.; Maurel, T. *Microelectronic Reliability* 2005, 45, 479-485.

In: MOSFETs: Properties, Preparations and Performance ISBN: 978-1-60456-762-5
Editors: N.T. Andre et al. pp. 217-251

Chapter 6

SOME MEDICAL APPLICATIONS OF MOSFETS IN RADIATION THERAPY: SURFACE DOSE AND ELECTRON BACKSCATTER MEASUREMENTS WITH MONTE CARLO SIMULATIONS

James Chun Lam Chow

Radiation Medicine Program, Princess Margaret Hospital, Toronto, ON M5G 2M9, Canada.
Department of Radiation Oncology, University of Toronto, Toronto, ON M5G 2M9, Canada.
Department of Physics, University of Waterloo, Waterloo, ON N2L 3G1, Canada.
Department of Physics, Ryerson University, Toronto, ON M5B 2K3, Canada.

Abstract

Metal-oxide semiconductor field effect transistor (MOSFET) dosimeter has been used to measure and monitor radiation dose for mega-voltage (MV) photon and electron beams in radiation therapy. The dosimeter has a very small physical size (active region of about $0.2 \times 0.2\ mm^2$), and the recorded dose can be read immediately after or during the treatment using a remote reader setup. This simplifies the cumbersome dosimetric procedure and minimizes the time of quality assurance in the treatment. Recently, the properties, preparations and performances of the MOSFET dosimeter are studied extensively in different aspects such as angular and energy dependence, lifespan and sensitivity. Monte Carlo simulation is also employed to predict dosimetric results under different conditions of measurements, and provides valuable information to improve the design of the dosimeter. In this chapter, some medical applications using the MOSFET dosimeter to measure the MV photon and electron beams were presented. Moreover, the properties and performances of the MOSFET dosimeter in determining the electron backscatter from the lead were studied using Monte Carlo simulations. The accuracy of using MOSFET dosimeter to predict the electron backscatter from the lead was also evaluated in this chapter.

For photon beam study in radiation therapy, surface doses for the 6 and 15 MV prostate intensity modulated radiation therapy (IMRT) plans were measured, and compared to those for typical prostate four-field box (4FB) plans. MOSFET dosimeters were used and placed on a transverse contour line along a Rando Phantom surface at the central beam axis in the measurement. The aims of measurement are to investigate how: (1) the moving multi-leaf collimator (MLC) leaves contribute to the surface dose during IMRT, and (2) the effects of both photon beam energy and number of beam used in the IMRT plan on the surface dose. Results showed that with the same number of beam in the IMRT plan, the 6 MV plan gave more surface dose than that of the 15 MV. However, when the number of beam was increased in the plan, the surface dose difference between the 6 and 15 MV photon beams became less. It was also found that the increase of surface dose in the IMRT plan compared to 4FB was predominantly due to the number of beam, and the calculated monitor units in the plans.

For electron beam study in radiation therapy, the electron backscatter, produced by clinical electron beams, was studied using the MOSFET dosimeter and Monte Carlo simulation (EGSnrc-based code). Electron backscatter factor (EBF), defined as the ratio of dose at the tissue-lead interface with and without the presence of backscatter, was calculated using the Monte Carlo simulation, and verified with the MOSFET measurement. The dependences of the (1) electron beam energy, (2) depth of the lead shield, (3) beam angle, and (4) aluminium sheet used to absorb electron backscatter, on the EBF were studied. It was found that for particular depth of lead shield, EBF decreases with increase of electron beam energy (4 – 16 MeV). In addition, within the electron practical range, R_p, and particular beam energy, EBF increases with the depth of lead. When the electron beam angle increases from 0 to 5 deg, EBF decreases only slightly (< 4%) for all beam energies. For low electron beam energy (4 MeV), 2 mm aluminium was found to effectively reduce the electron backscatter by 31%. While the electron beam energy increases, less electron backscatter were reduced by the same thickness of aluminium (only about 6% for the 16 MeV).

The performance of the MOSFET dosimeter in determining the electron backscatter from the lead was evaluated. The EBF was predicted by Monte Carlo simulation using the GEANT4-based code. Clinical electron beams with energies of 4, 6, 9 and 12 MeV were used in the simulations. It was found that with presence of the MOSFET body, EBF was underestimated to about 2% for the low electron beam energy of 4 MeV. However, such deviation decreases to 0.9% when higher electron beam energy of 12 MeV was used. This is because the small MOSFET dosimeter, mainly made of epoxy and silicon, not only attenuated the electron fluence from the upstream direction, but the electron backscatter generated by the lead underneath the dosimeter. Corresponding measurements were carried out and uncertainties compared to Monte Carlo results were within ± 2%. The electron energy spectra with and without the lead and MOSFET body were also determined by Monte Carlo simulations. It was found that the presence of the MOSFET body widened up the electron energy spectrum in the dosimetric volume. Moreover, no matter the MOSFET body is present or not in the simulation, deviations of electron energy spectra with and without the lead decrease with an increase of the electron beam energy. It is concluded that the MOSFET dosimeter performed well for measuring the electron backscatter from the lead using electron beams. The uncertainty of EBF is within 2% according to the Monte Carlo simulation, and decreases when higher electron beam energy is used (from 4 to 12 MeV).

I. Introduction

Metal-oxide semiconductor field effect transistor (MOSFET) is a miniature silicon field-effect transistor, which was proposed as a radiation dosimeter over 20 years ago (August 1984, Thomson et al.. 1984, Streubel et al.. 1988, Gladstone and Chin 1991). The physics principle of dose detection of the MOSFET is based on the measurement of threshold voltage, which is a linear function of absorbed dose (Holmes-Siedle 1974). Threshold, or device threshold voltage, is the potential difference necessary to initiate a current flow between the source and drain in the transistor. This current is based on a conduction channel with sufficient concentration of charge carriers between the source and drain. When the MOSFET dosimeter is exposed to ionizing radiation, charges will be generated and trapped permanently in the oxide layer. This results in a shift of threshold voltage, which can persist for years. This voltage shift can be measured during or after exposure and is proportional to the dose (Holmes-Siedle 1974).

The MOSFET dosimeter is well known for advantages of small size, light weight, minimal power requirement, immediate readout (remote possible), high sensitivity, and ease of use (Francescon et al. 1998, Kaplan et al. 2000, Lydon 2005). However, the dosimeter is also known to have energy and angular dependences for low energy photons. There are a lot of studies on the energy and angular dependences of the MOSFET using Monte Carlo simulations and measurements (Ramaseshan et al. 2004, Huang et al. 2003, Lanson et al. 1998, Soubra et al. 1994, Roshau and Hintenlang 1998, Wang et al. 2005). On the other hand, manufacturers such as Thompson and Neilson, Ottawa, Canada, are improving their products by introducing new MOSFET dosimeters such as the microMOSFET, claiming that this dosimeter is temperature, beam angular, dose rate and energy independent. The uncertainty of the microMOSFET is ± 2% at two standard deviations in measuring dose, and was verified elsewhere (Ramaseshan et al. 2004).

In recent years, MOSFET dosimeters are widely used in various medical applications such as dose verification and monitoring in radiation therapeutic and diagnostic applications (Huang et al. 2003, Lanson et al. 1998, Scalchi et al. 2005, Butson et al. 1996, Chuang et al. 2002, Rosenfeld 2002, Kron et al. 2002, Cheung et al. 2004). Moreover, with high sensitivity characteristic of the dosimeter, it becomes practical for the radiation protection measurement with low dose rates (Rosenfeld et al. 1995 and 2002, Peet and Pryor 1999, Belyakov et al. 1995). In this chapter, some medical applications of MOSFET in radiation therapy regarding photon and electron beams are presented. They are studies on (1) the surface dose generated in the prostate intensity modulated radiation therapy (IMRT) using the 6 and 15 MV photon beams, (2) the electron backscatter from the lead for clinical electron beams using Monte Carlo simulations (EGSnrc-based code), and (3) the influence of the MOSFET dosimeter on the electron backscattering measurement using Monte Carlo simulations (GEANT4-based code).

In the prostate radiation therapy, the conventional four-field box (4FB) technique is eventually replaced by IMRT, which not only provides a better conformal dose distribution in the prostate target, but spares the dose in the critical tissues such as rectal and bladder walls that need to be protected from complications (Burman et al. 1997, De Meerleer et al. 2000, Clark et al. 2002, Corletto et al. 2003, Price et al. 2003, Bauman et al. 2004, Jani et al. 2003, Bos et al. 2004, Klein et al. 2000, Kung et al. 2001). However, the extended treatment time, more monitor unit (MU) per each fraction of treatment, and specific patient quality assurance are extra clinical concerns in using IMRT (Niemierko et al. 2004, Ramsey et al. 2003, Ma et al. 1999, Cheng et al. 2002). The implementation of IMRT requires a multi-leaf collimator (MLC) to generate segmental fields per every treatment beam. One example of the MLC is the Varian 120-leaf Millennium system (Varian Medical System, Palo Alto, CA). This MLC contains two carriages, one located beneath the X1 jaw and the other directly beneath the X2 jaw. Each carriage holds 60 leaves with 40 inner leaves each having a width of 0.5 cm at the isocenter (source-to-axis (SAD) distance = 100 cm). The inner leaves can produce a maximum square field size of 20×20 cm^2. The width of the outer leaves is 1 cm and therefore all 60 leaves can generate a maximum square field size of 40×40 cm^2 at the isocenter. This MLC with collimator leaves would be a concern in the patient's surface dose. It is because the dynamic movement of the MLC varies with the configuration of the linear accelerator (linac) head during the treatment. The segmental fields generated by the MLC vary the collimations of the flattening filter, which is the main contributor to the "head scatter" to affect the surface and peripheral dose on the patient (Hounsell and Wilkinson 1999, Chow et al. 2006a). Moreover, the patient's surface dose is affected by the number of MUs and beams in the IMRT plan. It was found that patient's surface dose decreases in the IMRT with the same beam envelope compared to that of the conformal delivery. However, with a large increase of MU, number of beam and beam angle obliquity, it may contribute more surface dose (Dogan and Glenn 2003, Yokoyama et al. 2004). In this chapter, calibrated MOSFET dosimeters were used with an average-man Rando Phantom with bone anatomy to determine surface doses for the typical 5-, 7- and 9-beam prostate IMRT. Photon beam energies of 6 and 15 MV were used in the study.

The second medical application using the MOSFET dosimeter is to measure the backscattered electrons from the lead used as a shield in the electron radiation therapy. A thin layer of lead (2 – 3 mm) is applied during the electron beam irradiation treating superficial lesions on eyelid, ear and buccal mucosa. This layer of lead is to shield the underlying critical tissues such as eye and gum in the treatment (Sanuders and Peters 1974). However, electron backscatter from the lead is well known to enhance the dose at the tissue-lead interface in the upstream direction of the electron beam. This dose enhancement depends on the electron beam energy, thickness of the treatment volume above the lead, beam angle, and the presence of the low atomic number materials such as aluminium on top of the lead (Saunders and Peters 1974, Perez-Calatayud et al. 2000, Weatherburn et al. 1975, Khan et al. 1976, Gagnon et al. 1980, Lambert et al. 1982 and Das et al. 2004). Electron backscatter factor (EBF) is commonly used to quantify the electron backscatter for an electron beam, and is defined as the ratio of dose at the tissue-lead interface with to that without the presence of backscatter (Klevenhagen et al. 1982). It is difficult to measure the EBF because the physical size of the dosimeter should be very small so that: (1) the effective point of measurement can be as close to the tissue-lead interface as possible, and (2) the dosimeter body in the beam would not perturb the electron fluence in the experimental setup. MOSFET dosimeter was therefore employed, because it is physically very small compared to an ionization chamber, and was not expected to disturb the electron beam significantly in the measurement. In this study, the EBFs of the lead positioned at different depths from the treatment surface were determined systematically using clinical electron beams of 4, 6, 9, 12 and 16 MeV produced by a Varian 21 EX linac (Varian Medical System, Palo Alto, CA). The effect of an oblique electron beam on the EBF was also investigated because clinically, patient's surface contours such as eyelid, nose and ear are curved (Tabata 1967, Tabata and Ito 1992). This is different from our measured and calculated geometry (Chow et al. 2006b and 2007a). The MOSFET measurements were compared to Monte Carlo simulations using the EGSnrc-based code. The advantages of using Monte Carlo simulation to determine the EBF are: (1) the voxel size near the lead surface can be very small (< 1 mm) to fulfill the definition of the EBF, (2) any human uncertainties and variations during the measurement can be avoided (Chow et al. 2006c), (3) the perturbation due to the dosimeter body in the beam can be avoided in the measurement, and (4) once the Monte Carlo model has been verified by measurement, EBFs varying with different clinical treatment setups can be determined easily using computer. This saves time and human effort to carry out patient specific measurements case by case using linac (Chow 2007b). Moreover, it is well known that EGS4-based (previous version of EGSnrc) code underestimates the electron backscatter due to the used trajectory scheme compared to other advanced codes such as PENELOPE and MCNP (Jera et al. 1999). However, with the introduction of EGSnrc, the problem of electron backscatter in the simulation has been solved (Kawrakow 2000), and it is therefore worthwhile to evaluate the performance of the EGSnrc-based code in the electron backscatter calculation (Chow et al. 2008b).

The final study on MOSFET in this chapter is to evaluate the uncertainty and performance of this dosimeter in measuring the EBF. Monte Carlo simulation was used to model the presence of a MOSFET dosimeter on the lead to determine the electron backscatter. The GEANT4-based code was used in the simulation (Truscott et al. 2000, Carrier et al. 2004). In this study, the energy responses of electrons with and without presence of the MOSFET body were studied by Monte Carlo simulations. The influences due to the presence of the MOSFET body on the electron fluence with and without the dosimeter using simulations were compared to the related measurements, in order to find out the uncertainty and suitability of using MOSFET dosimeter to determine the EBF.

II. Methods and Materials

A. Surface Dose Measurements Using MOSFET

Unencapsulated TN-1002RD MOSFET dosimeters purchased from Thomson and Nielsen Electronic, Ottawa, Canada were used to measure surface doses in different positions on a Rando Phantom for the prostate 4FB and IMRT plan. The MOSFET dosimeter was calibrated with our local standard of Farmer-type ionization chamber (Capintec PR06C) and electrometer (Capintec 192) using solid water slabs. Figures 1(a) and 1(b) show percent depth dose (PDD) curves for the 6 and 15 MV photon beams measured using the MOSFET dosimeter. The calibrated MOSFET dosimeters were then used to measure surface doses in different specified points on the Rando Phantom for each beam in an IMRT/4FB plan with different photon beam energies. The measurement and isotopic uncertainty of the dosimeter is ± 3% and ± 2% for 360 deg reported by the manufacturer (Technical note #7: MOSFET dosimeter specifications. Thomson and Nielsen Electronics), respectively. During the measurement, the dosimeters were fixed on the Phantom surface using tape. The thickness of the tape was 0.13 mm. The ratio of the dose of each dosimeter with and without tape was determined in order to make correction in the measurement.

Prostate IMRT plans were carried out by the inverse planning routine (P^3IMRT) in the Pinnacle^3 system. Five-, 7- and 9-beam plans for the 6 and 15 MV photon beams were generated. The prescribed dose to the planning target volume (PTV) was 82 Gy in the volume of 95% contour (V_{95}) with 2 Gy per fraction. The dose constraints of plans were set as: 75 Gy for the bladder and rectum in V_{10}, 38 Gy maximum dose in V_0 to the left and right femur heads. The mean doses to the prostate PTV calculated by Pinnacle^3 for the 5-, 7- and 9-beam 6 MV IMRT plans were 83.8, 84.7 and 84.6 Gy, respectively, while for the 15 MV plans were 84.0, 85.7 and 84.5 Gy, respectively. The "step-and-shoot" segments of the MLC beams for each IMRT plan were then generated and sent through the DICOM RT to a Varian 21 EX linac with 120 leaves MLC. For comparison, a typical 15 MV 4FB plan was also generated with the same PTV. Four MLC beams conforming to the prostate PTV with gantry angles 0,

90, 180 and 270 deg were used. To compare further, all IMRT and 4FB plans were set at 2 Gy per fraction in the measurement.

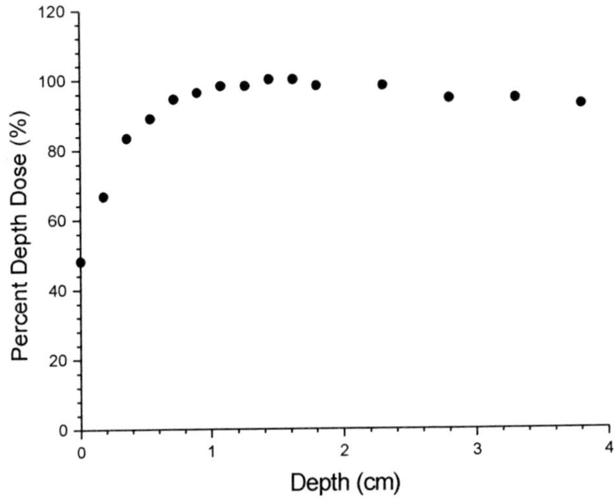

(a)

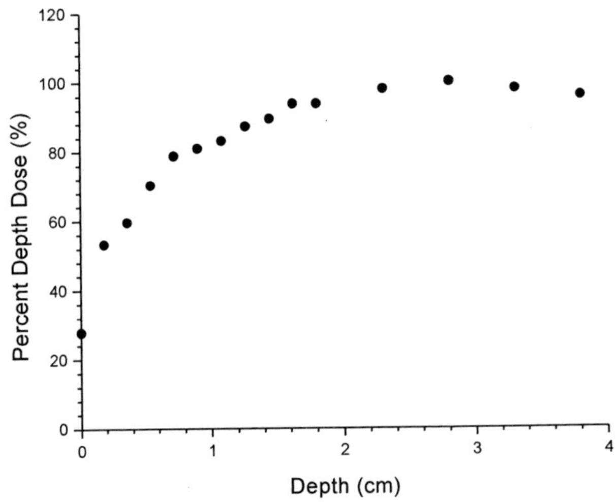

(b)

Figure 1. Percent depth dose curves for the (a) 6 MV and (b) 15 MV photon beam measured using the MOSFET dosimeter.

For the surface dose measurement on the Rando Phantom after the Pinnacle[3] planning, the Phantom was sent to the treatment room and set up on the couch for irradiation. Five MOSFET dosimeters were placed with equal distance of 3.75 cm along a transverse surface line around the Phantom external contour at the central beam axis (i.e. left, anterior and right surface). The length of the sampling line was from – 30 cm (left) to the central beam axis, and from the beam axis (i.e. 0 cm) to 30 cm (right) covering almost all the exposed surface area of the Phantom. There were 15 measuring points and the irradiation of each IMRT or 4FB beam was to be repeated three times for the five dosimeters in three different positions. Moreover, dosimeters on the transverse surface line were shifted towards the inferior to 5 and 10 cm from the beam central axis. Five sampling points along the surface contour with distances –15, -7.5, 0, 7.5 and 15 cm were set. The measurement was repeated for both IMRT and 4FB irradiations. For the MOSFET measurements of peripheral surface dose, dosimeters were placed from the central axis of the beam up to 30 cm inferior along the Phantom's sagittal surface line. The MU calculated by Pinnacle[3] for each IMRT/4FB beam was used in the measurement.

B. Electron Backscatter Measurement Using MOSFET with Monte Carlo Simulation

The experimental setup to determine the EBF is found in Fig. 2. Clinical electron beams with energies of 4, 6, 9, 12 and 16 MeV, and with a 10 × 10 cm^2 applicator were used at the source-to-surface distance (SSD) = 100 cm. The distance between the bottom of the applicator and the Phantom surface was 5 cm. A piece of lead sheet with thickness ranging from 1 – 6 mm was placed on top of a solid water slab, and a layer of tissue-equivalent bolus with thickness ranging from 0.5 – 2 cm was used to cover the lead. The area of the solid water

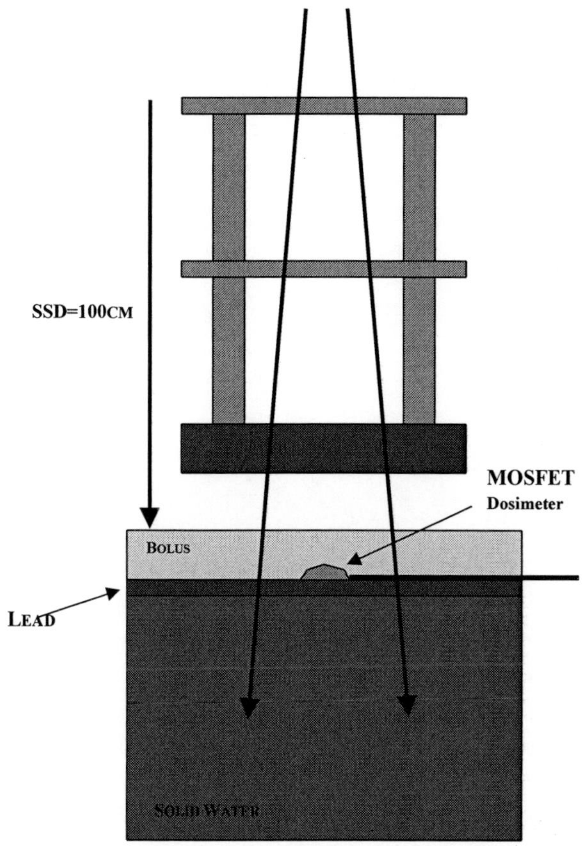

Figure 2. Schematic diagram (not to scale) of the experimental setup for the EBF measurement using the MOSFET dosimeter.

slab, lead sheet and bolus was 25 × 25 cm^2 which was larger than the electron field at SSD = 100 cm. To measure the dose due to the electron backscatter, MOSFET dosimeter was placed at the central beam axis between the bolus and lead. The clinical electron beams were produced by a Varian 21 EX linac. Doses were measured and calculated at the interface between the bolus and lead with beam angles varying from 0 – 15 deg.

For Monte Carlo simulation using the EGSnrc-based code (Nelson et al. 1965, Kawrakow and Rogers 2000), PDD curves for electron beams were determined using the EGSnrc-based DOSCTP and DOSXYZnrc (Chow and Leung 2007c and 2008a, Ma et al. 1995). The Phantom geometry as per Fig. 2 was input to the DOSXYZnrc with voxel volume set to 1 × 1 × 0.05 cm^3 for the x, y and z direction, respectively. The thickness of lead (1 – 6 mm) depended on the electron beam energy. The attenuations of electron beams due to different thicknesses of lead were investigated and found that 2 mm lead was enough to produce adequate electron backscatter for all electron beam energies (Perez-Calatayud et al. 2000). The thickness of bolus on top of the lead ranged from 0.5 – 2 cm, which is typical for the thickness of lip, buccal mucosa and ear. Fifty million histories were run for each calculation and the relative dose error (uncertainty as a fraction of dose in the voxel) was found around 1% for all electron beam energies.

For the measurement, MOSFET dosimeter was used to measure the dose due to the electron backscatter from the lead. The measured EBFs were then compared to those calculated using Monte Carlo simulations. The MOSFET dosimeter was calibrated with our local standard Farmer-type ionization chamber (Capintec PR06C) and electrometer (Capintec 192) using solid water slabs for all clinical electron beams with the 10 × 10 cm^2 applicator and cutout. High bias supply setting was used to produce a sensitivity of 9 mV/cGy for the MOSFET. The dosimeter is made of a silicon chip with a kapton base enclosed by a black epoxy. The thickness and length of the epoxy are 1.3 and 8 mm, respectively (Technical note #7: MOSFET dosimeter specifications. Thomson and Nielsen Electronics). One hundred cGy were given at the depth of maximum dose of the corresponding electron beam in each measurement. Although lower electron beam energies such as the 4 and 6 MeV are more likely to be used in the treatment of superficial tumour and lesion, all available electron beam energies were measured and calculated for a more thorough reference to results from other studies using higher energy beams (Klevenhagen et al. 1982, Das et al. 2004, Jera et al. 1999).

C. Monte Carlo Simulation on MOSFET for Electron Backscatter

To evaluate the performance and influence of the presence of the MOSFET body in the EBF measurement, MOSFET dosimeters were used with setup as shown in Fig. 2. A 2 mm lead was placed on a solid water slab (CNMC, Tennessee, USA; area = 25 × 25 cm^2) with 10 cm thickness. This solid water slab is designed for photon and electron beam calibration with its mass and volume electron densities, electron and photon transmission characteristics, and physical dimensions virtually the same as those in liquid water. A layer of tissue-equivalent SUPERFLAB plastic bolus (density = 1.02 g cm^{-3}) with thickness of 1 cm was placed on top of the lead. Electron beams with field size of 10 × 10 cm^2 and energies of 4, 6, 9 and 12 MeV produced by a Varian 21 EX linac were used in the measurement. The TN-1002RD MOSFET dosimeter (Thomson and Nielsen Electronic, Ottawa, Canada) was placed between the lead and bolus with the flat surface of the dosimeter lying on top of the lead as shown in Fig. 2. The center of the dosimeter was located at the central beam axis. The SSD was equal to 100 cm.

The TN-1002RD MOSFET is a high sensitivity dosimeter. The normal sensitivities under standard and high bias supply settings are 3 mV/cGy and 9 mV/cGy, respectively (Technical note #7: MOSFET dosimeter specifications. Thomson and Nielsen Electronics). A high bias supply setting was used, and the dosimeter was calibrated with our local standard Farmer-type ionization chamber (Capintec PR06C) and electrometer (Capintec 192) using solid water slabs for all electron beams. One hundred cGy were given at the depth ensuring maximum dose of the corresponding electron beam in each measurement. The doses with and without the lead underneath the dosimeter were measured as per the beam geometry in Fig. 2. The EBF was then calculated for comparison to the result of Monte Carlo simulation using the GEANT4-based code (Chow et al. 2008c).

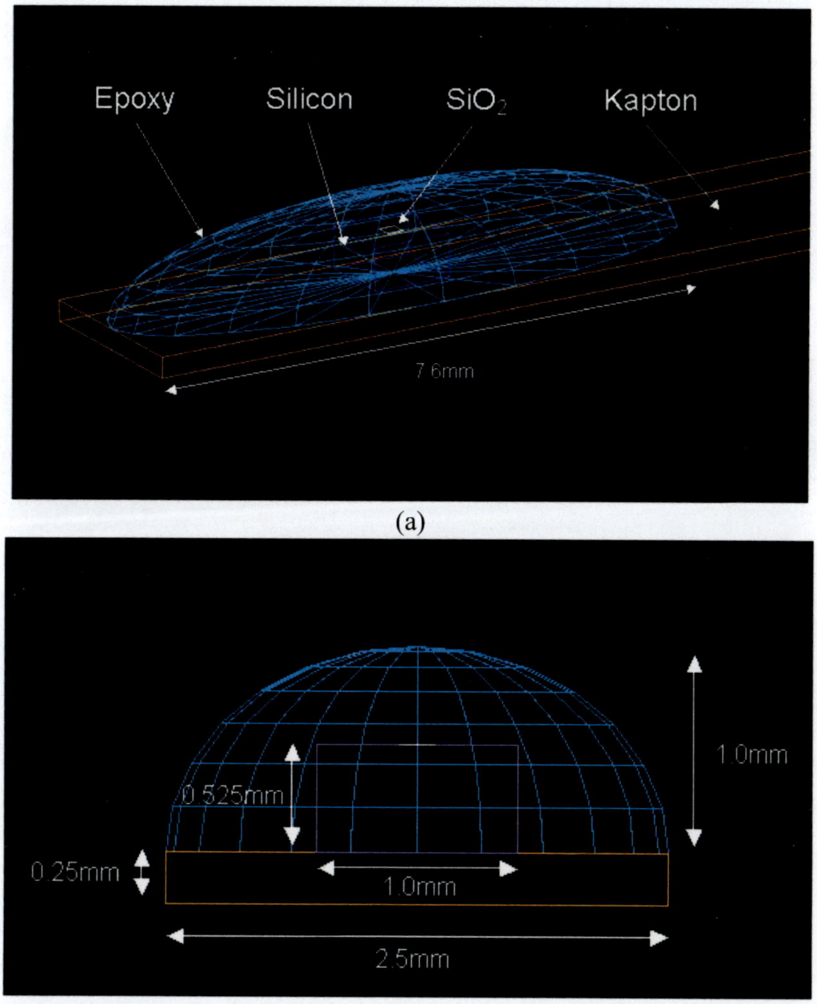

(b)

Figure 3. Schematic diagrams of the MOSFET dosimeter. (a) A 3D diagram showing the semi-ellipsoid epoxy bulb attached to the end of a kapton cable, and (b) the cross-section perpendicular to the cable axis showing the dimensions of different components.

The 3D and schematic diagram of the TN-1002RD MOSFET dosimeter is shown in Fig. 3. The structure and material was defined using the GEANT4 code. The epoxy bulb is modelled as a semi-ellipsoid attached to the end of a Kapton cable (0.25 mm thick and 2.5mm wide) as shown in Fig. 3(a). It encloses the silicon substrate (1×1 mm^2 in area, 0.525 mm thick as shown in Fig. 3(b)), and the thin SiO_2 sensitive volume (0.2×0.2 mm^2 in area, 1 μm thick).

Electron transport in a medium is characterized by many interactions, each of which contributes to small loss of energy. Secondary electrons produced in this manner are typically low in energy. For simulations involving small target volumes, physics models valid in the low energy ranges are required to ensure that the electron's trajectories are accurately modelled. This is important in this study given the small dimensions of the MOSFET dosimeter components. GEANT4 has two different low energy libraries for electromagnetic interactions. The first one, referred to as GEANT4 LowE, is a parameterized model based on the Livermore Evaluated Electrons and Photons Data Libraries (Rodrigues et al. 2004). The other is based on the set of processes originally developed from the FORTRAN-based Penelope code (Salvat et al. 2003). Both libraries describe interactions between photon and electron with matter down to 250 eV. In this study, the Penelope-based GEANT4 library is used, which has been verified to be in excellent agreement with the original PENELOPE code, and more suitable for microdosimetry applications (Spiga et al. 2007).

Four electron beam energies (4, 6, 9 and 12 MeV) were used in the simulation. The number of primary electron histories for each simulation was 20-50 million. In GEANT4, production thresholds are defined in terms of distance, which is then internally converted into an energy threshold, below which no secondary particles are generated. A universal threshold of 1 μm was used for all particle types. Electron interactions resulting in energy deposition occurring within the dosimeter volume were histogrammed to determine the energy spectrum, as well as accumulated to find the overall dose. Each simulation was 15 hours on average.

III. Results

A. Surface Dose Measured Using MOSFET for Prostate IMRT

Figures 4(a-e) were the percentage skin doses normalized to the prescribed dose at the isocenter, along the transverse line at the central beam axis for beam angles of 0, 80, 140, 220 and 280 deg in the 5-beam IMRT, respectively. Six and 15 MV photon beams were used in the prostate IMRT. The negative distance in the figures refers to the left hand side of the Phantom surface (viewing from the target to the gun position of the linac), while the positive distance the right hand side. The percentage surface dose (normalized to the dose at the isocenter with SAD = 100 cm) distributions around the Phantom surface of the 4FB, 5-, 7- and 9-beam IMRT plans for the 6 and 15 MV photon beams are shown in Figs. 5(a) and 5(b), respectively. Figure 6 shows the percentage surface doses measured from the central beam axis to 30 cm away along the Phantom's sagittal surface line (from superior to interior) for the 5-beam IMRT (6 MV) and 4FB (15 MV) plan.

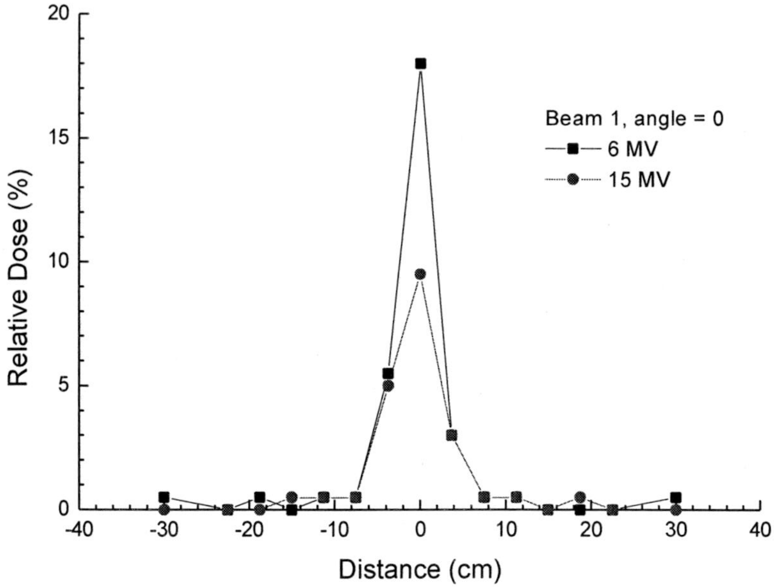

(a)

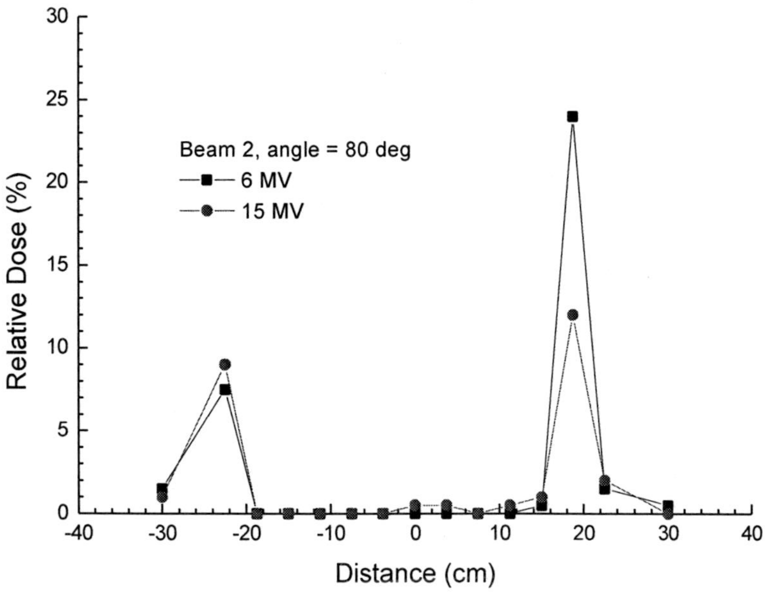

(b)

Figure 4. Continued on next page.

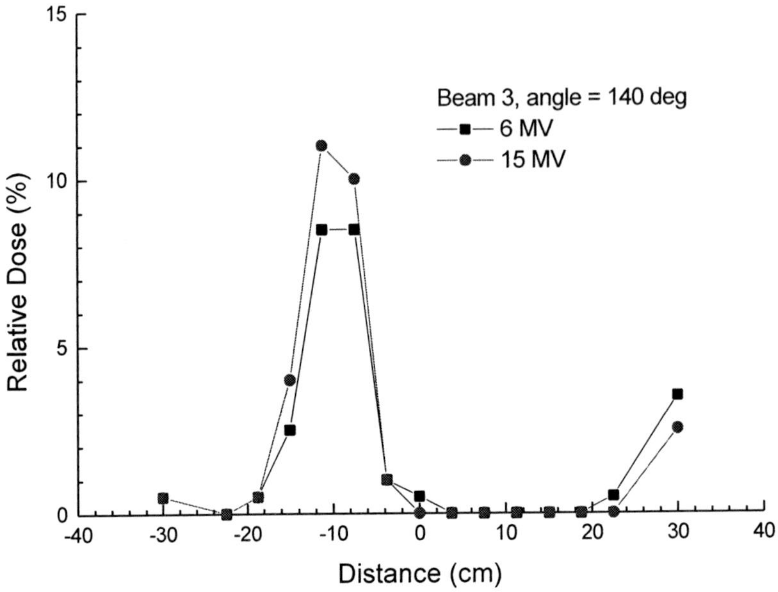

(c)

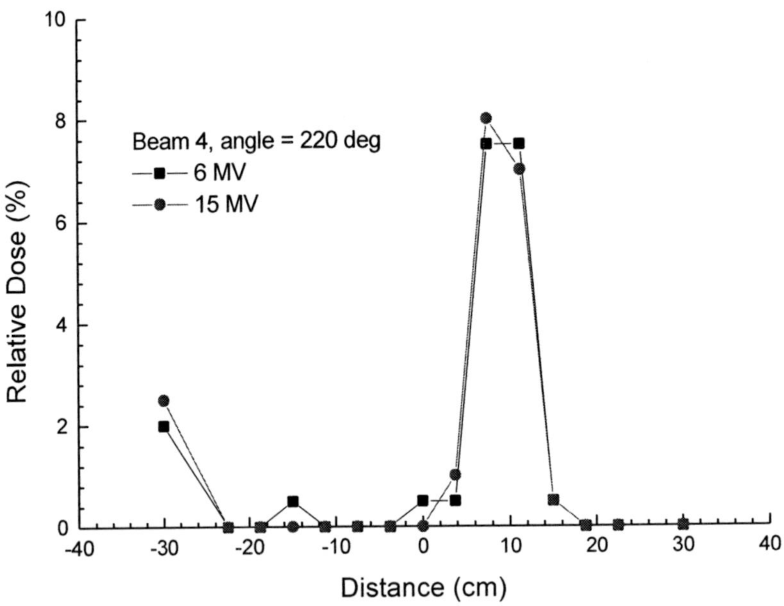

(d)

Figure 4. Continued on next page.

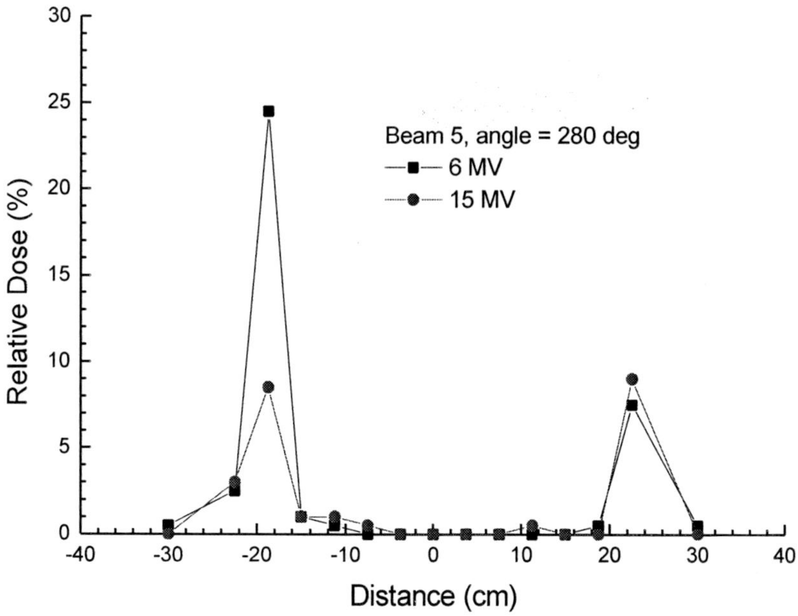

(e)

Figure 4. Percentage surface dose distributions along the transverse surface line at the beam central axis irradiated by each of the five beams: (a) beam 1, (b) beam 2, (c) beam 3, (d) beam 4, and (e) beam 5 in the 5-beam IMRT plans for the 6 and 15 MV photon beams, respectively. The surface dose is normalized to the prescribed dose at the isocenter.

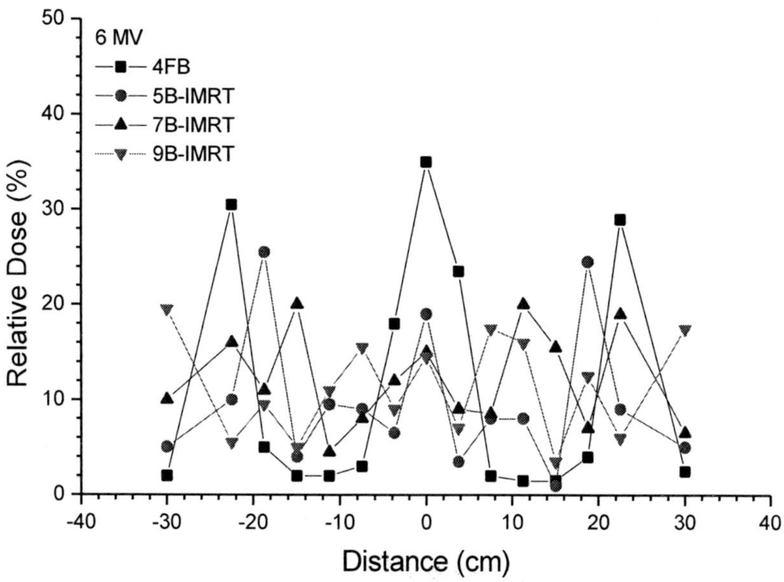

(a)

Figure 5. Continued on next page.

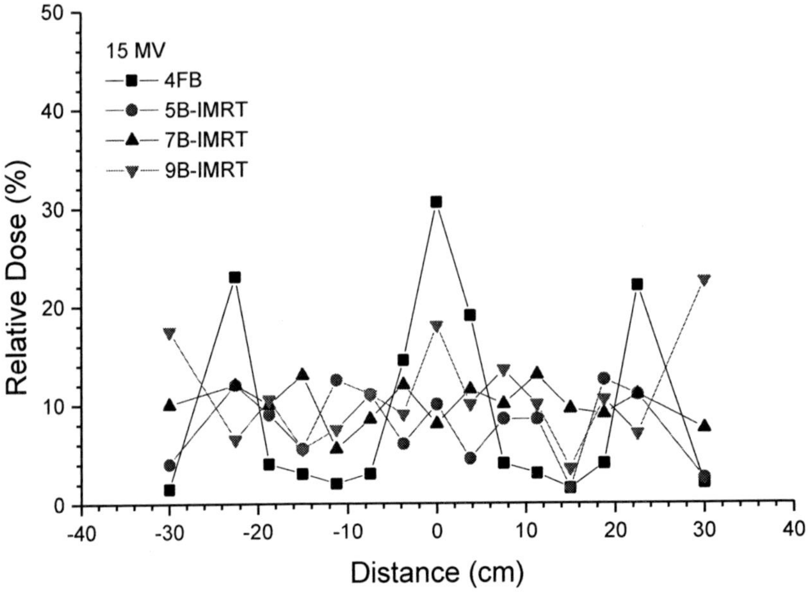

(b)

Figure 5. Percentage surface dose distributions around the Rando Phantom irradiated by the 4FB, 5-, 7- and 9-beam IMRT plans with the (a) 6 MV and (b) 15 MV photon beams. The dose is normalized to the dose at the isocenter.

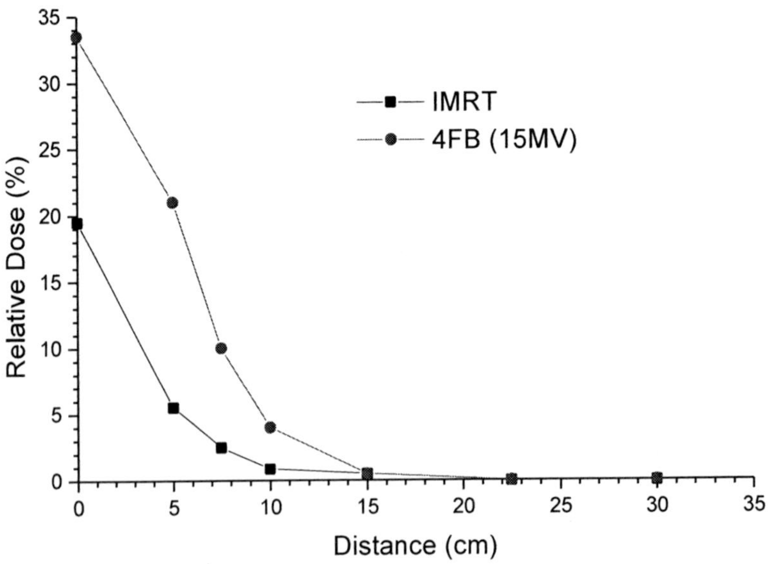

Figure 6. Percentage peripheral surface doses (normalized to the prescribed dose at the isocenter) measured from the beam central axis to 30 cm away along the Phantom's sagittal surface line for the 5-beam IMRT (6 MV) and 4FB (15 MV) plan.

B. EBF Measured Using MOSFET with Monte Carlo Simulation

Figure 7 shows the verification results of Monte Carlo simulations using the EGSnrc-based code compared to measurements. Figure 7(a) plots the measured and calculated EBFs against the electron beam energies for the 0.5 and 1 cm tissue-equivalent bolus on top of the lead.

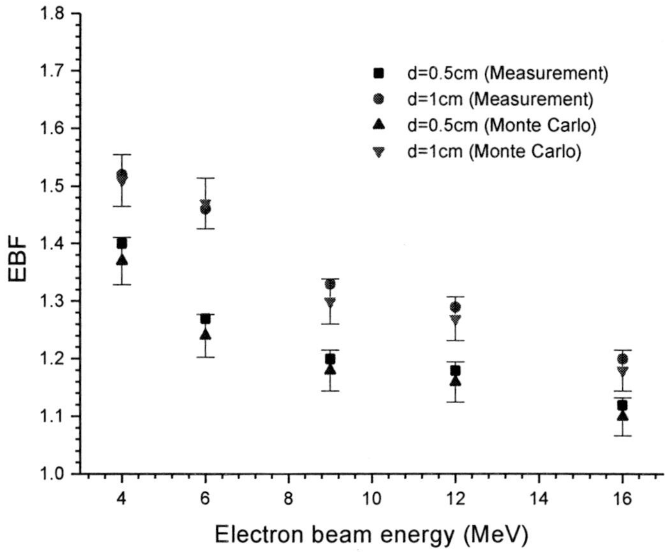

(a)

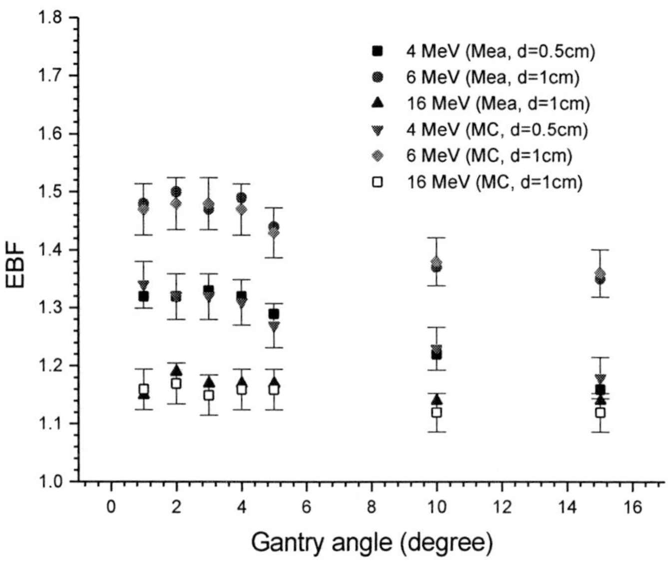

(b)

Figure 7. Measured and Monte Carlo simulated results of EBF (a) vs. electron beam energy with the lead at depths of 0.5 and 1 cm, and (b) vs. gantry angle for the 4, 6 and 16 MeV electron beams.

The relationships between EBFs (measured using MOSFET and calculated using Monte Carlo) and beam angles for the 4, 6 and 16 MeV electron beams are shown in Fig. 7(b). The thickness of lead used in Fig. 7 was 2 mm. The relationship between the calculated EBF and thickness of lead at depths of 0.5 and 1 cm using the 12 MeV electron beams is shown in Fig. 8.

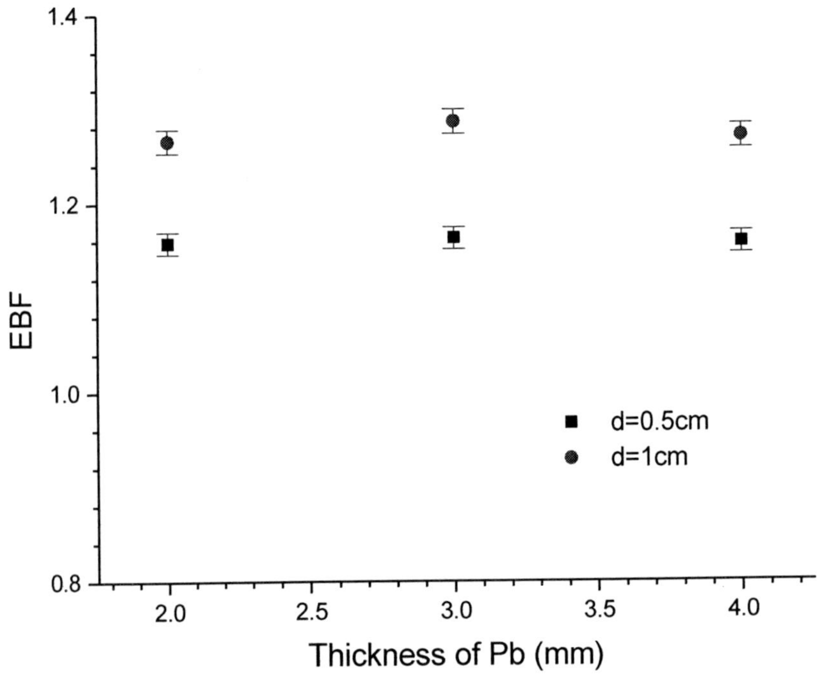

Figure 8. EBF plotted against the thickness of the lead at depths of 0.5 and 1 cm using the 12 MeV electron beams with the 10 × 10 cm^2 applicator and cutout.

In the figure, it can be seen that the EBF does not depend on the thickness of lead, and 2 mm lead is enough to produce electron backscatter for the calculation and measurement (Saunders and Peters et al. 1974). Figures 9(a-e) show the relative dose profiles with a 2 mm lead inserted at different depths in the Phantom for the 4, 6, 9, 12 and 16 MeV electron beams using the 10 × 10 cm^2 applicator and cutout, respectively. The PDD curve for each electron beam energy without the lead (labelled "No Pb") is also shown in the corresponding figure. Figures 10(a) and 10(b) show the relative dose profiles with 2 mm lead at 0.5 cm depth for the 4 MeV beam and 1 cm depth for the 6 MeV beam, respectively. The profiles of "black solid square" in Fig. 10 are PDD curves without the lead. The electron beam angles in Fig. 10 changed from 0 – 15 deg. Figures 11(a) and 11(b) show the relative dose profiles with 0.5 and 1 mm aluminium on top of the 2 mm lead at 1 cm depth for the 9 MeV electron beam, respectively.

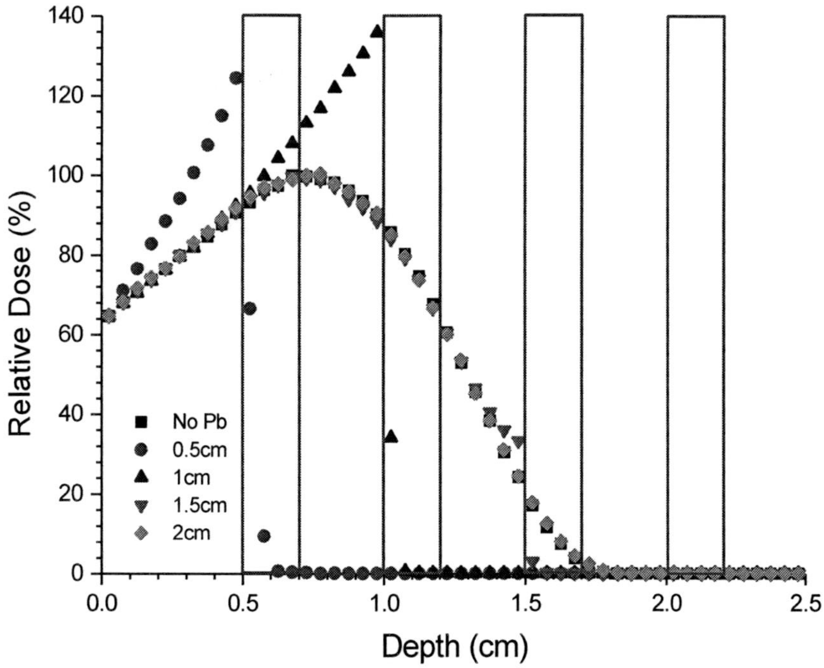

(a)

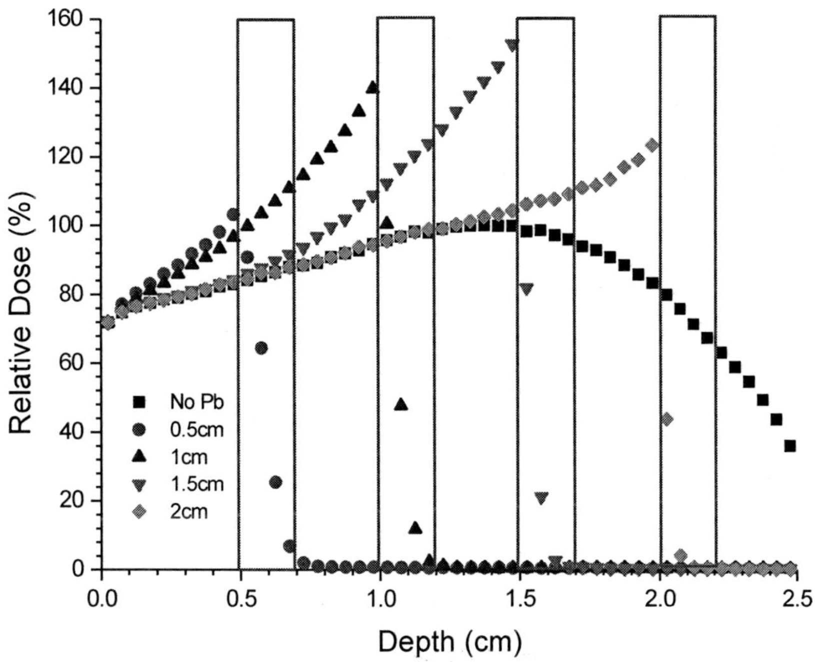

(b)

Figure 9. Continued on next page.

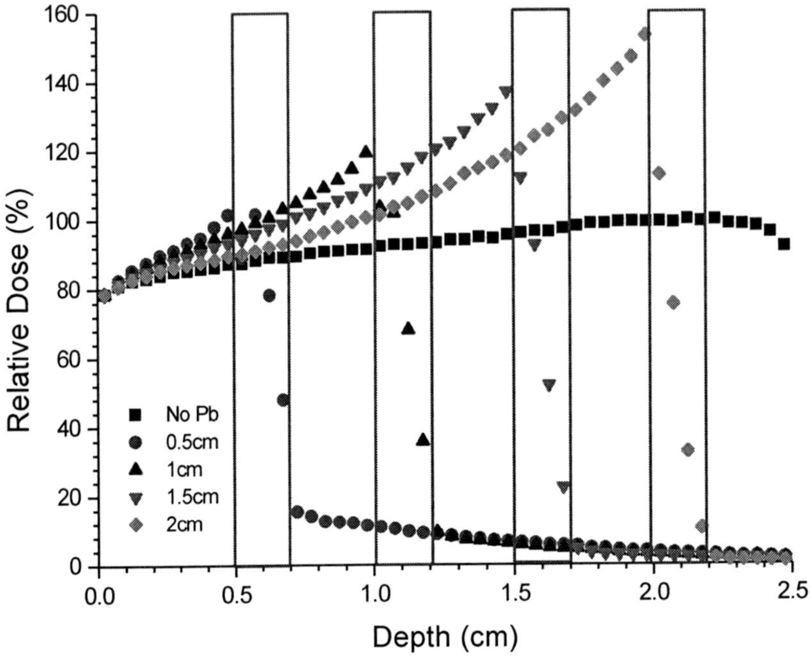

(c)

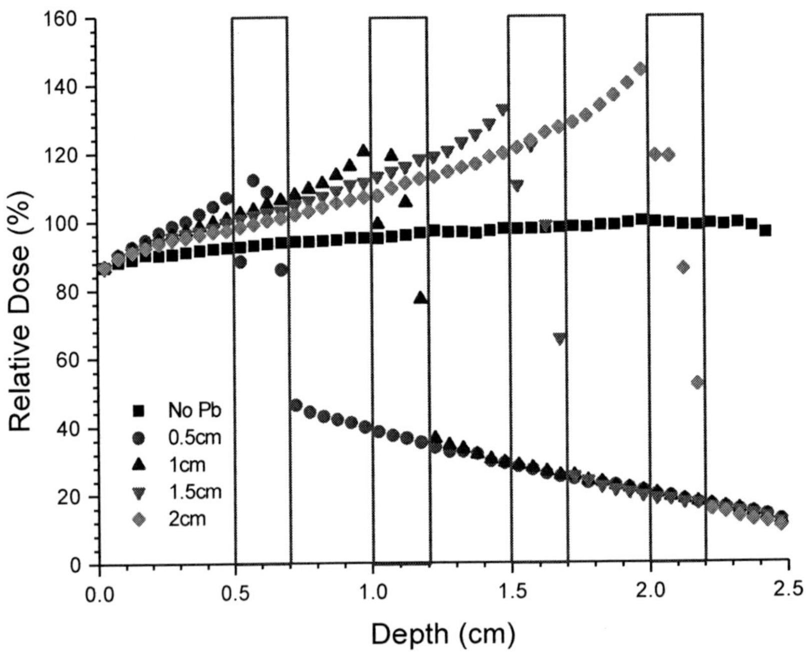

(d)

Figure 9. Continued on next page.

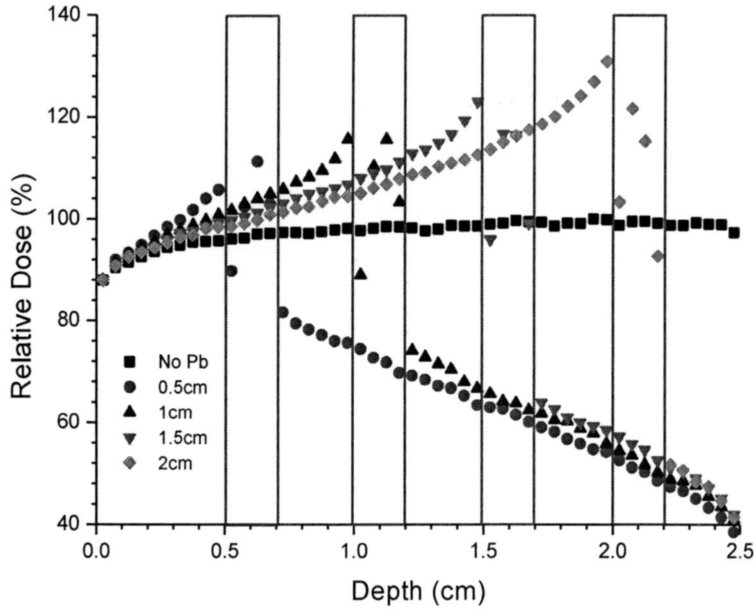

(e)

Figure 9. Relative dose profiles with a 2 mm lead sheet at depths of 0.5, 1, 1.5 and 2 cm for (a) 4 MeV, (b) 6 MeV, (c) 9 MeV, (d) 12 MeV and (e) 16 MeV electron beams. The percent depth dose curves of the above beam energies (No Pb) are also shown in the corresponding subfigures.

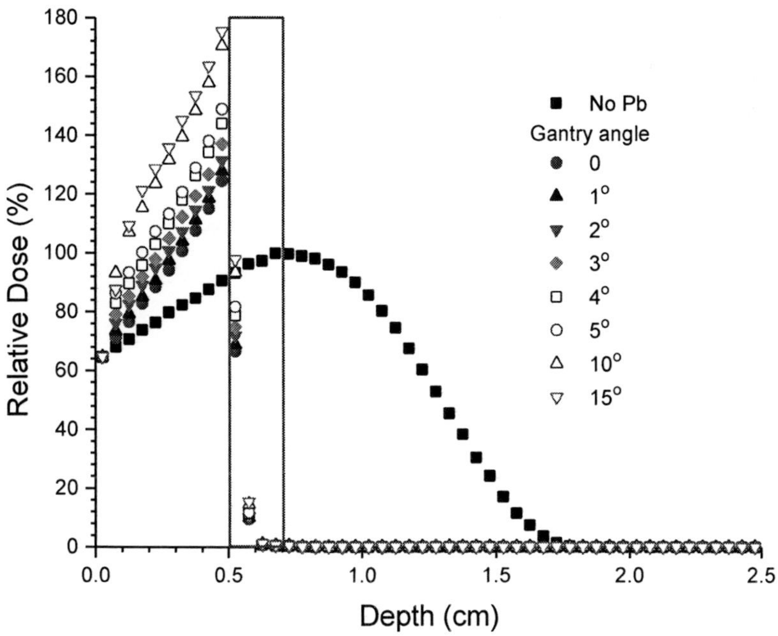

(a)

Figure 10. Continued on next page.

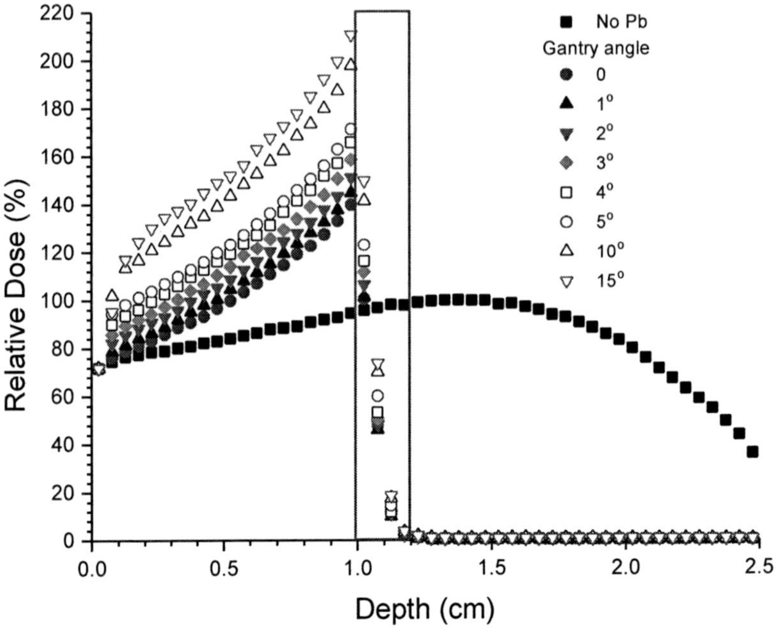

(b)

Figure 10. Relative dose profiles with a 2 mm lead sheet at (a) depth of 0.5 cm using the 4 MeV electron beam, and (b) depth of 1 cm using the 6 MeV electron beam with gantry angles varying from 1 – 15 deg.

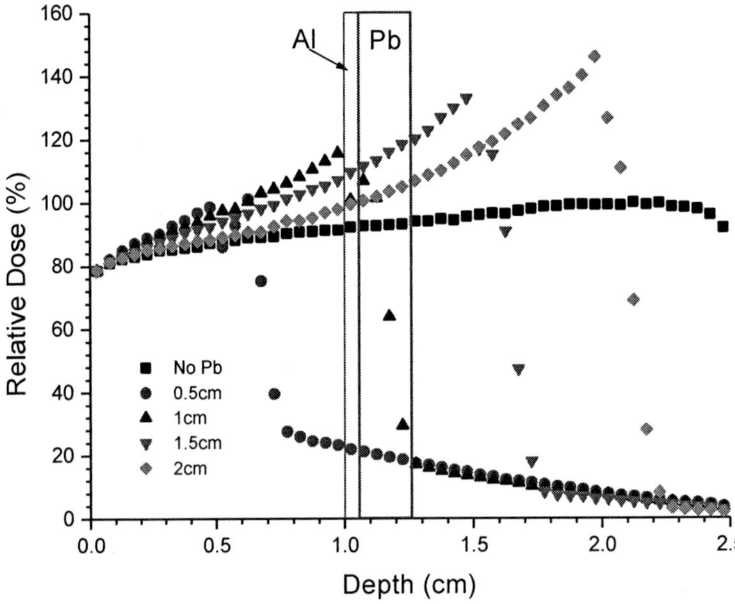

(a)

Figure 11. Continued on next page.

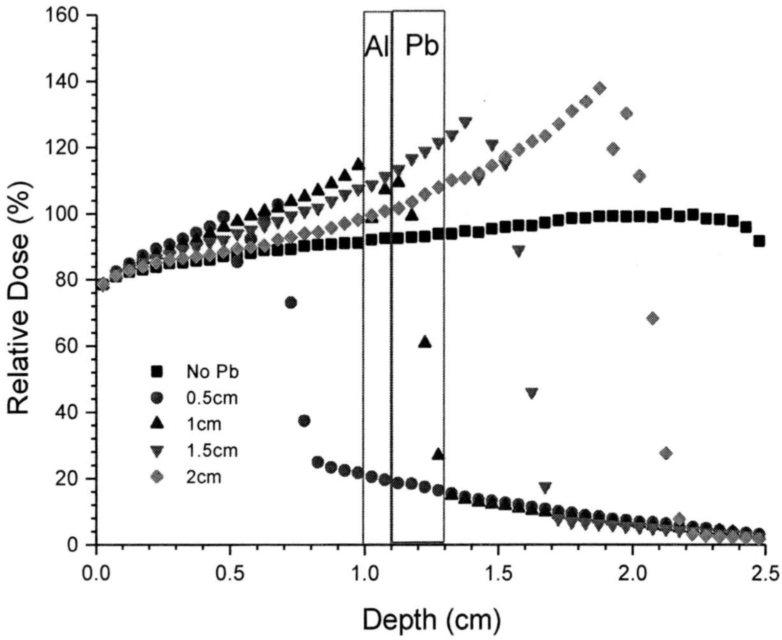

(b)

Figure 11. Relative dose profiles with a 2 mm lead sheet with (a) 0.5 mm and (b) 1 mm aluminium sheet on top of the lead at depths of 0.5, 1, 1.5 and 2 cm using the 9 MeV electron beams.

C. Monte Carlo Simulation on MOSFET for Electron Backscatter

The doses deposited in the MOSFET body (dosimetric volume) with and without the lead underneath were predicted by Monte Carlo simulations (GEANT4-based) with electron beam energies of 4, 6, 9 and 12 MeV according to the experimental setup in Fig. 2. On the other hand, the MOSFET body as shown in Fig. 3 was taken out and replaced by the bolus in the simulations, so that doses at the same effective point of measurement were predicted again with and without the lead. The results of absorbed doses in the above simulations were used to calculate EBFs with and without the MOSFET body for different electron beam energies. Figure 12 shows the EBF plotted against the electron beam energy with and without the presence of the MOSFET body. The results of MOSFET measurements were also shown in the figure for comparison. Figures 13(a-d) show the spectra of the energy deposited by the electrons in the dosimetric volume: (1) with and without the lead in the presence of the MOSFET, and (2) with and without the lead in the absence of the MOSFET, for electron beams with energies of 4, 6, 9 and 12 MeV, respectively. All energy spectra were normalized to the peak of the spectral curve without the lead and without the MOSFET body with respect to the electron beam energy.

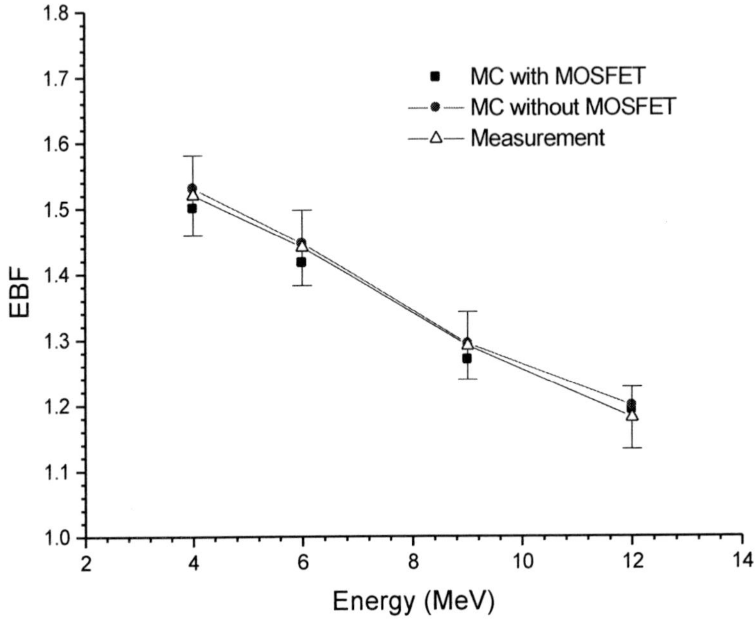

Figure 12. EBF vs. electron beam energy with and without the presence of the MOSFET body. EBFs were measured using MOSFET dosimeters and predicted by Monte Carlo simulations using the GEANT4-based code.

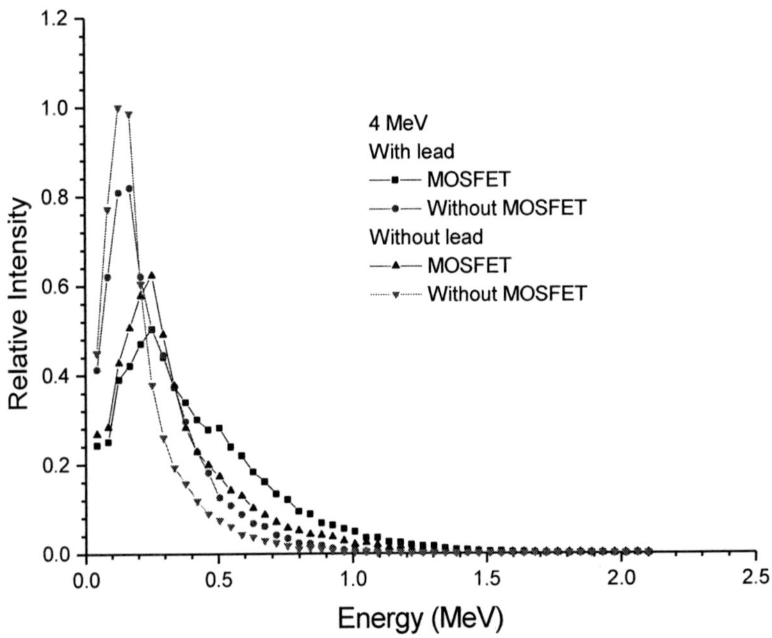

(a)

Figure 13. Continued on next page.

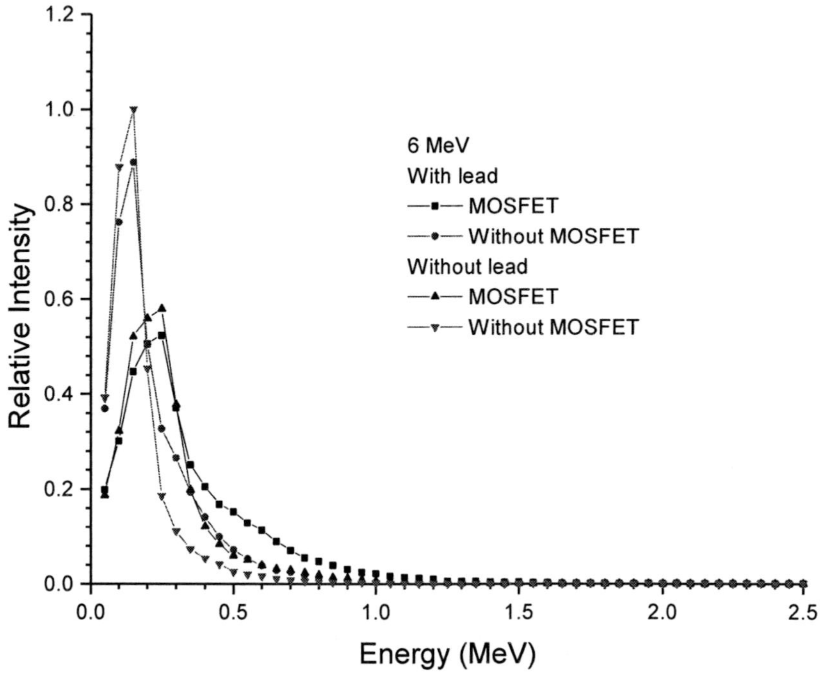

(b)

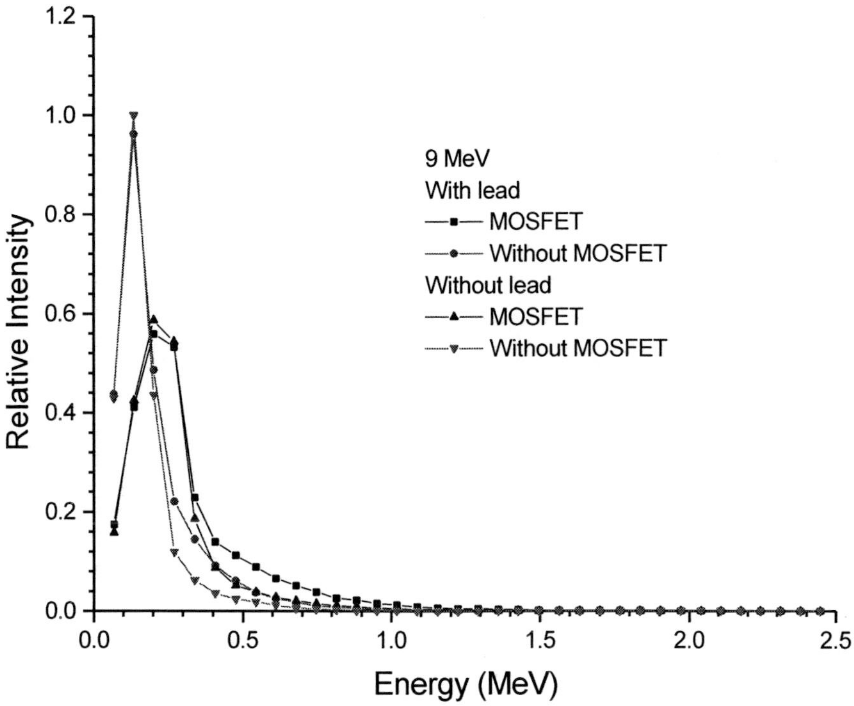

(c)

Figure 13. Continued on next page.

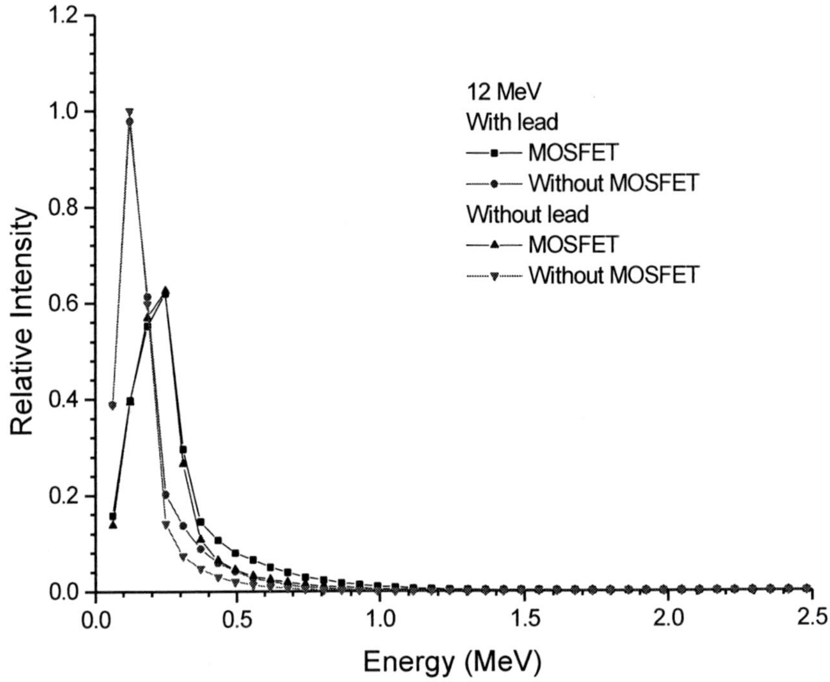

(d)

Figure 13. Spectra for the energy deposited by the backscattered electrons in the dosimetric volume of the MOSFET dosimeter for the (a) 4, (b) 6, (c) 9 and (d) 12 MeV electron beams. Energy spectra (1) with and without the lead in the presence of the MOSFET body, and (2) with and without the lead in the absence of the MOSFET body were plotted.

IV. Discussion

A. Surface Dose Measured by MOSFET for the Prostate IMRT

For every prostate IMRT using different number of beams and photon beam energies, the surface doses were measured around the Phantom's external contour with respect to every beam in the plan. In this study, only the percentage surface dose distributions normalized to the dose prescribed at the isocenter of each beam in the 5-beam IMRT are shown (Figs. 4(a-e)). It is found that the surface dose distribution for each beam around the Phantom surface depends on the IMRT dose intensity map generated by all segments (using step-and-shoot) in the beam, beam angle, photon beam energy, field size, and position of the dosimeter. As the abdomen part of the Rando Phantom is close to be cylindrical and MV photon beams are used, the directional dependence of the MOSFET dosimeter is neglected (Chuang et al. 2002).

It is reasonable that the surface dose at point zero (located at the central beam axis, beam angle = 0 deg) for the 6 MV photon beam is higher than that of the 15 MV as shown in Fig. 4(a). The surface dose curves are not exactly symmetric because of the non-symmetric dose intensity map in the plan, and the uncertainty of the Phantom setup, which also happens generally in the clinical situation. In Figs. 4(c) and (d), although beams with angles 140 and 220 deg are mirroring beams from the Phantom's anterior sagittal plane, the measured dose curves (6 and 15 MV) are not exactly mirrored. This is due to their different beam intensity maps and calculated MUs. The same situation can be found in Figs. 4(b) and (e) where they (80 and 280 deg) are another mirror pair.

In Figs. 5(a) and (b), the 4FB curves have peaks at about –24, 0 and 24 cm. They are all locations directly irradiated by the 270, 0 and 90 deg beam. The zero-angle beam has largest percentage dose peak because the beam passes through the thinnest part of the Phantom (anterior and posterior) and overlaps with the 180 deg beam. The sum of the surface (entrance) and exit dose of these two beams makes the maximum percentage dose peak in the irradiation. It can also be seen that for the 4FB irradiation, the Phantom surface directly under the beam has high dose, but low dose can be found between two of the irradiated fields. For the three IMRT plans (5-, 7- and 9-beam), dose peaks are found around the Phantom's irradiated surfaces. However, as the number of beam increases, the irradiated area coverage begins to spread out along the Phantom surface. Another reason why IMRT has a more even surface dose distribution is that the plan does not use parallel opposite beams. The specific IMRT beam arrangement leads to the field out of the Phantom surface sandwiched well by two directly irradiated IMRT fields over the Phantom. The result is to have a lower and more even surface dose distributed on the Phantom surface. Therefore, the chance of skin complication due to the treatment is low. From the two groups of IMRT surface dose distributions (6 and 15 MV), it can be seen that the group of curves for the 6 MV is more fluctuated along the Phantom surface than that of the 15 MV. This is because the 6 MV photon beam has a higher surface entrance dose and lower exit dose than the 15 MV. It means that the area, where the 6 MV surface entrance dose is dominant, has relatively higher dose compared to its neighbourhood area where the exit dose is dominant. When measuring the surface dose beyond the transverse surface line at the central axis (5 and 10 cm away interior), the surface doses are found to be reduced. The surface dose distribution depends mainly on the IMRT dose intensity map, MU irradiated and the photon beam energy. In Fig. 6, the 4FB irradiation has higher percentage peripheral surface dose than that of the IMRT. This is because the measurement was carried out on the anterior surface of the Phantom where the beam with angle zero was dominant. The higher percentage surface dose of 4FB is therefore due to the higher calculated MU of beam with angle zero compared to that in the IMRT plan.

B. EBF Measurement Using MOSFET with Monte Carlo Simulation

1. Verification of the Monte Carlo Model

Figure 7(a) shows the MOSFET measured and Monte Carlo predicted EBFs varying with electron beam energy. The EBF decreases with increase of electron beam energy for both 0.5 and 1 cm depth of lead. Moreover, the thicker of the bolus on top of the lead, the larger the EBF. A 2 mm lead was used here and was proved adequate according to Fig. 8 (Saunders and Peters et al. 1974). The measured and predicted EBFs agree well and match with each other within ± 5%. The relationship between the EBF and electron beam angle is shown in Fig. 7(b). Again, both measured and predicted EBF agree well within ± 5%. It is proved that

MOSFET dosimeter is suitable for the EBF measurement, and Monte Carlo simulation based on the EGSnrc code was verified in this study.

2. EBF Varying with the Depth of Lead

In electron radiation therapy for superficial lesion using internal shield, knowing the relationship between the EBF and lead shielding depth under the treatment surface is important. It is because radiotherapist can use such information of EBF to judge, based on the treatment volume thickness, how thick the tissue-equivalent bolus or aluminium sheet should be used to enclose the lead in order to minimize the electron backscatter. In Fig. 9(a) showing the relative dose profiles for the 4 MeV electron beam, it can be seen that only depths of 0.5, 1 and 1.5 cm are meaningful, because the result of the 2 cm depth of lead is determined beyond the practical range, R_p (1.63 cm) of the electron beam. It is also interesting to note that the dose profile was reduced to almost zero inside about half thick of the 2 mm lead. This in fact indicates that only 1 mm lead is enough for the shielding. The EBF increases initially in the build-up region and then decreases afterward, showing that the EBF is depth dependent. Figure 9(b) for the 6 MeV electron beam shows a relationship similar to Fig. 9(a), except the 2 cm depth of lead is now shallower than the R_p (2.98 cm). Again, the EBF increases with depth in the build-up region and then decreases afterward. In Fig. 9(c) for the 9 MeV electron beam, however, all calculated EBFs are within the build-up region of the PDD, and the EBF increases (from 1.18 - 1.54) with the thickness of tissue-equivalent bolus (0.5 – 2 cm) on top of the lead. For the relative beam profiles in Fig. 9(c), the 2 mm lead is not enough to shield the 9 MeV electron beam, especially when the lead is close to the surface. It is found that 3 mm lead is better in shielding, though the related profiles are not shown here. Figures 9(d) and 9(e) show the relative dose profiles for the 12 and 16 MeV electron beams. Although such energies are not likely to treat superficial lesions (unless sufficient thickness of lead shield can be put into the patient), the EBFs increase (1.16 – 1.44 for the 12 MeV and 1.10 – 1.31 for the 16 MeV), when the lead is positioned further away from the surface (from 0.5 to 2 cm). In addition, it is found that 6 mm lead should be used to shield the 16 MeV beam while 4 mm lead is enough for the 12 MeV beam to obtain about 95% of attenuation. As the EBF varying with the electron beam energy, it is difficult to compare the results in this study with the published data, because there is no publication concerning similar results using the same linac model, EGSnrc-based code and measurement geometry. However, from the data using linac model of Varian Clinac 1800 and EGS4 code (Perez-Calatayud et al. 2000), it was found that at least both set of EBF data decrease with increase of the electron beam energy.

3. EBF Varying with the Beam Obliquity

Clinically, the skin surfaces of treatment sites such as lip, buccal mucosa, eyelid and nose are not flat but curved. This makes the real treatment beam geometry different from that used in Section IV.B.2, in which the electron-backscattered dose for a phantom with flat surface was calculated and measured. In this Section, the relationship between the EBF and electron beam angle was investigated. To estimate the deviation due to the impact of the surface curvature on the EBF, the dose at the tissue-lead interface with tilted electron beam is normalized to the dose at the same point without the lead with beam angle equal to zero. Figures 10(a) and 10(b) show the relative dose profiles with different beam angles for the 4 and 6 MeV electron beams. It can be seen that the tilted electron beams have higher backscattered electron doses from the lead than those of beams with zero angle. Such electron backscattered dose increases with increase of the beam angle. Although the beam obliquity effect may not be significant on the dosimetry during treatment, the resulted uncertainty is a reason to be extremely cautious, when considering the thickness of tissue-equivalent bolus or aluminium sheet enclosed the lead to remove the electron backscatter.

4. EBF with Aluminium Sheet on the Lead

While tissue-equivalent bolus is used to enclose the lead shield in order to protect the patient from electron backscatter, adding aluminium on top of the lead would increase the efficiency of electron backscatter absorption, and was suggested in using internal shield for the electron radiation therapy (Saunders and Peters et al. 1974). As the degree of electron backscatter absorption is related to the electron beam energy and thickness of the aluminium sheet, it is worthwhile to investigate the EBFs by varying the electron beam energy and depth of the lead in the Phantom. Figures 11(a) and 11(b) show the change of relative dose profiles when 0.5 and 1 mm of aluminium sheet was added on top of the lead shield positioned at different depths, respectively. Inside the aluminium, it can be seen that aluminium is able to reduce the backscattered electron dose from the tissue-lead interface in the upstream direction of the beam. Moreover, although 2 mm lead is enough to produce precise yield of electron backscatter, such thickness is not adequate to shield 100% of the 6 MeV electron beam (Das et al. 2004). Three mm lead is needed to achieve more than 95% attenuation for a thickness of treatment volume equal to 2 cm. In Figs. 11(a) and 11(b), the reduction of backscattered electron dose does not change very much when increasing the thickness of aluminium sheet from 0.5 to 1 mm. The addition of aluminium is very effective to reduce the electron backscatter when the electron beam energy is low (4 MeV). For example, with 2 mm aluminium at 1 cm depth for the 4 MeV electron beam, the EBF can be reduced to 1.04 from 1.51 without aluminium (about 31% reduction). However, as the electron beam energy increases, the reduction of backscattered electron dose decreases. With the same beam geometry, thickness of aluminium and lead at 1 cm depth for the 16 MeV electron beam, the EBF can only be reduced to 1.11 from 1.18, which is about 6% reduction.

C. Monte Carlo Simulation on MOSFET for Electron Backscatter

1. Influence of the MOSFET Body on the EBF Prediction

Although MOSFET dosimeter has a small physical size compared to ionization chamber, the perturbation of the MOSFET body on the electron fluence is interested in studying. Unlike photon beam that has been studied thoroughly for the performance of the MOSFET (Panettieri et al. 2007, Lavallee et al. 2006, Briere et al. 2006, Jones et al. 2005, Briere et al. 2005, Chow et al. 2005, Chow et al. 2006d, Wang et al. 2004), electrons are charged particles hence they lose energy continuously via ionization. Therefore, electron fluence may be varied significantly under the influence of the small MOSFET body when measuring the electron backscatter. Studying EBFs calculated with and without presence of the MOSFET body would help to understand the effect and performance of the MOSFET dosimeter on disturbing the electron backscatter from the lead. In Fig. 12, the EBFs calculated without the MOSFET body on the lead were larger than those with presence of the MOSFET. The deviations of EBFs with and without presence of the MOSFET body vary with different electron beam energies. The deviation is around 2% when the electron beam energy is low (4 or 6 MeV), and decreases from 2% to 0.9% when the electron beam energy increases from 4 to 12 MeV. In Monte Carlo simulation, it was found that the presence of the MOSFET body would vary the electron fluence, and hence reduce the absorbed dose for the EBF calculation. However, such reduction of dose is more significant for the MOSFET with than without the lead. This is probably due to the fact that for the former case (MOSFET with the lead), both doses deposited from the incident electron beam in the upstream direction, and the electron backscatter from the downstream direction would be attenuated by the MOSFET body. This reduction of dose is larger for the electron beam with low energy of 4 MeV in which the beam is attenuated more by the MOSFET body. When the electron beam energy is higher, the dose difference due to the presence of the MOSFET with and without the lead becomes small. This is because the electron beam with higher energy becomes more penetrative to the MOSFET body, and it is the reason why the deviation of EBF decreases with an increase of the electron beam energy. In Fig. 12, the presence of the MOSFET body in the measurement underestimates the EBF to about 2% for the low energy beam (4 MeV), and the deviation becomes smaller when higher energy beam was used (e.g. 12 MeV). Comparing to measurements, the Monte Carlo results in Fig. 12 with the presence of the MOSFET body agree well with their corresponding measurements to within ± 2%.

2. Influence of the MOSFET Body on the Electron Energy Spectrum

Figures 13 show the electron energy spectra in the MOSFET body/dosimetric volumes with and without the lead and presence of the MOSFET in the Monte Carlo simulations. In Figs. 13, variations of the electron energy due to the presence of the MOSFET can be found with and without the lead underneath the dosimeter. The peaks in the energy spectra are roughly at the same location for all four incident electron energies. This is because the spectra recorded energies that were deposited by electrons crossing the dosimetric volume, which is the same for the four electron energies. As the deposited energy is related to the electron path length, which is restricted by the dimension of the dosimetric volume in the MOSFET, the peaks in the energy spectra should not change significantly with different electron energies. Generally, no matter whether the MOSFET is occupied in the dosimetric volume for the electron backscatter calculation, the electron energy spectral curve is shifted to higher energy when the lead is placed under the MOSFET. In another words, the relative intensity of the spectral curve with energy higher than the peak becomes larger with the lead present in the Monte Carlo simulation. This is because of the contribution of electron backscatter to the dosimetric volume, and results in the peak of the spectral curve lower than that of the curve without the lead. Moreover, when the MOSFET body is occupied in the dosimetric volume, the spectral curve was widened with or without the lead. This shows that the presence of the MOSFET body spreads out the energy distribution of electrons in the beam. In Figs. 13(a-d), deviations of relative intensities of the energy spectral curves decrease with an increase of electron beam energy. This is due to the fact that the intensity of the electron backscatter decreases with an increase of electron energy. Similar result is observed for spectral curves with the MOSFET occupying the dosimetric volume. In Figs. 13, it is also seen that deviations of relative intensities with and without the lead for the presence of the MOSFET body are slightly larger than those without the MOSFET. This effect is especially significant when the electron beam energy is low (Fig. 13(a)), and results in a larger uncertainty in determining the EBF. When the electron beam energy is higher, the contribution of the electron backscatter becomes lower, leading to a smaller deviation of relative intensity in the spectral curve with and without the lead.

V. Conclusions

Different prostate IMRT plans (5-, 7- and 9-beam) were generated for a Rando Phantom and the doses around its transverse surface line at the beam central axis were measured using MOSFET dosimeters. The MOSFET has been well recognized as an accurate and convenient dosimeter for the measurement of surface dose in MV photon beams. For IMRT plans using the same number of beam but different photon beam energies (6 and 15 MV), the 15 MV one has less surface dose measured than that of the 6 MV (about 5 to 30% depending on the number of beam). However, care should be taken regarding neutron dose and radiation safety brought by the 15 MV photon beam (Howell et al. 2005 and Kry et al. 2005). For the 15 MV IMRT plans using different number of beams, the surface dose measured increases with the number of beam in the plan. However, for the 6 MV IMRT plans, the measured surface dose increases from 5- to 7-beam plan, and then drops at the 9-beam. The decrease of surface dose using the 9-beam plan is due to the reduction of calculated MU compared to that in the 5- and 7-beam plan. Compared to the 4FB plans, the distribution of surface dose along the transverse surface contour around the Phantom is more even in the IMRT plans, especially when a greater number of beams are used. Such even surface dose distribution reflects a lower skin complication rate for the patient, though the total surface dose measured in the IMRT is still higher than that of the 4FB plan. Moreover, IMRT plans avoid very high surface dose peaks as seen in the 4FB case where certain surface areas were directly under the fields. For typical 6 MV IMRT plans with different number of beam used, surface doses for the IMRT plan measured were about 15% to 30% higher than those in the 15 MV 4FB. The increase of the surface dose in the IMRT plan is mainly affected by the beam energy, increase of calculated MU, number of beam and the IMRT dose intensity map (i.e. the dynamic MLC movement during delivery) compared to the 4FB.

The electron backscatter produced by clinical electron beam from a Varian 21 EX linac has been systemically studied using MOSFET dosimeter and Monte Carlo simulation. EBF was used to characterise the electron backscatter with and without the lead, and predicted using the Monte Carlo EGSnrc-based code. The Monte Carlo results were verified with measurements using the MOSFET dosimeter. The dependences of (1) the thickness of the lead on the beam attenuation, (2) the depth of the lead on the EBF, (3) the electron beam energy on the EBF, (4) the electron beam obliquity on the EBF, and (5) the presence of the aluminium sheet on top of the lead on the EBF were focused on. It was found that 2 mm lead was enough to give 95% beam attenuation at 0.5 and 1 cm depth for the 4 and 6 MeV electron beams, respectively. However, at 1 cm depth, 3, 4 and 6 mm leads are needed to shield the 9, 12 and 16 MeV electron beams with the same beam attenuation, respectively. Within the R_p, the EBF increases with the depth of the lead, and lower electron beam energy has larger EBF. When the beam angle increases, EBF decreases slightly for all electron beam energies.

However, if the effect of beam obliquity is neglected while still assuming the beam axis is perpendicular perfectly to the flat treatment surface, the EBF would be underestimated. Such deviation depends on the electron beam energy. Adding aluminium on top of the lead can effectively reduce backscattered electrons for relative low electron beam energy. The degree of electron backscatter reduction depends on the thickness of the treatment volume for particular electron beam energy. The Monte Carlo predicted EBFs characterised by the depth of the lead, electron beam energy, beam obliquity and presence of aluminium sheet are expected to provide important clinical information for the radiation oncology staff to estimate the electron backscatter, when internal shield is involved in electron radiation therapy to treat superficial lesions.

The influence of the MOSFET dosimeter on determining the EBF was studied by the Monte Carlo simulation using the GEANT4-based code. EBFs were calculated with and without a layer of lead (thickness = 2 mm) underneath a 1 cm tissue-equivalent bolus using electron beams with energies of 4, 6, 9 and 12 MeV. By comparing EBFs with and without presence of the MOSFET body, it was found that with the MOSFET, EBF was underestimated to be about 2% for the electron beam with low energy (4 MeV). Such underestimation decreases to 0.9% with higher electron beam energy of 12 MeV. This is due to the fact that the reduction of dose with presence of the MOSFET body is larger with the lead underneath the dosimeter. Compared to measurements, deviations with the Monte Carlo results (without the MOSFET body) were within ± 2%. Variations of electron energy spectra show that the electron energy distribution spreads out when the MOSFET body is occupied in the dosimetric volume. Moreover, the contribution of the electron backscatter to the energy spectrum decreases with an increase of the electron beam energy. According to the Monte Carlo simulations, it is found that with the small physical size of the MOSFET dosimeter, uncertainties with presence of the MOSFET body in determining the EBF are about 2 – 0.9% with the electron beam energy increases from 4 – 12 MeV. This agrees well with the measurement. The uncertainty of EBF determined by comparing the results of Monte Carlo simulations and measurements is well within the accuracy of the MOSFET dosimeter (< ±4.2%) provided by the manufacturer. It is concluded that the MOSFET dosimeter performed well in measuring the electron backscatter from the lead for the clinical electron beams.

Acknowledgements

The author would like to thank assistances from Dr Grigor Grigorov in the MOSFET dosimetric measurements in the Grand River Regional Cancer Center (GRRCC), and Michael Leung in programming the GEANT4 code in the Princess Margaret Hospital (PMH). The author would also like to thank supports from Dr Robert Barnett (GRRCC) and Dr David Jaffray (PMH) in the MOSFET dosimetric and Monte Carlo simulation studies, respectively. The GEANT4 Monte Carlo simulations were carried out in the Physics Department of the University of Toronto supported by Professor Martin Lee.

References

August L S 1984 Design criteria for high dose MOS dosimeter for use in space *IEEE Trans. Nucl. Sci.* **31** 801-803

Bauman G, Gete E, Chen J Z, Wong E 2004 Simplified intensity-modulated arc therapy for dose escalated prostate cancer radiotherapy *Med. Dosim.* **29** 18-25

Belyakov V V, Pershenkov V S, Shalnov A V, Shvetzov-Shilovsky I N 1995 Use of MOS structures for the investigation of low dose rate effects in bipolar transistors *IEEE Trans on Nuclear Sci.* **42** 1660-1666

Bos L J, Schwarz M, Bär Werner, Alber Markus, Mijnheer Ben J., Lebesque Joos V., and Damen Eugène M. F 2004 Comparison between manual and automatic segment generation in step-and-shoot IMRT of prostate cancer *Med. Phys.* **31** 122

Briere T M, Beddar A S, Gillin M T 2005 Evaluation of precalibrated implantable MOSFET radiation dosimeters for megavoltage photon beams *Med. Phys.* **32**, 3346-3349

Briere T M, Lii J, Prado K, Gillin M T, Beddar A Sam 2006 Single-use MOSFET radiation dosimeters for the quality assurance of megavoltage photon beams *Phys. Med. Biol.* **51** 1139-1144

Burman C, Chui C S, Kutcher G, Leibel S, Zelefsky M, LoSasso T, Spirou S, Wu Q, Yang J, Stein J, Mohan R, Fuks Z, Ling C C 1997 Planning, delivery, and quality assurance of intensity-modulated radiotherapy using dynamic multileaf collimator: a strategy for large-scale implementation for the treatment of carcinoma of the prostate *Int. J. Radiat. Oncol. Biol. Phys.* **39** 863-873

Butson M J, Rozenfeld A, Mathur J N, Carolan M, Wong T P Y, Metcalfe PE 1996 A new radiotherapy surface dose detector: The MOSFET *Med. Phys.* **23** 655-658

Carrier J F, Archambault L, Beaulieu L and Roy R 2004 Validation of GEANT4, an object-oriented Monte Carlo toolkit, for simulations in medical physics *Med. Phys.* **31**, 484-492

Cheng C W, Das I J, Ndlovu A M 2002 Suppression of dark current radiation in step-and-shoot intensity modulated radiation therapy by the initial pulse-forming network *Med Phys.* **29** 1974-1979

Cheung T, Yu P K N, Butson M 2004 Low-dose measurement with a MOSFET in high-energy radiotherapy applications *Radiat Mea* **39** 91-94

Chow J C L and Grigorov G N 2005 Dose measurements near a non-radioactive gold seed using radiographic film *Phys. Med. Biol.* **50** N227-234

Chow J C L, Grigorov G and Jiang R 2006a Improved peripheral dose calculation accuracy for a small MLC field brought by the latest commercial treatment planning system *J Radiother Pract.* **5** 121-128

Chow J C L and Grigorov G 2006b Peripheral dose outside applicators in electron beams *Phys. Med. Biol.* **51**, N231-240

Chow J C L and Grigorov G 2006c Electron radiotherapy: a study on dosimetric uncertainty using small cutouts *Phys. Med. Biol.* **52** N1-N11

Chow J C L, Grigorov G N and Barnett R B 2006d Study on the surface dose generated in the prostate intensity modulated radiation therapy treatment *Med. Dosim.* **31** 249-258

Chow J C L, Grigorov G N 2007a Effect of electron beam obliquity on lateral buildup ratio: a Monte Carlo dosimetry evaluation *Phys Med Biol.* **52** 3965-3977

Chow J C L 2007b Calculation of lateral buildup ratio using Monte Carlo simulation for electron radiotherapy *Med. Phys.* **34** 175-182

Chow J C L, Leung M K K 2007c Treatment planning for a small animal using Monte Carlo simulation. *Med. Phys.* **34** 4810-4817

Chow J C L, Leung M K K 2008a A graphical user interface for calculation of 3D dose distribution using Monte Carlo simulations. *JPCS* **102** 1-6

Chow J C L, Grigorov G N 2008b Monte Carlo simulation of backscatter from lead for clinical electron beams using EGSnrc. *Med. Phys.* **35** 1241-1250

Chow J C L, Leung M K K 2008c Monte Carlo simulation of MOSFET dosimeter for electron backscatter using the GEANT4 code. *Med. Phys.* **35** 2383-2390

Chuang C F, Verhey L J, Xia P 2002 Investigation of the use of MOSFET for clinical IMRT dosimetric verification *Med. Phys.* **29** 1109-1115

Clark C H, Mubata C D, Meehan C A, Bidmead A M, Staffurth J, Humphreys M E, Dearnaley D P 2002 IMRT clinical implementation: prostate and pelvic node irradiation using Helios and a 120-leaf multileaf collimator *J. Appl. Clin. Med. Phys.* **2** 273-284

Corletto D, Iori M, Paiusco M, Brait L, Broggi S, Ceresoli G, Iotti C, Calandrino R, Fiorino C 2003 Inverse and forward optimization of one- and two-dimensional intensity-modulated radiation therapy-based treatment of concave-shaped planning target volumes: the case of prostate cancer *Radiother. Oncol.* **66** 185-195

Das I J, Cheng C, Mitra R K, Kassaee A, Tochner Z and Solin L J 2004 Transmission and dose perturbations with high-Z materials in clinical electron beams *Med. Phys.* **31** 3213-3221

De Meerleer G O, Vakaet L A, De Gersem W R, De Wagter C, De Naeyer B, De Neve W 2000 Radiotherapy of prostate cancer with or without intensity modulated beams: a planning comparison *Int. J. Radiat. Oncol. Biol. Phys.* **47** 639-648

Dogan Nesrin and Glasgow Glenn P 2003 Surface and build-up region dosimetry for obliquely incident intensity modulated radiotherapy 6 MV x rays *Med Phys.* **30** 3091

Francescon P *et al.* 1998 Use of new type of radiochromic film, a new parallel plate micro-chamber, MOSFETs and TLD microcubes in the dosimetry of small beams *Med. Phys.* **25** 503-511

Gagnon W F and Cundiff B S 1980 Dose enhancement from backscattered radiation at tissue-metal interface irradiated with high energy electrons *Br. J. Radiol.* **53**, 466

Gladstone D J and Chin L M 1991 Automated data collection and analysis system for MOSFET radiation detectors *Med. Phys.* **18** 542-548

Holmes-Siedle A 1974 The space-charge dosimeter-General principles of a new method of radiation detection *Nucl. Instrum. Methods* **121** 169-179

Hounsell A R and Wilkinson J M 1999 Electron contamination and build-up doses in conformal radiotherapy fields *Phys Med Biol.* **44** 43-55

Howell R M, Hertel N E, Wang Z, Hutchinson J, Fullerton G D 2006 Calculation of effective dose from measurements of secondary neutron spectra and scattered photon dose from dynamic MLC IMRT for 6 MV, 15 MV and 18 MV beam energies *Med Phys.* **33** 360-368

Huang K, Bice W S Jr, Hidalgo-Salvatierra O 2003 Characteristion of an in vivo diode dosimetry system for clinical use *J Appl Clin Med Phys* **4** 132-142

Jani A B, Roeske J C, Rash C 2003 Intensity-modulated radiation therapy for prostate cancer. *Clin. Prostate Cancer* **2** 98-105

Jera R, Keall P J and Ostwald P M 1999 Comparisons between MCNP, EGS4 and experiment for clinical electron beams *Phys. Med. Biol.* **44** 705-717

Jones A K, Pazik F D, Hintenlang D E and Bolch W E 2005 MOSFET dosimeter depth-dose measurements in heterogeneous tissue-equivalent phantoms at diagnostic x-ray energies *Med. Phys.* **32**, 3209-3213

Kaplan G I, Rosenfeld A B, Allen B J, Booth J T, Carolan M G and Holmes-Siedle A 2000 Improved spatial resolution by MOSFET dosimetry of an x-ray microbeam *Med. Phys.* **27** 239-244

Kawrakow I 2000 Accurate condensed history Monte Carlo simulation of electron transport I EGSnrc, the new EGS4 version *Med. Phys.* **27** 485

Kawrakow I and Rogers D W O 2000 The EGSnrc code system: Monte Carlo simulation of electron and photon transport NRCC Report PIPRS-701, National Research Council of Canada, Ottawa

Klein E E , Low D A, Sohn J W, Purdy J A 2000 Differential dosing of prostate and seminal vesicles using dynamic multileaf collimation *Int. J. Radiat. Oncol. Biol. Phys.* **48** 1447-1456

Klevenhagen S C, Lambert G D and Arbabi A 1982 Backscattering in electron beam therapy for energies between 3 and 35 MeV *Phys. Med. Biol.* **27** 363-373

Khan F M, Moore V C and Levitt M D 1976 Field shaping in electron beam therapy *Br. J. Radiol.* **49** 883-886

Kron T, Rosenfeld A, Lerch M and Bazley S 2002 Measurements in radiotherapy beams using on-line MOSFET detectors *Radia Prot Dosim* **101** 445-448

Kry S F, Salehpour M, Followill D S, Stovall M, Kuban D A, White R A, Rosen I I 2005 Out of field photon and neutron dose equivalents from step-and-shoot intensity-modulated radiation therapy *Int J Radiat Oncol Biol Phys.* **62** 1204-1216

Kung J H, Reft H, Jackson W, Abdalla I 2001 Intensity-modulated radiotherapy for a prostate patient with a metal prosthesis *Med. Dosim.* **26** 305-308

Lambert G D and Klevenhagen S C 1982 Penetration of backscattered electrons in polystyrene for energies between 1 and 25 MeV *Phys. Med. Biol.* **27** 721-725

Lanson J H, Essers M, Meijer G J, Minken A W H, Uiterwall G J and Mijnheer B J 1998 In vivo dosimetry during conformal radiotherapy requirements for and findings of a routine procedure *Radiother. Oncol.* **52** 51-59

Lavallee M C, Gingras L and Beaulieu L 2006 Energy and integrated dose dependence of MOSFET dosimeter sensitivity for irradiation energies between 30 kV and 60Co *Med. Phys.* **33** 3683-3689

Lydon JM 2005 Theoretical and experimental validation of treatment planning for narrow MLC defined photon fields *Phys. Med. Biol.* **50** 2701-2714

Ma C M, Reckwerdt P, Holmes M, Rogers D W O and Geiser B 1995 DOSXYZ user manual NRC Report PIRS 509b

Ma L, Boyer A L, Ma C M, and Xing L 1999 Synchronizing dynamic multileaf collimators for producing two-dimensional intensity-modulated fields with minimum beam delivery time *Int. J. Radiat. Oncol. Biol. Phys.* **44**, 1147-1154

Nelson W R, Hirayama H, and Rogers D W O 1965 The EGS4 code system SLAC Report 265, Stanford Linear Accelerator Center, Stanford, CA

Niemierko A 2004 Quality assurance and validation of individual IMRT treatment plans before delivery *Med. Phys.* **31** 421

Panettieri V, Duch M A, Jornet N, Ginjaume M, Carrasco P, Badal A, Ortega X and Ribas M 2007 Monte Carlo simulation of MOSFET detectors for high-energy photon beams using the PENELOPE code *Phys. Med. Biol.* **52** 303-316

Peet D J and Pryor M D 1999 Evaluation of a MOSFET radiation senor for the measurement of entrance surface dose in diagnostic radiology *BJR* **72** 562-568

Perez-Calatayud J, Ballester F, Serrano M A, Lluch J L, Casal E, and Carmona V 2000 Dosimetric characteristics of backscattered electrons in lead *Phys. Med. Biol.* **45** 1841-1849

Price R A, Murphy S, McNeeley S W, Ma C M, Horwitz E, Movsas B, Raben A, Pollack A 2003 A method for increased dose conformity and segment reduction for SMLC delivered IMRT treatment of the prostate *Int. J. Radiat. Oncol. Biol. Phys.* **57** 843-852

Ramaseshan R, Kohli K S, Zhang T J, Lam T, Norlinger B, Hallil A and Islam M 2004 Performance characteristics of microMOSFET as an in vivo dosimeter in radiation therapy *Phys. Med. Biol.* **49** 4031-4048

Ramsey C, Dube S, and Hendee W R 2003 Moderator. It is necessary to validate each individual IMRT treatment plan before delivery *Med. Phys.* **30** 2271

Rodrigues P, Moura R, Ortigão C, Peralta L, Pia M G, Trindade A, Varela J 2004 Geant4 applications and developments for medical physics experiments *IEEE Trans. on Nuc. Sci* **51** 1412-1419

Rosenfeld A B 2002 MOSFET dosimetry on modern radiation oncology modalities *Radia Prot Dosim* **101** 393-398

Rosenfeld A B, Carolan M C, Kaplan G I, Allen B J, Khivrich V I 1995 MOSFET dosimeters: the role of encapsulation on dosimetric characteristics in mixed gamma-neutron and megavoltage x-ray fields *IEEE Trans on Nuclear Sci* **42** 1870-1877

Roshau J N and Hintenlang D E 1998 Characterization of the angular response of an "isotropic" MOSFET dosimeter *Health Phys.* **84** 376-379

Salvat F, Fernández-Varea J M, and Sempau J 2003 PENELOPE, a Code System for Monte Carlo Simulation of Electron and Photon Transport OECD Nuclear Energy Agency, Issy-les-Moulineaux, France

Saunders J E and Peters V G 1974 Back-scattering form metals in superficial therapy with high energy electrons *Br. J. Radiol.* **47** 467-470

Scalchi P, Francescon P and Rajaguru P 2005 Characterization of a new MOSFET detector configuration for in vivo skin dosimetery *Med. Phys.* **32** 1571-1578

Soubra M, Cygler J and Mackay G 1994 Evaluation of a dual bias dual metal oxide-silicon semiconductor field effect transistor detector as radiation dosimeter *Med. Phys.* **21** 567-572

Spiga J, Siegbahn E A, Bräuer-Krisch E, Randaccio P, Bravin A 2007 The GEANT4 toolkit for microdosimetry calculations: Application to microbeam radiation therapy (MRT) *Med. Phys.* **34** 4322-4330

Streubel G *et al.* 1988 Properties of MOS seniors for application in neutron field monitoring *Nucl. Instrum. Methods A* **274** 194-202

Tabata T 1967 Backscattering of electrons from 3.2 to 14 MeV *Phys Rev.* **162** 336-347

Tabata T and Ito R 1992 Simple calculation of the electron-backscatter factor *Med. Phys.* **19** 1423-1426

Thomson I, Thomson R E and Berndt L P 1984 Radiation dosimetry with MOS sensors *Radiat. Protect. Dosim.* **6** 121-124

Technical Note #7: MOSFET dosimeter specifications. Thomson and Nielsen Electronics

Truscott P, Fan L, Dyer C, Ferguson C, Gurriaran R, Nieminen P, Daly E, Apostolakis J, Giani S, Pia MG, Urbam L, Marie M 2000 Geant4 – a new Monte Carlo toolkit for simulating space radiation shielding and effects *Radiation Effects Data Workshop* 147-152

Wang B, Kim C H and Xua X G 2004 Monte Carlo modeling of a high-sensitivity MOSFET dosimeter for low- and medium-energy photon sources *Med. Phys.* **31** 1003-1008

Wang B, Xu X G, Kim C 2005 Monte Carlo study of MOSFET dosimeter characteristics: dose dependence on photon, energy, direction and dosemeter composition *Radiat Prot Dosim* **113** 40-46

Weatherburn H, McMillan K T P, Stedeford G and Durrant K R 1975 Physical Measurements and clinical observations on the backscatter of 10 MeV electrons from lead shielding *Br. J. Radiol.* **48** 229-230

Yokoyama S, Roberson P L, Litzenberg D W, Moran J M, and Fraass B A 2004 Surface buildup dose dependence on photon field delivery technique for IMRT *J Appl Clin Med Phys* **5** 71-81

In: MOSFETs: Properties, Preparations and Performance
Editors: N.T. Andre et al. pp. 253-273
ISBN: 978-1-60456-762-5

Chapter 7

SURROUNDING-GATE MOSFETs FOR TRANSISTOR SCALING: DEVICES, FABRICATION AND MODELING

Yijian Chen
Applied Materials, 3225 Oakmead Village Drive, M/S 1220, Santa Clara, CA 94086, USA

Abstract

Sustaining functional MOSFET device scaling encounters severe technological challenges as optical lithography and conventional transistor structure are approaching their physical limits. To break lithographic and structural barriers for continuous device scaling down to sub-10 nm half pitch, non-conventional transistors, fabrication processes, and their device physics have been extensively studied. In this article, recent research progress made in surrounding gate based nanowire and nanodot MOSFETs and a vertical integrated-gate CMOS (VIG CMOS) device is presented.

Vertical surrounding-gate MOSFET, compared with double-gate and tri-gate structures, proffers simultaneous gains in transistor density and device performance by consuming less space with better gate control and reduced leakage current. Device physics and modeling of surrounding-gate MOSFET are discussed, and the analytical solutions to the simplified Poisson's equation in both cylindrical and spherical coordinates are given to accurately calculate the electric potential and inversion charge concentration in the ultra-thin bodies. The stochastic characteristics of the device parameters of nanodot and nanowire MOSFETs are studied and several important process control issues are addressed.

For logic applications, a vertical integrated-gate CMOS (VIG CMOS) device is proposed to integrate vertical surrounding-gate PMOS and NMOS together to significantly increase the transistor density and improve the process control capability. Based on the VIG CMOS structure, novel NOR and NAND circuit architectures as well as their interconnect schemes are invented. Device physics of VIG CMOS is examined and we find that solving Poisson's equation in the domain of a Si shell under the single surrounding-gate control faces challenges related to device's boundary conditions and needs more research effort.

Index terms: nanowire, nanodot, surrounding gate, vertical integrated-gate CMOS, Poisson's equation

1. Introduction

The semiconductor industry has reached a critical stage as MOSFETs and lithography technologies are approaching their physical limits with increasing difficulties in sustaining functional device scaling. Leakage, patterning, and control of critical dimension (CD) variability are among the top challenges when the minimum feature of device shrinks to sub-10 nm half pitch.

To overcome these scaling barriers, non-conventional transistors such as nanowire and nanodot based MOSFETs, and vertical integrated-gate CMOS (VIG CMOS) have been proposed recently [1]-[6]. We first present a general review of surrounding-gate nanowire and nanodot MOSFET, and their advantages for deep nano-scale processing compared with conventional planar MOSFET. These two devices can be fabricated with "bottom-up" processes that do not require extremely high lithographic resolution. The devices remain functional even when nanodots and nanowires of various sizes are randomly distributed. Analytical solutions to 1-D Poisson's equation for cylindrical surrounding-gate MOSFETs are obtained. On the other hand, we show that Poisson's equation for spherical surrounding-gate MOSFET is mathematically more difficult and does not appear to have an analytical solution. To solve this problem, we use a boundary-condition coupled transformation technique to linearize the equation to obtain an analytical solution. It is shown that this analytical solution is accurate in calculating the electric potential in a spherical MOSFET.

Another vertical integrated-gate CMOS (VIG CMOS) device is proposed to integrate vertical surrounding-gate PMOS and NMOS together to significantly increase the transistor density. The gates of PMOS and NMOS are integrated into a narrow trench and the metal contact to this trench area is located on top of the device, thus without consuming extra space. The vertical channel/gate length is defined by the Epi-Si thickness while the channel/gate width is defined by the Si shell/cylinder circumference. Epi-Si thickness can be accurately controlled in the growth process with much less gate CD variation and line edge roughness (LER) than printed by lithography. The NOR and NAND circuit architectures as well as the interconnect schemes based on VIG CMOS are invented. Modeling device behaviour of a Si shell with single surrounding-gate control needs to overcome the difficulty in solving nonlinear Poisson's equation under relevant boundary conditions; and possible approaches to solve this problem are discussed.

2. Nanodot and Nanowire MOSFETs and Fabrication Processes

Nanodot- and nanowire-based Metal-Oxide-Semiconductor Field Effect Transistors (MOSFETs) for device scaling down to the half-pitch sub-10-nm generation have attracted extensive research interest recently owing to its efficient large-scale bottom-up fabrication process [2-5], yet a critical challenge is the uniformity control of the size and location of nanodots and nanowires in a production-worthy fabrication process. A cost-effective lithography technology to pattern sub-10-nm features for high-volume semiconductor manufacturing will be extremely difficult, if not impossible, in the near future. Even the self-aligned spacer double patterning, one of the main candidates for (high-volume) sub-10-nm line/space patterning, will face the daunting challenges of line/space profile and critical dimension (CD) control. Therefore, it is important to explore a MOSFET device architecture that can tolerate the statistical fluctuation of nanodot and nanowire size and location, and develop a fabrication process that does not need an impracticably high lithographic resolution capability.

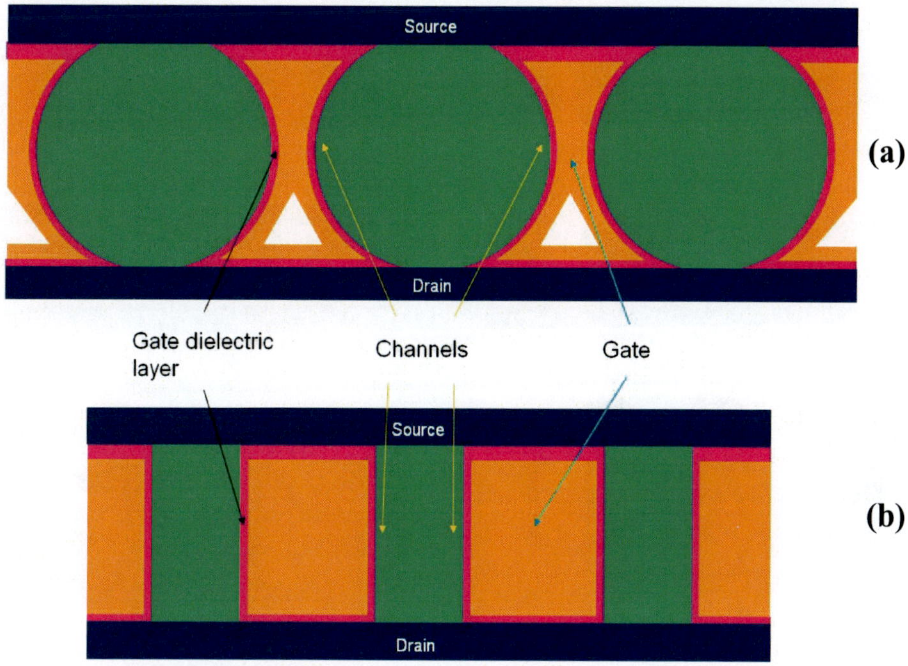

Figure 1. Concept of (a) nanodot MOSFET structure and (b) nanowire MOSFET structure.

Fig. 1 shows the concept of nanodot and nanowire MOSFETs (applicable to both nMOSFET and pMOSFET). Fig. 2 demonstrates a process flow to fabricate a nanodot nMOSFET (suitable for nanowire MOSFET as well), in which the effect of the random distribution of nanodot size and location is averaged out by embedding many dots in one MOSFET. We depict the process flow with Si body as an example, but any relevant semiconductor material can be used as the body/channel. First, we grow a cluster of Si nano-crystals on top of the n+ drain region (in a bottom-up manner) followed by a gate dielectric forming process. The nanocrystal growth process can be any cost-effective process such as self-assembly or ion implantation [5]. Then the gate material (e.g., poly Si or metal) is deposited and a chemical-mechanical polishing (CMP) process is used to planarize the surface, as shown in Fig. 2(3). The gate layer will then be etched down in step (4) (with highly selective wet etchant or dry etching) such that its post-etching surface is flat and slightly below the level of the top gate oxide. This step is important as the oxide deposited in the following step (5) will later (after the second CMP process) act as a dielectric layer to insulate the gate material from the top n+ source region as demonstrated in Fig. 2(6). Finally, an n+ source layer is deposited, followed by a conventional lithographic step to pattern the active area of MOSFET. Apparently, there is no need to resolve every individual nanodot, and the MOSFET devices remain functional regardless of whether all the nanodots are fully activated or not because of the statistical fluctuation of their radii and locations in the manufacturing process.

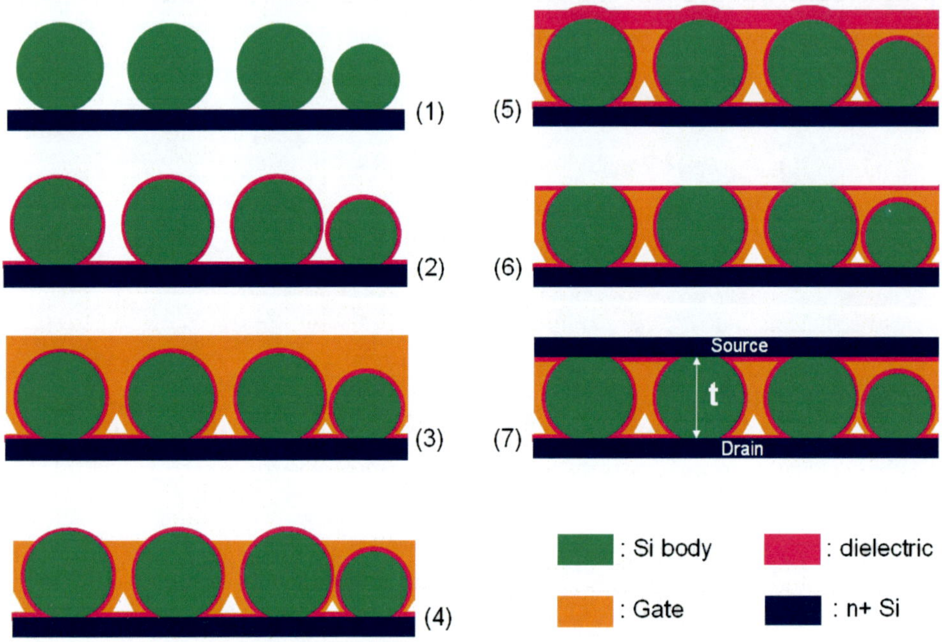

Figure 2. A process flow to fabricate a nanodot nMOSFET, which contains many parallel nanodots (individual MOSFETs) such that random effect can be statistically reduced. This process is also valid for nanowire MOSFET shown in Figure 1.

It is critical that the polished oxide in step (6) is thick enough to expose the Si nanocrystal body, but not too thick such that there is still some oxide left on top of the gate material to act as the insulation layer between the gate and source. Moreover, the polished oxide thickness in the CMP process will determine how many nanodots are activated (by opening the top oxide layer), as shown in Fig. 2. Some nanodots are not activated because of their smaller radii. Therefore, variations of the number of activated nanodots (or the total channel width of MOSFET) are directly related to the CMP polishing rate control.

3. Surrounding-Gate MOSFET Device Physics and Modeling

Both nanodot and nanowire MOSFETs involve surrounding-gate control of the channel and body to reduce the leakage current. Their electric behavior can be described by Poisson's equation [7, 9]:

$$\nabla^2\varphi = \frac{d^2\varphi}{dr^2} + \frac{m}{r}\frac{d\varphi}{dr} = -\frac{q}{\varepsilon_s}(-N_a + N_a e^{-q\varphi/kT} - \frac{n_i^2}{N_a}e^{q\varphi/kT}), \tag{1}$$

where m=1 is for the cylindrical nanowire structure, and m=2 is for the spherical nanodot structure. All the symbols here are following those widely used in device physics literature [7,

9]. When a nanowire MOSFET is biased into the inversion region, the first two terms on the right side of equation (1) can be neglected as below:

$$\frac{d^2 z}{dr^2} + \frac{1}{r}\frac{dz}{dr} = \delta e^z, \tag{2}$$

where $\delta = q^2 n_i^2 / \varepsilon_s kTN_a$ and $z = q\varphi / kT$.

An (analytical) particular solution to 1-D nonlinear Poisson's equation in the cylindrical coordinate was first given by Chambré in 1952 [8] to study the critical condition of inflammability in thermal explosion. For almost 50 years, this particular solution was unknown to the semiconductor device physics society. Consequently, limited progress was made in simplicity and accuracy of compact modeling of advanced transistor such as surrounding-gate MOSFET. In 2001, Chen et al. [9] first realized the potential impact of Chambré's solution on MOSFET microelectronic research. They successfully introduced Chambré's technique to analyze surrounding-gate MOSFET for a comparison with double-gate MOSFET performance. Since then, this particular solution to nonlinear Poisson's equation in the cylindrical coordinate has been frequently quoted in the literature of compact device modeling of surrounding-gate MOSFETs [10-15]. Nevertheless, the particular solution given by Chambré is valid only under certain boundary condition. It will be shown later that several different boundary conditions can be involved in surrounding-gate MOSFETs; therefore, a general solution to equation (2) must be solved. Following Chambré's technique, two transformation variables are defined as

$$\eta = r^2 e^z,\ \beta = r\frac{dz}{dr}. \tag{3}$$

Using above two transformation variables, equation (2) can be written as

$$\frac{d\beta}{d\eta} = \frac{\delta}{\beta + 2}. \tag{4}$$

Integration of equation (4) gives

$$\beta^2 + 4\beta + h = 2\delta\eta. \tag{5}$$

Here h is an integration constant. Substituting equation (3) into equation (5) yields

$$\frac{1}{2}\left(\frac{dz}{dr}\right)^2 + \frac{2}{r}\frac{dz}{dr} + \frac{h}{2r^2} = \delta e^z. \tag{6}$$

Equation (6) is then used to remove the term δe^z in equation (2)

$$\frac{d^2z}{dr^2} - \frac{1}{r}\frac{dz}{dr} - \frac{1}{2}(\frac{dz}{dr})^2 = \frac{h}{2r^2}. \tag{7}$$

The (analytical) general solution of equation (7) with arbitrary h is

$$z = A + (\sqrt{4-h} - 2)\ln r - 2\ln(r^{\sqrt{4-h}} + B), \tag{8}$$

where A and B are two integration constants. Equation (8) can be substituted back into equation (2) to solve the constant A as below

$$A = \ln[2B(h-4)/\delta]. \tag{9}$$

Combining equations. (8) and (9) gives the general solution of equation (2)

$$z = \ln[2B(h-4)/\delta] + (\sqrt{4-h} - 2)\ln r - 2\ln(r^{\sqrt{4-h}} + B), \tag{10}$$

where the integration constants B and h need to be determined from two boundary conditions. When h is zero required by the boundary condition of zero electric field at the body center, solution (10) becomes the same as the particular solution given by Chambré [8] and Chen et al. [9].

When a nanodot MOSFET is biased into the depletion and inversion regions, the hole term in equation (1) can be neglected as below

$$\frac{d^2\varphi}{dr^2} + \frac{2}{r}\frac{d\varphi}{dr} = \frac{qN_a}{\varepsilon_s} + \frac{q}{\varepsilon_s}\frac{n_i^2}{N_a}e^{q\varphi/kT}. \tag{11}$$

The boundary conditions are $d\varphi/dr(r=0)=0$ and $\varphi(r=r_0)=\varphi_0$ where $r=0$ is at the center of spherical nanodot body and $r=r_0$ is on the body surface with φ_0 being the surface potential. Equation (11) includes the contribution of both depletion and inversion charge and thus is suitable for a wide span of device behavior, even though it does not appear to have an analytical solution. A simplifying technique is to use the approximation $e^{q\varphi/kT} \approx 1 + q\varphi/kT$ to linearize equation (11). However, this is valid only when $q\varphi/kT$ is small (e.g., << 1); in reality, it is large. Direct scaling of $q\varphi/kT$ into a small variable (e.g., divided by a large constant) does not solve this problem because of the exponential term of equation (11). To overcome this difficulty, we apply a boundary-condition coupled transformation technique to rewrite equation (11) into a new differential equation that can be further simplified and analytically solved. Similarly, we set $z = q\varphi/kT$, then equation (11) can be written as

$$\frac{d^2z}{dr^2} + \frac{2}{r}\frac{dz}{dr} = (\frac{N_a}{n_i})^2 \cdot \delta + \delta \cdot e^z. \tag{12}$$

Two new variables are introduced: $y = z - z_0$ and $\xi = r(\delta e^{z_0})^{1/2}$, where $z_0 = z(r = r_0)$. Then equation (12) becomes

$$\frac{d^2 y}{d\xi^2} + \frac{2}{\xi}\frac{dy}{d\xi} = \mu + e^y. \tag{13}$$

Here, $\mu = N_a^2 / n_i^2 e^{z_0}$ is a parameter that can be used to determine whether the device is biased into the depletion or the inversion region. The transformation technique described above essentially is an approximation of z with the leading terms in its Taylor expansion about the boundary value z_0 instead of 0 as we just mentioned. Boundary value z_0 is usually given and it will enable a simple but accurate analytical solution for compact device modeling. To physically demonstrate how accurate this approximation is, the electric potential in the Si body of a nanowire cylindrical surrounding-gate MOSFET based on its exact analytical solution [9] is calculated and shown in Fig. 3(a). The potential drop from the outer surface ($r_0 = 5$ nm, $\varphi_0 = 0.85\,\text{V}$, $N_a = 10^{17}\ \text{cm}^{-3}$) to the center is about 0.0035 V, which corresponds to a z difference of 0.135. Intuitively, the spherical surrounding-gate control of a Si body in a nanodot MOSFET should be slightly better than the cylindrical surrounding-gate control in a nanowire MOSFET. Therefore, we expect that $z - z_0$ (or y) in the nanodot Si body will be smaller than 0.135. This indicates that $y \ll 1$ is satisfied in the nanodot MOSFET of our interest (e.g., $r \leq 5$ nm), and it is a good approximation to use $e^y \approx 1 + y$ to simplify equation (13) to such a form that an analytical solution can be obtained as below

$$z(\xi) = y(\xi) + z_0 = (1+\mu)\left(\frac{\sinh\xi}{\xi}\cdot\frac{\xi_0}{\sinh\xi_0} - 1\right) + z_0, \tag{14}$$

where $\xi_0 = r_0(\delta e^{z_0})^{1/2}$. The electric potential and inversion charge distribution using equation (14) are plotted in Figs. 3(b) and 3(c), which indicate an excellent body control by the spherical gate and a good agreement with our expectation. The electric field can be calculated from equation (14) as

$$\varepsilon(r) = \frac{d\varphi}{dr} = \frac{kT}{q}(\delta e^{z_0})^{1/2}(1+\mu)\frac{\xi_0}{\sinh\xi_0}\left\{\frac{\cosh[r(\delta e^{z_0})^{1/2}]}{r(\delta e^{z_0})^{1/2}} - \frac{\sinh[r(\delta e^{z_0})^{1/2}]}{[r(\delta e^{z_0})^{1/2}]^2}\right\}.$$

The surface electric field can be written as a function of the surface potential as

$$\varepsilon(r_0) = \left(\frac{d\varphi}{dr}\right)\Big|_{r=r_0} = \frac{kT}{q}(1+\mu)\left\{(\delta e^{z_0})^{1/2}\cdot\coth[r_0(\delta e^{z_0})^{1/2}] - r_0^{-1}\right\}.$$

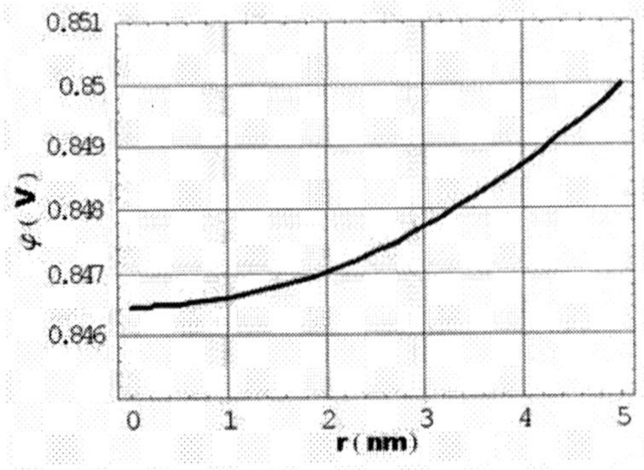

(a)

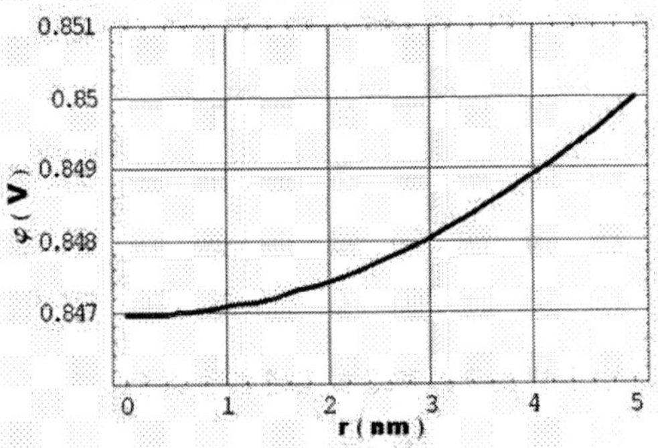

(b)

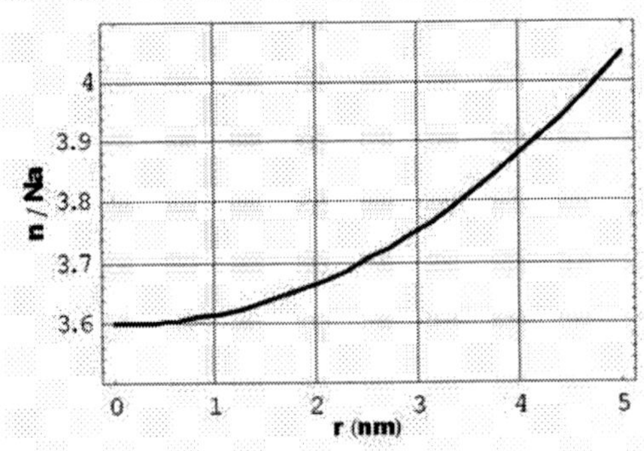

(c)

Figure 3. (a) Electric potential distribution in the Si body based on the exact solution for a nanowire MOSFET. (b) Electric potential as a function of *r* in the Si body of a nanodot MOSFET. (c) Inversion charge density (unit: *Na*) as a function of *r* in the Si body of a nanodot MOSFET. For all three curves, we have used the following parameters in our calculations: $r_0 = 5$ nm, $Na = 10^{17}$ cm^{-3}, $\varphi(5\,\text{nm}) = 0.85$ V.

The integrated charge density (per unit area) of the spherical surface can be obtained using Gauss's law $|Q_s| = |d\varphi / dr(r = r_0)| \cdot \varepsilon_s$, and consequently, the MOS capacitance (per unit area) is $C_{si} = |\frac{dQ_s}{d\varphi_0}|$.

3. Stochastic Analysis of Nanodot and Nanowire MOSFET Device Parameters

The radii of nanowire and nanodot bodies are random variables; therefore, calculation of the total effective channel width of MOSFET and its stochastic characteristics is important for device modeling and circuit design. What we show in Fig. 4 is the top view of a square MOSFET with a cluster of embedded nanodots/nanowires whose radii distribution follows a certain probability density function.

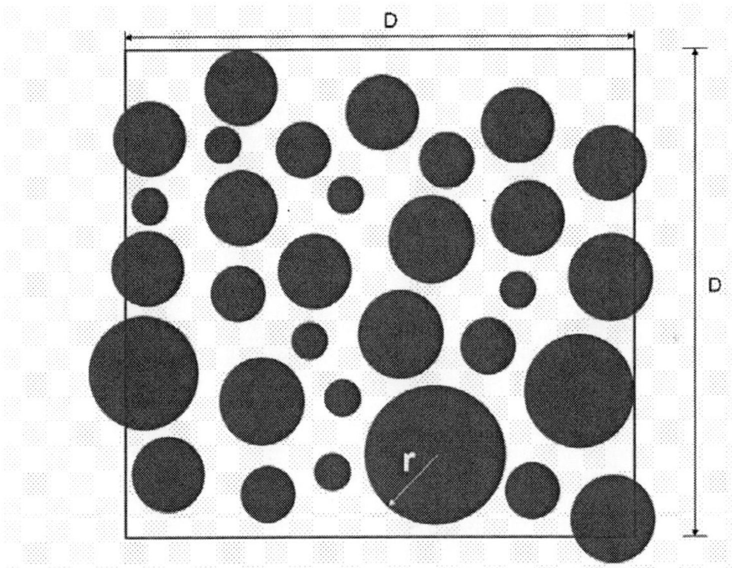

Figure 4. Top view of a cluster of nanodots or nanowires with random distributions of size and location. The square frame includes the active area of MOSFET.

The total channel width W_{nw} of a nanowire MOSFET can be estimated as below. First, we assume the length and width of the top source region D are much larger than the average radius of the embedded nanowires and thus ignore the statistical fluctuation of the total number of nanowires embedded underneath the source layer. This assumption apparently will fail when D is comparable to the average radius and a more complex stochastic model will be needed. If the mean number density (per unit area) of the nanowires is ρ_{nw}, the expected number of embedded nanowires is $N_{nw} = D^2 \rho_{nw}$. Given the probability density function of nanowires as a function of radius R, $p_{nw}(R)$, the (expected) total channel width of nanowire-based MOSFET is

$$W_{nw} = N_{nw} \cdot \int_0^{D/2} 2\pi R\, p_{nw}(R)\, dR \approx (D^2 \rho_{nw}) \cdot \int_0^{\infty} 2\pi R\, p_{nw}(R)\, dR. \qquad (15)$$

Here, we have assumed that the process control is good enough that the percentage of nanowires with radii larger than $D/2$ is negligible.

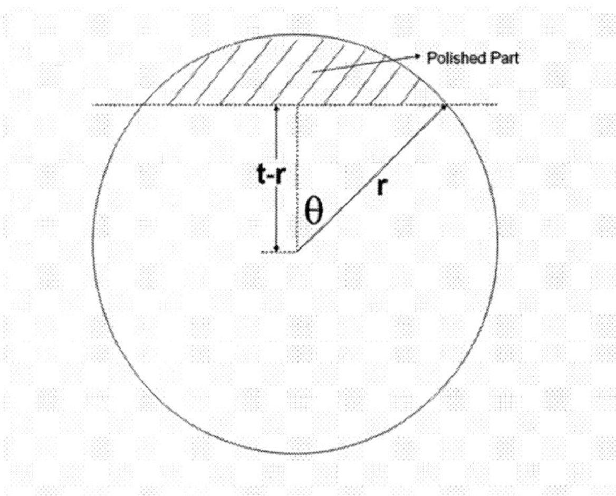

Figure 5. Diagram of a nanodot structure in which the shaded area indicates the polished part in the oxide CMP process. t is the vertical channel thickness, as indicated in Figure 2(7).

For a nanodot MOSFET, we first estimate the total channel area that is affected by the polished depth ($2R$-t). As shown in Fig. 5, the surface area of the polished part (in contact with gate oxide prior to polishing) is $2\pi R(2R-t)$ and the unpolished surface area (approximated as the channel area) is $2\pi Rt$. Since the nanodot thickness t can be considered to be the effective channel length of each individual nanodot MOSFET, the individual effective channel width is then $2\pi R$. It should be kept in mind that dots with thickness smaller than t will not be activated as shown in Fig. 2(7), because the source n+ region is not in contact with the body due to the oxide insulation layer. If we define the mean number density of the nanodots as ρ_{nd}, then the total number of embedded nanodots is $D^2\rho_{nd}$ while the number of activated nanodots is

$$N_{nd} = (D^2 \rho_{nd}) \int_{t/2}^{D/2} p_{nd}(R) dR \approx (D^2 \rho_{nd}) \int_{t/2}^{\infty} p_{nd}(R) dR,$$

where $p_{nd}(R)$ is the probability density function of nanodot radius distribution. Therefore, the (expected) total effective channel width W_{nd} can be estimated as

$$W_{nd}(t) = N_{nd}\int_{t/2}^{D/2} 2\pi R\, p_{nd}(R)dR \approx (D^2\rho_{nd})\int_{t/2}^{\infty} p_{nd}(R)dR \cdot \int_{t/2}^{\infty} 2\pi R\cdot p_{nd}(R)dR. \quad (16)$$

The stochastic characteristics of critical device parameters, such as the total effective channel width, can be obtained from the available data of random distribution of nanowire and nanodot radii.

Next, we shall calculate the variance of the effective channel width for the nanowire structure, and the same result can be applied to the nanodot structure. The average radius of embedded nanodots in each individual MOSFET is

$$\overline{R} = \frac{1}{N_{nw}}\sum_{i=1}^{N_{nw}} R_i\,, \quad (17)$$

and its variance is

$$\sigma^2[\overline{R}] = \sigma^2[\frac{1}{N_{nw}}\sum_{i=1}^{N_{nw}} R_i] = \frac{1}{N_{nw}^2}\sigma^2[\sum_{i=1}^{N_{nw}} R_i] = \frac{1}{N_{nw}}\sigma^2[R_i] = \frac{\sigma^2[R]}{N_{nw}}. \quad (18)$$

Here, we have naturally assumed that the radius R of each individual nanowire is an independent random variable, but with the same statistical characteristics such as variance. It is evident that the variance of the averaged nanowire radius is reduced by a factor of N_{nw} from the original individual variance. If the total channel width of a nanowire MOSFET is $W_{nw}^{m} = N_{nw}\cdot 2\pi\overline{R}$, wherein $\overline{R}$ is the average radius, then its variance is

$$\sigma^2[W_{nw}^{m}] = (2\pi N_{nw})^2\cdot\sigma^2[\overline{R}] = 4\pi^2 N_{nw}\cdot\sigma^2[R]. \quad (19)$$

This parameter is critical to nanowire MOSFET process control as it specifies the tolerable nanowire radius variance when the maximum allowed variance of the total channel width is provided based on the design rules. Similar results can be obtained for the nanodot MOSFET.

The ultrathin bodies of both nanowire and nanodot MOSFETs are completely inverted when biased into strong the inversion region. Owing to the spherical and cylindrical structures of the two devices, conventional concepts such as a planar channel width and inversion charge per unit area require some revision in order to describe the electrical characteristics of each individual nanowire/nanodot. The total conductance of a nanowire/nanodot MOSFET is the sum of (parallel) individual nanowire/nanodot conductances. The inversion charge in the whole body of the *ith* nanodot is the sum of the inversion charge in the periphery and the core regions as shown in Fig. 5

$$Q_{inv,i} = \int_{t-R_i}^{R_i} 2\pi r t\cdot n_{nd,i}(r)dr + \int_{0}^{t-R_i} 4\pi r^2\cdot n_{nd,i}(r)dr\,, \quad (20)$$

In equation (20), we consider the integration effect of the polished part of a nanodot on its total inversion charge, but ignore its influence on the boundary condition and on the final inversion charge distribution. Thus equation (20) can only be considered as a first-order approximation, and a more accurate model for predicting the I-V characteristics of MOSFET must be developed. Similarly, the inversion charge in the whole body of the *ith* nanowire is

$$Q_{inv,i} = \int_0^{R_i} 2\pi rt \cdot n_{nw,i}(r)dr . \tag{21}$$

Since the radii and the inversion charge distribution of nanodots/nanowires are both random variables, we can only calculate the expected inversion charge and the total conductance. The effects of stochastic variations of individual body radii and the inversion charge distribution on device performance are important issues that need further investigation.

4. Vertical Integrated-Gate CMOS

The challenges relate to sub-10 nm (half pitch) lithography include critical dimension (CD) patterning, CD variation and line edge roughness (LER). Moreover, in conventional logic design, NMOS (nMOSFET) and PMOS (pMOSFET) of a CMOS unit occupy separate space which limits our capability to further increase the transistor density.

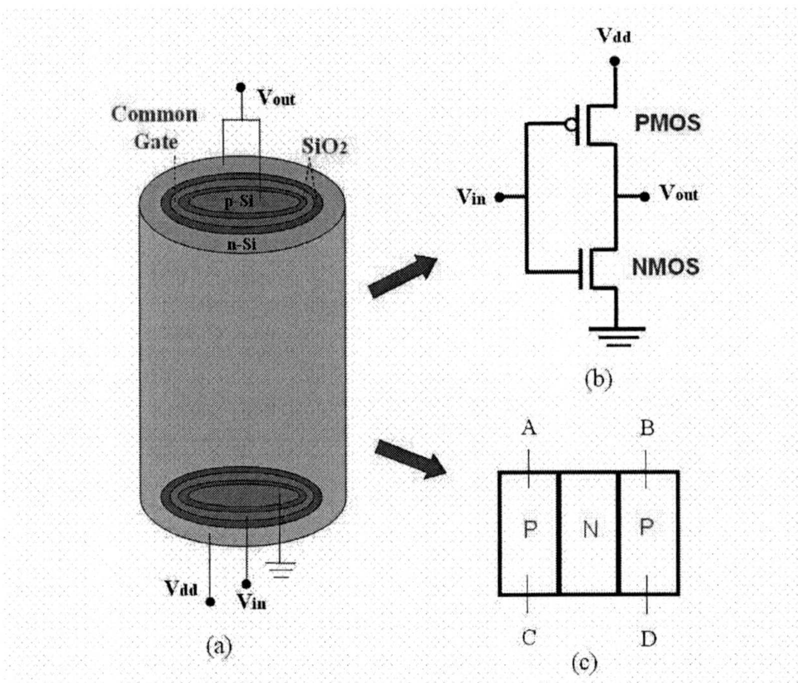

Figure 6. (a) The conceptual demonstration of a vertical integrated-gate (VIG) CMOS, (b) a conventional CMOS symbol, and (c) a VIG CMOS symbol.

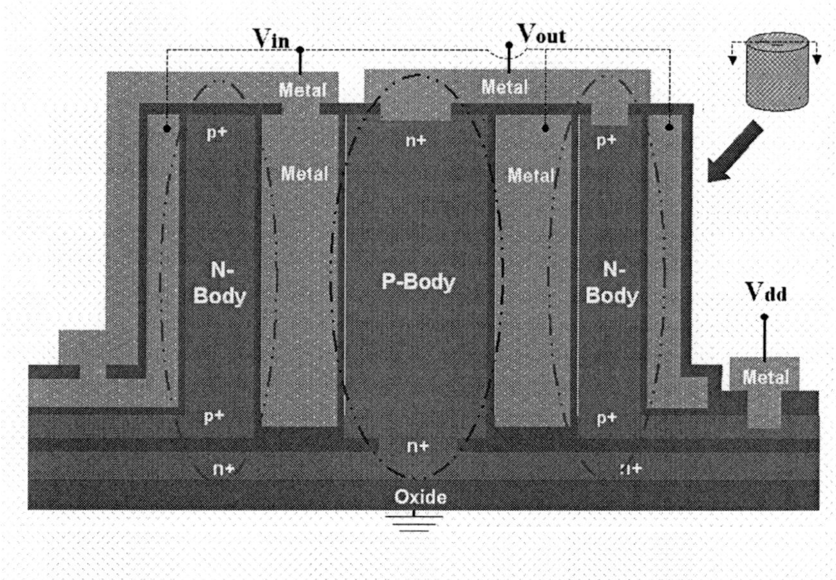

Figure 7. The cross section (cleaved as indicated at the right top of the figure) of a possible vertical integrated-gate (VIG) CMOS structure.

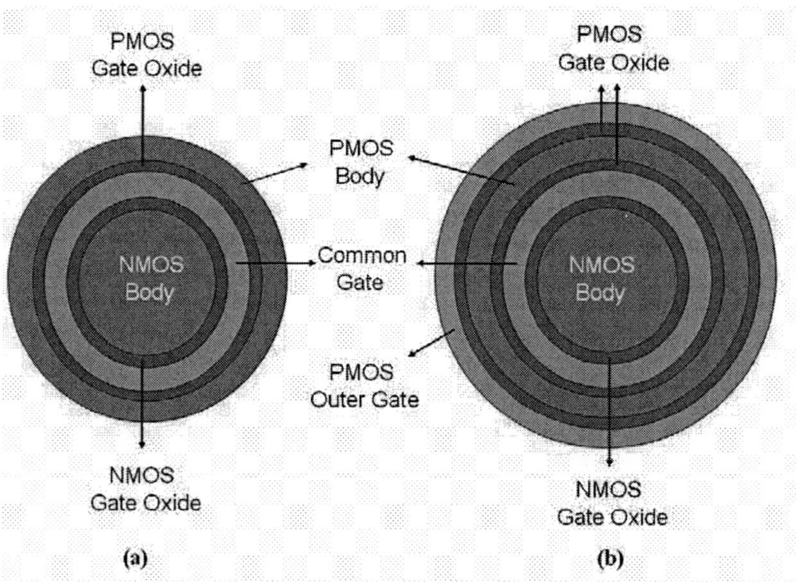

Figure 8. The top view (not to scale) of a (a) single-surrounding-gate PMOS, and (b) double-surrounding-gate PMOS in a VIG CMOS.

A novel vertical integrated-gate (VIG) CMOS has been proposed recently [4]. Fig. 6 is a conceptual demonstration of the VIG CMOS device, in which PMOS and NMOS, as well as their drain and source, are designed in the vertical way. In this device, NMOS is surrounded by a PMOS and they share a common gate which is integrated into a narrow trench between them. Metal contact to this trench area (gate) is located on top of the device (see Fig. 7) thus without consuming extra space. Again, Si is used as an example of body/channel material just to demonstrate the concept, it is not the only material that can be adopted in future devices. This CMOS architecture significantly increase the transistor density compared with the conventional structure in which PMOS and NMOS occupy separate space. The vertical channel/gate length is defined by the Epi-Si thickness while the channel/gate width is defined by the Si shell/cylinder circumference. Epi-Si thickness can be accurately controlled in the growth process with much less gate CD variation and LER than defined by lithography. Fig. 8 is the top view of two different VIG CMOS structures, one with a single-surrounding-gate control in PMOS, while the other with a double-surrounding-gate control. The NOR and NAND circuit architectures and practical interconnect schemes are shown in Fig. 9 (gate voltage inputs omitted for easier demonstration).

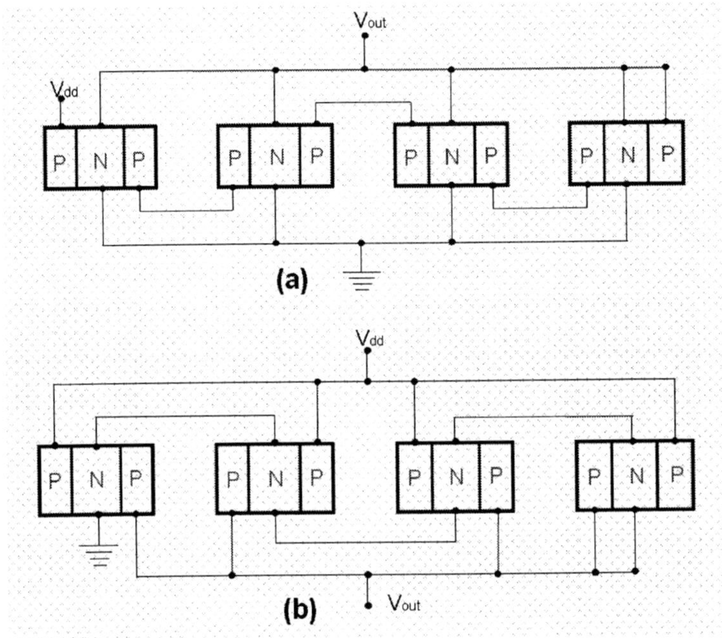

Figure 9. The circuit architecture and interconnect scheme of a 4-input (a) NOR and (b) NAND.

5. VIG CMOS Device Physics and Modeling

To study VIG CMOS device physics, we consider NMOS in strong inversion wherein the holes and depletion-charge terms can be neglected and equation (1) becomes:

$$\frac{d^2\varphi}{dr^2} + \frac{1}{r}\frac{d\varphi}{dr} = \frac{q}{\varepsilon_s}\frac{n_i^2}{N_a}e^{q\varphi/kT} \tag{22}$$

Three kinds of boundary conditions can be involved in a VIG CMOS device. First, the boundary conditions for the inner Si-cylinder (e.g., inner NMOS body in a VIG CMOS as shown in Fig. 6) structures are: $\frac{d\varphi}{dr}(r=0)=0$ and $\varphi(\mathrm{r}=\mathrm{r}_0)=\varphi_0$ where $r=0$ is at the center of Si body and $r=r_0$ is on the body surface with φ_0 as the surface potential. Secondly, for an outer Si-shell (e.g., outer PMOS body in a VIG CMOS as shown in Fig. 8(a)) structure with a single surrounding gate, the boundary conditions are $\varphi(\mathrm{r}=\mathrm{r}_1)=\varphi_1$ and $\frac{d\varphi}{dr}(r=\infty)=0$ where $r=r_1$ is the inner radius of the Si shell. This kind of boundary conditions is the same as that used for the planar MOSFET structure [7]. Finally, for an outer Si shell in a PMOS with double surrounding gates as shown in Fig. 8(b), the boundary conditions are $\varphi(\mathrm{r}=\mathrm{r}_1)=\varphi_1$ and $\varphi(\mathrm{r}=\mathrm{r}_2)=\varphi_2$ where $r=r_1$ and $r=r_2$ are the inner and outer radii of the Si shell, respectively. The last kind of boundary conditions results in a tighter control of the channel/body and consequently less leakage current.

If $r=0$ is at the center of the physical domain of device (e.g., the inner Si cylinder), both electric field and potential at the body center are real physical parameters and should be finite. Consequently, boundary condition $\frac{d\varphi}{dr}(r=0)=0$ requires $\beta(\mathrm{r}=0)=0$ and $\eta(\mathrm{r}=0)=0$, which result in the constant *h*=0 according to equation (5). However, it should be kept in mind that both electric potential and field calculated from the formula valid only in the domain excluding $r=0$ can be infinite when $r=0$ is substituted in. Since $r=0$ is out of the physical domain (which is the case for the second and third kinds of boundary conditions), it seems unnecessary for the solution to be meaningful at points out of its physical domain. An analytical solution to equation (22) with *h*=0 is given in [9]:

$$z=A_0-2\,ln(r^2+B_0) \tag{23}$$

To avoid confusion, symbols *A* and *B* used in the general solution for arbitrary *h* are replaced by A_0 and B_0 here. The zero electric field $\frac{d\varphi}{dr}(r=0)=0$ is automatically satisfied by (23). A relationship between A_0 and B_0 is found by substituting (23) into (2):

$$A_0=ln(-8\,B_0/\delta) \tag{24}$$

The other boundary condition $\varphi(\mathrm{r}=\mathrm{r}_0)=\varphi_0$ yields a formula to solve B_0:

$$exp(\frac{q\,\varphi_0}{kT})=-\frac{8\,B_0}{\delta(r_0^2+B_0)^2} \tag{25}$$

The surface charge density can be calculated with the surface electric field as below:

$$|\,Q_s\,|=|\,\frac{d\,\varphi}{dr}(r=r_0)\,|\cdot\varepsilon_s=\frac{kT}{q}\varepsilon_s\sqrt{\frac{-2\delta\,r_0^2}{B_0}exp(\frac{q\,\varphi_0}{kT})} \tag{26}$$

MOS capacitance can then be derived using: $C_{si} = |\frac{dQ_s}{d\varphi_0}|$.

As shown in Fig. 8, PMOS in a VIG CMOS device can have two different types of gate: double or single surrounding gate. From the processing point of view, it is easier to fabricate the double-surrounding-gate structure which also helps to reduce leakage because of its tighter control of Si body. Therefore, it is more interesting to study the double-surrounding-gate PMOS Si shell. Next, we shall examine device physics of a NMOS Si-shell with single and double surrounding gates. The results, when manipulated correctly, can be readily applied to analyze PMOS Si shell in a VIG CMOS.

The analytical solution to equation (2) with arbitrary *h* has been given in (10):

$$z = \ln[2B(h-4)/\delta] + (\sqrt{4-h} - 2)\ln r - 2\ln(r^{\sqrt{4-h}} + B)$$

where the integration constants B and h need to be determined from two boundary conditions. Based on the general solution (10), we can see that it is necessary to keep *h* in order to solve Poisson's equation for an outer Si shell with double surrounding gates (under the third kind of boundary conditions: $\varphi(r = r_1) = \varphi_1$ and $\varphi(r = r_2) = \varphi_2$). Otherwise we would have two (redundant) boundary conditions to solve for only one unknown parameter B. As mentioned before, it is possible that d*z*/d*r* (which is meaningful only within $r_1 < r < r_2$ in a shell structure) becomes infinity when *r* approaches zero resulting in the limit: $\beta(r \to 0) \neq 0$ and $h \neq 0$.

As in a planar MOSFET [7], zero electric field (or *dz/dr*) at infinity is usually taken as the boundary condition for an outer Si shell with a single surrounding gate as shown in Fig. 8(a). Unfortunately, this boundary condition is automatically satisfied by solution (10) for arbitrary *h* and does not provide a necessary equation to solve for *h*. Therefore, we may have to specify the voltages at both inner and outer radii even in an outer Si shell with a single surrounding gate in order to obtain the final solution. Nevertheless, the outer voltage is normally unknown and how to determine the integration constants remains an open question. The other approach is to assume a zero "virtual" electric field at *r*=0 which eliminates the unknown parameter *h*, and use the inner surface potential $\varphi(r = r_1) = \varphi_1$ to solve the parameter *B*, as described in reference [4]. However, the physical foundation of assuming a zero *dz/dr* at *r*=0 for solving the potential in a Si shell with a single surrounding gate is unclear.

It should be pointed out that only one root (*B*) of equation (25) is suitable for the device with a cylindrical Si body as shown in Fig. 6, in which a NMOS cylindrical Si body is surrounded by an all-around gate. To examine this, we assume:

$$N_a = 10^{17}\, cm^{-3}, \varphi_s = 1.0\,(V), r_0 = 15\, nm$$

From equation (25), we have two roots as below:

$$B_1 = -2.405 \times 10^{-16},\ B_2 = -2.105 \times 10^{-16}.$$

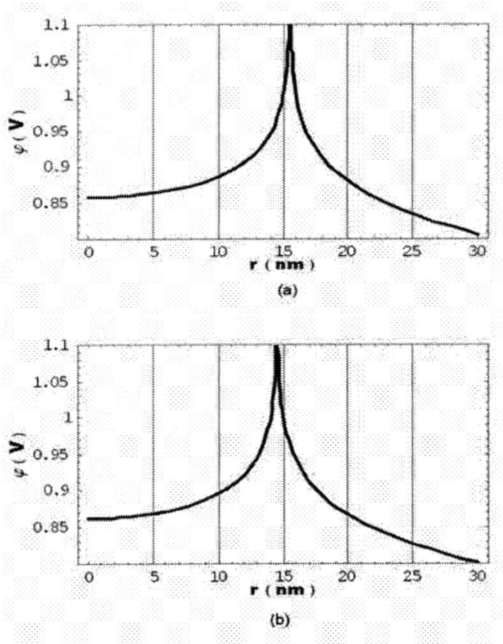

Figure 10. The electric potential in a NMOS, subject to a surface potential of 1V using two different roots (*B*) of equation (25).

The electric potential as a function of *r* for B_1 and B_2 are plotted in Figs. 10(a) and 10(b) respectively. For an inner Si body with $r_0 = 15nm$ in NMOS, solution (b) has a singular point at $r < r_0$ and the electric potential drops sharply when *r* approaches the boundary, which violates the physical intuition that φ should become higher when *r* increases. Therefore, (b) is a meaningless solution for an inner cylindrical Si body. The inversion charge concentration in this cylindrical Si body as a function of *r* is calculated and shown in Fig. 11(a). The ratio of the inversion charge concentration of a surrounding-gate NMOS to that of a double-gate NMOS is also shown in Fig. 11(b) for a comparison. Apparently, surrounding-gate MOSFET produces better body/channel control and less leakage current than double-gate MOSFET.

For a NMOS Si shell with double surrounding gates, surface potentials $\varphi(\mathrm{r} = \mathrm{r}_1) = \varphi_1$ and $\varphi(\mathrm{r} = \mathrm{r}_2) = \varphi_2$ are not necessarily the same when the same gate voltage is applied at the outer and inner gates, because the voltage drops in two thin oxide layers with different radii may vary slightly. Rigorous analysis need to numerically solve two surface potentials with the iteration method taking oxide voltage drop into account. In general, there are two solution sets of (B, *h*) to equation (10) provided that the outer and inner surface potentials are given. Our calculation shows that only one of them is physically meaningful. To demonstrate this, we assume that difference between φ_1 and φ_2 is small and arbitrarily choose:

$$\varphi_1(\mathrm{r} = 15\mathrm{nm}) = 0.82\,\mathrm{V} \text{ and } \varphi_2(\mathrm{r} = 30\mathrm{nm}) = 0.85\,\mathrm{V}$$

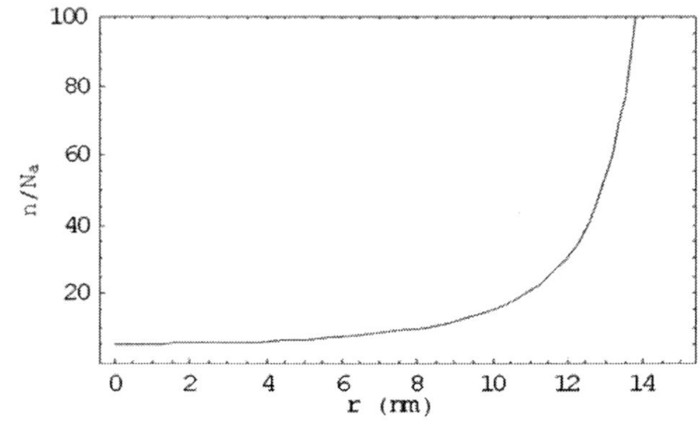

(a)

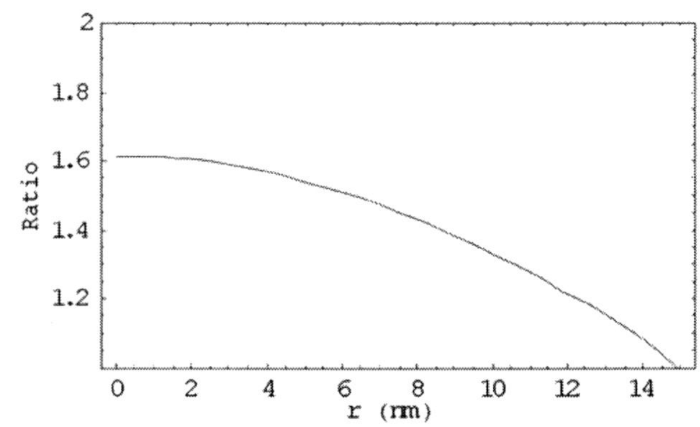

(b)

Figure 11. (a) The inversion charge concentration as a function of r calculated from the potential distribution shown in Figure 10(a). (b) Ratio of the inversion charge concentration of a surrounding-gate NMOSFET to that of a double-gate NMOSFET (subject to the same surface potential 1V).

Substituting above boundary conditions into (10) yields two equations which can be used to solve (B, h). What we show in Figs. 12(a) and 12(b) are the electric potential and the inversion charge density as a function of r based on the meaningful solution:

$$B = -2.08 \times 10^{-11} \text{ and } h = 1.912 .$$

The electric potential distribution for the other solution of (B, h) has also been plotted in Fig. 12(c), which is physically meaningless due to the observed singularity/spike. From Fig. 12(a), we can see that the body electric potential is locked between the surface potentials at $r = r_1$ and $r = r_2$. Namely, the potential across the whole Si body will be almost the same as the applied surface potential if we ignore the potential difference between two boundary surfaces. This characteristic indicates an improved control of Si body compared with the double-gate MOSFET in which we observed a more significant potential drop from the outer surface to the center [9].

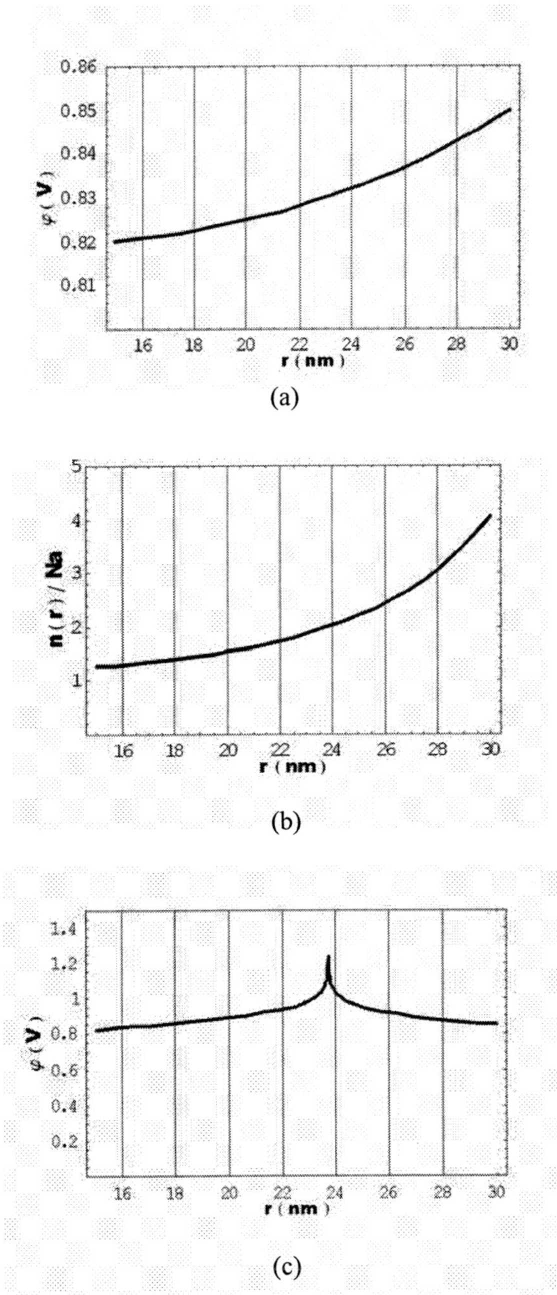

Figure 12. The electric potential (a), and inversion charge density (b) as a function of r in a NMOS with double surrounding gates based on one set of root (B, h) for equation (10). (c) The electric potential as a function of r based on another set of root (B, h), which however is an invalid solution. Here, we have used $\varphi(15\ nm) = 0.82\ V$, $\varphi(30\ nm) = 0.85\ V$, and $Na = 10^{17}\ cm^{-3}$.

6. Conclusion

Non-conventional nanodot/nanowire MOSFETs and a VIG CMOS device for continuous transistor scaling down to sub-10 nm have been proposed. We studied Poisson's equation in cylindrical and spherical coordinates under various boundary conditions. Analytical solution to the simplified Poisson's equation has been obtained to calculate the electric potential and inversion charge concentration within the ultra-thin bodies of nanodot and nanowire MOSFETs. The accuracy of calculating the induced electric potential and inversion charge in a nanodot MOSFET has been examined. The stochastic characteristics of the device parameters of nanodot and nanowire MOSFETs are studied, and several important process control issues as well as the impact of process variations (e.g., CMP polishing rate) on device performance have been discussed.

References

[1] S.-H. Oh et al., "Analytic description of short-channel effects in fully-depleted double-gate and cylindrical, surrounding-gate MOSFETs," *IEEE Electron Device Lett.,* vol. 21, No. 9, pp.397-399, September 2000.

[2] N. Singh, A. Agarwal, L. K. Bera, T. K. Liow, R. Yang, S. C. Rustagi, C. H. Tung, R. Kumar, G. Q. Lo, N. Balasubramanian, and D.-L. Kwong, "High-performance fully depleted silicon nanowire (diameter ≤ 5 nm) gate-all-around CMOS devices," *IEEE Electron Device Letters*, Vol. 27, No.5, pp. 383-386, May 2006.

[3] Y. Cui, Z. Zhong, D. Wang, W. U. Wang, and C. M. Lieber "High performance silicon nanowire field effect transistors," *Nano Letters*, Vol. 3, No. 2, pp.149-152, 2003.

[4] Y. Chen and A. Chu, "Vertical integrated-gate CMOS for ultra-dense IC," *Microelectronic Engineering*, Vol. 83, Issues 4-9, pp. 1745-1748, April-September, 2006.

[5] Y. Chen, "Nanodot and nanowire transistor device modeling and fabrication process," *Japanese Journal of Applied Physics*, Vol. 46, No. 9B, pp. 6213-6217, 2007.

[6] O. Moldovan, B. Iniguez, D. Jimenez, J. Roig, "Analytical charge and capacitance models of undoped cylindrical surrounding-gate MOSFETs," *IEEE Transactions on Electron Devices*, Vol. 54, No. 1, pp. 162-165, Jan., 2007.

[7] S. Wolf, *Silicon Processing,* vol. 3-The submicron MOSFET, Sunset Beach: Lattice Press, 1995, pp.106-112.

[8] P. L. Chambré, "On the solution of the Poisson-Boltzmann equation with application to the theory of thermal explosions," *J. Chem. Phys.*, Vol. 20, No. 11, pp.1795-1797, Nov. 1952.

[9] Y. Chen and J. Luo, "A comparative study of double-gate and surrounding-gate MOSFETs in strong inversion and accumulation using an analytical model," *Technical Proceedings of the Fourth International Conference on Modeling and Simulation of Microsystems*, pp. 546-549, 2001.

[10] T. Chiang, "New Current-Voltage Model for Surrounding-Gate Metal-Oxide-Semiconductor Field Effect Transistors," *Japanese Journal of Applied Physics*, Vol. 44, No. 9A, pp. 6446-6451, 2005.

[11] T. Chiang , "Concise Analytical Threshold Voltage Model for Cylindrical Fully Depleted Surrounding-Gate Metal-Oxide-Semiconductor Field Effect Transistors,"*Japanese Journal of Applied Physics*, Vol. 44, No. 5A, pp. 2948-2952, 2005.

[12] D. Jimenez, B. Iniguez, J. Sune, L. F. Marsal, J. Pallares, J. Roig, D. Flores, "Continuous analytic I-V model for surrounding-gate MOSFETs," *IEEE Electron Device Letters*, Vol. 25, Issue 8, pp.571 – 573, August, 2004.

[13] D. Jimenez, J. J. Saenz, B. Iniguez, J. Sune, L.F. Marsal, J. Pallares, "Modeling of Nanoscale Gate-All_Around MOSFETs,"*IEEE Electron Device Letters*, Vol. 25, No. 5, pp. 314-316, May, 2004

[14] B. Iniguez, D. Jimenez, J. Roig, H. A. Hamid, L.F. Marsal, J. Pallares, "Explicit continuous model for long-channel undoped surrounding gate MOSFETs," *IEEE Transactions on Electron Devices*, Volume: 52, Issue: 8, pp. 1868-1873, Aug. 2005.

[15] B. Yu, W. Lu, H. Lu, Y. Taur, "Analytic charge model for surrounding-gate MOSFETs," *IEEE Transactions on Electron Devices*, Vol. 54, No. 3, pp. 492-496, Mar., 2007.

In: MOSFETs: Properties, Preparations and Performance
Editors: N.T. Andre et al. pp. 275-299
ISBN: 978-1-60456-762-5

Chapter 8

QUANTUM, SELF-HEATING AND HOT ELECTRON EFFECTS OF SI-BASED DOUBLE-GATE MOSFET AND GAN-BASED MOS-HFET

Xiaoshuang Chen, Weida Hu and Wei Lu

National Laboratory for Infrared Physics, Shanghai Institute of Technical Physics, Chinese Academy of Sciences, Yutian Road 500, Shanghai 200083 China

Abstract

It is one of the most attractive devices for implementing nanoscale planar MOSFET because of its excellent characteristics such as ideal subthreshold swing, high transconductance, and reasonable short-channel effect. A kind of new sub-50 nm N channel self-aligned double gate MOSFETs has been studied by solving coupled Poisson-Schrödinger equations with a finite element method. A systematic simulation-based study has been presented on the quantum effects and the gate leakage current. The simulation results indicate that the deviation from the classical model becomes more important as the gate oxide, gate length and Fin channel width becomes thinner and the Fin channel doping increases. Gate tunneling current density is reduced with the body thickness decreasing. The excessive scaling increases the gate current below Fin thickness of 5nm. The gate current can be dramatically reduced beyond $10^{17}cm^{-3}$ with increasing the Fin body doping. Quantum mechanical simulation results are compared with the results from the classical approach indicating that quantum mechanical simulation is essential for the realistic optimization of the double gate structure.

Recent advances in GaN-based MOS-HFET have opened the way for their practical applications in high power and high frequency electronic devices. Undoped GaN-based MOS-HFETs with atomic-layer-deposited Al_2O_3 gate dielectrics are fabricated. Using a two-dimensional numerical calculation, we report the results of self-heating simulations of the GaN-based MOS-HFETs, including hot electron and quantum effects. The simulated electrical characteristics are in good agreement with the reported experimental data. The effect of the gate and source/drain extension lengths on both the output performance and self-heating is discussed in detail, allowing for device optimization. The dissipated Joule electric power causes the self-heating effects, which lead to negative differential output conductance. Our results have demonstrated that the hot electrons make a negligible contribution to the negative differential output conductance in our long channel MOS-HFETs. In order to investigate their joint interactions to the MOS-HFET's operation, the different static interface trap and charge densities created at the $AlGaN/Al_2O_3$ interface are considered in the output characteristics. Results show that the presence of the interface charges and traps are directly responsible for the observed current collapse and device switching in the GaN-based MOS-HFETs.

1. Introduction

CMOS designs below 0.1 um are severely constrained by lateral SCE and vertical gate insulator tunneling [1-3]. The double gate transistor is widely accepted as a highly promising candidate for fulfilling future CMOS device demands. Among the different concepts of double gate transistors which differ in their spatial arrangement [4-5], self-aligned double-gate MOSFETs (FinFETs) [6] have a straightforward process flow when starting from a conventional bulk transistor. It's one of the most attractive devices for implementing nanoscale planar MOSFET because of its excellent characteristics such as ideal subthreshold swing, high transconductance (G_m), and reasonable short-channel effect (SCE). [6-10]

The operation of deep-submicron FinFETs is now entering a regime in which quantum mechanical effects become noticeable. New methodologies and model for nanoscale FinFETs must be used for the devices simulations. In order to optimize the structure of FinFET, it is necessary to undertake a two-dimensional (2d) quantum mechanical (QM) simulation because of its inherent quantum and geometrical effects on the electronic properties of nanoscale semiconductor devices. The target oxide thickness (T_{ox}) for the FinFETs device is expected around 1.6–5nm for the gate length of 30 nm [3-6]. In such a thin gate oxide regime the gate direct tunneling current clearly appears, which can alter device characteristics significantly. To date, however, very few studies [8-10] on the gate leakage current and AC high frequency performance of FinFETs have been reported. Therefore , in this chapter we review the quantum mechanical and fully classical simulations of FinFETs. The self-consistent electron density is solved self-consistently [11] using the finite element method (FEM). The tunneling characteristics of electrons through ultra-thin oxides and AC high frequency performance are well simulated. Quantum mechanical effect on the electrical characteristics of FinFETs is also discussed.

Recent advances in GaN-based heterostructure field effect transistors (HFETs) or high-electron-mobility transistors (HFET) have opened the way for their practical applications in high power and high frequency electronic devices. However, the typically large gate leakage current reduces the breakdown voltage, power-added efficiency and device reliability, in addition to increasing the noise figure. The leakage current is thus an important roadblock to achieve the practical applications of these devices. To solve the problem, significant progress has been made on metal-insulator-semiconductor heterostructure field effect transistors (MIS-HFETs) and metal-oxide-semiconductor heterostructure field effect transistors (MOS-HFETs) using SiO_2, [12-17] Si_3N_4, [18-19] Al_2O_3 [20-21] and other oxides. [22] Recently, Ye et al. [23] have reported a GaN-based MOS-HFET with atomic-layer-deposited (ALD) Al_2O_3 as the gate dielectric, showing low leakage current, high breakdown voltage, strong accumulation, and high effective two-dimensional (2D) electron mobility under both low and high transverse fields. ALD Al_2O_3 has been widely studied as a gate dielectric and has become one of the leading candidates to replace SiO_2 in future-generation Si CMOS digital ICs. [24, 25] However, significant self-heating effects have been found in the MOS-HFETs when operating under high power.

The elimination of the self-heating effects requires an understanding of the physical phenomena responsible for them. The physics-based numerical simulation provides an efficient and economical way for complementing experimental investigations. However, there have only been a few reports on the self-heating simulation of the GaN-based HFET. Wu et al. have reported a thermal simulation using a 2D steady-state heat conduction model. [26, 27] Eastman et al. have carried out nonlinear three-dimensional heat spreading simulations as functions of the dissipated power and the device geometry. [28] Albrecht et al. have presented a gradual-channel-approximation-based Monte Carlo simulation of the GaN-based HFET. [29] Ahmad et al. have carried out a 2D finite element self-heating analysis of the GaN-based HFET. [30] Braga et al. have reported 2D steady-state simulations of the GaN-based HFET by considering the hot electron and the quantum effects. [31] The GaN-based HFET is a complicated device with multiple interfaces, which play an important role in the self-heating effects and the device operations. However, the reported simulations have not taken into account the interface charges or traps at all of the interfaces. All reported simulations, except for Braga et al., have neglected the quantum effects of the nanoscale GaN-based HFET. Furthermore, no publications have been dedicated to the self-heating simulation of the GaN-based MOS-HFETs. It is well recognized from the experimental data that it is necessary to optimize the geometry of the current devices, especially for the source/drain extension and the Al_2O_3/AlGaN interface. [31-33] Further , the radio-frequency (RF) drain-current (I_{ds}) collapse is the major factor limiting the output-power density at microwave frequencies in GaN-based HFETs [33-37]. Its detrimental consequences on device performance have been shown by using different experimental techniques, including measurements of pulsed I_{ds} versus drain-source voltage (V_{ds}) characteristics, gate- and drain-lag transients, and RF response [38-42]. These problems are being addressed by experimental studies that can produce epitaxy- and processing-related improvements, and simulations that can optimize device design. Numerical simulations, which can contain structural details such as layer thicknesses, doping profiles and trap concentrations, provide key insights into device operation and the degradation mechanisms of the reliability. A detailed assessment of the effect and characterization of the relevant trap states is crucial for optimization of nitride transistor performance. Therefore, in this chapter we make a review about the 2D simulations of the GaN-based MOS-HFETs taking into account the self-heating effects. The structural and interfacial optimizations of the MOS-HFETs are also discussed. Our results of numerical simulations that clearly show that the buffer-trapping could explain the observed feature of drain-current collapse and drain lag with a two-dimensional device simulator [43]. Pulse-on and pulse-down modes for drain lag simulations are presented and compared to the experimental observations respectively. In addition, it is shown that the quantum-well (QW) HFET structure is effective for suppressing the current collapse and drain lag.

2. Quantum Hot Electron Effects and Gate Leakage Current of Si-based Double-Gate Mosfet

2.1. Numerical Model for Simulation

The QM model for obtaining the band structure is based on the self-consistent electron density. 1) At equilibrium, solve the potential using flat-band approximation. The default uniform profile model is used for the density. This gives us the initial potential distribution. 2) At equilibrium, solve the potential again with potential obtained in 1) above. The self-consistent density model is used. A new density profile is obtained in this step. 3) We iterate 2) above until numerical self-consistency is achieved for density profile and potential distribution. 4) Once we achieve self-consistency in equilibrium after a number of numerical iterations, we increase the bias and repeat 2) and 3) above.

Direct tunneling model of current flow into the gate from first principle calculation is provided in this simulation. The tunneling transparency for a rectangular SiO_2 barrier is from Ref. [44]. The model takes into account the overlap of carrier wave function penetrating into the SiO_2 layer with current through the top of the SiO_2 barrier generating reasonable gate current over certain bias ranges. Because Assessment of high frequency performance of devices is a key reason of the modeling, an AC analysis using the deep level trap model developed by Li and Dutton [45-46] is discussed in this work. We also show the quantum mechanical effects on the threshold voltage, gate capacitance and channel conductance.

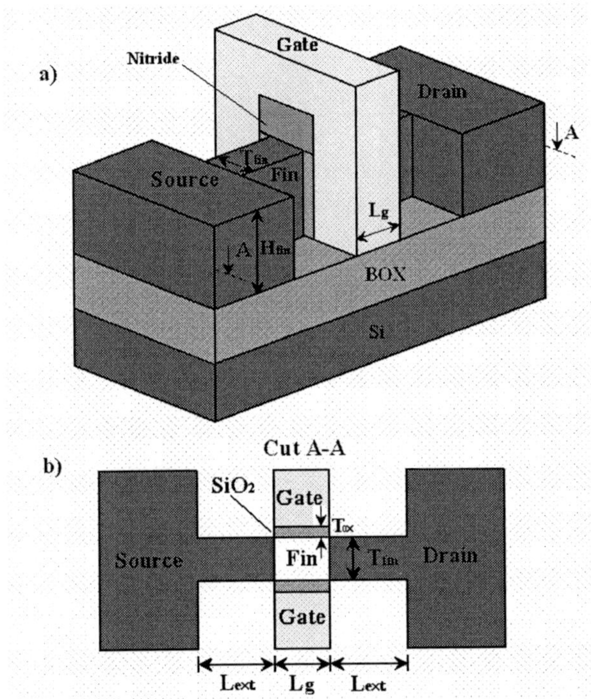

Figure 2.1. (a) Overview of 3-D FinFET (b) the cross-section of 2-D simulated FinFET under study.

Figure 2-1 shows 3-D view and cross-section of FinFET used in our work. The gate "wraps" over the thin Si Fin, yielding a quasi-planar symmetrical DG MOSFET with two inversion channels which are beneficially charge-coupled [47]. As a test vehicle for verifying the validity of our simulation, we chose N-type FinFET showing good short-channel performance down to a gate length of 30nm. Since the thickness of the Fin is desired to be smaller than two-thirds of the channel length in order to limit the SCEs [48-50], gate length L_g=30nm and Fin body width T_{fin}=20 nm are used in N-type FinFET. Our simulations are carried out for N-type FinFET at 300K with the (100) SOI layer.

2.2. Simulation Result and Discussion

2.2.1. Quantum-Mechanical Effects

In order to facilitate a fair comparison with single-gate MOSFETs, the currents are normalized by the effective device width (W) using $W = 2H_{fin}$ (fin height). We compared our results with the experimental results in the literature. The simulated I_d-V_d characteristics are reasonably well with experimental measurements.

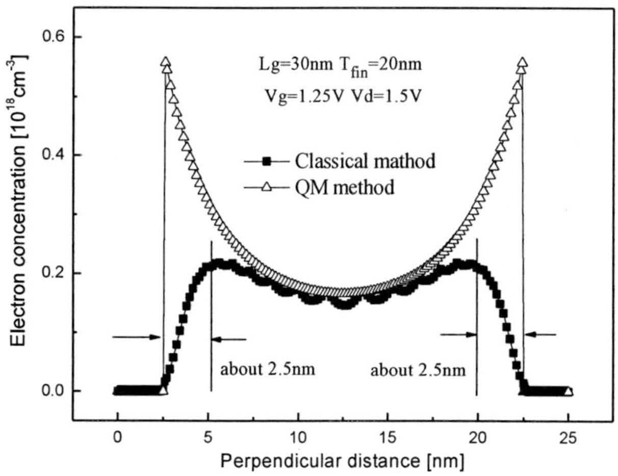

Figure 2.2. Comparison of the classical and QM electron density for a cut through the Fin channel.

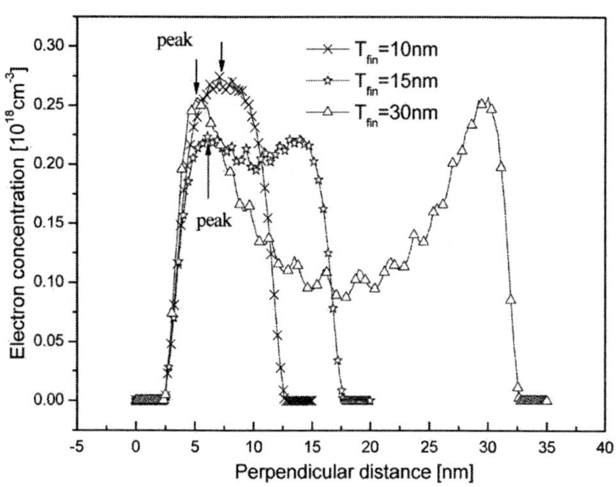

Figure 2.3. QM electron density at T_{fin}=10, 15 and 30nm with V_d=1. 5v, V_g=1.25 V and L_g=30 nm for cut through the Fin channels.

Figure 2-2 shows comparison of the classical and QM electron density for a cut through the Fin channel. However, QM simulation combined with gate bias creates a potential well between the two oxide boundaries. This well produces a quantization effect in the channel charge. The well splits the conduction band into discrete subbands, therefore the maximum of electron concentration is located approximately 2.5nm away from the oxide, which corresponds about 100% of the oxide thickness. This effect causes a deviation of effective oxide thickness from the physical oxide thickness which in turn reduces the gate capacitance and drain current with bias condition. Figure 2-3 demonstrates the distribution of electrons for T_{fin} = 10 , 15 and 30 nm. These simulations performed under conditions of L_g =30 nm, V_g = 1.25 V and V_d = 1.5 V exhibit clearly the quantum repulsion discussed earlier, as the peak of electron density is located farther from the Si/SiO_2 interface as the silicon channel becomes thinner.

The AC analysis of the classical drift-diffusion model predicts much higher capacitance in the accumulation and inversion regions in Fig. 2-4, because of the well in Fin channel occurs close to the surface. The energy levels in the quantum well are quantized, therefore, the electron concentration under the gate oxide peaks a few nanometers from the surface.

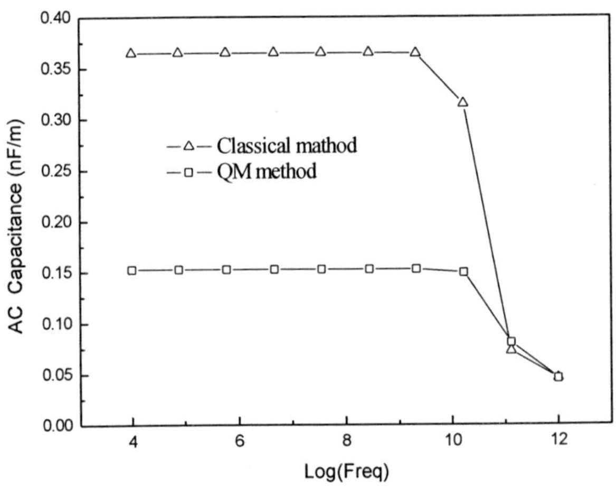

Figure 2.4. Comparison of the AC gate capacitance with and without quantum effects at V_d=1.5 V, V_g=1.25 V, L_g=30 nm and T_{fin}=20 nm.

2.2.2. Gate Leakage Current

In nanoscale n-type FinFETs, the gate current can be suppressed due to the reduced vertical effective electric field. The average vertical electric field in the channel inversion layer is described by an equation [51] Since the depletion charge densities is near 0 for low channel impurity concentration (2×10^{16} cm^{-3} in Fin channel), carriers in the inversion layer thus encounter a smaller average vertical electric field in thin-body devices than in standard bulk devices with heavy channel doping. The slope of the potential in the Fin silicon channel is dramatically reduced, particularly at points further away from the two Si/SiO_2 interfaces.

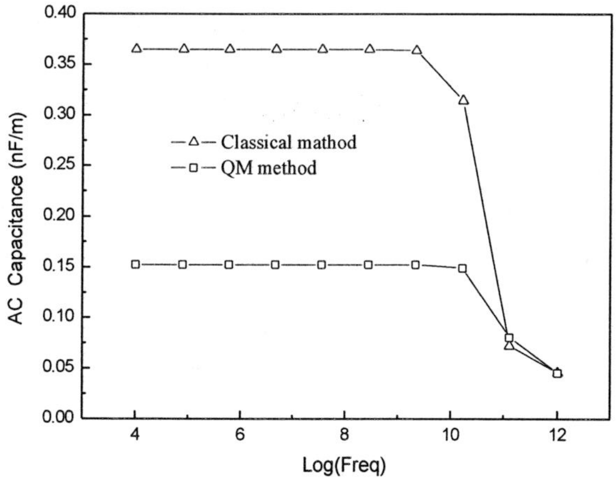

Figure 2.5. Vertical effective electric field reduction along the Fin channels in term of body thickness for N-type FinFETs.

As shown in Fig. 2-5, the electric effective field near the bottom of the inversion layer is dramatically reduced, as Fin body width (T_{fin}) decreases. This reduces the depth of the potential well (Fig. 2-6), which lowers the ground bound state energy (E_0) due to reduced carrier confinement and broadens the inversion charge distribution, thus resulting in a lower gate tunneling current density which is better than that of Ultrathin SOI MOSFETs [52] for the symmetry of the two gate electrodes.

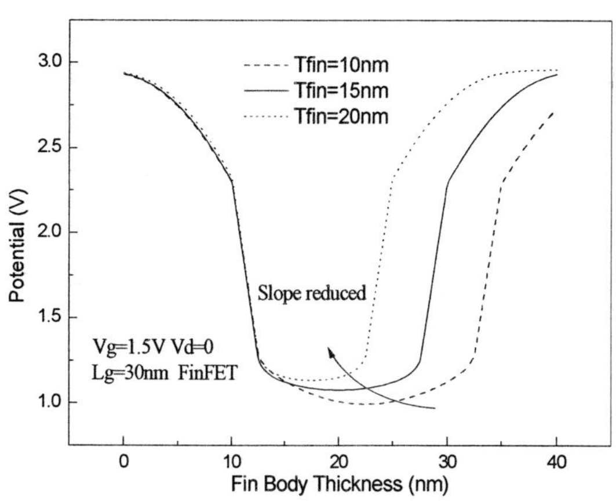

Figure 2.6 Potential slope change along the Fin channels in term of body thickness for N-type FinFETs.

Depending on the doping concentration, the electric field distribution in the Fin body can be affected as a significant amount of depletion charge can be formed when the channel doping is large. At same time, the electron concentration in the Fin body can be affected as a significant amount of conductive charge can be recombined when the channel doping is large. These impact the gate current. When the Fin doping is high the gate current can be dramatically reduced beyond $10^{17}cm^{-3}$. This occurs because conductive electron concentration in the well is reduced (Fig. 2-7), thus reducing the gate current. Though the electric field is raised because of increasing depletion charge when the Fin doping concentration is extremely high, conductive electron concentration is the major factor beyond the threshold doping $10^{17}cm^{-3}$.

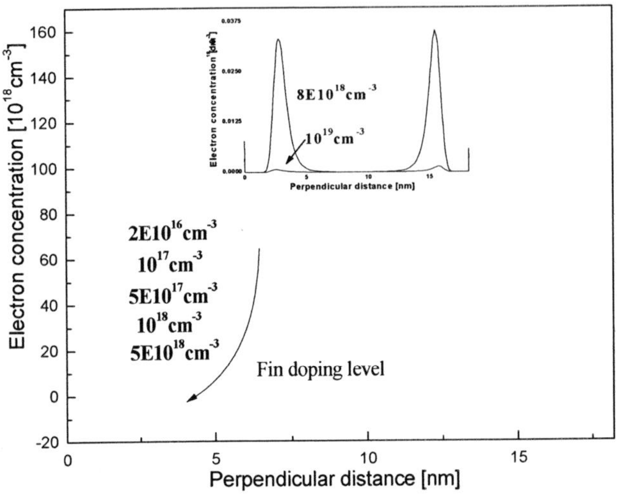

Figure 2.7. Inversion electron density at T_{fin}=15nm, Vd=0, V_g=0.8 V, T_{ox}=1.6 nm and L_g=30 nm with different Fin doping level for cut through the Fin channels. The inset shows the electron density with the other two Fin doping level.

3. Self-heating and Hot Electron Effects of GaN-based HFETS

3.1. Device Description and Simulation Model

Figure 3-1 shows the cross sections of the ALD Al_2O_3/AlGaN/GaN-based MOS-HFET. A 40 nm undoped AlN buffer layer, a 3 μm undoped GaN layer, and a 30 nm undoped $Al_{0.2}Ga_{0.8}N$ layer are sequentially grown by metal-organic chemical vapor deposition on a 2-in. sapphire substrate. After these layers are grown, the wafer is transferred via room ambient to an ASM Pulsar2000TM ALD module. A 16-nm-thick Al_2O_3 layer is deposited at 300°C, followed by annealing at 600°C per 60 s in oxygen ambient. Device isolation is achieved by nitrogen implantation. Using a wet etch in diluted HF, the oxide on the source and drain regions is removed while the gate region is protected by photoresist. Ohmic contacts are formed by electron-beam deposition of Ti/Al/Ni/Au and a lift-off process, followed by an 850°C anneal in nitrogen ambient, which also activates the previously implanted nitrogen. Finally, Ni/Au metals are e-beam evaporated and lifted off to form the gate electrodes. All four levels of lithography (alignment, isolation, ohmic, and gate) are done by using a contact aligner. The gate width (W_g) is 100 μm; the gate length (L_g) varies from 1 μm to 40 μm (most of the simulations are concentrated on 5 μm); the source and drain lengths (L_s and L_d) are 5μm for the simulations, and the source/drain extension (L_{ext}) is 2μm.

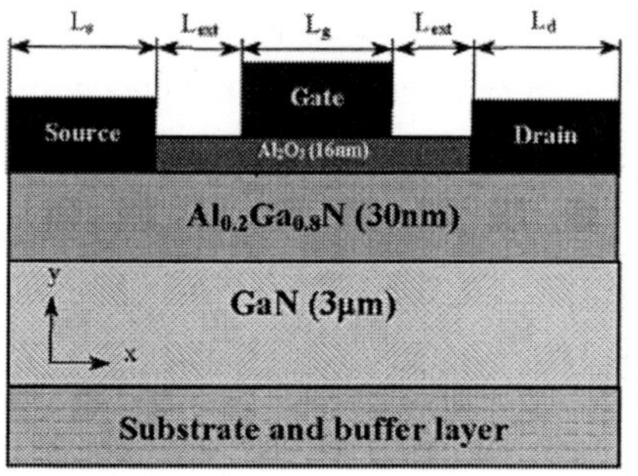

Figure 3.1. Schematic structure of GaN-based MOS-HFET with Al_2O_3 gate dielectric.

To account for the self-heating, the nonisothermal model available in DESSIS is utilized. The temperature gradient drift-diffusion transport equations and the temperature distribution equations are calculated self-consistently. Since the hot electrons have a significant effect on the vertical real space charge transfer and subsequent capture in bulk traps, [53] they are accounted for by the hydrodynamic transport model available in DESSIS. Mobility of GaN in the device grown on the sapphire substrate is approximately 1000 cm^2/Vs. [54] The electric fields induced by the high piezoelectric and spontaneous polarization lead to a significant increase in the sheet carrier concentration and a more narrow confinement of the 2DEG at the AlGaN/GaN heterointerface without doping in wurtzite group-III nitrides HFET[55-58]. In the absence of the external fields, the effect of the polarization leads to sheet interface charge accumulation on the end of the crystals. These interface charges are, of course, equal in magnitude and opposite in sign to maintain overall charge neutrality and will induce free carriers to compensate themselves. Theoretical computations of the interface charges of the GaN-based MOS-HFET due to the high piezoelectric and spontaneous polarization are given by Refs. [59, 60]. All of the parameters in our calculations can be found in Refs. [59] and [60].

Due to a large conduction band offset in AlGaN/GaN, some studies have indicated that the 2DEG induced by the polarization are subject to quantum confinement within 2~5nm [48, 50] at the AlGaN/GaN interface. For example, Ambacher et al. have shown that the confined carriers are within 2 nm of the AlGaN/GaN interface. [59] Based on the C-V-concentration experiments, Yu et al have found that the 2DEG arising from the polarization would be located within 2-5 nm of the AlGaN/GaN interface. [61] Furthermore, Braga et al. have shown using a self-consistent calculation that the 2DEG quantum confinement is within 2-5 nm of the AlGaN/GaN interface. [31] Therefore, the quantum effects become noticeable at the AlGaN/GaN interface. New methodologies and models for the GaN MOS-HFETs must be used for the device simulations to account for these effects. Since the density-gradient (DG) model is robust, fast, and can be applied to highly nonequilibrium situations, [53] the quantized electron gas has been accounted for by the DG transport model. [62] The DG approach is a self-consistent way to account for the quantum effects via the quantum potential correction to the continuity equation. [63]

The trap density and cross sections in III-Nitride materials have been shown in some experimental observations and theoretical calculations such as Ref. [64] and Ref. [65]. Here, we assume just a single acceptor type electron bulk trap level with a traps density of 1.0×10^{18} cm^{-3} and a cross section of $\sigma_{Tn}=1.0\times10^{-15}$ cm^{-2}, and position it 1eV above mid band gap. [31, 61, 64, 65] The thermal conductivity (κ) of GaN is known to depend on dislocation density

(ρ_D), which varies along the epitaxial growth direction. We divide the GaN layer into two layers in our simulation. The first layer is the 150 nm initial growth region, with κ=1.3 $Wcm^{-1}K^{-1}$; the remainder of the GaN layer has $\rho_D<10^{10}$ cm^{-2} and κ=1.5 W $cm^{-1}K^{-1}$. [66] Since the thin AlGaN layer plays a minor role in dissipating heat, we employ the properties of GaN. We use κ=0.28 $Wcm^{-1}K^{-1}$ for the sapphire [66, 67] and ignore its temperature dependence since the observed ΔT rise is small. Parameters of Al_2O_3, being very important to our MOS-HFET simulations, are extracted from Refs. [62] and [64]. The estimated oxide thickness (d_{OX}) is 10 nm, which is significantly less than the design value of 16 nm.

3.2. Quantization and Self-heating Effects

Figure 3-2 shows the simulated and experimental dc output characteristics (I_{ds}-V_{ds}) of the MOS-HFET for V_{gs} values varying from 0 to 3 V in 1 V steps. It is found that the simulation results are in good agreement with the experiment data. All of the curves display the distinct negative differential output conductance (NDC) due to the self-heating effects.

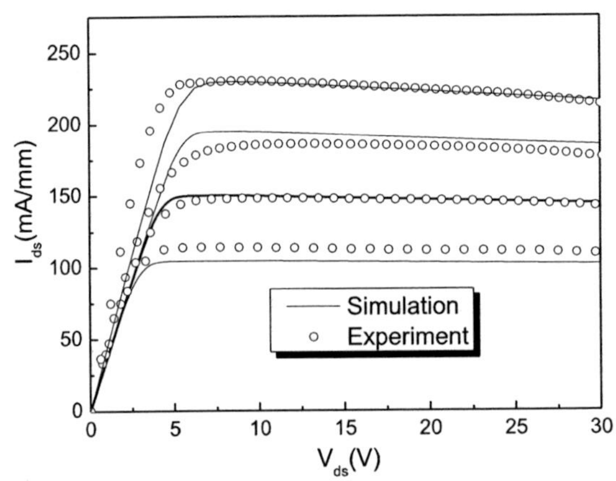

Figure 3.2. Simulated and experimental I_{ds}-V_{ds} of MOS-HFET. Experiment (circles) and simulation (solid lines) with thermodynamic and DG approaches for V_{gs}=0, 1, 2, and 3V.

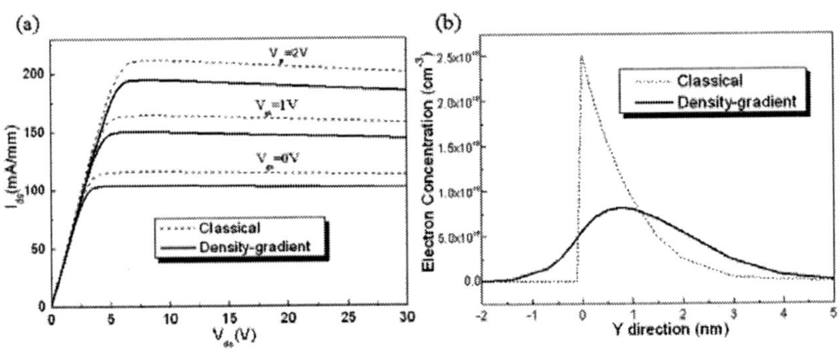

Figure 3.3. Comparisons of (a) simulated I_{ds}-V_{ds} results, and (b) electron density in the channel with DG quasi quantum mechanical method (solid lines) and classical method (dashed lines).

Figure 3-3 (a) shows the current-voltage (I_{ds}-V_{ds}) as predicted by a classical simulation and as predicted with the DG model. The simulation results reveal that the current predicted with the DG model is suppressed by approximately 9.6% due to the carrier confinement that arises when the quantum approach is chosen. We can also note that the data from the quantum method agree more closely to that of the experimental data. Figure 3-3(b) shows the corresponding distribution of electrons in the channel under the zero bias conditions. It is evident that the classical picture does not account for the quantum confinement of electrons. Both the quantum approaches and the classical simulations yield similar values for the sheet electron density $n_s \approx 6\times10^{12}$ cm^{-2}, which is in good agreement with the experimentally measured value. However, the classical approach does not account for the significant penetration of the electron wave function into the AlGaN barrier layer and the GaN bulk layer. The quantum calculations clearly show a significant penetration of the electron wave function from the 2D gas into three-dimensional states in AlGaN and GaN, thereby reducing the electron mobility and the drain current (Fig. 3-3(a)).

For a better description of the simulated behavior of the self-heating effects in the GaN-based MOS-HFETs, we use a simple theoretical model [68, 69] based on the heat dissipation and heat-transfer in the GaN-based MOS-HFETs. The drain current as a function of the field can be determined from the coupled equations. For simplicity, we take the thermal impedance θ as a constant (about 325 C·mm·W^{-1} on the sapphire). [69]

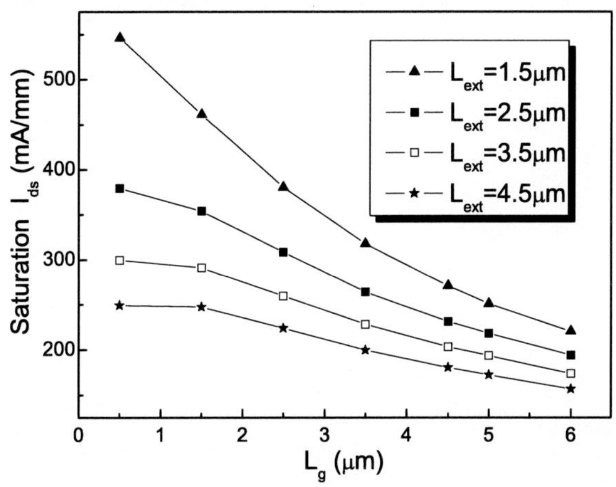

Figure 3.4. Dependence of saturation drain current (I_{ds}) on the gate lengths for the different source/drain extension (L_{ext}) lengths at V_{gs}=3V and V_{ds}=15V.

Figure 3-4 shows the dependence of the saturation drain current on the gate lengths for the different source/drain extension lengths. The shorter the channel is, the greater the reduction of the saturation I_{ds} will be. The current reduction is caused by the increasing intrinsic normalized channel resistance: $R_{ch}=V_{ds}\times W_g/I_{acc}\times L$, where I_{acc} is the accumulation current, and L is the channel length ($L=L_g+2L_{ext}$). Comparing the dependence of I_{ds} and ΔT_{max} on the gate lengths for the different L_{ext}, we can easily get the optimal gate length (about 2.5 μm) and source/drain extension (about 1.5 μm) considering the self-heating effects.

3.3. Hot Electron and Self-heating Effects in GaN-based MOS-HFETs

High carrier concentrations of 2DEG and large values of the electron mobility and operation voltages make it necessary to distinguish the hot electron and self-heating effects for better understanding of the transport properties aimed to use the GaN-based MOS-HFETs in high power and high temperature microelectronic applications. It is also very important to choose a suitable model for accurate and efficient (fast) physics-based device simulations to complement experiments. The hot electrons can overcome the potential barriers and be captured by the bulk traps in the barrier and bulk layers leading to the NDC in the output characteristics, [70] which is the so-called hot electron effects.

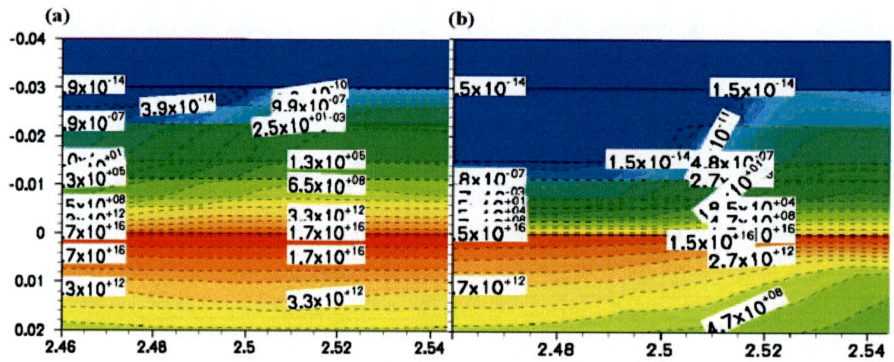

Figure 3.5. Cross section of GaN-based MOS-HFET with electron density contour maps comparing results predicted by (a) hot electron and (b) simple drift-diffusion simulations at the drain-side gate edge. The AlGaN/GaN interface is at y=0; the AlGaN/Al_2O_3 interface is at y=-0.03. Dimensions in μm.

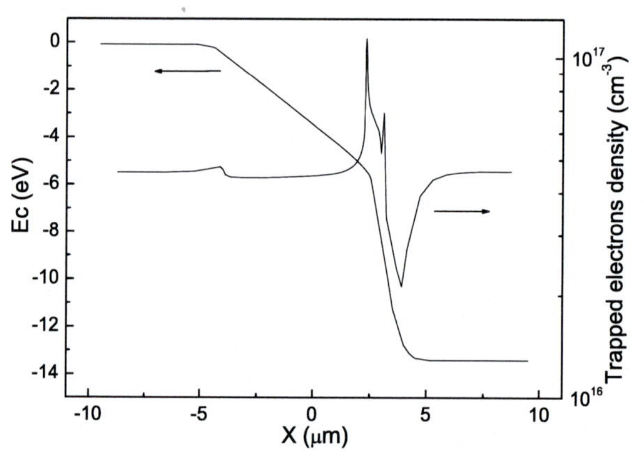

Figure 3.6. Conduction band (left y axis) and trapped electron density (right y axis) versus x direction position for V_{gs}=1V and V_{ds} =10V. L_g= 5μm L_{ext}=2μm.

Figure 3-5 compares the simulated electron density contour map with the drift-diffusion and hot electron models at zero gate bias and 15V drain bias. We can clearly see that in the drift-diffusion simulations electrons tend to be confined within the channel, while in the hot electron simulations the spreading of the hot electrons towards the AlGaN barrier and the GaN buffer is evident. At higher drain bias, the electrons become hotter; hence the wider flow spreading. This increase in drain bias also results in more trap levels in the bulk being occupied by the hot electrons. This extra negative charge subsequently lifts up the conduction band and decreases the 2DEG under the drain-side gate edge, leading to the formation of a potential barrier for the electron flow. However, for our long channel GaN-based MOS-HFETs, there is no evident potential barrier formed at the drain-side gate edge.

Figure 3-6 (left axis) shows the distribution of the trapped electrons at 5nm away from the AlGaN/GaN interface at 10V drain bias, which is where I_{ds} saturates. The average trapped electron density is about 7×10^{16} cm^{-3} smaller than that of the free 2DEG at the interface. Figure 3-6 (right axis) shows that no evident conduction band barrier is formed at the gate edge. Therefore, the trapped electrons make few contributions to the NDC in the long channel GaN-based MOS-HFETs.

3.4. Effects of Static Interface Charges and Traps

It is generally accepted that the empty donor like traps generated by the process damage positively charge the AlGaN interface under the gate. [71-73] As mentioned in Sec. II, the interface charges are induced by the polarization effects of the wurtzite group-III nitride heterointerface. In order to visualize the interface-related effects, we ran the static simulations with the different donor trap and charge densities (negative charge for the Al_2O_3/AlGaN interface with the theoretical computation). Unlike the conventional GaN-based HFET, our Al_2O_3/AlGaN/GaN-based MOS-HFET has an interface (Al_2O_3/AlGaN) under the gate contact. Therefore, it is very important to investigate the inner mechanisms of the interface-state-related effects on the MOS-HFET.

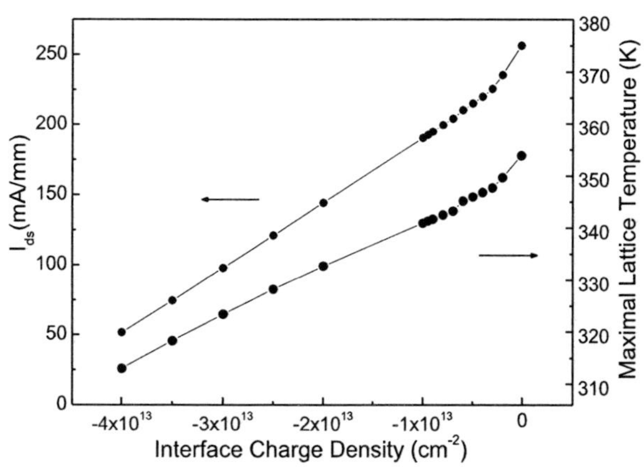

Figure 3.7. Drain current (left y axis) and maximal lattice temperature (right y axis) vs interface charge density for V_{gs}=0V and V_{ds}=15V. L_g=5μm and L_{ext}=2μm.

The interface states assumed in our model consist of a fixed interface charge and an interface trap of a uniform distribution at the Al_2O_3/AlGaN interface. As shown in Fig. 3-7, a reduction in the drain current is obtained when an order of magnitude increase of the negative interface charge is assumed at the Al_2O_3/AlGaN interface. The drastic reduction of the channel current occurs in the areas at the interface where the fixed interface charge is defined-

-the interface acts as a virtual gate located between the contacts, [74] as well as under the gate. Another reason for the channel current reduction is the increase in hole concentration next to the interface where the interface charge is introduced (we neglect the donor trap effects which are discussed later). As the magnitude of the interface charge density increases, the holes are attracted towards the interface, where they play an important role in neutralizing the negative interface charges. From the above discussions, we can conclude that the presence of the fixed charge at the interface can be directly responsible for the current collapse in the GaN-based MOS-HFETs. Figure 3-7 also shows that the maximal lattice temperature decreases down to 7.36% for the interface charge density of -4×10^{13}cm^{-2}. We also note that the self-heating effects are suppressed due to the neutralization of the attracted holes by the increasing negative interface charge density. However, the excessive increase of the charge density decreases the drain current because the virtual gate becomes too "thick".

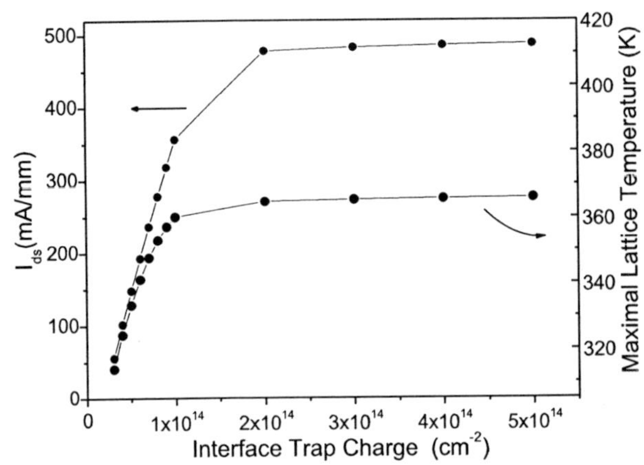

Figure 3.8. Drain current (left *y* axis) and maximal lattice temperature (right *y* axis) vs interface trap charge for V_{gs}=0V and V_{ds}=15V. L_g=5μm, L_{ext}=2μm.

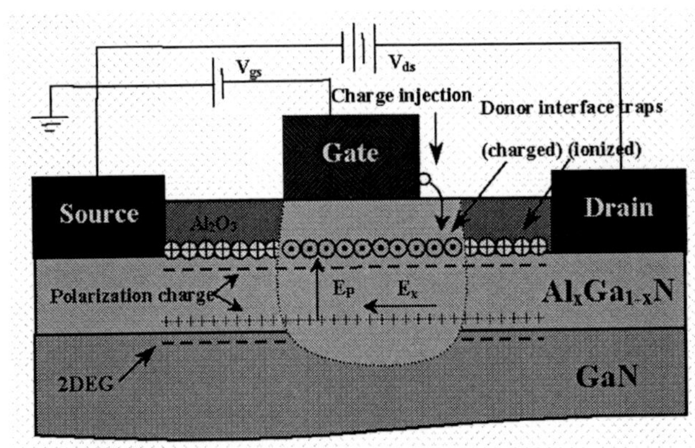

Figure 3.9. The pinch off state of GaN-based MOS-HFET under a positive drain bias.

Figure 3-8 shows the drain current and the maximal lattice temperature versus the interface trap charge for V_{gs}=3V and V_{ds}=15V. The drain current increases up to 92% for the interface trap charge $\geq 9\times10^{13}cm^{-2}$. The donor traps, which are partially ionized, create a positive step in the gate voltage, which further ionizes the donor traps and increases the drain current. Our results indicate that the drain current is sensitive to the interface donor trap charge, which reduces the net charges at the Al_2O_3/AlGaN interface. The net charges are a summation of the ionized donor traps and negative interface charges. [75, 76] As expected, a similarity in the drain current response is obtained when an increase in the trap charge is used in the model. We know that the increase of the drain current under the same electric field causes the correspondingly larger dissipated Joule electric power, thus increasing self-heating.

The above discussion gives a clear picture of the effects of the donor traps and polarization interface charges on the operation of the GaN-based MOS-HFET. When the gate is negatively biased and the drain is positively biased (as shown in Fig. 3-9), the interface traps at the Al_2O_3/AlGaN are filled with electrons injected from the gate during pinch off. [77] At the reverse gate bias, although the gate leakage current of the MOS-HFETs is lower than that of the conventional HFET, [23] it is reasonable to assume that some of the electrons can still be injected from the gate through Al_2O_3. With the surface traps filled with the injected electrons, the channel is vertically depleted to keep the system neutral, instead of being laterally depleted by the drain bias. Since there is no lateral depletion in the channel, the electric field peak in the x direction at the drain-side gate edge is alleviated. The electric field in the depletion region is a combination of the vertical polarization field (E_p) and the constant lateral bias electric field (E_x). When the gate and the drain are positively biased (which is the on state of the switching device), the trapped electrons escape from the traps leading to the disappearance of the vertical depletion and the recovery of density of the 2DEG in the channel. Finally, the current passes through the channel, and the device is turned on.

4. The Intrinsic Mechanism of Drain Lag and Current Collapse in GaN-based High- Electron-Mobility-Transistors

4.1. Device Description and Model

Figure 4-1(a) shows the sketch of the simulated structure. Two kinds of device parameters from [78] (device A) and [79] (device B) are employed in this simulation for the purpose of comparison with experimental results. For device A, the undoped GaN and $Al_{0.35}Ga_{0.65}N$ layer thicknesses are 1.5 μm and 29 nm, respectively. The gate length is 0.7 μm, whereas gate-drain and gate-source spacings are 2 and 0.7 μm, respectively. For device B, the undoped GaN and $Al_{0.3}Ga_{0.7}N$ layer thicknesses are 2 μm and 25 nm, respectively. The gate length is 1.0 μm, whereas gate-drain and gate-source spacings are 2.4 and 1.5 μm, respectively. The simulation tool is the two-dimensional device simulator Sentaurus-Device. Since the hot electrons play an important role in the vertical real space charge transfer and subsequent capture in bulk and surface traps, carrier temperature is calculated by the hydrodynamic transport (HT) equations [80]. The Schottky barrier tunneling is considered for the gate leakage current calculation.

The strain-induced and spontaneous polarizations cause interface charges at nitride heterointerfaces. These charges can be taken into account in our simulations by specifying the interface charges at the heterointerfaces. Theoretical computations of interface charges of GaN-based HFETs are given by Ref. [81]. All of the parameters in our calculations can be found in

Ref. [81]. An extended discussion on theoretical computations of the interface charge value, as well as on other model parameters used in our simulations, can be found elsewhere [82].

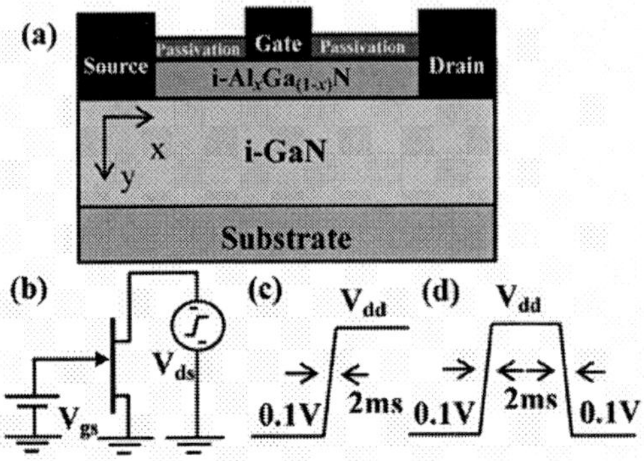

Figure 4.1. (a) Cross-sectional structure of simulated GaN-based HFET. The structure parameters of the device A and B are from [78] and [79], respectively; (b) Bias circuit used in simulations with a transient source-drain voltage to obtain the drain current response versus time; (c) Drain turn-on pulsing mode for device A. A fixed voltage is applied to the gate and the drain is pulsed from 0 V to a V_{dd}; (d) Drain turn-off pulsing mode for device B. A fixed voltage is applied to the gate. The drain is pulsed from 0 V to a V_{dd} and changed back to 0 V. And then the drain voltage is left at 0 V for the rest of the simulation.

4.2. Results and Discussions

Fig. 4-2(a) shows the simulated and experimental $I_{ds}(t)$ transients in response to V_{ds} turn-on pulse at V_{gs} = 0 V for device A. It is found that the simulation results are in good agreement with the reported experiment data. Note that there is a quick fall in Ids during the initial 0.05 s, where I_{ds} decreases exponentially ($I_{ds}(t)$~exp($-t/\tau$)) with the characteristic time τ=0.03 s. Then I_{ds} decreases gradually and reaches the steady-state value with a collapse of about 8%. The simulated characteristic time is comparable to that of the reported [78]. In order to analyze the dynamic pictures of hot electrons accounting for current collapse and drain lag, the simulated trapped charge at the drain-side gate edge is investigated. Fig. 4-3(a) shows the spatial distributions of trapped charge density under the pulse-on drain bias measured at time t=0.1, 0.2, and 0.6 s for device A. It is well known that acceptor-type traps can be either negative or neutral like the acceptor. Acceptor traps are neutral when empty and negatively charged (ionized) when filled with electrons. Filled acceptor traps can emit electrons or capture holes called de-trapping. Empty acceptor traps can capture electrons or emit holes called trapping. There is a dynamic equilibrium between the trapping and de-trapping at different time stages in the active region. At t=0.1 s, V_{ds} =V_{gs}=0, the system is at the steady state. The deep traps in the buffer are under the dynamic equilibrium--trapping and de-trapping. When V_{ds} is positively biased from t=0.2 to 0.6 s, the equilibrium conditions determine the de-trapping process. The excess channel electrons can spill over in all directions while the drain bias is ramped up and acquire enough energy at the gate edges to reach far deep into the buffer where a number of traps are initially unoccupied. The deep traps in the buffer need certain time to capture the electrons. Finally, I_{ds} decreases gradually to the steady-state value at another equilibrium.

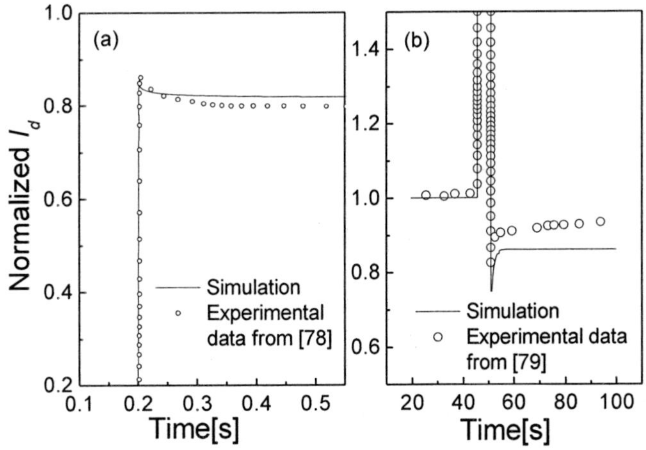

Figure 4.2. Simulated (solid line) and experimental (circles) $I_{ds}(t)$ transients in response to (a) V_{ds} turn-on pulse at V_{gs} = 0V for device A, and (b) V_{ds} turn-down pulse at V_{gs} = 0V for device A. For comparison, the drain currents have been normalized.

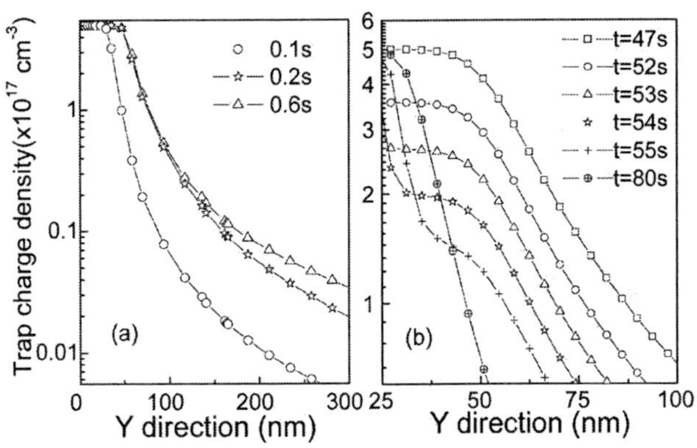

Figure 4.3. (a) Trapped charge density as a function of *Y* direction (cut at the drain-side gate edge, x = 0.4 μm. The origin of the X axis is at the middle of the gate, while the origin of the Y axis is at the AlGaN/GaN interface.) under pulse drain bias measured at time t = 0.1, 0.2, and 0.6 s (as shown in Fig. 2 (a)) for device A. (b) Trapped charge density as a function of y direction (cut at the drain-side gate edge, x = 0.2 μm) under pulse drain bias measured at time t = 47, 52, 53, 54, 55, and 80 s (as shown in Fig. 2 (b)) for device B.

Fig. 4-2(b) shows the simulated and experimental $I_{ds}(t)$ transients in response to V_{ds} turn-off pulse at V_{gs} = 0 V for device B. It is found that the simulation results are comparable with the reported experiment data [79], however, the simulated results are smaller than that reported at the pulse-off duration. We attribute the reduction of I_{ds} to the shallow donors, which are not considered in our simulations. A small number of traps in the buffer are already filled by compensating the shallow donors before being occupied by hot electrons [79], [83]. In order to analyze the dynamic pictures of hot electrons during the pulse-off mode, the simulated trapped charge density at the drain-side gate edge is investigated. Fig. 4-3(b) shows the spatial distributions of trapped charge density under the pulse-off drain bias measured at

t=47, 52, 53, 54, 55, and 80 s for device B. Similar to the former discussions, there is a dynamic equilibrium between trapping and de-trapping at the different time stages in the active region. When V_{ds} is ramped up to 20 V from 0 V while keeping V_{gs}=0, the equilibrium conditions determine the de-trapping process. A number of traps are occupied by the hot electrons, and the trapped charge density is maximal at t=47 s as shown in Fig. 4-3(b). When the gate voltage is switched to the initial level, the trapped charges remain nearly the same as that of V_{ds}=20 V. At the same time, the system of the trapped electrons finds itself far from equilibrium—de-trapping and trapping. With the trap filling mechanism being interrupted, the de-trapping process alone dominates the dynamics of the trap population until a new equilibrium is reached. Finally, I_{ds} recovers to the steady state value at the equilibrium.

Although growth at low pressure by introducing more defects or antidoping by carbon may be used to increase the resistivity of GaN, the dopants also increase the defects density in the GaN bulk, which has been shown to further enhance the current collapse effect [84]. To minimize the drain current collapse effect and the parasitic conduction, quantum-well HFET structures are proposed and demonstrated on the basis of the simulations.

Since AlN (or SiC) with a band gap of 6.1 eV (7 eV) is highly resistive, and high-quality AlN epilayers can be grown and bulk AlN substrates are also available, we substitute the GaN bulk layer with highly resistive AlN bulk layer [85-87] to build the AlGaN/GaN/AlN (AlGaN/GaN/ SiC) QW HFETs. The channel of the narrower band gap GaN between the buffer layer of the wider band gap AlN and the AlGaN barrier layer increases the quantum confinement of the two-dimensional electron gas (2DEG) and forms a double-heterostructure (DH) band structure. The large net polarization charges, inducing the two-dimensional hole gas (2DHG) at the GaN/AlN interface, could possibly deplete the 2DEG. A backside-doping scheme is adopted to overcome the depletion for device B.

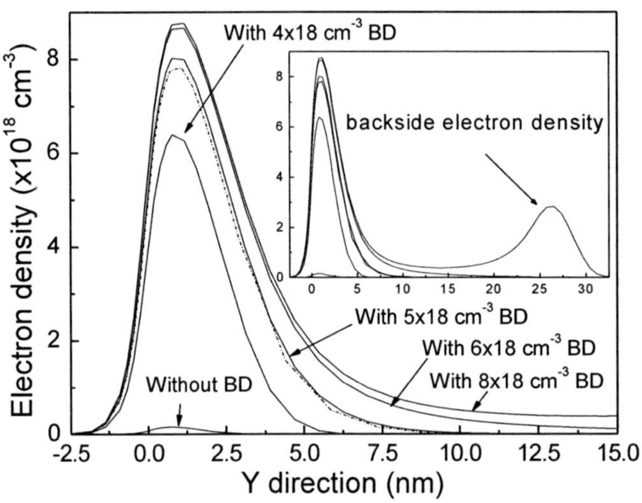

Figure 4.5. Simulated electron density as a function of distance from AlGaN/GaN (cut at x=0 μm) with different backside doping for the proposed QW-HFET. Dashed line shows the electron density for the conventional HFET under the same condition. Inset shows the whole electron density throughout the device. The origin of the Y axis is at the AlGaN/GaN interface.

Figure 4-5 shows the simulated electron density as a function of the distance from AlGaN/GaN with and without the backside doping (BD) for the proposed QW-HFET. Comparing the electron density with and without the BD, our simulations indicate that the electron density without the backside doping is about 80% lower, which is clearly not acceptable. The simulations demonstrate that a 5×10^{18} cm^{-3} BD in the QW-HFET has the similar sheet electron density as the conventional structure. An excess BD causes an undesired backside electron density in the GaN channel as shown in the inset of Fig. 4-5. This backside electron density can significantly increase the parasitic conduction in GaN. Taking into consideration the parasitic conduction in the bulk and the electron density in the channel, we demonstrate that the doping value around 4×10^{18} cm^{-3} is much better than others at this narrow range. So, in a wide range of doping value, the device has the good performance, and the best doping value may be around 4×10^{18} cm^{-3}.

If the GaN layer is thick enough, the 2DHG avoids the shunting of the 2DEG by background electrons in the structure bulk. In this case, a reduction in the channel thickness can improve the electron confinement and prevent the electron spillover towards the bulk. However, the co-existence of the 2DEG and 2DHG in a thin GaN channel may affect negatively the QW-HFET performance. Indeed, the ohmic source and drain contact regions formed by annealing may overlap both the 2DEG and 2DHG channels. This is expected to result in shunting the electron conductivity by the parasitic holes. In addition, the electrons heated in the electric field induced by the drain-source voltage are capable of penetrating into the 2DHG region, leading to carrier losses through the electron-hole recombination [87]. Finally, the existence of the 2DHG at the GaN/AlN interface makes the application of our backside-doping concept necessary.

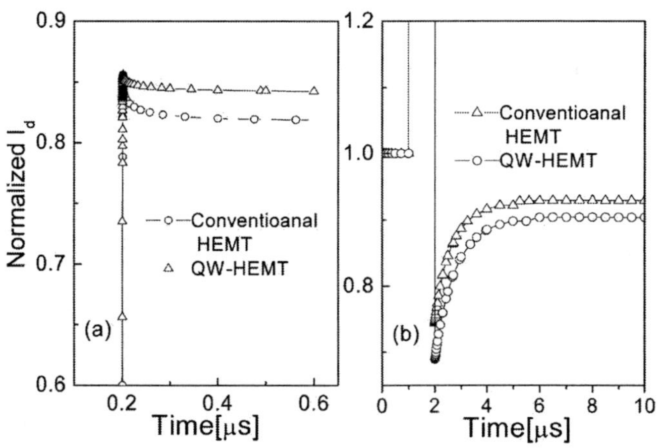

Figure 4.6. (a) Simulated $I_{ds}(t)$ transients in response to Vds turn-on pulse at $V_{gs} = 0$ V for conventional HFET (circles) and QW HFET (triangles). (b) Simulated $I_{ds}(t)$ transients in response to V_{ds} turn-off pulse at $V_{gs} = 0$ V for conventional HFET (circles) and QW HFET (triangles). For comparison, the drain currents have been normalized.

To verify the advantage of our proposed QW HFET, we perform transient simulations for the current collapse effect. The gate pulse as transient signal is such that the minimum gate voltage is close to threshold, and the maximum is at zero. The gate voltage is held at 0V for 0.

1μs, and is changed to −5V (pinch-off voltage) for 1μs pulse width, with the fall and rise time of 250ns, and then is left at 0V for the rest of the simulation. Compared with the conventional HFET, the proposed QW HFET shows a decreased current collapse as shown in Fig. 4-6. The designed quantum-well has a large band edge discontinuity and effective quantum confinement, which limit the electron spillover into the bulk, and finally result in a reduction of the drain current collapse (5~10%). Our simulation results have provided guidance for the device structural optimization.

5. Conclusion

AC high frequency performance was well simulated. The results indicate that the deviation from the classical model becomes more important as the gate oxide, gate length and Fin channel width becomes thinner and the Fin channel doping increases. The energy quantization effect becomes evident in the quantum well, which considerably affects electron tunneling significantly. The gate current can be dramatically reduced beyond $10^{17}cm^{-3}$ with the Fin body doping increasing. The mechanism of gate direct tunneling in very thin silicon layer nanoscale FinFETs must be assessed based on a quantum-mechanical approach to determine the design parameters of future nanoscale FinFETs.

High performance GaN-based MOS-HFETs have been fabricated with Al_2O_3 gate dielectrics. The theoretical results are found to be in good agreement with the experimental data for ALD Al_2O_3/AlGaN/GaN MOS-HFETs on a sapphire substrate with a 5μm gate length. Both the simulated and experimental results show that the transfer characteristics are not greatly affected by the self-heating effects when operating below a 10V drain voltage. The quantum calculations clearly show a significant penetration of the electron wave function from the 2D gas into three-dimensional states in AlGaN and GaN. The presence of the interface charges and traps are directly responsible for the observed current collapse and device switching in the GaN-based MOS-HFETs. Our results allow us to identify the nature of the donor traps affecting the interface net charge. The self-heating is also strongly affected due to the fluctuation of the interface states.

2DEG can significantly heat and spread toward the GaN bulk. Those “hot electrons” are transferred in the vertical real space and subsequently captured by bulk traps at the drain-side gate edge causing a notable current collapse.-Trapping and de-trapping processes determine the collapse and recovery of drain current under drain pulse. Quantum-well HFET structures have been proposed and demonstrated to obtain the optimized performance. The simulation results also link the current collapse with double-heterostructure effect of AlGaN/GaN/AlN structure. A possible approach for removing the current collapse that follows from our simulations is field-plate over the drain-side gate electrode, which can lessen the electric field changes at the drain-side gate edge.

Acknowledgement: This work was supported in part by Chinese National Key Basic Research Special Fund (2007CB613206), National Natural Science Foundation of China (10725418, 60576068), Key fund of Chinese National Natural Science Foundation (10734090) and Creative Foundation of SITP (C2-14).

References

[1] Y. Taur, D. A. Buchanan,W. Chen, D. J. Frank, K. E. Ismail, S.-H. Lo, G. A. Sai-Halasz, R. G. Viswanathan, H.-J. C.Wann, S. J.Wind, and H.-S. P. Wong, CMOS scaling into the nanometer regime, *Proc. IEEE* **85**(4) (1997) 486-504.

[2] D. J. Frank, R. H. Dennard, and E. Nowak, Device scaling limits of Si MOSFETs and their application dependencies, *Proc. IEEE* **89** (3) (2001) 259-288

[3] W. D. Hu, X. S. Chen, Z. J. Quan, X. C. Zhou, and W. Lu Quantum and short-channel effects study for sub-50 nm FinFETs, *J. Infrared Millim. Waves* **25**(2) (2005) 90-93

[4] S. L. Jang, M. C. Hu, and S. S. Liu, An Analytical Symmetric Double-Gate Silicon-on-Insulator Metal-Oxide-Semiconductor Field-Effect-Transistor Model, *Jpn. J. Appl. Phys.* **36** (10) (1997) 6250-6253.

[5] T. Schulz, W. Rösner, E. Landgraf, L. Risch, U. Langmann, Planar and vertical double gate concepts, *Solid State Electron* **46** (7) (2002) 985-989.

[6] D. Hisamoto, W. C. Lee, J. Kedzierski, H. Takeuchi, K. Asano, C. Kuo, E. Anderson, T. J. King, J. Bokor and C. Hu, FinFET-a self-aligned double-gate MOSFET scalable to 20 nm, *IEEE Trans. Electron Devices* **47** (12) (2000) 2320-2325.

[7] W. D. Hu, X. S. Chen, X. C. Zhou, Z. J. Quan, and W. Lu, "Quantum-mechanical effects and gate leakage current of nanoscale n-type FinFETs: A 2d simulation study, *Microelectronics Journal* **37** (2006) 613-619.

[8] G. Pei, J. Kedzierski, P. Oldiges, M. Ieong, and E. C. Kan, FinFET Design Considerations Based on 3-D Simulation and Analytical Modeling, *IEEE Trans. Electron Devices*, **49** (8) (2002) 1411-1419.

[9] G. A. Kathawala, U. Ravaioli, *3-D Monte Carlo Simulation of FinFETs*, IEDM'03 (2003) 683-686.

[10] K. Kim, O. Kwon, J. Seo and T. Won, Nanoscale Device Modeling and Simulation: Fin Field-Effect Transistor (FinFET), *Jpn. J. Appl. Phys.* **43** (6) (2004) 3784-3789.

[11] Svizhenko, M. P. Anantram, T. R. Govindan and B. Biegel, Two-dimensional quantum mechanical modeling of nanotransistors, J. Appl. Phys. **91** (4) (2002) 2343-2354.

[12] U. K. Mishra, P. Parikh, and Y. F. Wu, *Proc. IEEE* **90**, 1022 (2002).

[13] M. Asif Khan, X. Hu, G. Sumin, A. Lunev, J. Yang, R. Gaska, and M. S. Shur, *IEEE Elect. Dev. Lett.* **21**, 63 (2000).

[14] M. Asif Khan, X. Hu, A. Tarakji, G. Simin, J. Yang, R. Gaska, and M. S. Shur, *Appl. Phys. Lett.* **77**, 1339 (2000).

[15] G. Simon, X. Hu, N. Ilinskaya, A. Kumar, A. Koudymov, J. Zhang, M. A. Khan, R. Gaska, and M. S. Shur, *Electron Lett.* **36**, 2043 (2000).

[16] Koudymov, X. Hu, K. Simin, G. Simin, M. Ali, J. Yang, and M. Asif Khan, *IEEE Elect. Dev. Lett.* **23**, 449 (2002).

[17] G. Simin, A. Koudymov, H. Fatima, J. Zhang, J. Yang, and M. Asif Khan, X. Hu, A. Tarakji, R. Gaska, and M. S. Shu, *IEEE Elect. Dev. Lett.* **23**, 458 (2002).

[18] G. Simin, X. Hu, N. Ilinskaya, J. Zhang, A. Tarakji, A. Kumar, J. Yang, M. Asif Khan, R. Gaska, and M. S. Shur, *IEEE Elect. Dev. Lett.* **22**, 53 (2001).

[19] X. Hu, A. Koudymov, G. Simon, J. Yang, M. Asif Khan, A. Tarakji, M. S. Shur,and R. Gaska, *Appl. Phys. Lett.* **79**, 2832 (2000).

[20] S. Ootomo, T. Hashizume, and H. Hasegawa, *Phys. Status Solidi B* **1**, 90 (2002).

[21] T. Hashizume, S. Ootomo, and H. Hasegawa, *Appl. Phys. Lett.* **83**, 2952 (2003).

[22] R. Mehandru, B. Luo, J. Kim, F. Ren, B. P. Gila, A. H. Onstine, C. R. Abernathy, S. J. Pearton, D. Gotthold, R. Birkhahn, B. Peres, R. Fitch, J. Gillespie, T. Jenkins, J. Sewell, D. Via, and A. Crespo, *Appl. Phys. Lett.* **82**, 2530 (2003).

[23] P. D. Ye, B. Yang, K. K. Ng, J. Bude, G. D. Wilk, S. Halder, and J. C. M. Hwang, *Appl. Phys. Lett.* **86**, 063501 (2005).

[24] G. D. Wilk, R. M. Wallace, and J. M. Anthony, *J. Appl. Phys.* **89**, 5243 (2001).

[25] S. W. Huang; J. G. Hwu, *IEEE Trans. Elect. Dev.* **50**, 1658 (2003).

[26] Y. F. Wu, B. P. Keller, S. Keller, D. Kapolnek, S. P. Denbaars, and U. K. Mishra, *IEEE Elect. Dev. Lett.* **17**, 455 (1996).

[27] Y. F. Wu, B. P. Keller, S. Keller, D. Kapolnek, P. Kozodoy, S. P. Denbaars, and U. K. Mishra, *Solid State Elect.* **41**, 1569 (1997).

[28] L. F. Eastman, V. Tilak, J. Smart, B. M. Green, E. M. Chumbes, R. Dimitrov, H. Kim, O. S. Ambacher, N. Weimann, T. Prunty, M. Murphy, W. J. Schaff, and J. R. Shealy, *IEEE Trans. Elect. Dev.* **48**, 479 (2001).

[29] J. D. Albrecht, P. P. Ruden, S. C. Binari, and M. G. Ancona, *IEEE Trans. Elect. Dev.* **47**, 479 (2001).

[30] Ahmad, V. Kasisomayajula, M. Holtza, J. M. Berg, S. R. Kurtz, C. P. Tigges, A. A. Allerman, and A. G. Baca, *Appl. Phys. Lett.* **86**, 173503 (2005).

[31] W. D. Hu, X. S. Chen, Z. J. Quan, C. S. Xia, W. Lu, P. D. Ye, *J. Appl. Phys.*, **100**, 074501 (2006).

[32] W. D. Hu, X. S. Chen, Z. J. Quan, X. M. Zhang, Y. Huang, C. S. Xia, W. Lu, P.D. Ye, *J. Appl. Phys.,* **102**, 034502 (2007).

[33] W. D. Hu, X. S. Chen, Z. J. Quan, C. S. Xia, W. Lu, H. J. Yuan, *Applied Physics Letters*, **89**, 243501 (2006).

[34] Tarakji, G. Simin, N. Ilinskaya, X. Hu, A. Kumar, A. Koudymov, J. Zhang, M. A. Khan, M. S. Shur and R. Gaska, "Mechanism of radio-frequency current collapse in GaN-AlGaN field-effect transistors," *Appl. Phys. Lett.*, vol. 78, pp. 2169-2171, 2001.

[35] Nguyen, N. X. Nguyen and D. E. Grider, "Drain current compression in GaN MODFETs under large-signal modulation at microwave frequencies," *Electron Lett.*, vol. 35, pp. 1380-1382, 1999.

[36] G. Simin, A. Koudymov, A. Tarakji, X. Hu, J. Yang, M. A. Khan, M. S. Shur and R. Gaska, "Induced strain mechanism of current collapse in AlGaN/GaN heterostructure field-effect transistors," *Appl. Phys. Lett.*, vol. 79, pp. 2651-2653, 2001.

[37] Daumiller, D. Theron, C. Gaquiere, A. Vescan, R. Dietrich, A.Wieszt, H. Leier, R. Vetury, U. K. Mishra, I. P. Smorchkova, N. X. Nguyen, and E. Kohn, "Current instabilities in GaN-based devices," *IEEE Electron Device Lett.*, vol. 22, pp. 62–64, Feb. 2001.

[38] W. D. Hu, X. S. Chen, Z. J. Quan, C. S. Xia, W. Lu, H. J. Yuan, *Applied Physics Letters*, **89**, 243501 (2006).

[39] R. Vetury, N. Q. Zhang, S. Keller, and U. K. Mishra, "The impact of surface states on the dc and RF characteristics of AlGaN–GaN HFETs," *IEEE Trans. Electron Devices,* vol. 48, pp. 560–566, Mar. 2001.

[40] T. Kikkawa, M. Nagahara, N. Okamoto, Y. Tateno, Y. Yamaguchi, N. Hara, K. Joshin, and P. M. Asbeck, "Surface-charge controlled AlGaN–GaN power HFET without current collapse and G dispersion," in *IEDM Tech. Dig.*, 2001, pp. 585–588.

[41] E. Kohn, I. Daumiller, M. Kunze, M. Neuburger, M. Seyboth, T. J. Jenkins, J. S. Sewell, J. Van Norstand, Y. Smorchkova, and U. K. Mishra, "Transient characteristics of GaN-based heterostructure field-effect transistors," *IEEE Trans. Microwave Theory Tech.*, vol. 51, pp. 634–642, Feb. 2003.

[42] M. Green, V. Tilak, V. S. Kaper, J. A. Smart, J. R. Shealy, and L. F. Eastman, "Microwave power limits of AlGaN–GaN HEMTs under pulsed-bias operation," *IEEE Trans. Microwave Theory Tech.*, vol. 51, pp. 618–623, Feb. 2003.

[43] N. Braga, R. Gaska, R. Mickevicius, M. S. Shur, X.Hu, M. A. Khan, G. Simin, and J. Yang, "Simulation of gate lag and current collapse in gallium nitride field-effect transistors," *Appl. Phys. Lett.*, vol. 85, pp. 4780-4783, 2004.

[44] Yariv, An introduction to theory and applications of quantum mechanics, John Wiley & Sons, 1982.

[45] S. E. Laux, Techniques for small-signal analysis of semiconductor devices, *IEEE Trans. Electron Devices*, **ED-32** (10) (1985) 2028-2037.

[46] Q. Li, R.W. Dutton, Numerical small-signal AC modeling of deep-leveltrap related frequency-dependent output conductance and capacitance for GaAs MESFET's on semi-insulating substrates, *IEEE Trans. Electron devices* **38** (6) (1991) 1285-1288.

[47] H. K. Lim and J. G. Fossum, Threshold voltage of thin-film silicon-oninsulator (SOI) MOSFETs, *IEEE Trans. Electron. Devices* **ED-30** (10) (1983) 1244-1251.

[48] X. Huang, W. C. Lee, et al, Sub-50nm FinFET: PMOS, in *Proc. IEDM* (1999) 67-70.

[49] L. Chang, S. Tang, T. J. King, J. Bokor, and C. Hu, Gate Length Scaling and Threshold Voltage Control of Double-Gate MOSFETs, in *Proc. Intl. Electron Devices Meeting (2000)* 719-722.

[50] R. Chau, et al, Advanced Depleted-Substrate Transistors: Single- Gate, Double-Gate and Tri-Gate, in *Proc. SSDM*, Nagoya, Japan, Tokyo, Japan (2002) 68-69.

[51] G. Sabnis and J. T. Clemens, Characterization of the electron mobility in the inverted <100> Si surface, in *IEDM Tech. Dig.* (1979) 18-21.

[52] H. Iwata, Fully quantum-mechanical modeling of tunneling current in ultrathin gate oxide metal–oxide-semiconductor devices, *Jpn. J. Appl. Phys.* **40** (7) (2001) 4496-4500.

[53] DESSIS ISE TCAD Manual, Release 10.06 (ISE Integrated Systems Engineering AG, Zurich, 2005).

[54] D. Bykhovski, R. Gaska, and M. S. Shur, *Appl. Phys. Lett.* **73**, 3577 (1998).

[55] P. M. Asbeck, E. T. Yu, S. S. Lau, G. J. Sullivan, J. Van Hove, and J. M. Redwing, *Electron Lett.* **33**, 1230 (1997).

[56] T. Yu, G. J. Sullivan, P. M. Asbeck, C. D. Wang, D. Qiao, and S. S. Lau, *Appl. Phys. Lett.* **71**, 2794 (1997).

[57] M. B. Nardelli, K. Rapcewicz, and J. Bernholc, *Appl. Phys. Lett.* **71**, 3135 (1997).

[58] T. Takeuchi, H. Takeuchi, S. Sota, H. Sakai, H. Amano, and I. Akasaki, *Jpn. J. Appl. Phys.* **36**, 177 (1997).

[59] O. Ambacher, J. Smart, J. R. Shealy, N. G. Weimann, K. Chu, M. Murphy, W. J. Schaff, L. F. Eastman, R. Dimitrov, L. Wittmer, M. Stutzmann, W. Rieger, and J. Hilsenbeck, *J. Appl. Phys.* **85**, 3222 (1999).

[60] O. Ambacher, J. Smart, J. R. Shealy, N. G. Weimann, K. Chu, M. Murphy, W. J. Schaff, L. F. Eastman, R. Dimitrov, L. Wittmer, M. Stutzmann, *J. Appl. Phys*. **87**, 334 (2000).

[61] T. Yu, G. J. Sullivan, P. M. Asbeck, C. D. Wang, D. Qiao, and S. S. Lau, *Appl. Phys. Lett*. **71** 2794 (1997)

[62] M. G. Ancona and H. F. Tiersten, *Phys. Rev. B* **35**, 7959 (1987).

[63] D. K. Ferry and J. R. Zhou, *Phys. Rev. B* **48**, 7944 (1993).

[64] X. H. Wu, L. M. Brown, D. Kapolnek, S. Keller, B. Keller, S. P. DenBaars, and J. S. Speck, *J. Appl. Phys*. **80**, 3228 (1996).

[65] P. B. Klein, S. C. Binari, K. Ikossi, A. E. Wickenden, D. D. Koleske, and R. L. Henry, *Appl. Phys. Lett.* **79**, 3527 (2001).

[66] J. Zou, D. Kotchetkov, A. A. Balandin, D. I. Florescu, and F. H. Pollak, *J. Appl. Phys*. **92**, 2534 (2002).

[67] R. Gaska, Q. Chen, J. Yang, A. Osinsky, M. Asif Khan, M. S. Shur, *IEEE Elect. Dev. Lett*. **18**, 492 (1997).

[68] S. A. Vitusevich, S. V. Danylyuk, N. Klein, M. V. Petrychuk, A. Yu. Avksentyev, V. N. Sokolov, V. A. Kochelap, A. E. Belyaev, V. Tilak, J. Smart, A. Vertiatchikh, and L. F. Eastman, *Appl. Phys. Lett*. **82**, 748 (2003).

[69] R. Gaska, A. Osinsky, J. W. Yang, and M. S. Shur, *IEEE Elect. Dev. Lett*. **19**, 89 (1998).

[70] W. D. Hu, X. S. Chen, Z. J. Quan, C. S. Xia, W. Lu, H. J. Yuan, *Applied Physics Letters*, **89**, 243501 (2006).

[71] P. Smorchkova, C. R. Elsass J. P. Ibbetson and R. Vetury B. Heying, P. Fini, and E. Haus S. P. DenBaars J. S. Speck U. K. Mishra, *J. Appl. Phys*. **86**, 4520 (1999).

[72] H. Hasegawa, T. Inagaki, S. Ootomo, and T. Hashizume, J. *Vac. Sci. Technol. B* **21**, 1844 (2003).

[73] J. Kotani, T. Hashizume, and H. Hasegawa, *J. Vac. Sci. Technol. B* **22**, 2179 (2004).

[74] R. Vetury, N. Q. Zhang, S. Keller, UK Mishra, *IEEE Trans. Elect. Dev*. **48**, 560 (2001).

[75] J. M. Tirado, J. L. Sanchez-Rojas and J. I. Izpura, *Semicond. Sci. Technol.* **20**, 864 (2005).

[76] W. Saito, M. Kuraguchi, Y. Takada, K. Tsuda, I. Omura, and T. Ogura, *IEEE Elect. Dev. Lett*. **52**, 159 (2005).

[77] N. Q. Zhang, B. Moran, S. P. DenBaars, U. K. Mishra, X. W. Wang, and T. P. Ma, in *Proceeding of International Electron Devices Meeting,* 2001 p. 589.

[78] Meneghesso, G. Verzellesi, R. Pierobon, F. Rarnpazzo, A. Chini, U. K. Mishra, C. Canali, and E. Zanoni, "Surface-related drain current dispersion effects in AlGaN/GaN HEMTs", *IEEE Trans. Electron Devices,* vol. 51, pp. 1554-1561, 2004.

[79] S. C. Binari, K. Ikossi, J. A. Roussos, W. Kruppa, D. Park, H. B. Dietrich, D. D. Koleske, A. E. Wickenden, and R. L. Henry, "Trapping effects and microwave power performance in AlGaN/GaN HEMTs," *IEEE Trans. Electron Devices*, vol. 48, pp. 465-471, 2001.

[80] Sentaurus Device Manual (former ISE-DESSIS), Ver. Z-2007. 03, Synopsys Inc., USA, 2006.

[81] O. Ambacher, J. Smart, J. R. Shealy, N. G. Weimann, K. Chu, M. Murphy, W. J. Schaff, L. F. Eastman, R. Dimitrov, L. Wittmer, and M. Stutzmann, "Two dimensional electron gases induced by spontaneous and piezoelectric polarization in undoped and doped AlGaN/GaN heterostructures," *J. Appl. Phys.,* vol. 87, pp. 334-344, Jan. 2000.

[82] W. D. Hu, X. S. Chen, Z. J. Quan, C. S. Xia, W. Lu, and P. D. Ye, "Self-heating simulation of GaN-based metal-oxide-semiconductor high-electron-mobility transistors including hot electron and quantum effects," *J. Appl. Phys.*, vol. 100, 074501: 1-9, 2006.

[83] S. C. Binari, P. B. Klein, and T. E. Kazior, "Trapping effects in GaN and SiC microwave FETs," *Proc. IEEE 99*, 2002, pp. 201–204.

[84] P. B. Klein, S. C. Binari, K. Ikossi, A. E. Wickenden, D. D. Koleske, and R. L. Henry, "Current collapse and the role of carbon in AlGaN/GaN. HEMTs grown by MOVPE," *Appl. Phys. Lett.*, vol. 79, no. 21, pp. 3527–3529, Nov. 2001

[85] Z. Y. Fan, J. Li, M. L. Nakarmi, J. Y. Lin, and H. X. Jiang, "AlGaN/GaN/AlN quantum-well field-effect transistors with highly resistive AlN epilayers," *Appl. Phys. Lett.*, vol. 88, pp. 073513 1-3, 2006.

[86] X. Hu, J. Deng, N. Pala, R. Gaska, M. S. Shur, C. Q. Chen, J. Yang, G. Simin, M. Asif Khan, J. C. Rojo, and L. J. Schowalter, "AlGaN/GaN heterostructure field-effect transistors on single-crystal bulk AlN, " *Appl. Phys. Lett.*, vol. 82, pp. 1299-1301, 2003.

[87] V. F. Mymrin, K. A. Bulashevicha, N. I. Podolskaya, and S. Y. Karpova, "Bandgap engineering of electronic and optoelectronic devices on native AlN and GaN substrates: A modelling insight, " *J. of Crystal Growth*, vol. 281, pp. 115-124, Jul. 2005.

In: MOSFETs: Properties, Preparations and Performance ISBN: 978-1-60456-762-5
Editors: N.T. Andre et al. pp. 301-353

Chapter 9

BULK FINFETS : FABRICATION AND THRESHOLD VOLTAGE

Jong-Ho Lee*

School of EECS, Kyungpook National University, 1370 Sangyuk-Dong, Buk-Gu
Daegu, 702-701, Korea

Abstract

There are two types of FinFETs: SOI FinFETs and bulk FinFETs. SOI FinFETs have an excellent scalability and high performance, but require expensive SOI wafer and has heat dissipation problems. To solve these problems, bulk FinFETs was devised and demonstrated firstly by our group. Since the bulk FinFETs have also excellent scalability and high performance, they can be applied to future high speed logics, scaled memory cells like sub-50 nm DRAM and flash memory cells. First, SOI and bulk FinFETs were compared in terms of scalability and heat dissipation. Then we focused on the bulk FinFETs by covering from fabrication to threshold voltage modeling. The bulk FinFETs have no back-bias effect differently from conventional planar channel MOSFETs. For the fabrication of the bulk FinFETs, we explained key fabrication steps and resultant structures with many drawings and pictures. The first 40 nm bulk FinFET was briefly introduced with device structure and basic I-V characteristics. SEM views of the bulk FinFETs fabricated at industry were shown. The bulk FinFETs showed much better DIBL than that of conventional planar channel devices, and also shown very small back-bias effect. Key device characteristics such as threshold voltage, maximum transconductance, subthreshold swing, and saturation drain current were compared with those of the control devices by changing fin body width. Especially, p-type bulk FinFET shown much better key characteristics compared to those of planar channel p-type MOSFETs. The threshold voltage (V_{th}) behaviors of the bulk FinFETs were modeled based on charge-sharing. Threshold voltage modeling of the double/triple-gate bulk FinFETs was performed by considering the potential lowering at the channel surface of fin body at $V_{GS}=V_{th}$ condition. With decreasing gate length and/or fin width, the surface band-bending (theoretically $2\psi_B$) at $V_{GS}=V_{th}$ in fully depleted bulk FinFETs decreases since actual surface potential decreases at the V_{th} condition. Based on the $2\psi_B$ correction, charge-sharing length (x_h) and corner factor (α_c) were extracted. The V_{th} behaviors of bulk FinFETs were modeled based on the surface potential lowering, 3-D charge-sharing, narrow-width effect, and corner factor. The V_{th} model predicted well the V_{th} behavior with fin body width, body doping

* E-mail address: jongho@ee.knu.ac.kr. (Phone & Fax) +82-53-950-6561

concentration, gate length, gate height, and corner shape of the fin body. The compact models predicted accurately V_{th} of the devices with the gate length up to 20 nm and the fin body width up to 5 nm.

1. Introduction

The basic concept of the field effect transistor (FET) was patented by Lilienfeld [1] in 1930, but was not implemented at that time because of no available practical means. The first oxides grown in high pressure steam were used by Kahang and Atalla [2] to fabricate the first MOSFET structures in 1960. In 1963, Grove, Deal, and Snow developed the first commercially available process for the construction of MOSFETs with stable characteristics. [3]. Since around 1960, MOSFETs have been applied to integrated circuits (ICs) which can perform functions such as digital and mixed signal processing, signal amplification, computing, and information storage efficiently and cheaply. Thus silicon MOSFETs have been emerged over at least the last 30 years as the predominant technology of the microelectronics industry. MOSFETs have been scaled down consistently over many manufacturing technology generations, resulting in steady improvements in circuit performance (speed), integration density, and cost per function (or bit). The gate lengths of MOSFETs were reduced from μm range to several tens nm. According to the forecast of 2006 International Technology Roadmap for Semiconductors (ITRS06) [4], production year for high-performance (HP) MOSFETs with 18 nm gate length is 2010. Conventional nano-scale bulk-Si MOSFETs are suffering from the increase of off-current (I_{off}) as one of the short channel effects (SCE). A method to suppress the short channel effect is to make the gate oxide thickness thin, but thin gate oxide increases the gate leakage current. Requirements in the nano-scale devices are low-resistance ultra-shallow source/drain junction depth, ultra-thin equivalent gate oxide thickness using high-κ dielectrics, and low-resistance metal (or silicide) gate electrodes to meet the specifications of the ITRS [4]. One of the technology boosters for future technology generations is device structure. Advanced device structures are ultra-thin body silicon on insulator (SOI) single-gate transistor [5],[6] and double/triple-gate (or multiple gate) transistor [7]-[9], and those devices have better device scalability than classic bulk-Si transistors [10]. Thus, better device architecture will be necessary to continue the benefits obtained from the device scaling in the past. Double-gate (DG) MOSFETs which have top and bottom gates or left and right gates have long been recognized for their potential to better control SCEs. Triple-gate MOSFETs show behaviors similar to the DG devices, but have three gate electrodes on the three surfaces of the rectangular shape body. Several different structures for DG MOSFETs have been proposed or demonstrated experimentally. The DG MOSFETs can be classified into three basic categories [11] labeled types I, II, and III as shown in Fig. 1 and Table 1. Among three types of DG MOSFET structures, type III FinFET is the most popular because fabrication process of the device structure is relatively easy and compatible with that of conventional planar channel MOSFETs. Therefore, in this chapter, we focus on the FinFETs, and deal with first device fabrication and characterization of the FinFETs. Then we describe in detail the threshold voltage behavior through modeling of the FinFETs to understand the devices well.

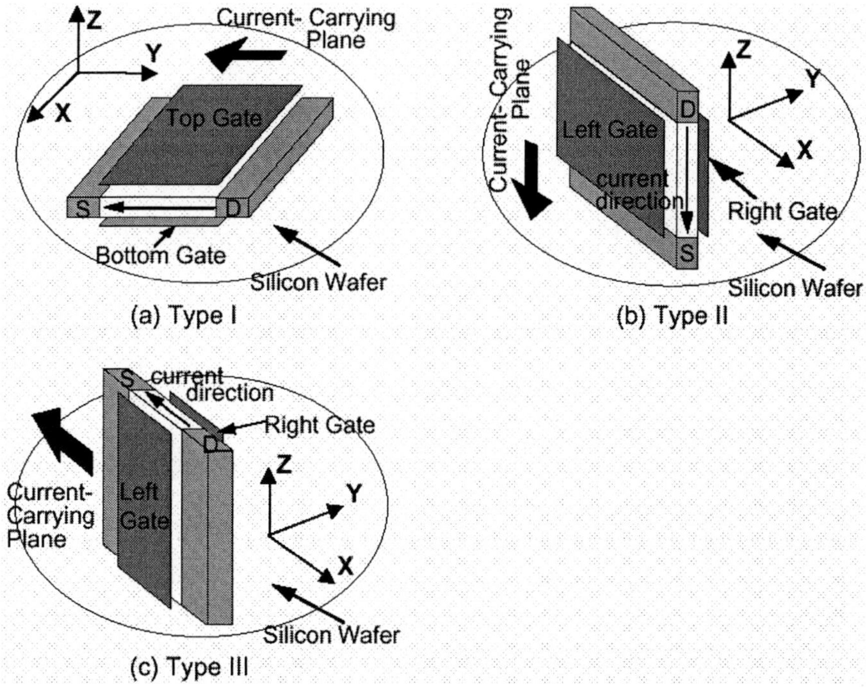

Figure 1. Three types of double-gate MOSFETs. [11] (a) Planar type double-gate FET. (b) Vertical type double-gate FET. (c) FinFET.

Table 1. Key geometrical factors of three types of a double-gate (DG) MOSFET on a silicon wafer.

Type Key Geometry	Type I (Planar DG FETs)	Type II (Vertical DG FETs)	Type III (FinFETs)
Gate Position	Top/Bottom	Left/Right (or Cylinder)	Left/Right
Body Shape	Horizontal	Vertical	Vertical
Current Carrying Plane	Horizontal Surfaces	Side Surfaces	Side Surfaces
Current Flow Direction	Horizontal	Vertical	Horizontal

All directions and positions are set on the basis of the plane surface of wafers

2. FinFETs: SOI and Bulk FinFETs

Si-based bulk planar MOSFETs have been the building blocks for modern VLSI circuits. However, as the design rule shrink down beyond 50 nm, the conventional planar MOSFETs are facing many challenges. For decades, various kinds of three-dimensional (3-D) transistors have been reported [7],[8],[12]-[35]. Now, however, FinFETs [7],[8],[15]-[35] seem to be very promising for the future CMOS technology including logic and memory applications. As mentioned in Introduction, FinFETs have basically double-gate structure. Double-gate MOSFETs have shown excellent scaling-down characteristics and high performance among devices structures. FinFETs have been developing for various applications such as high speed digital integrated circuits (ICs)[17]-[21], analog ICs [22],[23], SRAMs [24]-[26], flash memory[27]-[31], DRAMs [32]-[35].

The FinFETs are known to most of us as the ones fabricated on SOI wafers, and they are basically SOI MOSFETs. So far, double-gate transistors have been demonstrated on SOI substrates to overcome problems associated with short channel effect (SCE) [7],[9],[11],[15]-[20]. We call the FinFETs built on SOI wafers SOI FinFETs. These devices have the advantages of easy fabrication and excellent scalability. These SOI devices built on SOI wafers are normally applied to high speed circuits due to their low parasitic capacitances and high current drivability. In general, floating body SOI devices including single- and double-gate devices may have floating body problem [36] depending on doping, Si film thickness, and bias condition. SOI wafers have higher wafer cost and higher defect density than bulk Si wafers, and heat transfer problem. Switching from the conventional planar MOSFETs to the SOI FinFETs means not only the active splitting [16] for narrow and depleted channels, but also changing from conventional four-terminal devices to three-terminal (body floating) ones. It consequently narrows the circuit operation windows because of the three-terminal characteristics. Therefore, it is rather more appropriate to consider four-terminal FinFETs that are achieved by connecting the fin body directly to the Si substrate. These four-terminal FinFETs are built on bulk Si wafers and called body-tied FinFETs or more preferably bulk FinFETs [26]. When we first reported the body-tied FinFETs, we called them as Omega (Ω) MOSFETs because the cross-section of the body resembles Greek letter Ω. F.-L. Yang *et al.* called their MOSFET as omega FET when they report it in 2002 IEDM because the gate structure looks like Ω [16]. To differentiate the body-tied FinFET from the SOI FinFET and the omega FET, we call the body-tied FinFET as bulk FinFET. Fig. 2 briefly explains the classification of the FinFETs and shows cross-sectional views of SOI and bulk FinFETs along the gate electrode. The fin bodies of SOI and bulk FinFETs are floated and tied to the substrate, respectively. "G" means gate electrode.

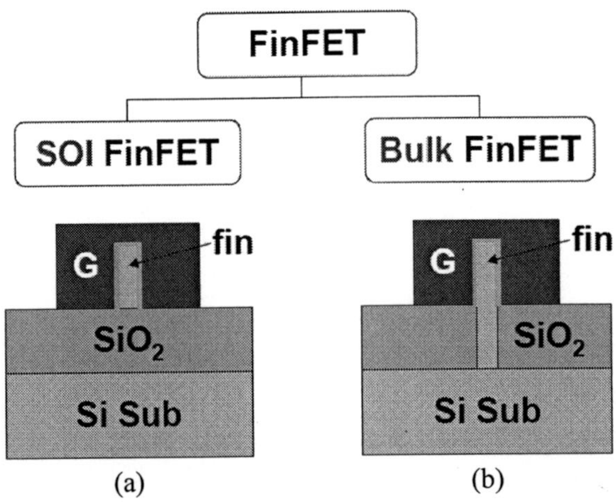

Figure 2. FinFET classification, and cross-sectional views of SOI (a) and bulk (b) FinFETs.

The bulk FinFETs are the more familiar to IC design engineers and the fabrication steps of the devices are compatible with those of the conventional planar channel CMOS devices fabricated on bulk Si wafers. The schematic 3-D view of the bulk FinFET is shown in Fig. 3. Here, W_{fin} and H_{fin} represent fin body width and fin height, respectively. The source/drain junction depth is represented by x_{j} and the x_{j} is generally similar to H_{fin}.

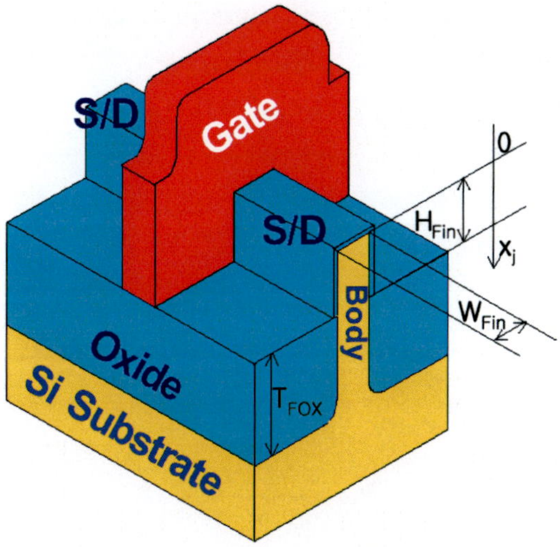

Figure 3. 3-dimensional schematic view of bulk FinFET. H_{Fin} and W_{Fin} represent fin height and width, respectively. T_{FOX} and x_j stand for the field oxide thickness and the source/drain junction depth, respectively.

The bulk FinFETs eliminate problems such as wafer cost, defect density, floating body effect, and heat accumulation associated with SOI FinFETs. The bulk FinFETs keep nearly the same scalability with the SOI FinFETs and better heat dissipation characteristics [37]. To see the scalability of the SOI and bulk FinFETs, Fig. 4 is prepared. Electrical characteristics of the n-type SOI and bulk FinFETs were studied through the extensive 3-D device simulation using ATLAS [38]. The drift-diffusion transport model was adopted in this study, because the drift-diffusion model mostly accounts for the salient features of the scaling properties. Physical gate length (L_{g}) and gate oxide thickness (T_{ox}) are 25 nm and 1.5 nm, respectively. The p-type body concentration is 1×10^{19} cm^{-3}. The H_{fin} and the T_{FOX} are fixed at 70 nm and 100 nm, respectively. The W_{fin} is a variable and the x_{j} is 66 nm. The length of LDD (lightly doped drain) region is about 17.5 nm and concentration is 6×10^{19} cm^{-3}. The distance between n^+ poly-Si gate edge and S/D contact is approximately 60 nm and the HDD (heavily doped drain) concentration is 2×10^{20} cm^{-3}. The corner regions of the fin were rounded to remove any possible leakage path. For comparison, the SOI FinFETs were designed to have exactly the same structure and parameters as those of the bulk FinFETs. The only difference is the existence of the buried oxide. Fig. 4 shows the threshold voltage (V_{th}) and drain induced barrier lowering (DIBL) versus the W_{fin}. The bulk FinFETs (solid circles) have nearly the same V_{th} and DIBL characteristics as those of SOI FinFETs (open circles), which guarantees a good scalability of the bulk FinFETs. The subthreshold swing (SS) characteristics are not shown in this figure, but nearly the same for both devices. The relation between V_{th} and W_{fin} can be explained by following equation.

$$V_{th} = V_{FG} + 2\psi_B + \frac{qN_b W_{fin}}{2C_{ox}} \quad (1)$$

where V_{FB}, ψ_B, N_b, t_b, and C_{ox} are flat ban voltage, Fermi potential, fin body doping, and gate oxide capacitance, respectively. In Fig. 4, as W_{fin} increases, V_{th} increases for fully depleted fin body. For W_{fin} thicker than ~25 nm, the V_{th} is saturated because the body is not fully depleted and depletion width is nearly constant.

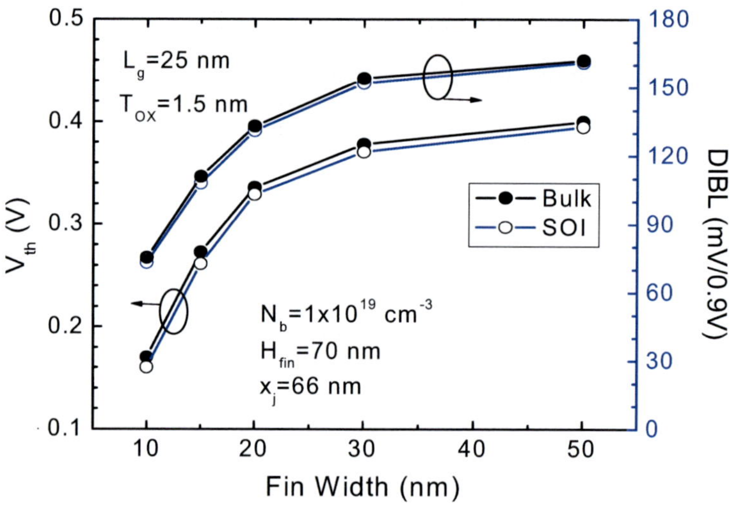

Figure 4. Threshold voltage (V_{th}) and DIBL characteristics of 25 nm SOI and bulk FinFETs versus fin body width. The n^+ poly-Si gate was applied.

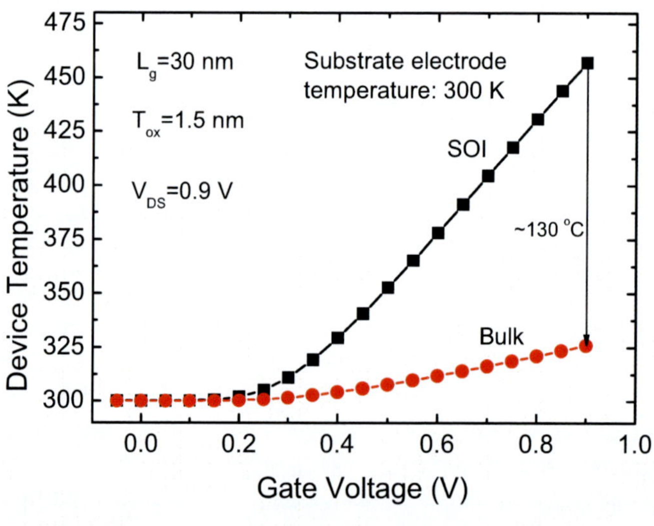

Figure 5. Temperature characteristics of bulk (solid circle symbols) and SOI (solid square symbols) FinFETs versus gate bias. Both devices have nearly the same drain current with gate bias. The results were obtained through 3-D device simulation considering lattice temperature.

Fig. 5 shows device temperature of 30 nm devices versus gate bias at a given V_{DS} of 0.9 V. The bulk FinFET with a W_{fin} of 20 nm shows lower device temperature by ~130 °C than the SOI FinFET at a gate bias of 0.9 V.

Thus it is very cost effective and useful to implement double-gate transistors on bulk Si wafer (bulk FinFETs) while keeping excellent scalability and performance of SOI double-gate devices (SOI FinFETs). Also, controlling body bias in bulk FinFETs widens the device application window and provides the memory and circuit design engineers with more familiarity.

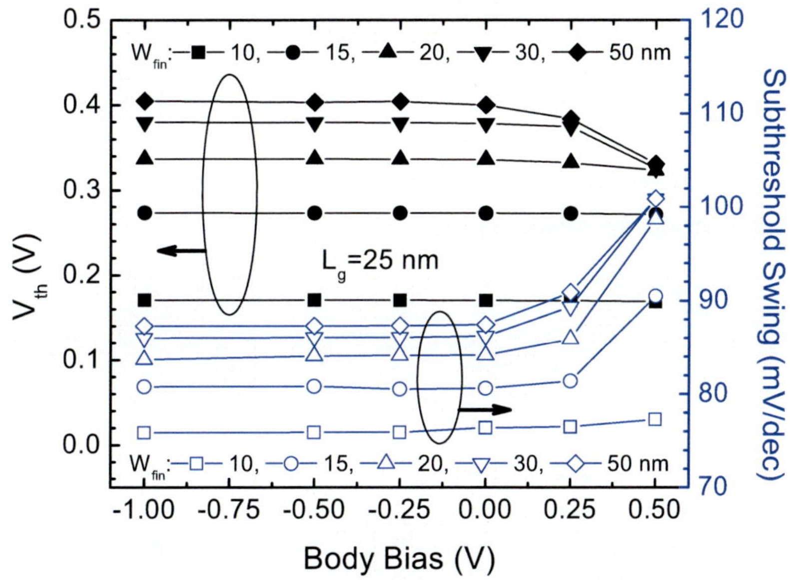

Figure 6. Simulated V_{th} and subthreshold swing (SS) of bulk FinFETs versus body bias as a parameter of fin body width. Solid and open symbols represent V_{th} and SS, respectively.

The bulk FinFETs have much less back bias effect than conventional planar channel MOSFETs [37]. Fig. 6 shows the V_{th} and the SS versus body bias as a function of W_{fin} (10, 15, 20, 30, and 50 nm). For the negative body bias, the V_{th}'s of the bulk FinFETs with various W_{fin}s keep nearly constant, which alleviates delay problem due to V_{th} increase in planar channel MOS devices connected in series. The SS characteristics for various W_{fin}'s are nearly constant for negative body bias but increase with positive body bias due to the increase of DIBL. As the W_{fin} decreases, the increase of the SS is less significant because the DIBL is suppressed effectively.

3. Fabrication and Characterization of Bulk FinFET

3.1. World 1st bulk FinFET

We briefly introduce world 1st bulk FinFET in terms of device fabrication and main characteristics. Key fabrication steps of the 1st bulk FinFET are as follows. Conventional g-line stepper was used and the nano-scale patterning was achieved by using spacer technology.

On p-type Si wafers, 30 nm thick thermal SiO_2 was grown and B^+ (35 keV and 3×10^{13} cm^{-2}) ions were implanted. LPCVD SiN, TEOS CVD SiO_2, and LPCVD SiN were deposited successively, with the thicknesses of 25, 30, and 70 nm, respectively (Fig. 7 (a)). The bottom SiN/SiO_2 stack was used as a stopping layer during CMP and the top SiN/SiO_2 was used for a sacrificial layer while making nano-scale sidewall spacers. A photolithography step and the SiN etching were performed, and a poly-Si film of 35 nm was deposited and anisotropically etched. Spacers were made on every side of the original pattern (Fig. 7 (b)). The SiN inside the spacers was removed in a phosphoric acid solution (Fig. 7 (c)). Anisotropic etching of the $SiO_2/SiN/SiO_2$ (Fig. 7 (d)), and another anisotropic etching of the Si substrate with the depth of 230 nm were performed, and consequently, the active fin profile was obtained as shown in Fig. 7 (e). TEOS CVD SiO_2 of 350 nm were deposited and densified at 950 °C for 30 min in nitrogen ambient (Fig. 7 (f)). CMP planarization (Fig. 7 (g)) and isotropic etching in a diluted

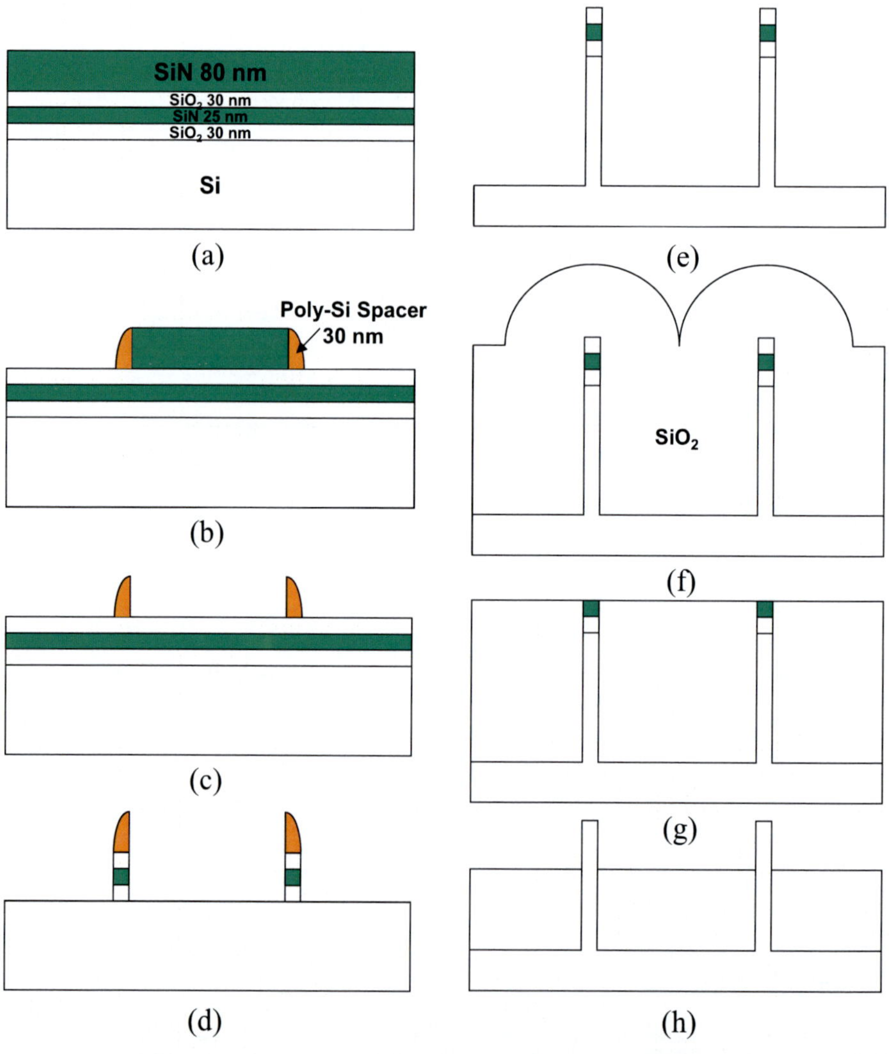

Figure 7. Key process steps for the fabrication of world 1st bulk FinFET.

HF solution were performed to expose upper part of the fin (Fig. 7 (h)). The remaining field SiO_2 was about 80 nm. A gate SiO_2 of 2.8 nm was grown and a poly-Si film of 100 nm was deposited. Again, the sidewall spacer patterning mentioned above were carried out to make 40 nm gate. After the gate etching, a TEOS CVD SiO_2 spacer of 30 nm width was formed. As^+ (20 keV and 3×10^{15} cm^{-2}) ions were implanted for source/drain and gate doping. RTA at 1000 °C for 10 sec in nitrogen ambient was performed for the dopant activation. Interconnection was made by a conventional sequence, such as passivation, contact opening, aluminum sputter deposition, and etching.

Fig. 8 shows SEM views of key fabrication process steps of the bulk FinFETs. In figure (a), SEM view after oxide CMP and partial wet etching of oxide is shown. Part of the fin body is protruded beyond the oxide and both vertical surfaces of the protruded body become the channel of the bulk FinFETs. Top and bottom widths of the fin body are 25 nm and 100 nm, respectively. Fin height is 230 nm. The shape of the fin body may affect device characteristics. So the etching process of the fin body needs to be controlled carefully. Figure (b) shows SEM views after defining polysilicon gate. The profile is nearly vertical and the physical gate length is 40 nm. By using 1.0 μm CMOS technology, the distance between gate and source/drain contact about 2.5 μm so that the parasitic source/drain resistance is very high.

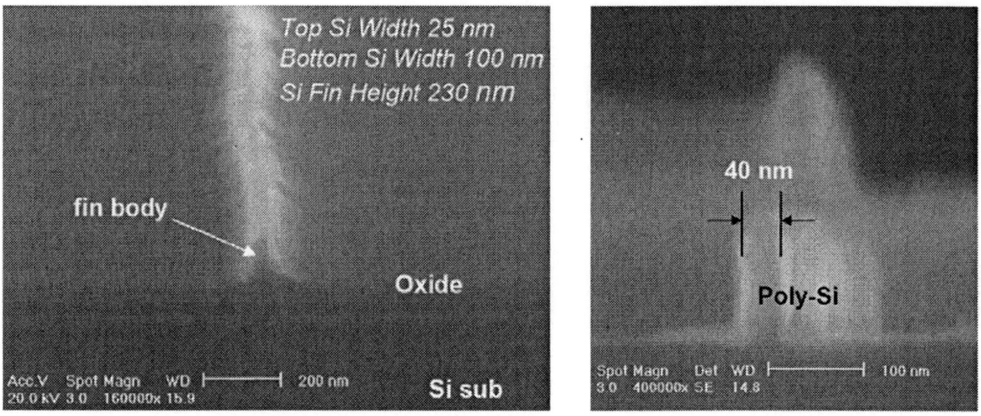

(a) (b)

Figure 8. SEM photographs of the key steps for the fabrication of the bulk FinFETs. (a) SEM view after oxide CMP and partial wet etching of the oxide. (b) SEM view after poly-Si gate definition.

Fig. 9 illustrates I_D-V_{GS} (a) and I_D-V_{DS} (b) characteristics of the world 1st bulk FinFET with a gate length of 40 nm. In Fig. 9 (a), the insert shows schematically the fin body where the channel is formed. Substrate bias dependences are shown in Fig. 9 (a), and the reverse body bias suppresses significantly the leakage current level. The leakage current is relatively high because parasitic channel along the side-wall of the body butting to the field oxide was not suppressed. By controlling implantation dose to dope the body, the leakage current can be suppressed. As the body bias increases negatively, the leakage decreases due to the suppression of the parasitic channel. The threshold voltage has no dependence on the body bias while conventional planar channel MNOSFETs have significant back bias effect. We think the fin body is fully depleted by the electric field from the gate so that the body bias cannot change effectively the channel potential as explained in Fig. 6. As the fin width decreases, the body bias effect is suppressed significantly. Although the body where the channel and the source/drain are formed is tied to the substrate, the proposed device has the no V_{th} change with negative body bias, which is very helpful in low voltage/high speed circuit design. Fig. 9 (b) shows that the drain current can be modulated by gate bias although SCE and large S/D resistance problems exist. Further optimization of channel doping and body geometry will lead to high performance devices.

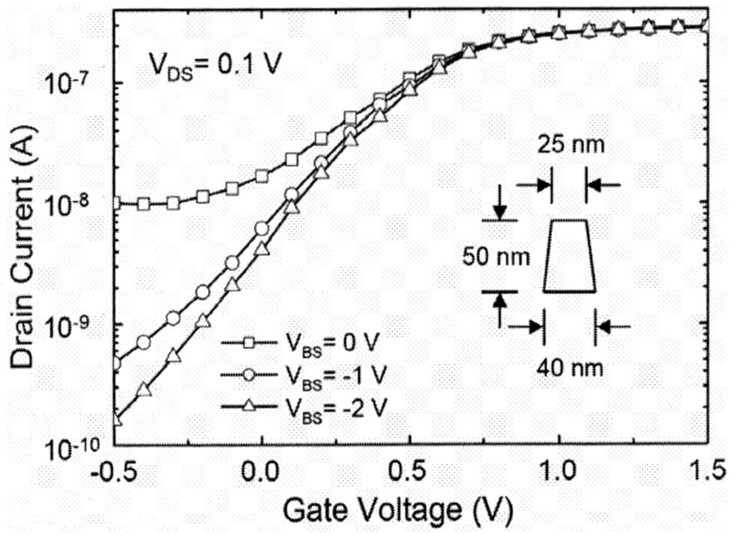

(a)

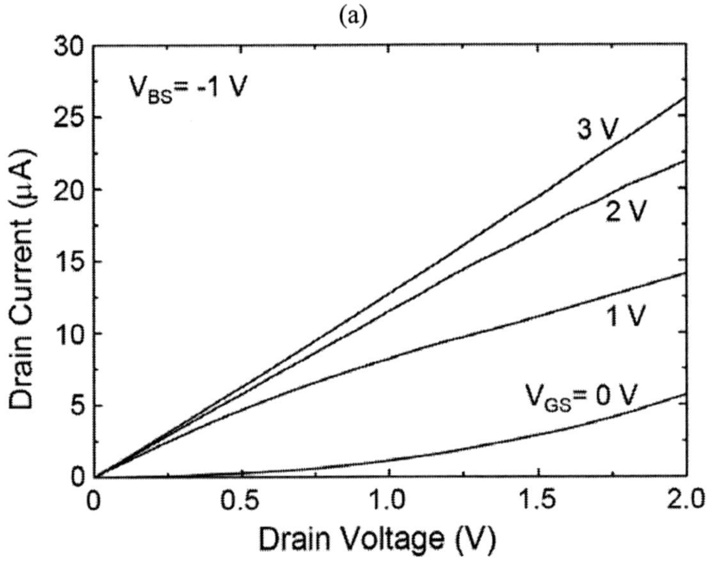

(b)

Figure 9. I-V characteristics of world 1st 40 nm bulk FinFET. (a) NMOS I_d-V_{gs} characteristics with respect to the body biases. The inset shows the fin channel region tied to the Si substrate. (b) NMOS I_d-V_{ds} characteristics as a parameter of the gate bias.

3.2. General Characteristics of Bulk FinFETs

3.2.1. Device Fabrication

In this section, various characteristics of the bulk FinFETs that were fabricated at industry are described. Full process steps to fabricate the bulk FinFETs are very similar to those of conventional planar channel CMOS FETs built on bulk Si wafers mentioned above. The key process steps to fabricate the bulk FinFETs are as follows. Active fin body is defined by using KrF lithography and can be thinned with the thermal oxidation and stripping (Fig. 10 (a)). After fin body formation by RIE, it should be needed to remove slightly Si surface damaged by plasma etching so that we can obtain low leakage and high performance. A layer of oxide growth and strip can be a method to remove the damaged Si surface. A layer of thin SiO_2 is formed on the whole Si wafer surface including the fin bodies and followed by sequential deposition of 50 nm thick SiN liner (Fig. 10 (b)) and a layer of thick SiO_2 for STI (Fig. 10 (c)). The SiN layer is optionally used as both CMP (Chemical Mechanical Polishing) stopping layer and sacrificial layer for the selective opening of the fin sidewall. Fig. 10 (d) shows the cross-sectional view after CMP using the SiN as a stopping layer. Fin channel height is determined by the extent of SiN recess after the surface planarization using CMP. The SiN layer is etched away from the top of the flat surface to a depth vertically in hot phosphoric acid by controlling the etch time (Fig. 10 (e)). If we skip the SiN recess etching below the top of the active level, then the devices in this case become planar transistors with

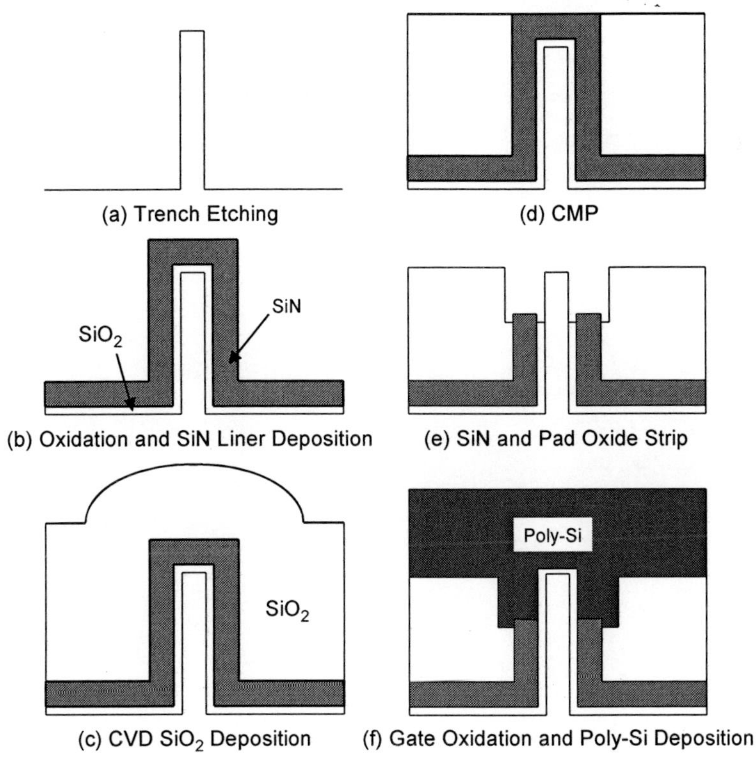

Figure 10. Cross-sectional views of key process steps for the fabrication of the bulk FinFETs.

nano-scale channel width. Using this selective recess etch technology, global field SiO_2 level is sustained at the level equal to that of the active fin top, and this is advantageous for subsequent photo-lithography and reduction of parasitic capacitance related to the isolation region. The thin oxide layer on top and the recessed side surfaces of the fin bodies is removed and gate oxide is formed. A layer of poly-Si was deposited and patterned (Fig. 10 (f)). Here we should be careful in etching away the poly-Si in the recessed region since the thickness of the filled poly-Si in the recessed region is thicker than the poly-Si on the other region by the recessed depth. We need high etch selectivity between poly-Si and SiO_2 during the gate etch process by considering the thickness of the filled poly-Si and process margin. Instead of normal gate electrode formation, gate damascene process can be applied to the fabrication of the bulk FinFETs to control fin body doping more effectively [33], [34].

Other important processes are SiN sidewall spacer formation on both sidewall of the patterned gate, ion implantation for halo, spike RTA for activation, salicide formation, multiple W-damascene processes for local interconnection, and Al sputtering for metallization.

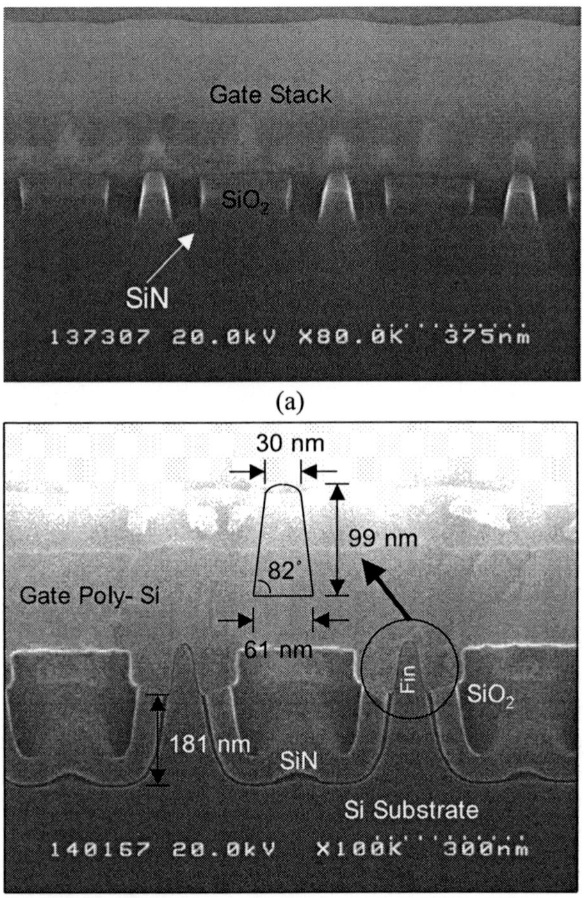

Figure 11. Cross-sectional SEM views after gate poly-Si etching (a) and final process (b) [41].

Fig. 11 (a) and (b) show SEM micrographs taken after the gate poly-Si etching and the final step, respectively. As shown in Fig. 11 (a), no poly-Si residue was remained at either side of the recessed region after poly-Si etching. The fin bodies are clearly seen and the SiN besides the fin bodies is recessed from the top of the fin body. Fig. 11 (b) shows cross-sectional view along the gate electrode after final process. Top corners of the fin bodies are rounded because of the repetitive oxidation and isotropic etching (chemical dry etching), which lead to high reliability by spreading the electric field from the gate and uniform threshold voltage control. Such rounded top corner can be also implemented through the annealing in hydrogen ambient. The insert shows simple cross-section of the fin body used for the channel. Although there are recessed regions by recessing the SiN layer before poly-Si gate deposition, the surface of the poly-Si gate is nearly flat as shown in Fig. 11 (b). The flat top of the gate poly-Si obtained by filling the narrow SiN recess regions helps subsequent gate photo-lithography and etching. The oxide thickness on the top and the sidewall of the fin bodies was nearly the same (no crystal orientation dependence), because In-Situ Steam Generated (ISSG) radical SiO_2 was grown.

Fig. 12 shows cross-sectional TEM micrographs along the fin body in the DRAM cell as an example. It seems that process induced defects were not generated. The region notated by ‘Si substrate’ is actually fin body. There are STI isolation oxide on left and right sides of the region. Since the fin body is so thin, it looks like semitransparent layer in the TEM view. We can see blurredly the regions backside of the fin body.

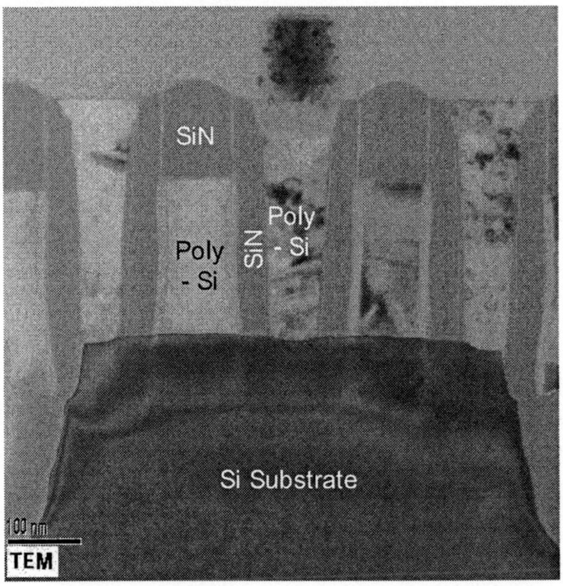

Figure 12. Cross-sectional TEM view along the fin body in the DRAM cell region [41].

Fig. 13 shows the TEM image of the fin body of the fabricated bulk FinFET with thinned active fin width. The slope of the side surfaces of the fin body is almost vertical. The thickness and height are about 12 nm and 100 nm, respectively. This TEM image is not the one used for actual device fabrication, but from this image, we can see that further fin thickness scaling is possible for next generation transistors. Here, an advantage of the bulk

FinFET needs to be noted. In gate-all-around [12]-[14] and Omega type [16] transistors, it is difficult to etch out residues of the gate electrode materials that are deposited inside the undercut region of the fin body, and if not etched clearly, gate electrode may be shorted to source/drain. However, the body-tied FinFET never have undercut regions because of the fin bodies tied to the Si substrate, resulting in easier etching of the gate poly-Si.

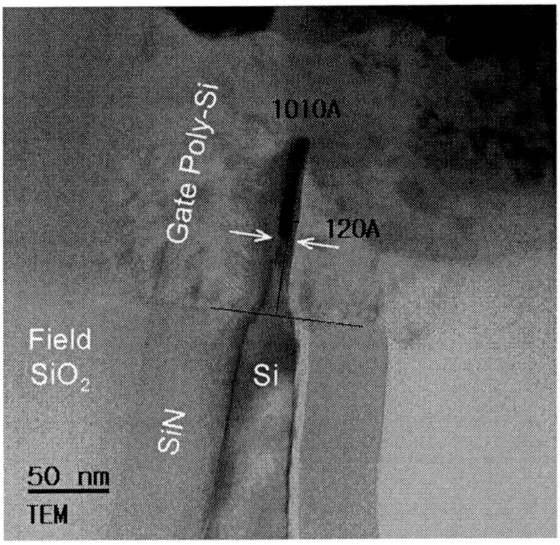

Figure 13. A cross-sectional TEM micrograph of thinned fin body. The minimum width of the fin is about 12 nm. This view is prepared to show the scalability of the fin body.

3.2.2. Characterization

In Fig. 14, subthreshold swing distribution of the buk FinFETs is compared with that of the conventional DRAM cell transistors. Both devices have a gate length of 120 nm. It is shown that the bulk FinFET has much lower value and smaller deviation. The bulk FinFETs and conventional cell devices have the gate oxide thicknesses of 4 nm and 5.4 nm, respectively. The gate electrode is n^+ poly-Si.

For the devices mentioned in Fig. 14, body bias dependencies of the transistors are compared as shown in Fig. 15. The bulk FinFET shows very small dependency, whereas the dependency is significant in conventional DRAM cell transistor at a fixed gate length. Since the top and both side surfaces of the fin body is wrapped by the gate, the fin body is fully depleted and the channel is controlled well by the gate. Therefore the effect from the gate is more dominant than the back bias effect and the threshold voltage is changed very slightly with the back bias as shown in Fig. 16. The results are well corresponded to simulated results shown in Fig. 6. Effective channel width (W_{eff}) of the bulk FinFETs can be estimated as $W_{eff} = W + 2H_{fin}$.

As shown in Fig. 16, Drain Induced Barrier Lowering (DIBL) of 24 mV/V is obtained with the bulk FinFET while the conventional DRAM cell transistor shows 108 mV/V. The threshold voltage of the bulk FinFETs is much lower than that of conventional planar channel devices, because the fin body is fully depleted and a half of fin body is smaller than the depletion width under the gate of the planar channel devices. If the gate electrode is changed from n^+ poly-Si to p^+ poly-Si, the threshold voltage of n-type bulk FinFETs can be ~1 V. The p^+ poly-Si gate bulk FinFET as a DRAM cell could be useful in reducing off-current at V_{GS} of 0 V. However, the vertical electric field in the surface of n-type drain overlapped by the p^+ poly-Si gate is very high, resulting in high Gate-Induced Drain Leakage (GIDL) current [34],[39].

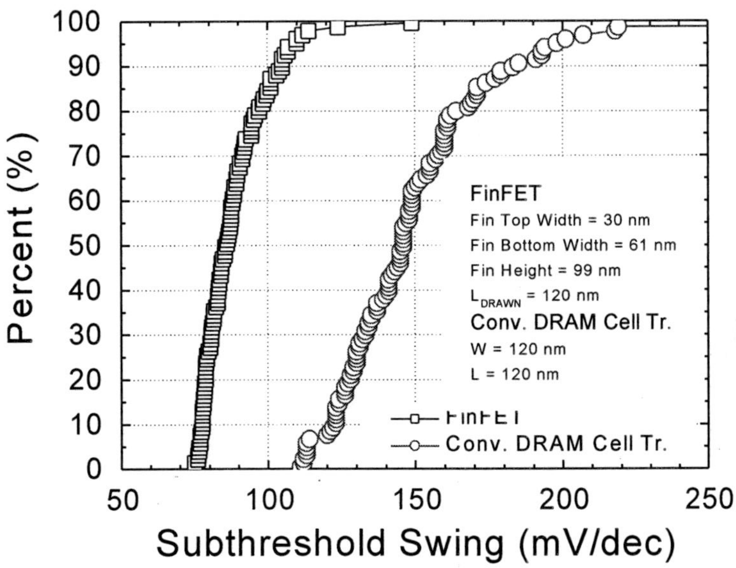

Figure 14. Comparison of subthreshold swing distributions. The gate oxide thicknesses are 4 nm and 5.4 nm for the bulk FinFET and conventional DRAM cell transistors, respectively [41].

In Figs. 14-16, 1st bulk FinFETs for DRAM cell application were demonstrated and compared. The off-current of the bulk FinFETs for DRAM cell is now much improved. We briefly discuss characteristics of bulk FinFETs for SRAM with the fin body width in the following Figs. 17 and 18. We firstly applied the bulk FinFETs to SRAM technology.

Three different device structures are compared by changing active width (equivalent to fin body width for bulk FinFETs). They are bulk FinFET, nano width MOSFET (planar channel), control devices. The nano-scale width planar channel FETs are fabricated by using identical process steps as the bulk FinFET process until the CMP step, but with less SiN recess etch by the phosphoric acid solution. Therefore, the nano-scale width MOSFETs have planarized active (no side channel) region. Finally, the control planar FETs are the state-of-the-art SRAM cells with 0.79 μm^2 and 90 nm technology [40]. Both nano-scale width FETs and control planar FETs are compared with the bulk FinFETs. SRAM cell using the bulk FinFETs shows SNM of 280 mV at a V_{CC} of 1.2 V, which is the similar level as that (300 mV) of the state-of-the-art 90 nm node conventional SRAM cell although the bulk FinFETs are not optimized [26]. The cell consisted of the nano-scale width planar CMOS SRAM shows poor transfer characteristics and much smaller SNM (170 mV).

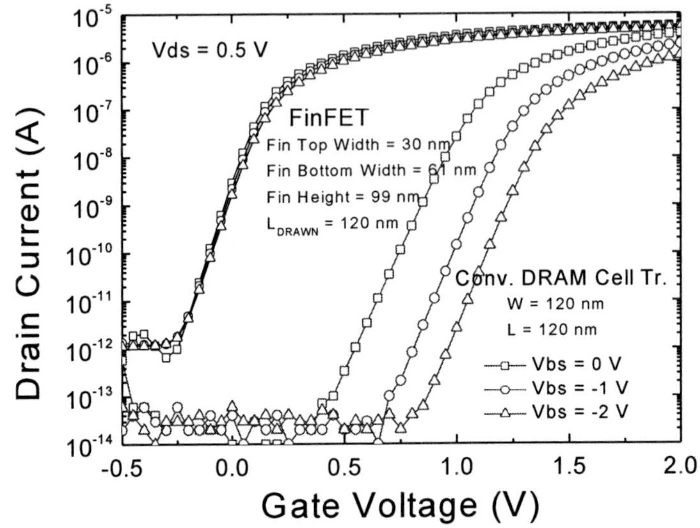

Figure 15. Comparison of the body bias dependency of the bulk FinFET with that of the conventional DRAM cell transistor [41].

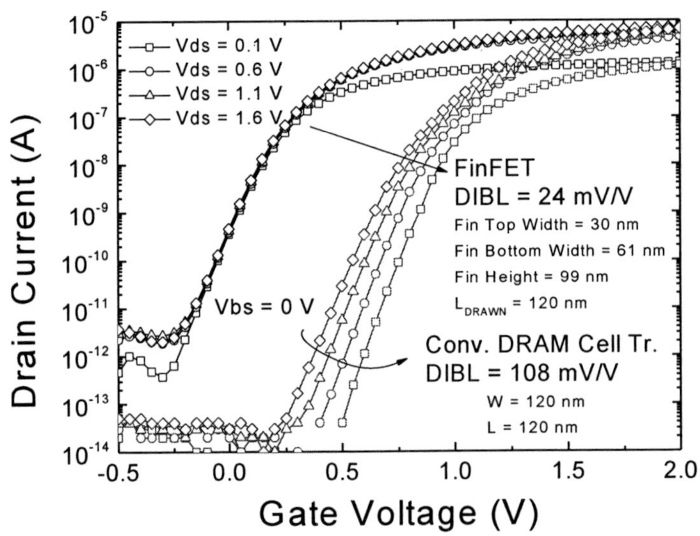

Figure 16. Drain induced barrier lowering (DIBL) comparison [41].

Now we investigate device characteristics by changing active width. As shown in Fig. 17, V_{th}s of the bulk NFinFETs are lower than those of the nanoscale width planar nMOSFET and also than those of the control group nMOSFET. It is well known that the top corner regions of a fin body have lower V_{th} than top and side channel regions because the electric field from the gate electrode concentrates at the corner region. It is noted that the fin width (>50 nm in this case) of the bulk NFinFET is too thick at a given gate length of 100 nm, which degrades SS and DIBL. The large DIBL decreases V_{th}. Fig. 17 (c) shows the SS with the active width. As expected, the bulk NFinFET showed the worst SS (>~100 mV/dec) among there different devices. By thinning the fin body width, we can improve the SS further. The maximum transconductance (g_{mmax}) and drain current in saturation region (I_{Dsat}) of each device are shown with active width (or fin body width) in Fig. 17 (b) and (d), respectively. Although the effective channel width (50+2×60=170 nm) of the bulk NFinFET is wider than that (115 nm) of the control group nMOSFET, the bulk NFinFET shows similar I_{Dsat} to that of control group nMSOFET. However, the g_{mmax} of the bulk NFinFET is nearly half of that of the control nMOSFET, because the electron mobility on the side surfaces of the fin body is poor due to (110) plane and untreated surface.

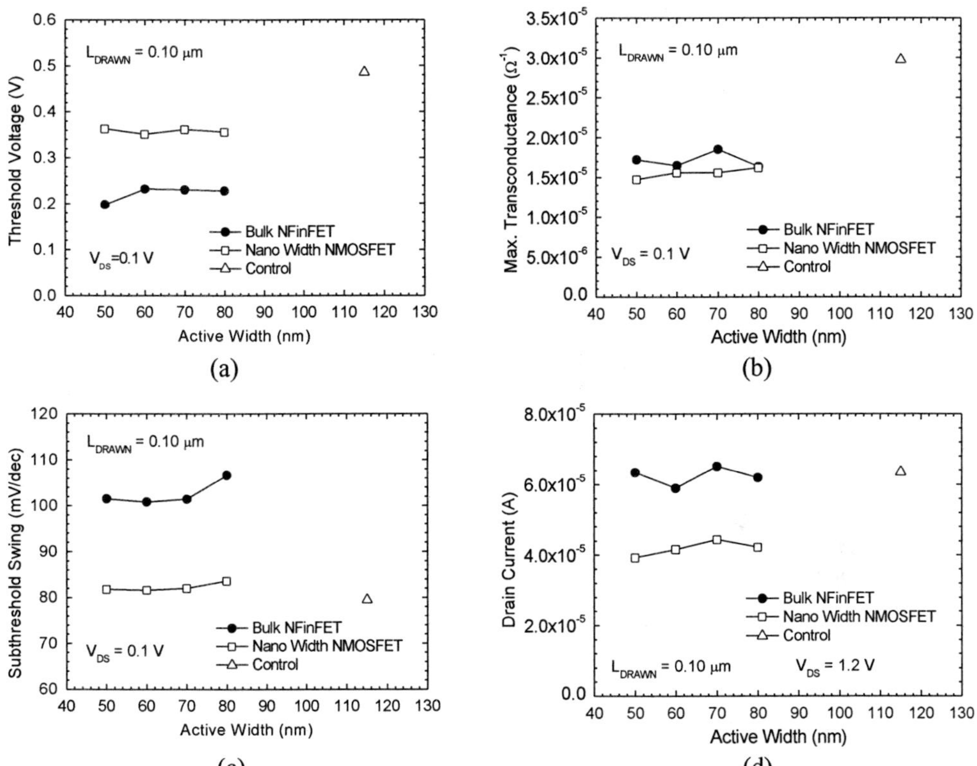

Figure 17. V_{th}, maximum transconductance (g_{mmax}), subthreshold swing (SS), and saturation current comparison of the bulk NFinFET, the nano-scale width planar nMOSFET, and state-of-the-art nMOSFET (control). Gate length is 100 nm, and H_{fin} and T_{ox} are 60 nm and 1.5 nm, respectively. Active width of the nano-scale width planar nMOSFET is the same as the fin top width of the bulk NFinFET, because the planar transistor was fabricated using the same process steps without SiN recess etching after the CMP. Each datum is the median of the 16 data points over a wafer [26].

Fig. 18 shows the device characteristics of three different pMOS devices with active width. Here pMOS devices have p^+ poly-Si gate. The V_{th} comparison is shown in Fig. 18 (a) and the V_{th} difference is relatively small for three different devices. The nano-scale width planar PMOSFETs show increase of the magnitude of the V_{th} ($|V_{th}|$) with decreasing active channel width (narrow width effect) by the increase of the channel doping due to the segregation of phosphorus ions for n-well. The bulk PFinFETs show slight decrease and very slight increase of the $|V_{th}|$ with decreasing fin body width. According to the $|V_{th}|$ expression for PFinFET with p^+ poly-Si gate, the $|V_{th}|$ decreases with decreasing fin body width. The segregation effect of the phosphorus ions acts as a key factor which increases the $|V_{th}|$ in the bulk PFinFET. It seems the segregation effect is more significant in the top corner regions of the fin body, resulting in the higher phosphorus concentration in the corner regions. As mentioned in the explanation of the V_{th} of NFinFET, there should be electric field concentrated at the corner regions, which lowers the $|V_{th}|$. It also should be noted that the positive oxide charge inside the gate oxide along the channel surface increases the $|V_{th}|$ in the bulk PFinFET. Differently from the NFinFET case, the segregation and the positive oxide charge increase the $|V_{th}|$ of the bulk PFinFET. The $|V_{th}|$ of the bulk PFinFET shows relatively small change with decreasing the active width, which is very positive effect for future nano-scale CMOS devices. Fig. 18 (c) shows subthreshold swing comparison for three different devices. The bulk PFinFETs show the best SS among them, and improvement as the active

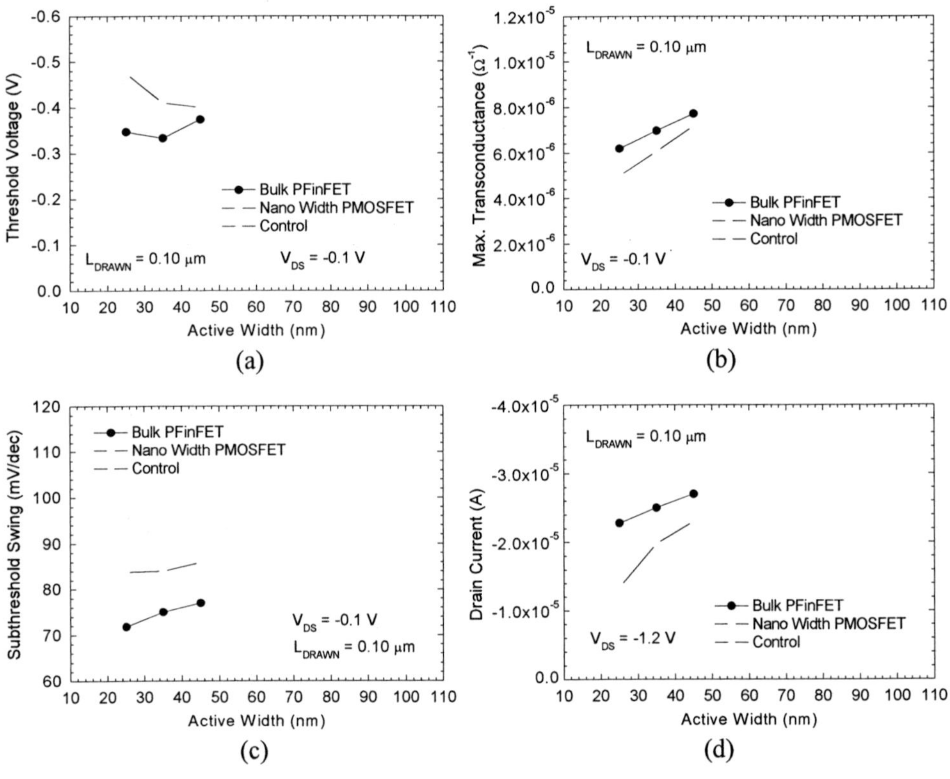

Figure 18. V_{th}, maximum transconductance (g_{mmax}), subthreshold swing (SS), and saturation current comparison of the bulk PFinFET, the nano-scale width planar pMOSFET, and state-of-the-art nMOSFET (control). Gate length is 100 nm, and H_{fin} and T_{ox} are 60 nm and 1.5 nm, respectively. Active width of the nano-scale width planar pMOSFET is the same as the fin top width of the bulk PFinFET, because the planar transistor was fabricated using the same process steps without SiN recess etching after the CMP. Each datum is the median of the 16 data points over a wafer [26].

width (or fin body width) decreases because of the increased channel controllability. We compared the g_{mmax} and the $|I_{Dsat}|$ for the devices in Fig. 18 (b) and (d). Although the effective channel width (35+2×60=155 nm) of the bulk PFinFET is wider than that (100 nm) of the control device, the bulk PFinFETs show lower $|I_{Dsat}|$ and g_{mmax} than those of the control device. We think that the degradation of the $|I_{Dsat}|$ and g_{mmax} is attributed to mainly higher parasitic source/drain resistance and the poor side channel surfaces. It is still required to improve the surface before growing the gate oxide and to find a method to reduce the parasitic resistance. As the active width decreases, the nano-scale width planar devices show more significant $|I_{Dsat}|$ degradation than the bulk PFinFETs. The PFinFETs described here show very good device characteristics compared to nano width PMOSFETs. So we examine into details of PFinFETs starting with body bias dependency in the following.

Body bias dependencies of the PMOSFETs are compared as shown in Fig. 19. The bulk PFinFET shows very small dependency, whereas a significant one with the nano-width planar PMOSFET at a fixed gate length of 100 nm. Both devices have p^+ poly-Si gate. According to our simulation results, the potential in the channel region is dominated by both side-gates of the fin body, and not dependent on the body bias. The channel potential around the bottom of the side-gates shows a small change with the body bias, resulting in slightly higher V_{th} in the I_D-V_{GS} curve with positive body bias. With the back bias of 1 V, the bulk PFinFET shows higher drain current over the positive gate bias because the n-side doping concentration of source/drain p^+-n junction in the bulk PFinFET is higher than that in planar devices. The retrograde doping profile of the n-well needs to be optimized to reduce the leakage current in the bulk PFinFETs. Here, the planar PMOSFET has the same gate width as the fin top width of the bulk PFinFET (about 40 nm). It is well known that the back-bias effect becomes less significant as the channel width decreases. The planar channel device in this work has relatively thick SiN layer (50 nm) as a liner, the fringing field from the gate electrode penetrates more easily into the channel near the surface due to higher dielectric constant of the SiN than that of SiO_2, which may make the back-bias effect less significant. The inset in the Fig. 19 is the I_{SUB}/I_D comparison of PMOSFETs. The bulk PFinFET shows lower I_{SUB}/I_D characteristic than planar PMOSFET, since it seems that the I_{SUB} of the triple-gate devices is low due to the drain field perturbed by the normal gate field and the on-current of the device is higher than that of planar device [43]. We can say hot carrier generation in bulk FinFETs is lower than that of planar channel MOSFETs at the same drain current.

As shown in Fig. 20, drain induced barrier lowering (DIBL) of 34 mV/V was obtained with the parallel arrayed 30,000 bulk PFinFETs while that of the planar PMOSFETs show 92 mV/V at the same gate length of 100 nm. Also, maintaining comparably low I_{OFF} level in the transistors connected in parallel is a quite affirmative result and means that the proposed device can be a good candidate for production level devices. However, it should be mentioned that, as the fin width and gate length decrease down to sub 100 nm, threshold voltage reduces as well [44], so higher off-current is inevitable. Therefore, new approaches such as workfunction engineering using combination of various metals and poly-Si [45] should be considered. The bulk PFinFETs show better subthreshold swing of 74 mV/dec compared with 92 mV/dec in the planar channel devices, and higher on-current.

It is noted that in the bulk PFinFETs the positive oxide charge at the interface between gate oxide and Si channel region induce the increase of the V_{th}, which compensates the V_{th} lowering due to fully depleted thin body. Here, the bulk PFinFET shows only slightly lowered V_{th} by about 50 mV in absolute value compared to planar PMOSFET. The bulk

NFinFEETs for DRAM application showed intolerably high leakage current because the thin body and the positive oxide charge significantly lower the V_{th}.

Fig. 21 shows I_D-V_{GS} characteristics of 100 nm N and PFinFETs as a function of V_{DS}. The devices in Fig. 21 were designed and fabricated to improve the device characteristics shown in Figs. 15-18. Both N and PFinFETs show very good I_D-V_{GS} characteristics and quite symmetric. As mentioned above, effective channel width (W_{eff}) of the FinFETs is approximately given by $2H_{fin}+W_{fin}$ and 225 nm in these devices. Extracted I_D of NFinFET considering the W_{eff} is ~200 μA/μm at V_{GS} of 1.5 V, which is still not reasonable and needs further optimization in the channel surface and doping profile (channel and source/drain). PFinFET shows reasonable I_D of ~200 μA/μm at $|V_{GS}|$ of 1.5 V. It seems that the NFinFET has relatively low doping of source/drain extension because the substrate current has a peak around V_{GS} of 0.7 V and another peak at a bias larger than 2 V. Of course, the low substrate current is clear evidence of the low doping of the extension region. Such low extension doping gives high parasitic source/drain resistance, resulting in the degradation of the drain current. Low doping of source/drain region suppresses hot carrier generation. The I_{SUB}/I_D ratio at a peak I_{SUB} and a V_{DS} of 1 V is less than 10^{-7} which is quite low value. Off-current of both devices at V_{GS}=0 V is very small and less than 0.2 nA/μm at a V_{DS} of 1 V. It is well known that the crystal orientation of the vertical surface of the fin body becomes (110) if the surface is parallel or orthogonal to the primary flat zone (or notch) of a (100) plane wafer. The (110) vertical surface plane of the fin body degrades electron mobility and enhances hole mobility [46].

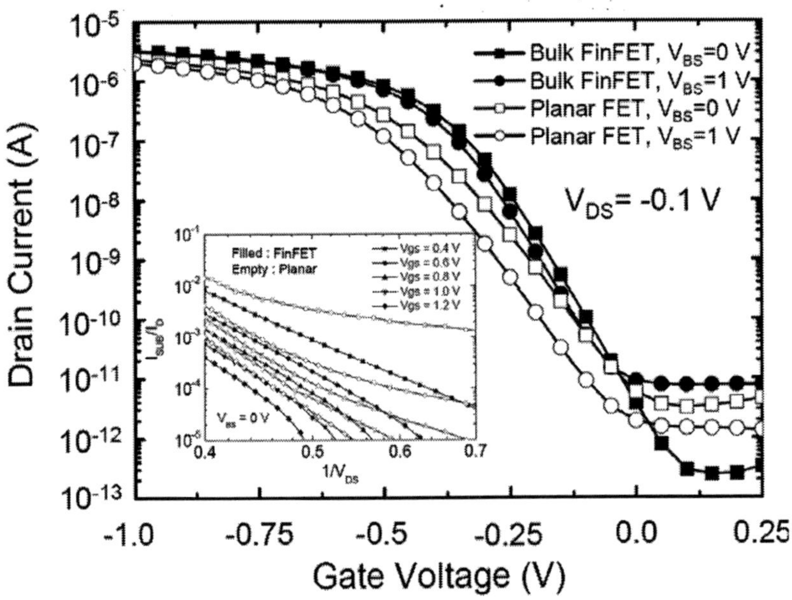

Figure 19. Comparison of the body bias dependency of the bulk PFinFET with that of the planar pMOSFET. L_g and T_{ox} are 100 nm and 1.8 nm, respectively. 40 nm active width of the planar pMOSFET was the same as the fin top width of the bulk PFinFET. Fin height was about 60 nm for the bulk FinFET. The inset is I_{SUB}/I_D comparison of discrete pMOSFETs according to the inverse of the drain bias. Here, V_{GS} is changed from -0.6 to -1.2 V with -0.2 V step in the measurement [42].

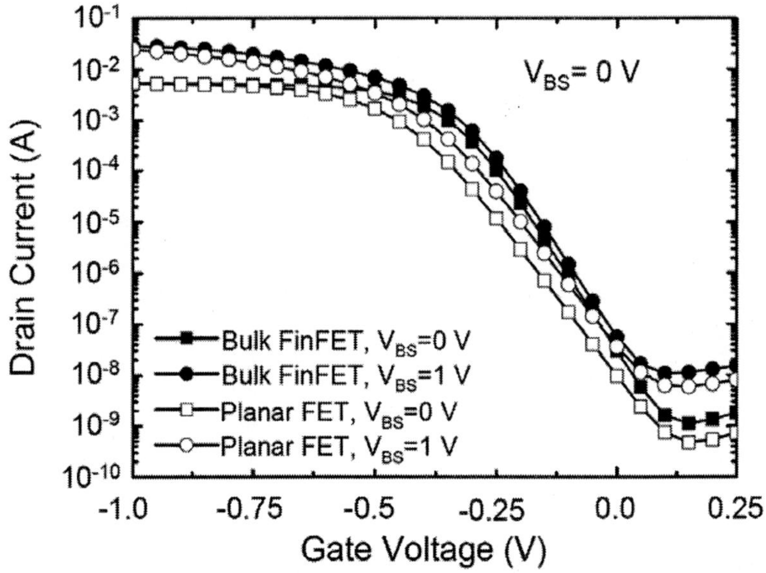

Figure 20. DIBL comparison of parallel arrayed 30,000 pMOSFETs. L_g and T_{ox} are 100 nm and 1.8 nm, respectively. 40 nm active width of the planar pMOSFET was the same as the fin top width of the bulk PFinFET. Fin height is about 60 nm for the bulk FinFET [42].

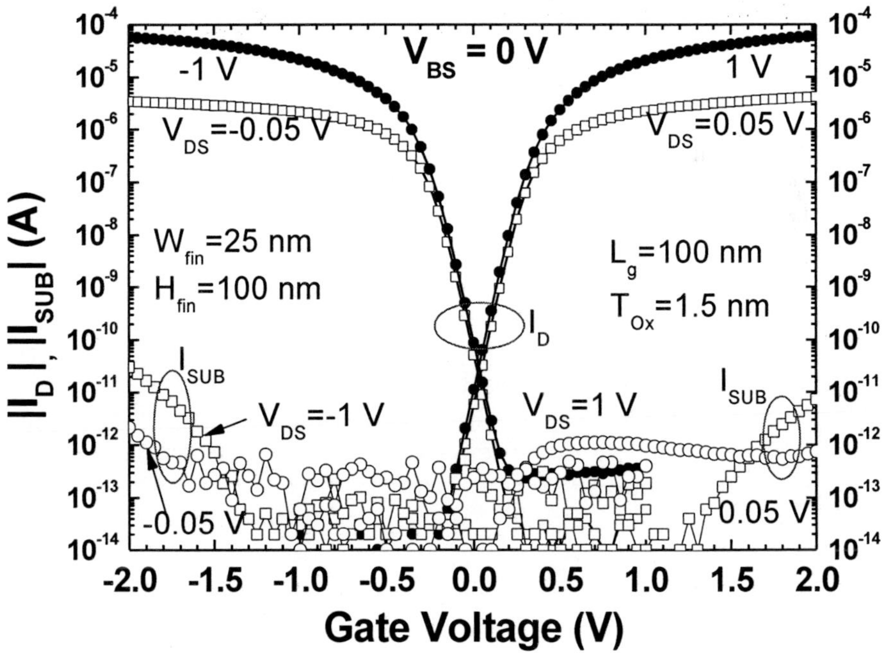

Figure 21. Drain and substrate currents versus gate bias for 100 nm NFinFET and PFinFET as a parameter of drain bias. The width (W_{fin}) and height (H_{fin}) of channel fin body are 25 nm and 100 nm, respectively. Substrate bias is 0 V.

It is meaningful to check the reliability of the bulk FinFETs, but this is beyond the scope of this chapter. We can find device degradation by hot-carrier [47], Negative Bias Temperature Instability (NBTI) of PFinFETs [48], and impact ionization behavior [49]. It is interesting to check temperature dependence of substrate and drain currents in the bulk FinFETs with nitride liner as shown in Fig. 11, because the nitride gives large stress to the fin body with increasing temperature [50].

4. Threshold Voltage Modeling of Bulk FinFETs

4.1. Threshold Voltage Modeling of the Bulk FinFET without Corner

In the bulk FinFETs, double-gate (DG) [51] nature is a key part in FinFETs and needs to be understood well. Understanding on threshold voltage (V_{th}) behavior with device geometry and channel doping concentration is very important in accurate analysis, device design, and circuit design. To apply the devices to integrated circuits (ICs), the behavior of V_{th} should be characterized and the parameters for V_{th} should be extracted. However, it is difficult to model the V_{th} of this device, because of corner effect and 3-dimensional (3-D) device structure. Therefore, we perform V_{th} modeling of the bulk FinFETs without corner (hereafter called as an ideal bulk FinFET) based on short-channel effect (SCE), narrow-width effect (NWE), and 3-D charge-sharing at low drain bias (V_{DS}=0.05 V). Then, the V_{th} modeling of the bulk FinFETs with corner is achieved based on SCE, NWE, 3-D charge-sharing, and corner factor (α_c) [52]. In this chapter, threshold voltage is extracted by using $g_{m,max}$ for a given V_{DS} of 0.05 V.

Fig. 22 (a) shows 3-D schematic view of the bulk FinFET. The H_g, W_{fin} and T_{FOX} represent gate height, fin width and field oxide thickness in STI region, respectively. The x_{jSDE} is the junction depth of source/drain extension (SDE) defined from the top of the fin body. The LDD concentration is 7×10^{19} cm^{-3} and has simple Gaussian profile of dec/5nm, and the doping concentration at source (or drain) region is 1.4×10^{20} cm^{-3}. Gate oxide thickness (T_{ox}) is fixed at 1.5 nm and n$^+$ poly gate is applied. For the fin body with the 90° corner shown in Fig. (a), there are three different channels (W_{tc}, W_{sc}, and W_s) where W_{tc}, W_{sc}, and W_s are channel widths of top-corner, side-corner, and side region of a half fin body, respectively. Fig. 22 (b) shows 2-D schematic view of DG MOSFET cut along A-A' in Fig. 22 (a). Metallurgical channel length (L) is given by the difference between L_g and two lengths (L_{ov}) overlapped by source and drain. L_c(=$L-2x_m$) is the channel length based on the box channel doping profile. Although channel doping is uniform, the channel doping profile near source and drain junction is nonuniform because of the Gaussian doping profile of S/D extension region. To approximate this channel doping profile to box-type channel doping profile, x_m is introduced in the modeling. We ignore oxide charge in this work to simplify the modeling and simulation.

Fig. 23 (a), (b), and (c) show the 2-dimension (2-D) cross-sectional views of different fin bodies along the gate width direction, and show the cross-sectional views of the bulk FinFETs with or without corner: no corner, 90° corner, and half-circle corner. We start with the V_{th} modeling for the device without corner (called as an ideal bulk FinFET) to avoid complicated corner effect.

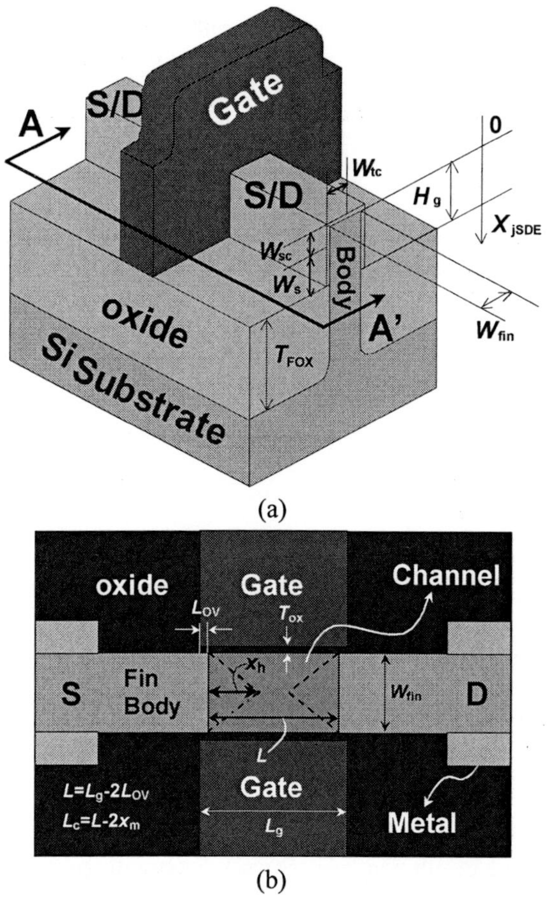

Figure 22. (a) 3-D schematic view of bulk FinFET. H_g and W_{fin} represent fin height and width, respectively. (b) 2-D cross-sectional view of the bulk FinFET cut along A-A' in Fig. 22 (a). The T_{ox} and x_h represent gate oxide thickness and charge-sharing length, respectively.

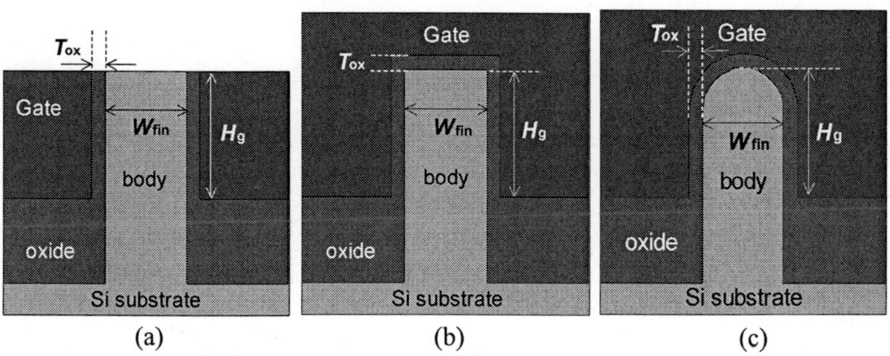

Figure 23. 2-D schematic cross-sectional views of the fin body: without corner (a), with 90° corner (b), with a half-circle corner (c).

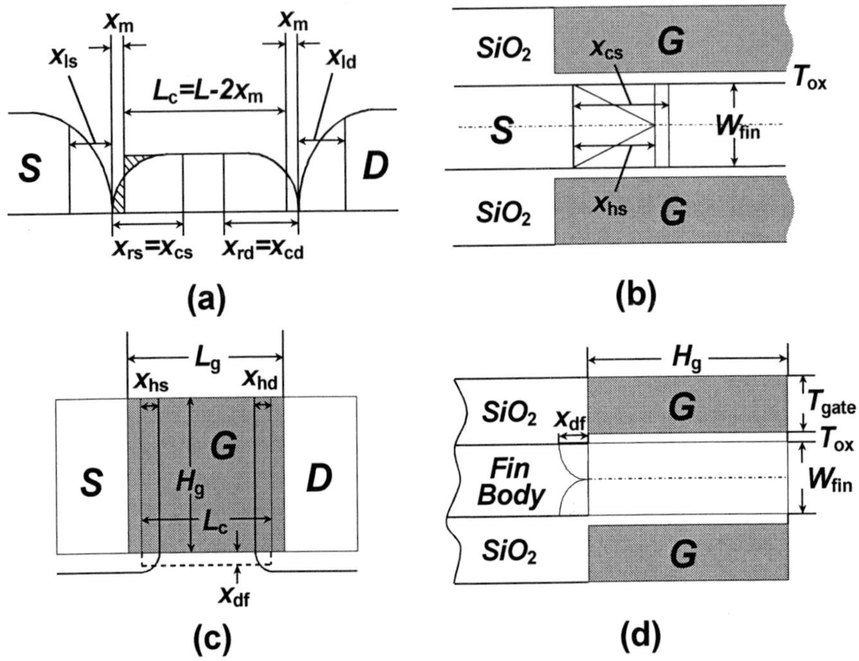

Figure 24. Schematic 2-D views of structure and doping profile. (a) Net doping profile in the fin body along channel length direction. (b) Top cross-section along channel length direction. (c) Side view. (d) Cross-section along the center of gate electrode.

Fig. 24 shows doping profile and 2-D schematic views of the ideal bulk FinFET. Fig. 24 (a) represents the net doping profile along the channel length direction. The $L_c(=L-2\ x_m)$ is the channel length based on the box channel doping profile. Metallurgical channel length (L) is given by the difference between L_g and two lengths (L_{ov}) overlapped by source and drain. The x_m is introduced to approximate actual Gaussian doping profile to the channel doping box profile. In Fig. 24 (a), each of two hatched regions in the channel has same area. The V_{th} model of the bulk FinFET is based on 3-D charge-sharing. In the V_{th} model of conventional planar channel MOSFETs, the SCE and NWE were given by Yau et al. [53] and Akers et al. [54], respectively. The conventional SCE and NWE models are widely used to understand the physics of the device intuitively. In Fig. 24 (a), the amount of charge in the left region (from junction to x_{ls}), Q_L, is equal to the amount of charge in the right region (from junction to x_{rs}), Q_R. The x_{ls} (or x_{ld}) is a depletion width in the source (or drain) region, corresponding to the x_{rs} (or x_{rd}) in the channel. And the x_{ls}, x_{rs}, x_{ld}, and x_{rd} are used to calculate x_{cs} (or x_{cd}). Fig. 24 (b), the cross-sectional top view of fin body, explains short channel effect due to channel charge-sharing. Actually, the doping concentration near junction is changed with a gradient. The x_{hs} (or x_{hd}) is a fitting parameter to represent the charge-sharing length at source-side (or drain-side), and used to reflect the SCE. The W_{fin} represents fin body thickness. Note that the shaded region in Fig. 24 (b) represents the depletion region depleted by the source at low drain bias under $V_{GS}=V_{th}$ condition. Fig. 24 (c) shows side view along fin body direction to explain the NWE. We first consider the NWE in the bulk FinFETs. As shown in Fig. 24 (c) and (d), the fringing field goes from the bottom of the gate to the fin body along the channel length direction. The fringing field makes depletion in non-channel region (see the x_{df}), which increases V_{th}. Fig. 24 (d) shows the cross-sectional view along gate electrode direction. We assumed that the fringing field from the gate makes depletion region like hatched regions, and each of hatched regions looks like a quarter of an oval. The x_{df} is an effective depletion width depleted by the fringing field from the gate.

The V_{th} model of the conventional planar MOSFETs with long channel length and wide channel width is given by [55]

$$V_{th} = V_{FB} + 2\psi_B + \frac{qN_b x_{dep}}{C_{ox}} \tag{2}$$

where V_{FB} and ψ_B are flat band voltage and Fermi potential, respectively. The N_b, x_{dep}, and C_{ox} are body doping, channel depletion width under the gate, and gate capacitance, respectively. We ignore oxide charge in this work to simplify the modeling and simulations. Through the modification of the last term in (2), we can reflect the V_{th} change in small geometry planar channel devices with the SCE and NWE. Basically, (2) can be used for V_{th} of the ideal bulk FinFET with a simple modification. As the device becomes smaller, the V_{th} equation of the ideal bulk FinFET, denoted as $V_{th,s}$ (side-gate V_{th}), should be modified to reflect SCE and the NWE too. We assumed that the fin body is fully depleted. Then the x_{dep} in the bulk FinFETs is limited by a half of the fin body width (W_{fin}). In Fig. 24 (b), since the hatched triangle region is symmetric for dash-dot line, we consider a half of them (the upper or lower part of the dash-dot line) to consider SCE. Because there are two hatched right triangular depletion regions in the channel near the source and the drain, these two right triangular depletion regions have area of

$$\frac{1}{2} x_{dep} \left(x_{hs} + x_{hd} \right) \tag{3}$$

where x_{dep} is fixed to $W_{fin}/2$ under the assumption that the fin body is fully depleted, and x_{hs} and x_{hd} are fitting parameters at source and drain sides, respectively. The V_{th} reduction ($\Delta V_{th,SCE}$) due to the SCE is given by

$$\Delta V_{th,SCE} = -\frac{qN_b x_{dep}}{C_{ox}} \left(\frac{x_{hs} + x_{hd}}{2L_c} \right) \tag{4}$$

where the minus sign (-) represents V_{th} reduction. If the drain bias is very small compared to the built-in potential at the drain junction, the x_{hd} is nearly the same as the x_{hs}. It is quite general to set the drain bias less than (or equal to) 0.1 V to measure the V_{th}. From this bias condition, we can use only one fitting parameter (x_h) instead of using two fitting parameters, x_{hs} and x_{hd} by putting $x_h=(x_{hs}+x_{hd})/2$.

The V_{th} equation of the ideal bulk FinFET including only the SCE is given by

$$V_{th,SCE} = V_{FB} + 2\psi_B + \frac{qN_b x_{dep}}{C_{ox}}\left(1 - \frac{x_h}{L_c}\right). \tag{5}$$

In (5), the term considering SCE is simple, but much different from that of the V_{th} equations for conventional planar MOSFETs and thin-film SOI MOSFETs [56]. The x_h is a fitting parameter depending mainly on N_b and W_{fin}. For a case of fully depleted fin body having high N_b, the x_{hs} and x_{hd} have similar values to the x_{cs} and x_{cd}, respectively. The x_h extraction procedure is as follows. For given N_b and W_{fin}, we obtain a set of $V_{th,SCE}$ for different L_g from the simulation, and then fit the V_{th}'s from our model ($V_{th,model}$) to a set of V_{th} from the simulation ($V_{th,simulation}$) by changing the x_h. With a reasonable x_h, we can achieve the best fitting, and finally an optimum x_h is determined for the given W_{fin} and N_b.

Now we show how to select initial $x_h(=(x_{hs}+x_{hd})/2)$ by plotting the relation between the x_h and the $x_c(=(x_{cs}+x_{cd})/2)$. The x_c can be calculated simply by using doping profile. Once the initial x_h is set, then we can easily extract a reasonable x_h by using the $V_{th,SCE}$ data from a set of L_g.

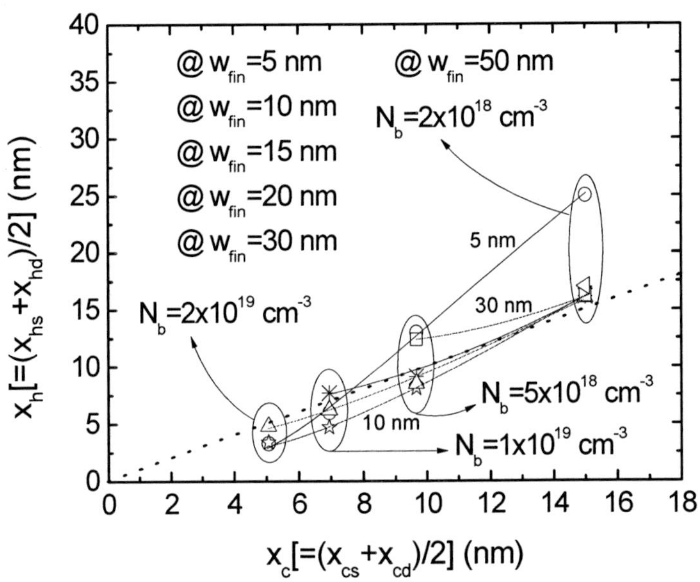

Figure 25. Relation between $x_c=(x_{cs}+x_{cd})/2$ and a fitting parameter, $x_h=(x_{hs}+x_{hd})/2$. The dotted line represents the points for $x_h=x_c$.

Fig. 25 shows the relation between a fitting parameter x_h and an x_c (channel-side depletion width) for given N_b and W_{fin}. The dotted line stands for the line for $x_h=x_c$. In Fig. 25, when we draw a line for the fitting parameter x_h values for various N_b's at a fixed W_{fin}, the line shape for each fin body width looks like parabola. As the W_{fin} is increased, the slope of the parabola is decreased. Anyway, we can roughly guess the x_h from the figure.

As shown in Fig. 24 (a), we calculate the left-side charge (Q_L) after choosing a depletion length (x_{ls}) arbitrarily in the source region from the junction between the source and the channel. We can calculate the right-side charge (Q_R) in the channel that has the same value as the Q_L. By calculating the channel depletion charge (Q_R) equal to the Q_L, the x_{rs} can be obtained. Then we calculate the potential (V_x) by integrating the space charges from x_{ls} to x_{rs} twice. If the calculated V_x is different from the built-in potential (Φ_{bi}), the x_{ls} is changed until the V_x is equal to the built-in potential (Φ_{bi}). The built-in potential (Φ_{bi}) is given by

$$\Phi_{bi} = \frac{kT}{q} ln\left(\frac{N_d(x_{ls})N_b(x_{rs})}{n_i^2} \right). \tag{6}$$

In this way, the x_{rs} (or x_{rd}) is extracted and becomes the channel-side depletion width (x_{cs} or x_{cd}). Starting with an initial x_{hs} similar to an x_{cs}, a final x_{hs} for given device structure and parameters can be determined by fitting the calculated V_{th}'s to the simulated V_{th}'s obtained by changing L_g only.

Figs. 26 and 27 are the results of 2-D double-gate device simulation. Fig. 5 shows $V_{th,SCE}$ versus W_{fin} as a parameter of the body doping at a fixed L_g of 50 nm. Fig. 27 shows $V_{th,SCE}$ versus L_g as a parameter of the W_{fin} at a fixed body concentration of 5×10^{18} cm^{-3}. The $V_{th,SCE}$ model explains well the simulated V_{th} behavior. Until now, the corner effect and the NWE were not included in the model because 2-D double-gate device structure was used.

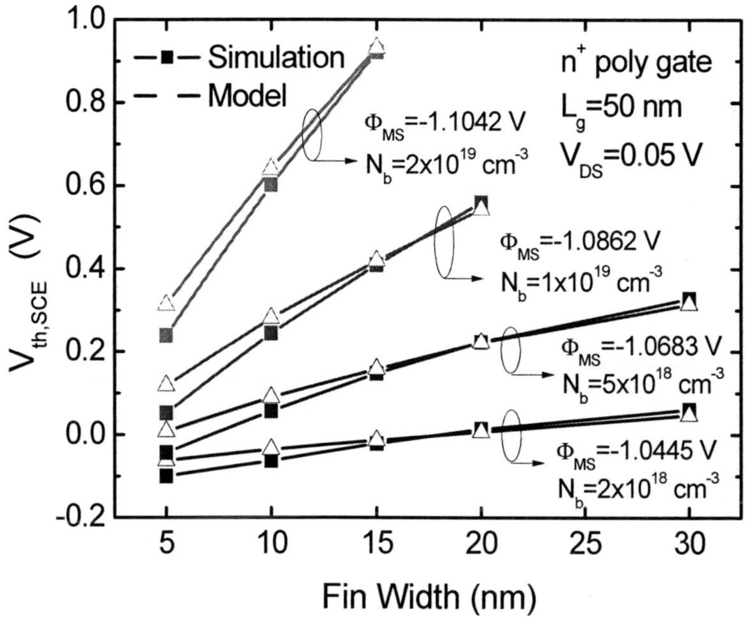

Figure 26. $V_{th,SCE}$ versus fin width as a parameter of the body doping. Corner effect and NWE are not considered. The gate length (L_g) is fixed at 50 nm.

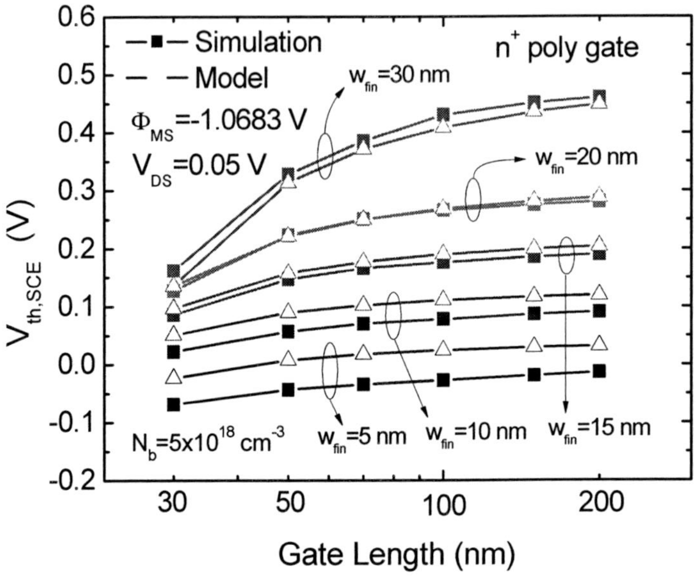

Figure 27. $V_{th,SCE}$ versus gate length as a parameter of the fin width at a fixed body concentration of 5×10^{18} cm^{-3}. The model includes only short channel effect. The fin width changes from 5 nm to 30 nm. The x_h's corresponding to the W_{fin}'s of 5 nm to 30 nm are 13, 8.1, 8.5, 9.1, and 12.4, respectively.

In Fig. 24 (d), the NWE is attributed to the charge (a quarter oval area) in the fin body depleted by the fringing field from the gate. The area of a quarter oval is given by

$$\frac{1}{4}\pi x_{dep} x_{df} \tag{7}$$

where x_{df} is an effective width in the body below the channel region affected by gate fringing field as shown in Fig. 24 (d). Then the V_{th} increase ($\Delta V_{th,NWE}$) due to the NWE is given by

$$\Delta V_{th,NWE} = \frac{qN_b x_{dep}}{C_{ox}}\left(\frac{\pi x_{df}}{4H_g}\right) \tag{8}$$

where the H_g is equivalent to the channel width of planar MOSFET, and the x_{df} is shown in Fig. 24 (c) and (d). In (8), the terms in the parenthesis was obtained by considering the volume depleted by the fringing field and the channel volume. The $V_{th,s}$ model including the SCE in (5) and the NWE in (8) can be derived for the device without the corner region. The threshold voltage is expressed as $V_{th,s}$ and given by

$$V_{th,s} = V_{FB} + 2\psi_B + \frac{qN_b x_{dep}}{C_{ox}}\left(1 - \frac{x_h}{L_c} + \frac{\pi x_{df}}{4H_g}\right). \tag{9}$$

In (9), $\Delta V_{th,SCE}$ and $\Delta V_{th,NWE}$ terms can be canceled out each other at a given L_c of $(4H_g x_h)/(\pi x_{df})$. Here the x_{df} is still an unknown. The x_{df} can be calculated using fringing capacitance C_{of} reported by Shrivastava et al. [57]. The C_{of} by gate fringing field is given by

$$C_{of} = \frac{2\varepsilon_{ox}}{\pi} ln\left(1 + \frac{T_{gate}}{T_{ox}}\right) \quad (10)$$

where the T_{gate} and T_{ox} represent gate thickness and gate oxide thickness, respectively. The ε_{ox} is permittivity of silicon dioxide (3.45×10^{-13} F/cm). We assumed that the depleted body charge by the fringing field is given by $C_{of}V_{th,s}$ because we are focusing on the $V_{th,s}$ among various gate biases. If we set the depleted charge as Q_{fring}, the relation is $Q_{fring} = C_{of}V_{th,s}$. From the relation, the x_{df} is analytically given by

$$x_{df} = V_{th,s} \frac{8\varepsilon_{ox}}{\pi^2 x_{dep} q N_b} ln\left(1 + \frac{T_{gate}}{T_{ox}}\right). \quad (11)$$

The x_{df} and the $V_{th,s}$ are obtained by solving simultaneously (9) and (11).

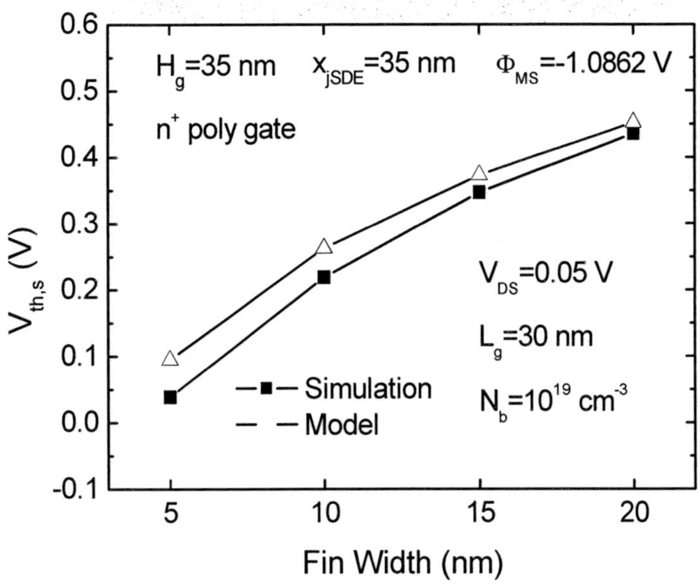

Figure 28. $V_{th,s}$ versus fin width at given H_g=35 nm, L_g=30 nm, and $N_b=10^{19}$ cm^{-3}. The model shows a good agreement with the simulation. Here, the devices have no top corners.

Using (9), we calculate first the $V_{th,s}$ including the SCE and NWE for the device without top corner. At fixed H_g =35 nm and L_g =30 nm, the model predicts well the simulated $V_{th,s}$ with the W_{fin} as shown in Fig. 28. The increase of the W_{fin} increases the $V_{th,s}$. Fig. 29 shows the validity of the $V_{th,s}$ model with body doping at fixed H_g=35 nm, W_{fin}=20 nm, and L_g=30 nm. We can observe a good agreement between the model and the simulation. In this figure we used different Φ_{MS} for different body doping, because the Φ_{MS} depends on the body doping.

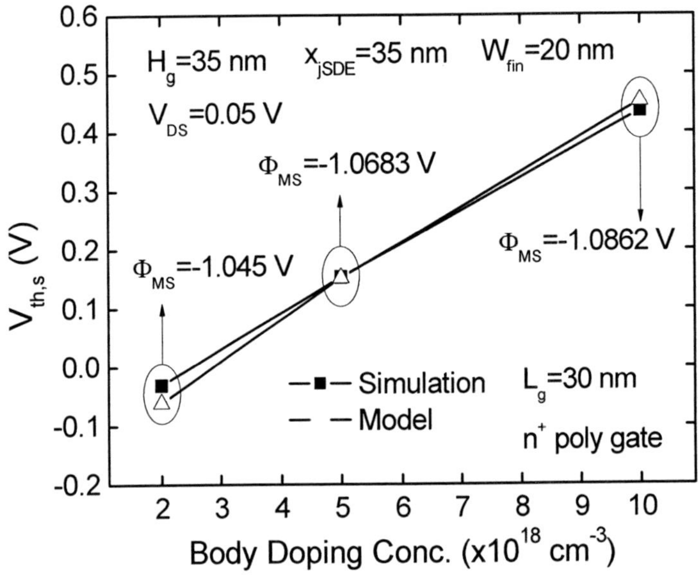

Figure 29. $V_{th,s}$ versus body doping concentration at given H_g=35 nm, W_{fin}=20 nm, and L_g=30 nm. Since the body doping affects work-function difference (Φ_{MS}), we used different Φ_{MS}'s for different body doping concentrations. Here, the devices have no top corners.

4.2. Threshold Voltage Modeling of the Bulk FinFETs with Corner

Now we derive the threshold voltage in the corner region ($V_{th,c}$). Fig. 30 shows schematic 3-D view (a), cross-section view of the fin body with 90° corner (b), and cross-section view of the fin body with half-circle corner (c) of the fin body to model the $V_{th,c}$. Firstly, we will study the $V_{th,c}$ in the device with the fin body where the corner angle is 90° as shown in Fig. 23 (b). The model for the half-circle corner as shown in Fig. 23 (c) can be derived similarly in a simple form. In Fig. 30 (a), the W_{tc} is the channel width in the top corner and is assumed to be $0.5 \cdot W_{fin}$ at the top if the fin body is fully depleted. The W_{sc} is the channel width in the side corner of the fin body and is also $0.5 \cdot W_{fin}$ for the fully depleted fin body. In Fig. 30 (b), the H_g is equal to the summation of W_{sc} and W_s. In this work, we considered that the corner region of the fin body is the region defined by the W_{tc} and W_{sc}. Under the condition of the same oxide thickness on top and side surfaces of the fin body, the W_{sc} is equal to the W_{tc}.

In deriving the $V_{th,c}$ model, we consider the SCE and the shape of the corner region. The SCE in the corner region decreases the $V_{th,c}$. The corner shape changes the V_{th} for the corner channel which is lower than that of side channel of the fin body.

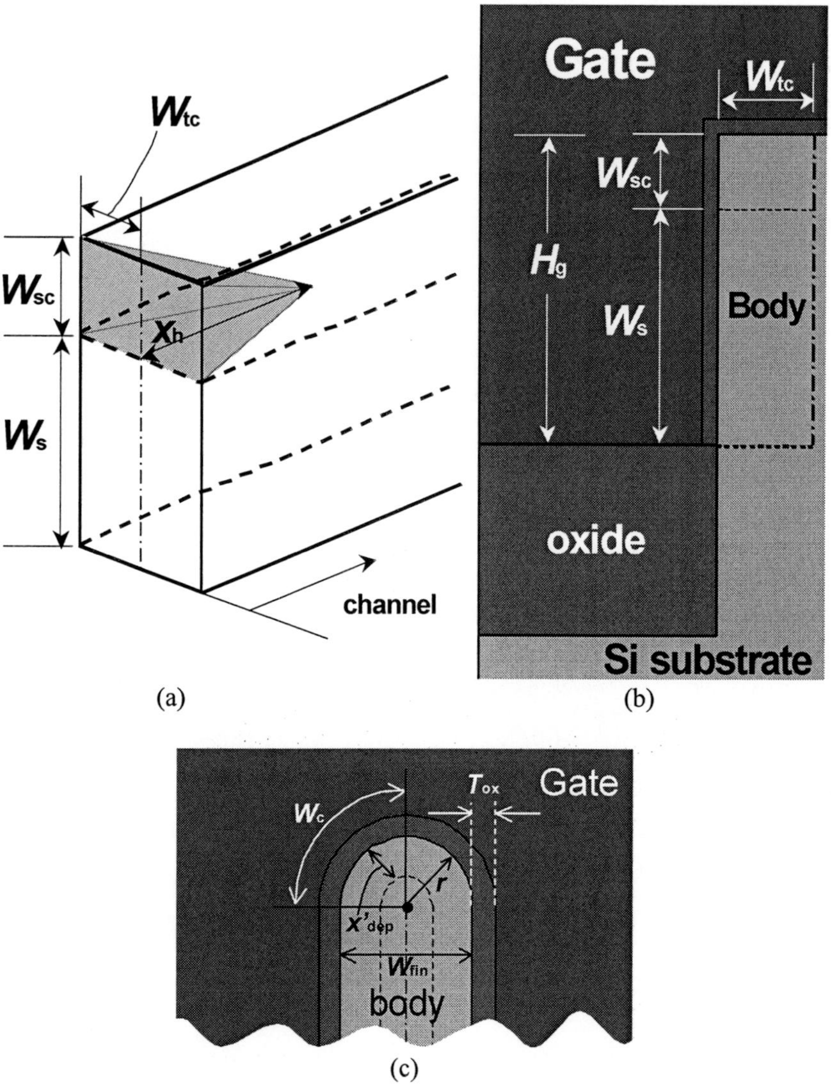

Figure 30. (a) Schematic 3-D view for considering of the corner effect. (b) Cross-section view of the fin body with 90 ° corner. (c) Cross-sectional view of the fin body with half-circle corner.

Before deriving the SCE in the corner region, we defined the corner region as shown in Fig. 30. For a fully depleted fin body, the channel formed on the fin body can be classified into 3 regions: top-corner, side-corner, and side channels. We think that the corner channel can be simply defined by the W_{tc} and the W_{sc} if the fin body is fully depleted. Then the $V_{th,c}$ of the top-corner channel is equal to that of the side-corner channel, and the $V_{th,c}$ modeling of the corner region is simplified. It is only required to consider corner and side channels for the fully depleted fin body. If the fin body is not fully depleted, then we should consider each contribution from top-corner, side-corner, top, and side channels in modeling V_{th} of the bulk FinFET. In Fig. 30 (a), the volume charge shared by source (or drain) in the fin body with 90° corner is shaded. The shaded region in the corner region is depleted by source or drain, and the shaded volume shown in Fig. 30 (a) is given by $(1/3)(W_{tc}\cdot x_h\cdot W_{sc})$. We multiply $(1/3)(W_{tc}\cdot x_h\cdot W_{sc})$ by 2 to take account of both contributions from source and drain. The volume of the corner in the bulk FinFET is given by $W_{tc}\cdot L_c\cdot W_{sc}$. Therefore, the SCE term in corner region with 90° corner is given by

$$1-\frac{2}{3}\frac{x_h}{L_c}. \tag{12}$$

The SCE term for the fin body with the half-circle was also derived and given by the same as (12).

Now we derive the $V_{th,c}$ expression for the corner region. The volume charge shared by the source or the drain in the corner region is less than that in the body region excluding the corner region because the electric field from the top and side gates of the corner region is stronger than that from the side gate only in the fin body excluding the corner region. Based on the 3-D charge-sharing in the corner region, we obtained the $V_{th,c}$. The channel charge including inversion and depletion charges is induced by the electric field from the gate electrode. Since the volume of the charge per unit width to be depleted in the corner region is smaller by comparing with that of side channel and the gate area where gate electric field gets through is wider in the corner channel region than in the side channel region of the fin body, we introduce a corner factor (α_c) to take into account the effect in the $V_{th,c}$ model. For a half-circle corner, several parameters for $V_{th,c}$ modeling should be derived based on cylindrical coordinate and we derived expressions for gate capacitance (C'_{ox}), depletion charge (Q'_D), and surface potential (ψ'_s) as given by

$$C'_{ox}=\frac{\varepsilon_{ox}}{r\ln(1+T_{ox}/r)} \tag{13}$$

$$Q'_D=-qN_b x'_{dep}\left(1-\frac{x'_{dep}}{2r}\right) \tag{14}$$

$$\psi'_s=\frac{qN_b}{2\varepsilon}\left\{(r-x'_{dep})^2\ln\left(1-\frac{x'_{dep}}{r}\right)+x'_{dep}\left(r-\frac{x'_{dep}}{2}\right)\right\} \tag{15}$$

where ε_{ox}, r, and T_{ox} represent dielectric constant of oxide, the radius in the half-circle corner of fin body, and gate oxide thickness, respectively, as shown in Fig. 30 (c). The r is the same as $0.5\cdot W_{fin}$ for a half circle corner. The x'_{dep} represents the depletion width in the half-circle corner region. In (15), we can calculate the x'_{dep} when the ψ'_s is $2\psi_B$. By substituting the x'_{dep}

into (14), the $Q_D^{'}$ is obtained. For a fully depleted body, the $x_{dep}^{'}$ is equal to *r* or larger than *r* theoretically. Then we cannot calculate the $\psi_s^{'}$ in (15). Therefore, the (13)-(15) can be applied to calculate the $V_{th,c}$ for non-fully depleted fin body with half-circle corner. Thus we need to find reasonable approach to obtain the $V_{th,c}$ for fully depleted half-circle corner. In this case, we can use the α_c to take into account the corner effect. The α_c is a fitting parameter and has the extracted value of 0.25 regardless of the corner shape. Finally we express the $V_{th,c}$ by considering the α_c and the 3-D charge-sharing as follows

$$V_{th,c} = V_{FB} + 2\psi_B + \frac{q\alpha_c N_b x_{dep}}{C_{ox}}\left(1 - \frac{2}{3}\frac{x_h}{L_c}\right). \tag{16}$$

Now we calculate the V_{th} of a bulk FinFET. There are three different channels (W_{tc}, W_{sc}, and W_s) for 90° corner and two different channels (W_c and W_s) for half-circle corner. The W_c is a channel width corresponding to 0.5 of half-circle corner of the fin body. Each channel may have the same or different V_{th}. Since the channels are in parallel in the channel length direction, the V_{th} of the device is calculated by considering contribution from each channel as follows

$$V_{th} = \left((W_{tc} + W_{sc})/W\right)V_{th,c} + (W_s/W)V_{th,s} \qquad \text{for } 90^{\circ} \text{ corner} \tag{17}$$

$$V_{th} = (W_c/W)V_{th,c} + (W_s/W)V_{th,s} \qquad \text{for half-circle corner} \tag{18}$$

where W, W_{tc}, W_{sc}, and W_s are channel widths of total, top-corner, side-corner, and side regions, respectively, as shown in Fig. 30 (b). The total channel widths (W) for 90 ° corner and half-circle corner are $W_{tc}+W_{sc}+W_s$ and W_c+W_s, respectively. Here, the $V_{th,s}$ is given by (9). The $V_{th,c}$ is given by (16) for uniform gate oxide thickness. As shown in (17) and (18), the combined threshold voltage (V_{th}) of the bulk FinFET with corner can be calculated as the perimeter-weighted sum [58] of the threshold voltage equations ($V_{th,s}$, $V_{th,c}$).

Fig. 31 shows V_{th} extraction procedure using derived model. Using parameters given in the input card, the $V_{th,s}$ including SCE and NWE is first calculated for the case without the corner region as mentioned before. Here, the x_{df} (a distance depleted by the fringing electric field from the bottom of the gate) of the ideal bulk FinFET is calculated by (9) and (11). For the bulk FinFET with a corner, the final V_{th} is considered to calculate the x_{df} in (11). The $V_{th,c}$ is calculated with the α_c(=0.25). The V_{th} calculation routine was coded by using MATLAB.

Fig. 32 shows V_{th} versus radius ratio as parameters of body doping, gate height, and gate workfunction. The V_{th}'s are modeled with the α_c of 0.25. The radius *r* of a quarter circle in the corner region is normalized to a half of the fin body width. The *r* is 0 for the 90° corner and $0.5{\cdot}W_{fin}$ for a half-circle corner. We can observe a good agreement between the model and the simulation as the given parameters. Now we check validity of the V_{th} model by changing gate length, gate height, fin body width, body doping, and corner shapes.

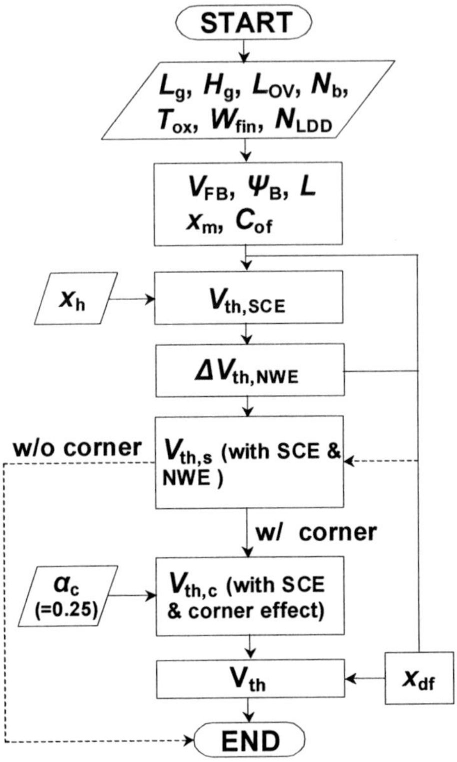

Figure 31. Procedure of the V_{th} modeling and extraction of the bulk FinFET. Here, x_h is a fitting parameter for charge-sharing. The corner factor α_c is fixed at 0.25.

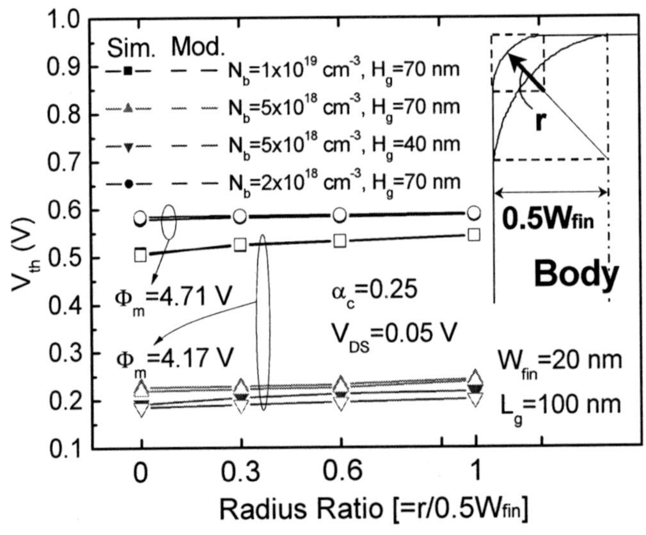

Figure 32. V_{th} versus radius ratio as parameters of body doping, gate height, and gate workfunction. The ratio is 0 for the 90° corner and 1 for a half-circle corner. The model shows a good agreement with the simulation at the given parameters.

Fig. 33 shows V_{th} versus L_g for devices as shown in Fig. 23 (a) and (b). The data for the devices with the 90° corner are compared with those of devices without the corner. Using only (9), the $V_{th,s}$'s of the devices without the corner are calculated. As the L_g increases, the $V_{th,s}$ increases because the short channel effect becomes small. The V_{th}'s of the devices with the corner are calculated by using (9), (16), and (17). Judging from the comparison with the data from 3-D device simulation, the model predicts well the V_{th} decrease due to the corner region for several L_g's. For given H_g of 70 nm and L_g of 100 nm, the V_{th} including the contribution from the corner shows about 20 % lower value than that ignoring the corner contribution.

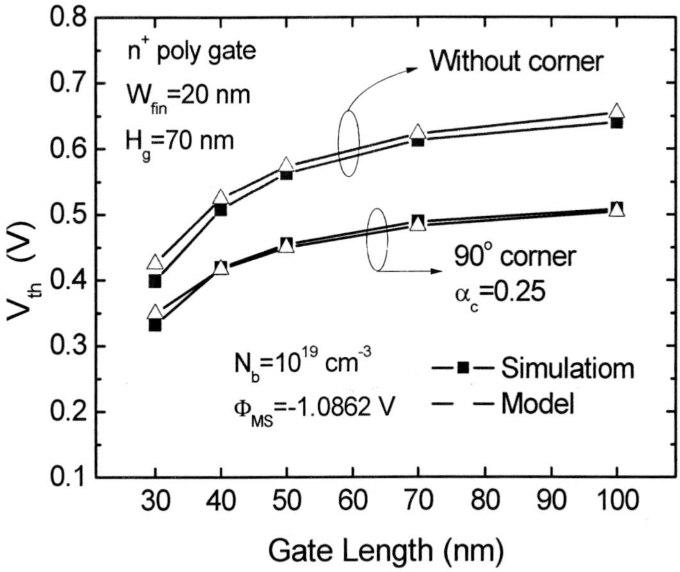

Figure 33. V_{th} versus gate length as a parameter of device structure. The $V_{th,s}$'s in the upper part were obtained from the device without corner in Fig. 23 (a). The data in the lower part were obtained from the device with 90° corner in Fig. 23 (b).

Fig. 34 shows V_{th} versus H_g for the devices with no corner and 90° corner at given W_{fin}= 20 nm and L_g=100 nm. The model shows a good agreement with 3-D simulation. The devices with the corner show lower V_{th}'s than those of the device without the corner as mentioned in Fig. 33. As the gate height decreases, the V_{th} of the bulk FinFETs with the corner is decreased, but the $V_{th,s}$ for the device without corner shows opposite trend. For the devices without the corner, the $V_{th,s}$ increases with decreasing H_g since relative contribution from the NWE increases. The V_{th} of devices with the corner decreases with the decrease of the H_g since the V_{th} lowering due to the corner channel is more dominant than the V_{th} increase due to the NWE.

Fig. 35 shows the V_{th} versus W_{fin} for the device with 90° corner at given L_g=100 nm and H_g=70 nm as a function of the N_b. The slope of the V_{th} with the W_{fin} is steeper for a higher body doping since the higher N_b in (9) and (16) increases the slope. The model explains well the V_{th} behavior.

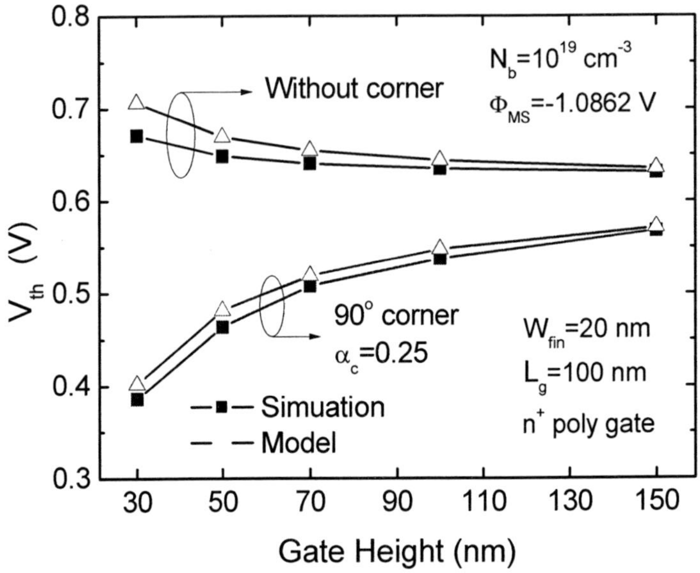

Figure 34. V_{th} versus gate height as a parameter of device structure. The $V_{th,s}$ for the devices without corner increases with decreasing the gate height. However, the device with 90° corners shows decreasing V_{th} with decreasing gate height.

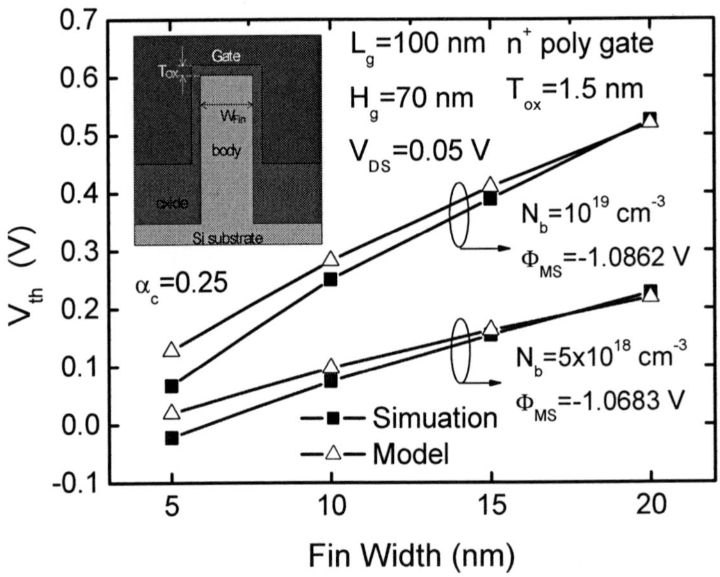

Figure 35. V_{th} versus fin width for the device with 90° corner as a function of the N_b.

Fig. 36 shows the V_{th} versus L_g for the device with a half-circle corner as a function of the N_b. Given W_{fin} and H_g are 20 nm and 70 nm, respectively. The model shows a good agreement with the simulation data.

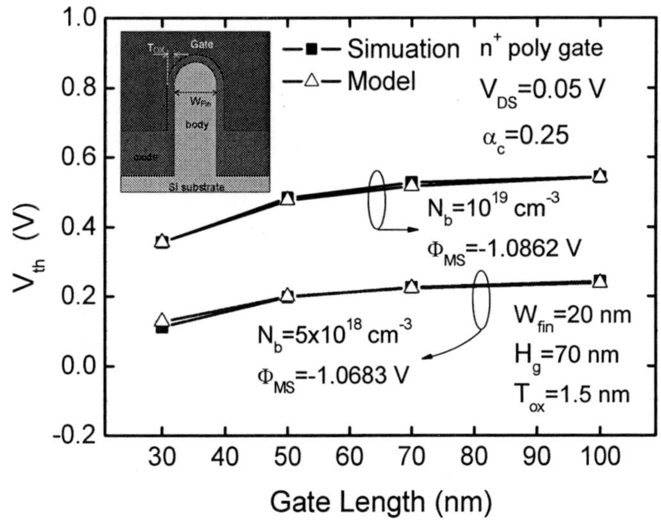

Figure 36. V_{th} versus gate length for the device with a half-circle corner as a function of the N_b.

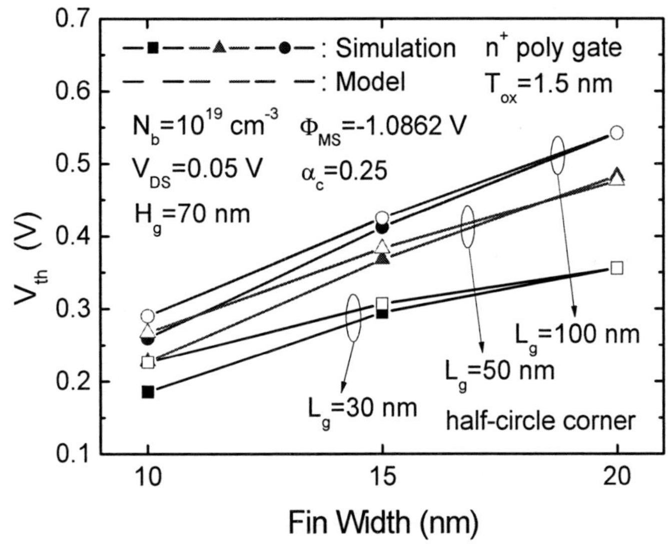

Figure 37. V_{th} versus fin width as a parameter of L_g. These data were obtained from the devices with a half-circle corner as shown in Fig. 23 (c).

Fig. 37 shows the V_{th} versus W_{fin} for the device with a half-circle corner as a function of the L_g. The model predicts well the V_{th}'s at each gate length. As the W_{fin} increases, the V_{th} increases since the x_{dep} in (9) and (16) increases. The slope of the V_{th} with W_{fin} is smaller in shorter L_g than in longer L_g because the value in the parenthesis in (9) and (16) becomes small due to the SCE.

The compact model predicts well the V_{th} behavior of the bulk FinFETs. It should be noted that final V_{th} model has only one fitting parameter (x_h). In this work, quantum effect due to thin fin body was not included, since the quantum effect is very small for the device with fin body thicker than about 10 nm [59]. Of course the quantum effect from the triangular potential well in the channel needs to be included in the V_{th} model to be more accurate.

4.3. Threshold-Voltage Modeling of Bulk FinFETs by Considering Surface Potential Lowering

The V_{th} models in sections 4.1 and 4.2 have predicted well the V_{th} behavior of the bulk FinFETs by using only one fitting parameter (charge sharing length, x_h). But these V_{th} models have shown relatively big difference with simulation data in L_g shorter than 40 nm and/or W_{fin} narrower than 10 nm because the surface band-bending ($2\psi_B$) in bulk FinFETs with fully depleted fin body decreases significantly at $V_{GS}=V_{th}$ condition for the small geometry.

In this section, new V_{th} model by considering the surface potential lowering is proposed in order to obtain better accuracy. First, parameters δ_W and δ_L are introduced to consider the surface potential lowering with the scaling-down of W_{fin} and L_g, respectively. By adopting δ_W and δ_L, we can predict accurately functional form of the V_{th} behavior with the geometry variation. However, there is still constant error between modeled and simulated data. So another fitting parameter (δ_B) in V_{th} modeling is introduced to correct the error. Finally, we drive compact V_{th} model of bulk FinFETs by considering 3-D charge-sharing, NWE, surface potential, and the shape of the top corner region of the fin body. Extensive 2-D and 3-D device simulations [38] are performed to verify the validity of derived V_{th} model in terms of gate length, gate height, fin body width, fin body doping concentration, and corner shape.

The $V_{th,s}$ model (side-gate V_{th}) of ideal bulk FinFETs with fully depleted fin body was given by (9).

In (9), Fermi potential, ψ_B is given by

$$\psi_B = V_t \, ln\left(\frac{N_b}{n_i}\right) \tag{19}$$

where V_t and n_i are the thermal voltage and intrinsic carrier concentration, respectively. Here, ψ_B in (19) depends on only N_b at a fixed temperature. However, the surface band-bending in fully depleted body at V_{th} decreases with decreasing L_g and/or W_{fin} since actual surface potential at V_{th} condition decreases appreciably with the geometry scaling-down. We can observe the surface potential lowering with the decrease of L_g at given W_{fin} and N_b through conduction band energy diagrams in Fig. 38 (a). If the conduction band energy diagrams are turned over, we obtain potential profiles for different L_gs.

Fig. 38 (a) and (b) shows the conduction band energy diagrams cut along the channel length direction in the surface and the center of fin body, respectively, at given V_{DS}=0.05 V, $V_{GS}=V_{th}$, and $N_b=2\times10^{18}$ cm^{-3}. Fig. 38 (c) shows the conduction band energy diagrams cut along the channel width direction at a point showing maximum-conduction-band-energy (MCBE) over the channel length. MCBE at a given L_g increases gradually with the decrease of L_g as shown in Fig. 38 (a). As devices scale down, such appreciable surface potential decrease affects V_{th} behavior. In Fig. 38 (b), MCBE starts to decrease in the devices with L_g shorter than 100 nm. In Fig. 38 (a) and (b), we need to notice MCBE at an L_g of 20 nm. In

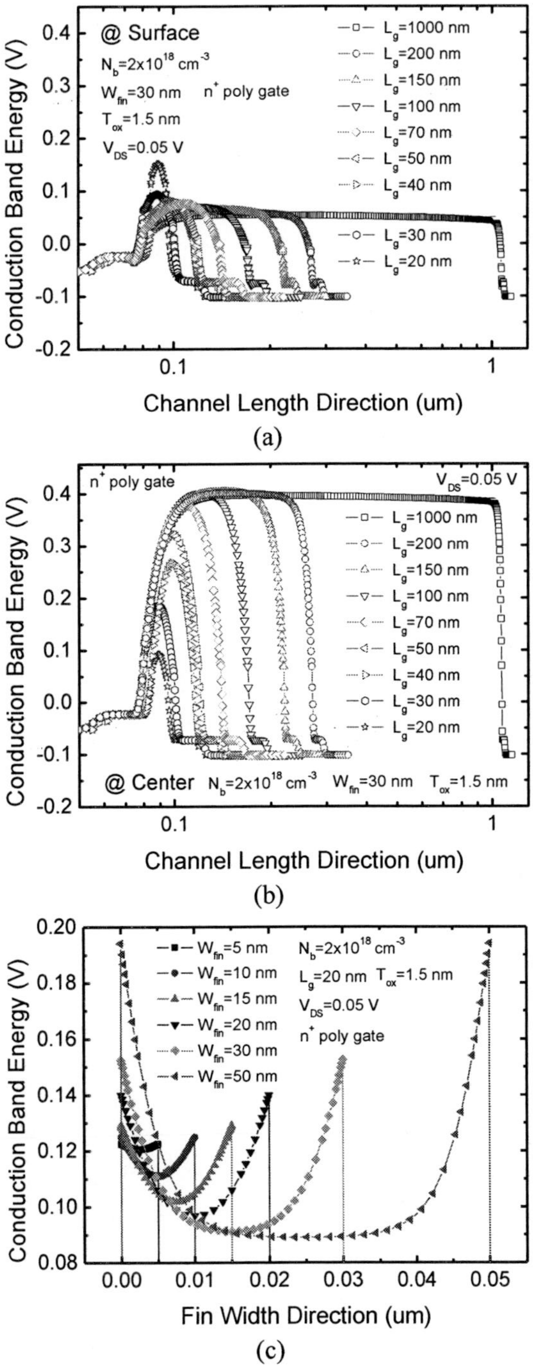

Figure 38. Conduction band energy along the channel length direction as a parameter of L_g at the surface (a) and the center (b) of fin body. (c) Conduction band energy along the fin width direction as a parameter of W_{fin}. Figure (c) shows conduction band energy diagrams cut along the channel width direction at a point showing maximum-conduction-band-energy over the channel length at given V_{DS}=0.05 V and $V_{GS}=V_{th}$. Here, N_b is 2×10^{18} cm^{-3}.

this case, the MCBE in the surface is higher than that in the center, which means that bulk punchthrough occurs at an L_g of 20 nm. The bulk punchthrough causes a large drop of V_{th} and degrades significantly gate controllability to the channel. Fig. 38 (c) shows bulk punchthrough at $V_{GS}=V_{th}$ as a parameter of W_{fin} for given $N_b=2\times10^{18}$ cm^{-3} and L_g=20 nm. The conduction band energy in the center is lower than that in the surface, which is more significant in wide fin body.

Fig. 39 shows δ_W versus W_{fin} of long channel (1 μm) DG devices as a parameter of N_b at a given V_{DS}=0.05 V. Here, L_g of 1 μm is used to exclude SCE. The decrease of W_{fin} causes the lowering of V_{th} in DG devices with fully depleted fin body. The decrease of W_{fin} lowers the surface potential at V_{th}, resulting in the decrease of the band-bending from $2\psi_B$. So δ_W is introduced to correct $2\psi_B$ with W_{fin}, and defined as the surface potential difference obtained by subtracting the potential of a fully depleted fin body from that of a fin body with the neutral region at $V_{GS}=V_{th}$. Here, δ_W is a sort of physical parameter to consider the surface potential lowering with decreasing W_{fin}. The slope of δ_W at a given N_b is proportional to increment of N_b.

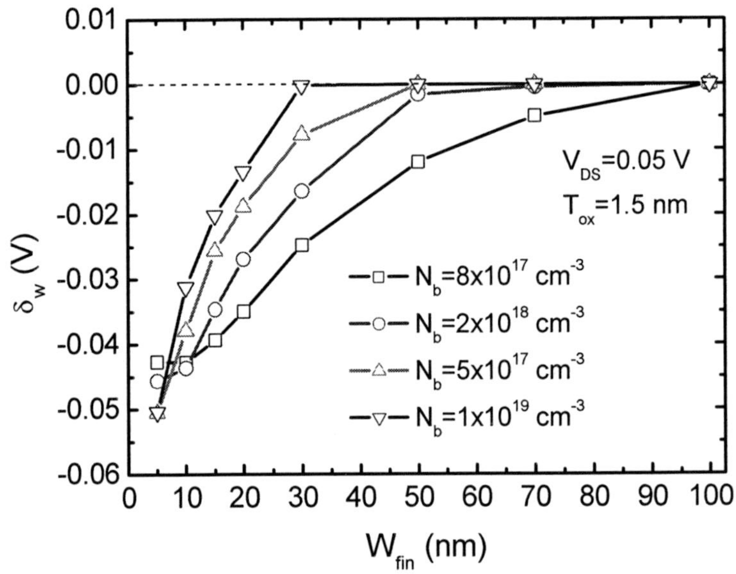

Figure 39. δ_W versus W_{fin} as a parameter of N_b. The δ_W is a sort of physical parameter to consider the surface potential lowering with decreasing W_{fin}, and 0 for non-fully depleted fin body. Here, the L_g of devices are fixed at 1 μm to exclude SCE.

Fig. 40 shows δ_L versus L_g of DG devices at a given V_{DS}=0.05 V. δ_L is introduced to correct $2\psi_B$ with L_g, and defined as the surface potential difference obtained by subtracting the potential at a given L_g from that of device with an L_g of 1 μm at $V_{GS}=V_{th}$ condition. It is interesting δ_L has very small dependence (see the distribution of δ_L at a given L_g in Fig. 40) on W_{fin} and N_b variation if the fin body is fully depleted. Therefore, we can get easily δ_L depending on only L_g by averaging the surface potential change with various N_b and W_{fin} at a given L_g. δ_L is also a sort of physical parameter to consider the surface potential lowering with decreasing L_g.

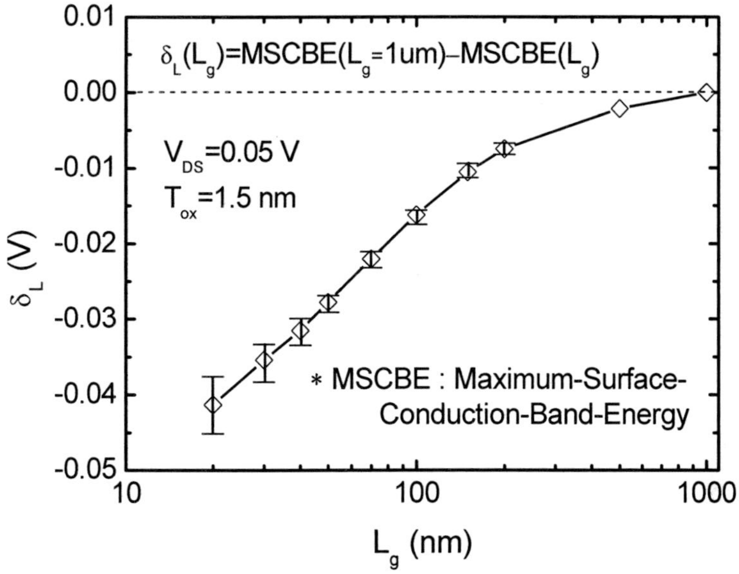

Figure 40. δ_L versus L_g at given V_{DS}=0.05 V and T_{ox}=1.5 nm. Here, δ_L is also a sort of physical parameter to consider the surface potential lowering with decreasing L_g, and in proportion to L_g.

Table 2 shows the gate length that bulk punchthrough happens at given N_b and W_{fin} in DG devices. When bulk punchthrough occurs, the surface potential of DG device decreases and the center potential increases rapidly, which leads to a large drop of V_{th} in DG devices. Under the punchthrough, our V_{th} model is not correct.

Table 2. The gate length that bulk punchthrough happens at given N_b and W_{fin} in DG devices.

W_{fin} / N_b	5 nm	10 nm	15 nm	20 nm	30 nm	50 nm
8×10^{17} cm^{-3}	< 40 nm	< 40 nm	< 40 nm	< 40 nm	< 40 nm	< 50 nm
2×10^{18} cm^{-3}	< 30 nm	< 30 nm	< 30 nm	< 30 nm	< 30 nm	< 40 nm
5×10^{18} cm^{-3}	< 20 nm	< 20 nm	< 20 nm	< 20 nm	< 30 nm	
1×10^{19} cm^{-3}	< 20 nm	< 20 nm	< 20 nm	< 20 nm	Non-fully depleted	

Table 3 shows the surface potential difference (δ_L) when bulk punchthrough happens at given N_b and W_{fin}. Actually, such surface potential difference greatly deviates from the tendency of δ_L shown in Fig. 40.

Table 3. The surface potential difference (δ_L) under bulk punchthrough.

N_b / W_{fin}	8×10^{17} cm^{-3} L_g (nm)	δ_L (V)	2×10^{18} cm^{-3} L_g (nm)	δ_L (V)	5×10^{18} cm^{-3} L_g (nm)	δ_L (V)	1×10^{19} cm^{-3}
5 nm	30	-0.0385	20	-0.0439	δ_L shown in Fig. 40 is applied to L_gs from 20 nm to 1μm		
	20	-0.0407					
10 nm	30	-0.0353	20	-0.0485			
	20	-0.0449					
15 nm	30	-0.0356	20	-0.0617			
	20	-0.0576					
20 nm	30	-0.0408	20	-0.0803			
	20	-0.0863					
30 nm	30	-0.0586	20	-0.1031	20	-0.0529	
	20	-0.1305					
50 nm	40	-0.0437	30	-0.0489	Non-fully depleted		
	30	-0.0771					
	20	-0.2285	20	-0.1594			

Fig. 41 shows δ_B versus N_b of DG devices at given V_{DS}=0.05 V. To improve V_{th} accuracy for small geometry devices, δ_W and δ_L were introduced to consider the surface potential lowering in the V_{th} modeling, but a constant error between the modeled and the simulated data was found nevertheless. The error lineally depends on only N_b regardless of W_{fin} and L_g. So we took into account another fitting parameter δ_B in the V_{th} modeling to correct the error. We can roughly guess δ_B for various N_bs shown in Fig. 41.

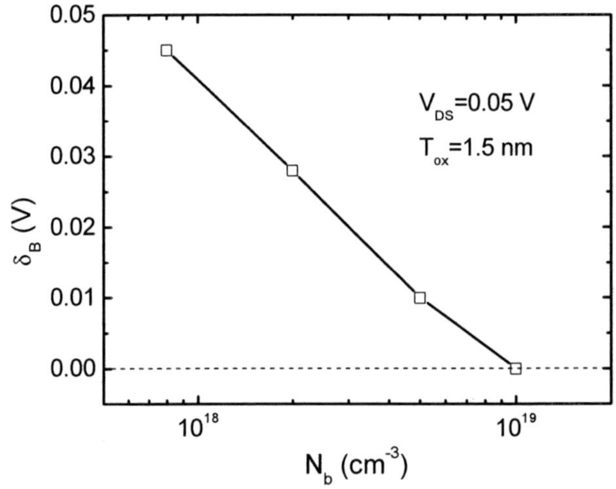

Figure 41. δ_B versus N_b. The δ_B is a fitting parameter to correct a constant difference between the model and the simulation data, and changes lineally with only N_b regardless of W_{fin} and L_g.

According to the above explanation, the surface potential in (2) is corrected as follows:

$$2\psi_B' = 2\psi_B + \delta_W + \delta_L + \delta_B . \tag{20}$$

Finally, the $V_{th,s}$ model of ideal bulk FinFETs is given by

$$V_{th,s} = V_{FB} + 2\psi_B' + \frac{qN_b x_{dep}}{C_{ox}}\left(1 - \frac{x_h}{L_c} + \frac{\pi x_{df}}{4H_g}\right). \tag{21}$$

In (21), the second and third term in the parenthesis reflect SCE and NWE, respectively.

Fig. 42 (a) shows the relation between a fitting parameter $x_h(=(x_{hs}+x_{hd})/2)$ corrected by considering (4) for the given N_b and W_{fin} and $x_c(=(x_{cs}+x_{cd})/2)$. Here, x_c is channel side-depletion width. The dashed line represents the line for $x_h=x_c$. In Fig. 42 (a), when we draw a line for the fitting parameter x_h values for various N_bs at a fixed W_{fin}, the line shape for each fin body width looks like a parabola. As W_{fin} is increased, the slope of the parabola is decreased. Since the x_h is a charge-sharing length under $V_{GS}=V_{th}$ condition, it is physically reasonable that the x_h is less than or equal to x_c with various W_{fin} and N_b. Anyway, we can roughly guess x_h from Fig. 42 (a). Fig. 42 (b) shows x_h versus W_{fin} as a function of N_b. Here, x_h(old) (open symbol) decreases with decreasing W_{fin} and then increases around W_{fin} of 5 nm because the effect from the charge-sharing and the surface potential lowering is reflected in only one fitting parameter (x_h(old)). The x_h decreases physically with decreasing W_{fin} since decreasing W_{fin} suppresses effectively the charge-sharing from the S/D. x_h(new) (solid symbol) decreases with decreasing W_{fin} regardless of N_b by separating the surface potential lowering from x_h(old) in this work.

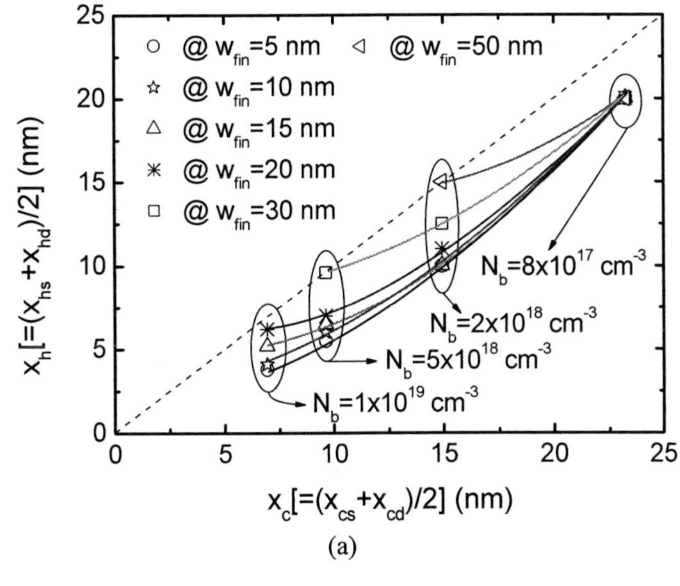

(a)

Figure 42. Continued on next page.

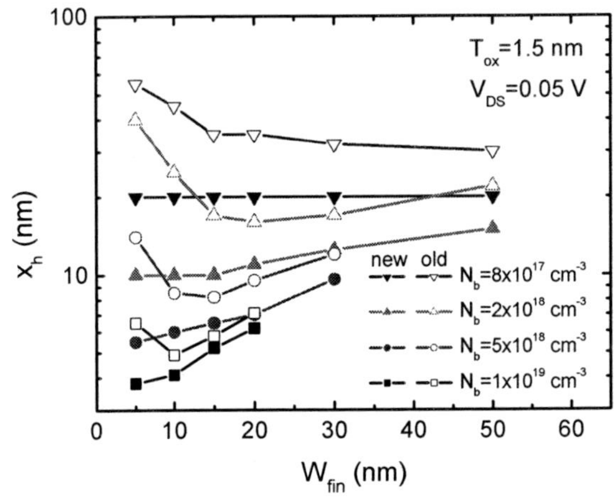

(b)

Figure 42. (a) Relation between x_c and charge-sharing length x_h after $2\psi_B$ correction. The dashed line represents the points of x_h=x_c. (b) x_h versus W_{fin} as a parameter of N_b. Here, new x_hs represented by solid symbols were newly extracted by considering the surface potential lowering.

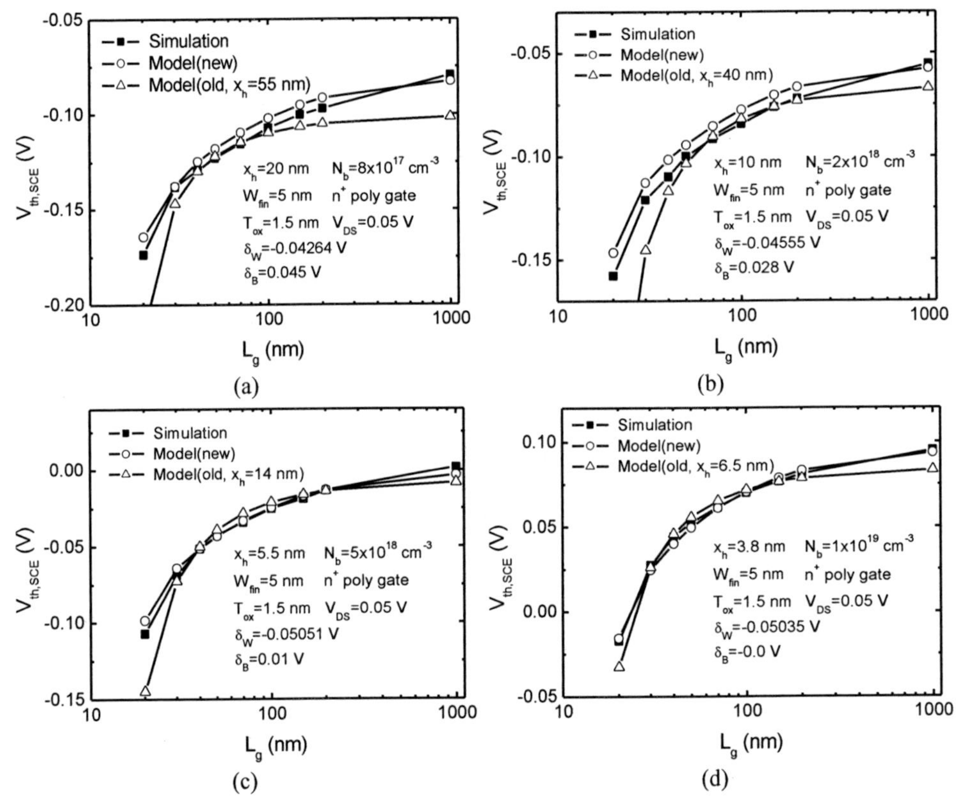

Figure 43. $V_{th,SCE}$ versus L_g as a function of N_b for DG devices with a W_{fin} of 5 nm. Here, corner effect and NWE are not included. Model(new) (open triangle symbols) considering the surface potential lowering shows a good agreement with the simulation data with various N_bs for given a W_{fin} of 5 nm.

Fig. 43 shows $V_{th,SCE}$ versus L_g as a function of N_b for DG devices with a W_{fin} of 5 nm. Here, $V_{th,SCE}$ does not include NWE term in (21) because we consider only 2-D DG devices. Quantum effect is not considered because the quantum effect influences significantly when W_{fin} is narrower than 5 nm [60]. In this figure, the model(old) represents our old V_{th} model with x_h(old). The model(new) including the correction of the surface potential lowering shows a good agreement with the simulation regardless of N_b for a given W_{fin} of 5 nm.

Fig. 44 shows $V_{th,SCE}$ versus L_g as a function of N_b for DG devices with a W_{fin} of 10 nm. The model(old) has some error at L_g of 20~30 nm for N_b of 8×10^{17} cm^{-3} and 2×10^{18} cm^{-3}. But the model(new) shows a good agreement with the simulation regardless of N_b for a given W_{fin} of 10 nm. The comparison between simulation and model(old and new) for W_{fin} of 15~50 nm as a function of N_b is not shown, but both old and new models show a good agreement with the simulation regardless of N_b, L_g, and W_{fin}. Because the surface potential lowering at $V_{GS}=V_{th}$ condition becomes small with the increase of W_{fin} for various N_bs as shown in Fig. 39, the results from both models coincide well with simulation data. Since the results of 2-D DG devices are shown in Figs. 43 and 44, corner effect and NWE are not considered in $V_{th,SCE}$ model.

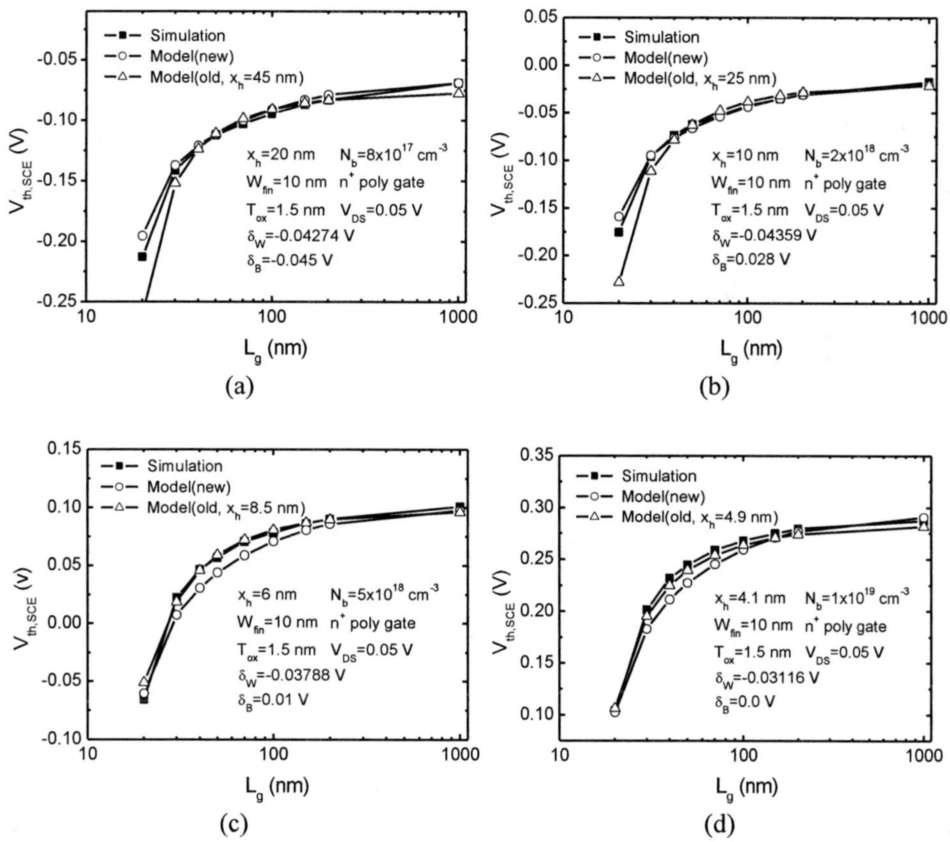

Figure 44. $V_{th,SCE}$ versus L_g as a function of N_b for DG device with a W_{fin} of 10 nm. Here, corner effect and NWE are not included. Model(new) (open triangle symbols) considering the surface potential lowering shows a good agreement with the simulation data with various N_bs for given a W_{fin} of 10 nm.

We now discuss the V_{th} of bulk FinFETs with top corner. Bulk FinFET with the corner as shown in Fig. 22 (a) has the side-channel and corner-channel components. The $V_{th,c}$ (corner-channel V_{th}) model have been derived based on the SCE and the shape of corner region without considering surface potential lowering [52]. The SCE of the corner-channel was derived to be 2/3 of the SCE in the side-channel regardless of the corner shape by considering 3-D charge-sharing, because the electric field from the top-and side-gate of the corner region is stronger than that from the side-gate only in the fin body excluding the corner region. To consider the electric field focusing in the corner region, we introduced a corner factor α_c to take into account the effect in the $V_{th,c}$ model. Before the correction of $2\psi_B$, α_c was 0.25 [52] and had no dependence on the corner shape, body geometry, and N_b. After the correction, α_c is extracted to 0.4 regardless of the corner shape. In the corner-channel $V_{th,c}$ modeling of bulk FinFETs with corner, the surface potential lowering at V_{th} also has to be considered as in the side-channel $V_{th,s}$ modeling of ideal bulk FinFET. Finally we express $V_{th,c}$ using $2\psi'_B$ and corrected α_c as follows:

$$V_{th,c} = V_{FB} + 2\psi'_B + \frac{q\alpha_c N_b x_{dep}}{C_{ox}}\left(1 - \frac{2}{3}\frac{x_h}{L_c}\right). \tag{22}$$

We now calculate the V_{th} of a bulk FinFET with corner. There are two different channels (W_c and W_s) for the half-circle corner. W_c is the channel width corresponding to 0.5 of the half-circle corner of the fin body. Since the channels are in parallel along the channel length direction, the V_{th} of the device is calculated by using (17) and (18).

As shown in (17) and (18), the combined V_{th} of the bulk FinFET with corner is calculated as the perimeter-weighted sum [58] of the $V_{th,s}$ and $V_{th,c}$ models. The V_{th} calculation routine was coded by using MATLAB.

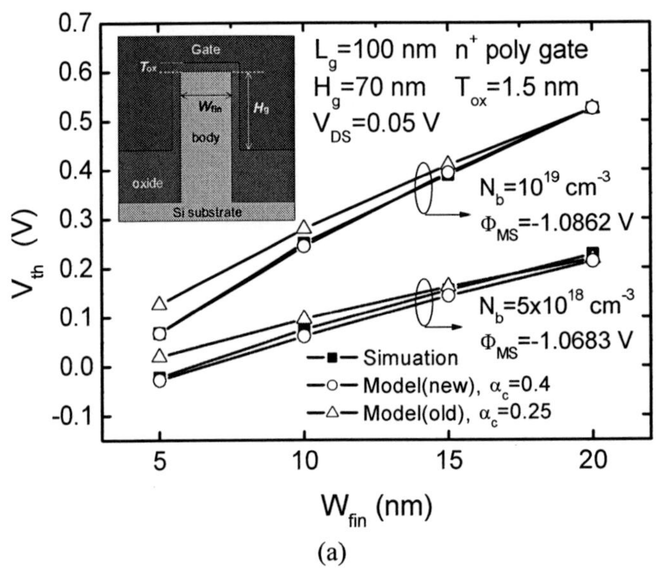

(a)

Figure 45. Continued on next page.

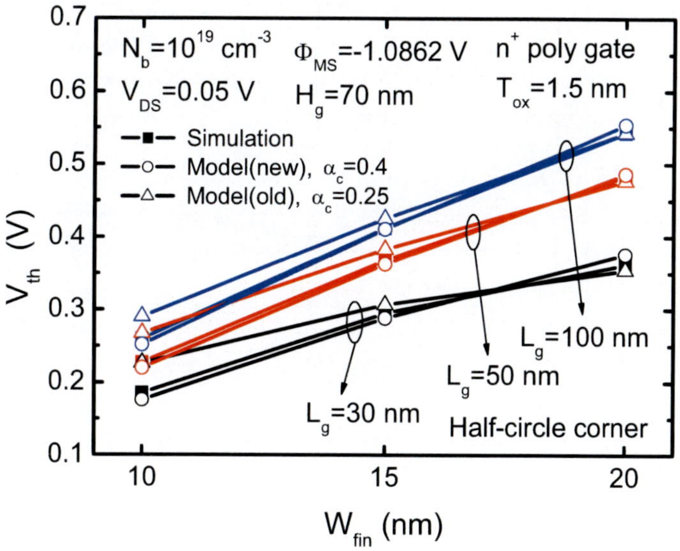

(b)

Figure 45. V_{th} versus W_{fin} for bulk FinFETs with 90° corner (a) and half-circle corner (b). Parameters in figures (a) and (b) are body doping N_b and gate length L_g, respectively. Corner factor, α_cs are 0.4 for model(new) and 0.25 for model(old). Model(new) shows a good agreement with 3-D simulation data regardless of L_g and N_b.

Fig. 45 (a) shows V_{th} versus W_{fin} as a parameter of N_b for bulk FinFET with 90° corner. Open circle and triangle symbols represent V_{th} data obtained by new and old models, respectively, and these data are compared with simulation data. Corner factor α_cs for new and old models are 0.4 and 0.25, respectively. The model(old) shows an increasing error with decreasing W_{fin} for given N_bs. But the model(new) considering the surface potential lowering and corrected α_c shows a good agreement with 3-D simulation. Fig. 45 (b) shows V_{th} versus W_{fin} as a parameter of L_g for bulk FinFET with half-circle corner. As the W_{fin} decreases, the model(old) represented by open triangle symbols shows an increasing deviation from the simulation results, and the deviation is more significant for shorter L_g. But the model(new) represented by open circle symbols explains well V_{th} behavior obtained by 3-D device simulation.

Fig. 46 shows V_{th} versus H_g for bulk FinFET with 90° corner. Channel doping is uniform and 1×10^{19} cm^{-3}. Data of old and new V_{th} models are represented by open triangle and circle symbols, respectively, and compared with simulation data (square symbols). Both old and new models show a good agreement with the simulation at a given W_{fin} of 20 nm, but the model(old) shows a deviation from the simulated value by about 10 % at an H_g of 20 nm. As the H_g decreases, device V_{th} decreases since the contribution of corner-channel V_{th} ($V_{th,c}$) which is lower than that of side-channel V_{th} ($V_{th,s}$) becomes large. In other words, $(W_s/W)V_{th,s}$ term in (17) becomes small with decreasing H_g. Deducing from the data in Fig. 45, it seems that the model(old) shows an increasing error with decreasing W_{fin} in the range less than 20 nm.

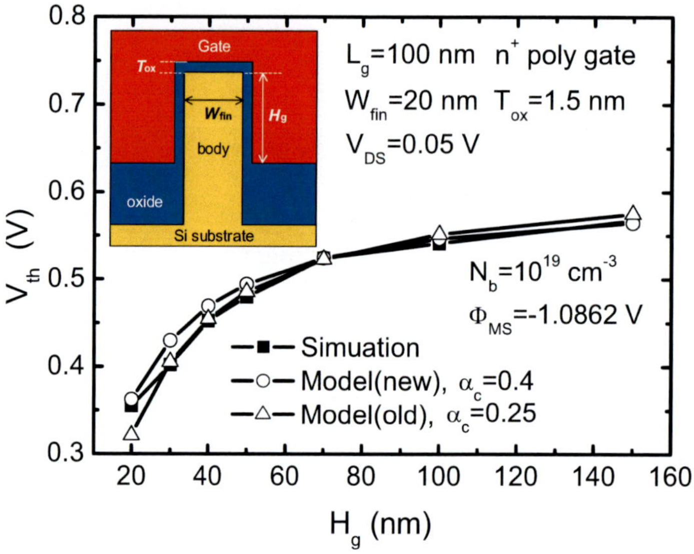

Figure 46. V_{th} versus H_g for bulk FinFET with 90° corner. Channel doping is uniform and 1×10^{19} cm^{-3}. Both new (open circle symbols) and old (open triangle symbols) show a good agreement with the simulation data for a given W_{fin} of 20 nm. Model(new) give better accuracy than model(old) at an H_g of 20 nm.

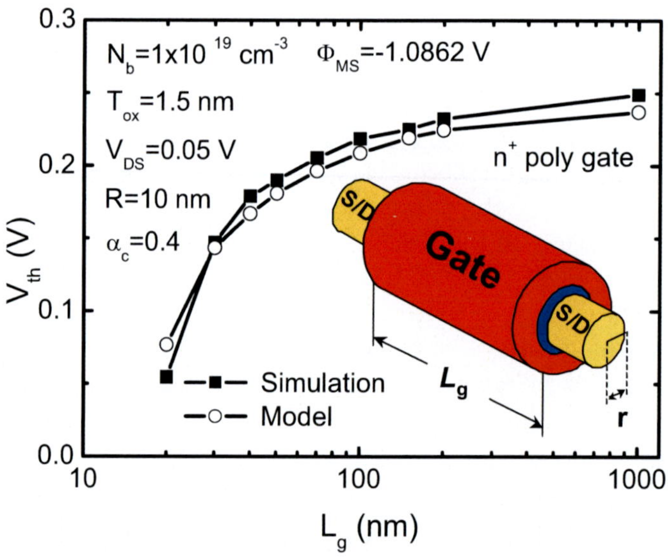

Figure 47. V_{th} versus L_g for wire-MOSFET. Here, r is the radius of Si wire and 10 nm. Here, α_c is 0.4. The model shows a good agreement with 3-D simulation data.

Our corner V_{th} model ($V_{th,c}$) can be easily extended to the V_{th} modeling of wire transistors. Fig. 47 shows V_{th} versus L_g of wire-MOSFET. The insert shows 3-D schematic view of wire-MOSFET. The V_{th} model of wire-MOSFET is given by (22), and verified by comparing the modeled with simulated data. In Fig. 47, the wire-MOSFET has an r (radius of wire) of 10 nm, a T_{ox} of 1.5 nm, an N_b of 1×10^{19} cm^{-3}, and n$^+$ poly gate. Here, corner factor α_c is 0.4. The modeled V_{th} data shows a good agreement with 3-D simulation data up to an L_g of 20 nm.

References

[1] J. E. Lilienfeld, "Method and apparatus for controlling electric currents," U.S. Patent 1,745,175 (filed in 1926, issued in 1930), 1930.

[2] D. Kahng and M. M. Atalla, "Silicon-silicon dioxide field induced surface devices," presented at the IRE Solid-State Device Res. Conf., Pittsburgh, PA, June 1960.

[3] B.E. Deal, U.S. Patent no. 3,426,422, "Method of Making Stable Semiconductor Devices," Feb. 11, 1969.

[4] Semiconductor Industry Association (SIA), International Technology Roadmap for Semiconductors, 2006 ed. 2006.

[5] J. Kedzierski, P. Xuan, V. Subramanian, E. Anderson, J. Bokor, T.-J. King, and C. Hu, "A 20 nm gate-length ultra-thin body p-MOSFET with silicide source/drain," *Superlattices and Microstructures*, vol. 28, no. 5, pp. 445-452, Nov. 2000.

[6] T. Low, F. F. Li, C. Shen, Y.-C. Yeo, Y. T. Hou, C. Zhu, A. Chin, and D. L. Kwong, "Electron mobility in Ge and strained-Si channel ultrathin-body metal-oxide semiconductor field-effect transistors," *Appl. Phys. Lett., vol.* 85, no. 12, pp. 2402-2404, Sep. 2004.

[7] Y.-K. Choi, N. Lindert, P. Xuan, S. Tang, D. Ha, E. Anderson, T.-J. King, J. Bokor, C. Hu, "Sub-20 nm CMOS FinFET technologies," in *IEDM Technical Digest*, pp. 421-424, 2001.

[8] Tai-su Park, Euijoon Yoon, and Jong-Ho Lee, "A 40 nm body-tied FinFET (OMEGA MOSFET) using bulk Si wafer", *IEE Physica E*, vol. 19, no. 1, pp. 6-12, 2003..

[9] B. Doyle , B. Boyanov, S. Datta, M. Doczy, S. Hareland, B. Jin, J. Kavalieros, T. Linton, R. Rios, and R. Chau, "Tri-gate fully-depleted CMOS transistors: fabrication, design, and layout," in *Tech. Dig. of Symposium on VLSI Tech.*, 2003, pp. 133-134.

[10] C. H. Wann, K. Noda, T. Tanaka, M. Yoshida, and C. Hu, "A comparative study of advanced MOSFET concepts," *IEEE Trans. Electron Devices*, vol. 43, pp. 1742-1753, Oct. 1996.

[11] H.-S. P. Wong, K. K. Chan, and Y. Taur, "Self-aligned (top and bottom) double-gate MOSFET with a 25 nm thick silicon channel," in *IEDM Tech. Dig.,* 1997, pp. 427-430.

[12] J. P. Colinge, M. H. Gao, A. Romano-Rodriguez, H. Maes, and C. Claeys, "Silicon on insulator 'gate-all-around device'," in *IEDM Tech. Dig*, 1990, pp. 595-598.

[13] Effendi Leobandung, Jian Gu, Lingjie Guo, and Stephen Y. Chou, "Wire-channel and wrap-around-gate metal-oxide-semiconductor field-effect transistors with a significant reduction of short channel effects", *J. Vac. Sci. Technol. B*, vol. 15, no. 6, pp. 2791-2794, 1997.

[14] Christopher P. Auth and James D. Plummer, "Scaling theory for cylindrical, fully-depleted, surrounding-gate MOSFET's", *IEEE Electron Device Letters*, vol. 18, no. 2, pp. 74-77, 1997.

[15] J. Kedzierski, E. Nowak, T. Kanarsky, Y. Zhang, D. Boyd, R. Carruthers, C. Cabral, R. Amos, C. Lavoie, R. Roy, J. Newbury, E. Sullivan, J. Benedict, P. Saunders, K. Wong, D. Canaperi, M. Krishnan, K.-L. Lee, B. A. Rainey, D. Fried, P. Cottrell, H.-S. Philip Wong, M. Ieong, and W. Haensch, "Metal-gate FinFET and fully-depleted SOI devices using total gate silicidation", in *IEDM Tech. Dig.*, 2002, pp. 247-250.

[16] Fu-Liang Yang, Hao-Yu Chen, Fang-Cheng Chen, Cheng-Chuan Huang, Chang-Yun Chang, Hsien-Kuang Chiu, Chi-Chuang Lee, Chi-Chun Chen, Huan-Tsung Huang, Chih-Jian Chen, Hun-Jan Tao, Yee-Chia Yeo, Mong-Song Liang, and Chenming Hu, "25 nm CMOS omega FETs", in *IEDM Tech. Dig.*, 2002, pp. 255-258.

[17] J. Kedzierski, D. M. Fried, E. J. Nowak, T. Kanarsky, J. H. Rankin, H. Hanafi, W. Natzle, D. Boyd, Y. Zhang, R. A. Roy, J. Newbury, C. Yu, Q. Yang, P. Saunders, C. P. Witlets, A. Johnson, S. P. Cole, H. E. young, N. Carpenter, D. Rakowski, B. A. Rainey, P. E. Cotrell, M. Ieong, and H.-H. P. Wong, "High-performance symmetric-gate and CMOS-compatible Vt asymmetric-gate FinFET devices," in *IEDM Tech. Dig.*, 2001, pp. 437-440.

[18] W. Rosner, E. Landgraf, J. Kretz, L. Dreeskornfeld, H. Schafer, M. Stalele, T. Schulz, F. Hofmann, R. J. Luyken, M. Specht, J. Hartwich, W. Pamler and L. Risch, "Nanoscale FinFETs for low power applications," *Solid-State Electronics*, vol. 48, no.10-11, pp. 1819-1823, 2004.

[19] H. Shang, L. Chang, X. Wang, M. Rooks, Y. Zhang, B. To, K. Babich, G. Totir, Y. Sun, E. Kiewra, M. Ieong, W. Haensch, "Investigation of FinFET devices for 32 nm technologies and beyond," in *Symp. on VLSI Technology Dig.*, 2006, pp. 66-67.

[20] K. von Arnim, E. Augendre, C. Pacha, T. Schulz, K. T. San, F. Bauer, A. Nackaerts, R. Rooyackers, T. Vandeweyer, B. Degroote, N. Collaert, A. Dixit, R. Singanamalla, W. Xiong, A. Marshall, C. R. Cleavelin, K. Schrufer, and M. Jurczak, "A low-power multi-gate FET CMOS technology with 13.9 ps inverter delay, large-scale integrated high performance digital circuits and SRAM," in *Symp. on VLSI Technology Dig.*, 2007, pp. 106-107.

[21] K. Okano, T. Izumida, H. Kawasaki, A. Kaneko, A. Yagishita, T. Kanemura, M. Kondo, S. Ito, N. Aoki, K. Miyano, T. Ono, K. Yahashi, K. Iwade, T. Kubota, T. Matsushita, I. Mizushima, S. Inaba, K. Ishimaru, K. Suguro, K. Eguchi, Y. Tsunashima, and H. Ishiuchi, "Process integration technology and device characteristics of CMOS FinFET on bulk silicon substrate with sub-10 nm fin width and 20 nm gate length," in *IEDM Tech. Dig.*, 2005, pp. 721-724.

[22] P. Wambacq, B. Verbruggen, K. Scheir, J. Borremans, V. De Heyn, G. Van der Plas, A. Mercha, B. Parvais, V. Subramanian, M. Jurczak, S. Decoutere, S. Donnay, "Analog and RF circuits in 45 nm CMOS and below: planar bulk versus FinFET," in *Proc. European Solid-State Device Research Conf. Tech. Dig.*, 2006, pp. 53-56.

[23] J.-P. Raskin, T. M. Chung, V. Kilchytska, D. Lederer, D. Flandre, "Analog/RF performance of multiple gate SOI devices: wideband simulations and characterization," *IEEE Trans. on Electron Devices*, vol. 53, no. 5, pp. 1088-1095, 2006.

[24] Z. Guo, S. Balasubramanian, R. Zlatanovici, T.-J. King, B. Nikolic, "FinFET-based SRAM design," in *Proc. International Symp. on Low Power Electronics and Design Tech. Dig.*, 2005, pp. 2-7.

[25] H. Kawauka, K. Okano, A. Kaneko, A. Yagishita, T. Izumida, T. Kanemura, K. Kasai, T. Ishida, T. Sasaki, Y. Takeyama, N. Aoki, N. Ohtsuka, K. Suguro, K. Eguchi, Y. Tsunashima, S. Inaba, K. Ishimaru, and H. Ishiuchi, "Embedded bulk FinFET SRAM cell technology with planar FET peripheral circuit for hp32 nm node and beyond," in *Symp. on VLSI Technology Dig.*, 2006, pp. 86-87.

[26] T. Park, H. J. Cho, J. D. Chae, S. Y. Han, D. Park, K. Kim, E. Yoon, and J. H. Lee, "Characteristics of the full CMOS SRAM cell using body-tied TG MOSFETs (bulk FinFETs)," *IEEE Trans. on Electron Devices*, vol. 53, no. 3, pp. 481-487, March 2006.

[27] I. H. Cho, T.-S. Park, S. Y. Choi, J. D. Lee, J.-H. Lee, "Body-tied doube-gate SONOS flash (omega flash) memory device built on bulk Si wafer," in *Proc. Device Research Conf. Tech. Dig.*, 2003, pp. 133-134.

[28] T.-H. Hsu, H.-T. Lue, E.-K. Lai, J.-Y. Hsieh, Z.-Y. Wang, L.-W. Yang, Y.-C. King, T. Yang, K.-C. Chen, K.-Y. Hsieh, R. Liu, and C.-Y. Lu, "A high-speed BE-SONOS NAND flash utilizing the field-enhancement effect of FinFET," in *IEDM Tech. Dig.*, 2007, pp. 913-916.

[29] C. Gerardi, S. Lombardo, G. Cina, E. Tripiciano, D. Corso, V. Ancarani, G. Iacono, C. Bongiorno, C. Garozzo, P. Barbe G. A. Costa, C. Coccorese, M. Vecchio, E. Rimini and M. Melanotte, "Highly manufacturable/low aspect ratio Si nano floating gate FinFET memories: high speed performance and improved reliability," in *Proc. Non-Volatile Semiconductor Memory Workshop Tech. Dig.*, 2007, pp. 44-45.

[30] J.-R. Hwang, T.-L. Lee, H.-C. Ma, T.-C. Lee, T.-H. Chung, C.-Y. Chang, S.-D. Liu, B.-C. Perng, J.-W. hsu, M.-Y. Lee, C.-Y. Ting, C.-C. Huang, J.-H. Wang, J.-H. Shieh, and F.-L. Yang, "20 nm gate bulk-FinFET SONOS flash," in *IEDM Tech. Dig.*, 2005, pp. 154-157.

[31] E. S. Cho, T.-Y. Kim, B. K. Cho, C.-H. Lee, J. J. Lee, A. Fayrushin, C. Lee, D. Park, and B.-I. Ryu, "Technology breakthrough of body-tied FinFET for sub 50 nm NOR flash memory," in *Symp. on VLSI Tech. Dig.*, 2006, pp. 110-111.

[32] C. H. Lee, J. M. Yoon, C. Lee, H. M. Yang, K. N. Kim, T. Y. Kim, H. S. Kang, Y. J. Ahn, D. Park, and K. Kim, "Novel body tied FinFET cell array transistor DRAM with negative word line operation for sub 60 nm technology and beyond," in *Symp. on VLSI Tech. Dig.*, 2004, 130-131.

[33] C. Lee, J.-M. Yoon, C.-H. Lee, J. C. Park, T. Y. Kim, H. S. Kang, S. K. Sung, E. S. Cho, H. J. Cho, Y. J. Ahn, D. Park, K. Kim, and B.-I. Ryu, "Enhanced data retention of damascene-finFET DRAM with local channel implantation and <100> fin surface orientation engineering," in *IEDM Tech. Dig.*, 2004, pp. 61-64.

[34] Y.-S. Kim, S.-H. Lee, S.-H. Shin, S.-H. Han, J.-Y. Lee, J.-W. Lee, J. Han, S.-C. Yang, J.-H. Sung, E.-C. Lee, B.-Y. Song, D.-J. Chung, K. Kim, and W. Lee, "Local-damascene-FinFET DRAM integration with p+ doped poly-silicon gate technology for sub-60 nm device generations," in *IEDM Tech. Dig.*, 2005, pp. 315-318.

[35] D.-H. Lee, S.-G. Lee, J. R. Yoo, G.-H. Buh, G. H. Yon, D.-W. Shin, D. K. Lee, H.-S. Byun, I. S. Jung, T.-S. Park, Y. G. Shin, S. Choi, U.-I. Chung, J.-T. Moon, and B.-I. Ryu, "Improved cell performance for sub-50 nm DRAM with manufacturable bulk FinFET structure," in *Symp. on VLSI Tech. Dig.*, 2007, pp. 164-165.

[36] Jean-Pierre Colinge, Silicon-on-Insulator Technology : Materials to VLSI, Kluwer Academic Publishers, Dordrecht, 1991.

[37] J.-H. Lee, T.-S. Park, E. Yoon, and J. J. Park, "Simulation study of a new body-tied FinFETs (Omega MSOFETs) using bulk Si wafers," in *Proc. Si Nanoelectronics Tech. Dig.*, 2003, pp. 102-103.

[38] SILVACO International, ATLAS.

[39] K.-H. Park and J.-H. Lee, "Gate workfunction engineering in bulk FinFETs for sub-50 nm DRAM cell transistors," *IEEE Electron Device Letters*, vol. 28, no. 2, pp. 148-150, 2007.

[40] S.-M. Jung, H. Kwon, J. Jeong, W. Cho, S. Kim, H. Lim, K. Koh, Y. Rah, J. Park, H. Kang, G. Lyu, J. Park, C. Chang, Y. Jang, D. Park, K. Kim, M. Y. Lee, "A novel 0.79 μm^2 SRAM cell by KrF lithography and high performance 90 nm CMOS technology for ultra high speed SRAM", in *IEDM Tech. Dig.*, 2002, pp. 419-422.

[41] T.-S. Park, S. Choi, B.-H. Lee, U.-I. Chung, J. T. Moon, E. Yoon, and J.-H. Lee, "Body-tied triple-gate NMOSFET fabrication using bulk Si wafer," *Solid-State Electronics*, vol. 49, pp. 377-383, 2005.

[42] T.-S. Park, H. J. Cho, J. D. Choe, I. H. Cho, D. Park, E. Yoon, and J.-H. Lee, "Characteristics of body-tied triple-gate pMOSFETs," *IEEE Electron Device Letters*, vol. 25, no. 12, pp. 798-800, Dec. 2004.

[43] T. Park, D. Park, J. H. Chung, E. J. Yoon, S. M. Kim, H. J. Cho, J. D. Choe, J. H. Choi, B. M. Yoon, J. J. Han, B. H. Kim, S. Choi, K. Kim, E. Yoon, and J. H. Lee, "PMOS body-tied FinFET (Omega MOSFET) characteristics", *IEEE Device Research Conference (DRC)*, Salt Lake City, Utah, pp. 33-34, June 2003.

[44] R. Katsumata, N. Tsuda, J. Idebuchi, M. Kondo, N. Aoki, S. Ito, K. Yahashi, T. Satonaka, M. Morikado, M. Kim, M. Kido, T. Tanaka, H. Aochi, and T. Hamannoto, "Fin-Array-Fet on Bulk Silicon for sub-100 nm Fin Trench Capacitor DRAM", *JSAP and IEEE Symposium on VLSI Technology*, Kyoto, Japan, Technical Digest, pp. 61 ~ 62, June 2003.

[45] J. Kedzierski, D. Boyd, Y. Zhang, Michelle Steen, Fen F. Jamin, John Benedict, Meikei Ieong, Wilfred Haensch, "Issues in NiSi-gated FDSOI device integration", *IEEE International Electron Devices Meeting (IEDM), Technical Digest*, pp. 441 ~ 444, December 2003.

[46] B. Yu, L. Chang, S. Ahmed, H. Wang, S. Bell, C.-Y. Yang, C. Tabery, C. Ho, Q. Xiang, T.-J. King, J. Bokor, C. Hu, M.-R. Lin, and D. Kyser, "FinFET scaling to 10 nm gate length," in *IEDM Tech. Dig.*, Dec. 2002, pp. 251–254.

[47] S.-Y. Kim and Jong-Ho Lee, "Hot carrier-induced degradation in bulk FinFETs," *IEEE Electron Device Letters*, vol. 26, no. 8, pp. 566-568, August 2005.

[48] S.-Y. Kim, K.-R. Han, B.-K. Choi, S.-H. Kong, J.-S. Lee, J.-H. Lee, "Negative bias temperature instability of bulk fin field effect transistor," *Japanese J. of Applied Physics*, vo. 45, no. 3A, pp. 1467-1470, 2006.

[49] S.-Y. Kim, Y. M. Kim, K.-H. Baek, K.-H. Park, K.-R. Han, B.-K. Choi, and J.-H. Lee, "Impact ionization behavior in bulk fin field effect transistors with fin body width and bias conditions," *Japanese J. of Applied Physics, vo.* 46, no. 6A, pp. 3347-3350, 2007.

[50] S.-Y. Kim, Y. M. Kim, K.-H. Baek, B.-K. Choi, K.-R. Han, K.-H. Park, and J.-H. Lee, "Temperature dependence of substrate and drain-currents in bulk FinFETs," *IEEE Trans. on Electron Devices*, vol. 54, no. 5, pp. 1259-1264, May 2007.

[51] Q. Chen, E. M. Harrel, and J. D. Meindl, "A physical short-channel threshold voltage model for undoped symmetric double-gate MOSFETs," *IEEE Trans. Electron Devices*, vol. 50, no. 7, pp. 1631–1637, July 2003.

[52] B.-K. Choi, K.-R. Han, Y. M. Kim, Y. J. Park, J.-H. Lee, "Threshold-voltage modeling of body-tied FinFETs (bulk FinFETs)," *IEEE Trans. on Electron Devices*, vol. 45, no. 3, pp. 537-545, March 2007.

[53] L. D. Yau, "A simple theory to predict the threshold voltage of short channel IGFET's," *Solid-State Electronics*, vol. 17, pp. 1059–1063, 1974.

[54] L. A. Akers and J. J. Sanchez, "Threshold voltage models of short, narrow and small geometry MOSFET's: A review," *Solid-State Electronics*, vol. 25, pp. 621–641, 1982.

[55] Richard S. Muller and Theodore I. Kamins, "Device Electronics for Integrated Circuits," *John Wiley & Sons*, Inc., second edition, pp. 443, 1986.

[56] Surya Veeraraghavan and Jerry G. Fossum, "A physical short-channel model for the thin-film SOI MOSFET applicable to device and circuit CAD," *IEEE Tran. Electron Devices*, vol. 35, no. 11, pp. 1866–1875, Nov. 1988.

[57] R. Shrivastava and K. Fitzpatrick, "A simple model for the overlap capacitance of a VLSI MOS device," *IEEE Trans. Electron Devices* ED-29, pp. 1870, 1982.

[58] Christoper P. Auth and James Plummer, "A Simple Model for Threshold Voltage of Surrounding-Gate MOSFET's," *IEEE Trans. Electron Device*, vol. 45, no. 11, pp. 2381-2383, Nov. 1998.

[59] E. Lyumkis, R. Mickevicius, O. Penzin, B. Polsky, K. El Sayed, A. Wettstein, and W. Fichtner, "Simulations of ultrathin, ultrashort double-gated MOSFETs with the density gradient transport model," *SISPAD 2002*, pp. 271–274, Sept. 2002.

[60] Y. Omura, T. Ishiyama, M. Shoji, and K. Izumi, "Quantum mechanical transport characteristics in ultimately miniaturized MOSFETs/SIMOX," in *Proc. Int. Symp. Silicon-on-Insulator Tech.*, 1996, pp. 199–211.

In: MOSFETs: Properties, Preparations and Performance ISBN: 978-1-60456-762-5
Editors: N.T. Andre et al. pp. 355-379

Chapter 10

DISCUSSION ON 1/F NOISE IN CMOS TRANSISTORS: MODELLING-SIMULATION AND MEASUREMENT TECHNIQUES

T. Noulis [a,1], ***S. Siskos*** [a,2], ***L. Bary*** [b,3] ***and G. Sarrabayrouse*** [b,4]

[a] Electronics Laboratory of Physics Department, Aristotle University of Thessaloniki, Aristotle University Campus, 54124 Thessaloniki, Greece.

[b] LAAS – CNRS, University of Toulouse, 7 Avenue du colonel Roche 31077 Toulouse cedex 4, France

Abstract

In this chapter the most important flicker noise models that have been proposed in the literature are analytically presented and discussed. The available flicker noise simulation models in software packages like SPICE and HSPICE and SPECTRE are also examined and selection criteria in relation to the type of MOSFET and the operating regime are proposed. The impact of scaling down technologies is discussed and extra analysis is presented in relation to the type of the bias in CMOS transistors. In particular the switched biasing effect in the MOSFET low frequency noise performance is presented. Additionally, low frequency noise measurement set-ups are presented and measurement techniques are suggested.

1.1. Introduction

CMOS transistors are required as "building blocks" of analog applications circuits. They are preferred for dense, low-voltage and low-power applications. While the advantages of CMOS technology are well established, there is a major drawback for analog applications: MOSFETs are "noisy". On the other hand, the conventional bipolar semiconductor devices such as BJT or HBT with their superior noise characteristics are commonly used for such analog circuit

---

[1] E-mail address: tnoul@physics.auth.gr. (Corresponding author.)
[2] E-mail address: siskos@physics.auth.gr.
[3] E-mail address: bary@laas.fr.
[4] E-mail address: sarra@laas.fr.

applications requiring satisfactory noise performance. However, with the recent progress in noise reduction for MOS transistors (MOS-FETs), analog circuit configuration using MOSFETs is becoming possible in areas traditionally regarded as difficult.

In particular, the noise behaviour of CMOS devices is primarily dominated by two sources, thermal and flicker noise. Other sources that are sometimes present in the noise spectrum are shot noise, generation/recombination noise, and "popcorn" noise. Of these noise sources, thermal and shot noise are physically fundamental to the operation of the device and are always present. Thermal noise is the voltage fluctuations due to the random Brownian motion of electrons in the resistive medium. It is broadband white noise, and it gets larger with increasing resistance and temperature. Shot noise is caused by the non smooth current flowing across a junction and it is comprised of individual electrons arriving at random times. This non-uniform flow gives rise to broadband white noise. Generation/recombination (G/R) noise is due to trapping of carriers in the bulk or at the interfaces of the device. The trapping of carriers causes fluctuation in their number, and thus fluctuation in the resistance. The G/R noise creates a Lorentzian noise spectrum. Popcorn noise, sometimes called burst noise or random-telegraph-signal (RTS) noise is due to a discrete modulation of the channel by the capture and emission of a channel carrier.

Regarding flicker noise phenomenon, it appears through both quality-dependent and fundamental noise processes. It seems to be large in MOSFETs and therefore sets the lower limit of the signal level that can be processed by VLSI devices and circuits. It is also commonly called $1/f$ noise because the noise spectrum varies as $1/f^{\alpha}$, where f is the signal frequency and α, an exponent very close to unity (α=1±0.2) (see Figure 1). In contrast to the channel thermal noise and the shot noise which are quite understood, flicker noise is not yet fully known, although its presence is surprisingly universal in all types of semiconductor devices [1]. A lot of studies have been done so far in order to explain and describe the flicker noise phenomenon [2]-[19]. However, controversy still exists in the modelling of $1/f$ noise in CMOS transistors.

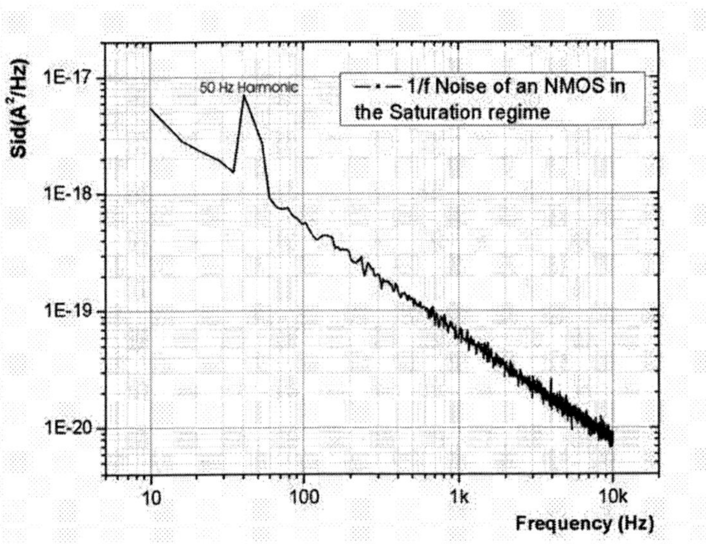

Figure 1. Power Spectral Density of Flicker Noise is inversely proportional to frequency.

In this chapter, the most important 1/*f* noise models proposed in the literature are presented and discussed. The available flicker noise simulation models in software packages like SPICE and HSPICE are examined and selection criteria in relation to the type of MOSFET and the operating regime are also proposed. Extra analysis is presented in relation to the type of the bias in CMOS transistors. In addition, low-frequency noise measurement set-ups are also presented and measurement techniques are suggested.

1.2. Flicker Noise Modelling and Simulation

In integrated circuits design, noise simulation is a key point of the process. To evaluate circuit noise performance, the designer uses one of the noise models available in the simulator. The selection of the noise model is very important in flicker noise evaluation depending upon the type of MOSFET (n or p) and the operating mode. So any information or rule about the model to select is very essential and useful to the designer.

1.2.1. The Empirical and the Typical 1/*f* Noise Models

A. Simple Empirical Model

The most common flicker noise model used in the IC analog designer's community was proposed by Tsividis [20]. Tsividis says that "in a common bulk CMOS fabrication process," one may find an input-referred voltage due to flicker noise of:

$$v_f^2(f) = \frac{K_F}{C_{OX}^{2} \cdot W \cdot L \cdot f} \tag{1}$$

where K_F is the empirical flicker noise coefficient equal to 5×10^{-9} $fC^2/\mu m^2$ for NMOS devices, and 2×10^{-10} $fC^2/\mu m^2$ for buried channel PMOS devices. W and L represent the transistor's channel width and length respectively and C_{ox} *the* gate capacitance per unit area and f the frequency.

Many of the empirical models, including the above model, given by textbooks are simple and easy for hand calculations. Unfortunately they are also questionable. In particular, this equation is an empirical relation, and K_F is not entirely independent of the operating point (and is not the same in weak and strong inversion). This model was implemented in early versions of SPICE, and is generally considered to be oversimplified.

B. Number Fluctuation Model and Mobility Fluctuation Model

According to the Number Fluctuation model (or McWhorter Model) the flicker noise phenomenon is primarily a surface effect [2]. In particular, the McWhorter number fluctuation (Δn) theory states that flicker noise is generated by fluctuations in the number of carriers due to charge trapping in surface states. McWhorter obtained the necessary 1/*f* spectrum by assuming that the time constant τ of the surface states varied with a $1/\tau$ distribution. Christensson et al. [3, 4] were the first to apply the McWhorter theory to

MOSFETs, using the assumption that the necessary time constants are caused by the tunnelling of carriers from the channel into traps located inside the oxide.

A number of other applications of the McWhorter theory to MOSFETs have been done. Das and Moore [5] reviewed and compared a number of these theories and found that the basic assumptions behind any explanation, affects the interpretation of the experimental results. They said, "theoretical calculations of the MOSFET drain noise current have been performed by many investigators, which have led to different results, mainly due to the difference in the method of attack and nature of assumptions".

Reimbold [6] developed the McWhorter model further, taking into account all the capacitive components of the small-signal equivalent circuit, so as to account for all transistor operating regimes. This development was done to fit Reimbold's measurements in weak inversion. Ghibaudo [7] presented a shortcut through Reimbold's work and came up with the same result.

According to the above model the spectral density of the drain current I_{DS} is:

$$i^2{}_f = \frac{K_F q^4 I^2{}_{DS}}{n^2 kTWLf} \cdot \frac{N_{ST}}{(C_{OX} + C_{SS} + C_D + C_I)^2} \qquad (2)$$

where N_{ST} is the surface trap density, C_{SS} is the interface state capacitance, C_D is the depletion capacitance, and C_I is the inversion layer capacitance. T is the temperature, k the Boltzman constant, q the electron charge.

On the other hand according to the mobility fluctuation model (or Hooge model), $1/f$ noise is not associated to surface effects. F. N. Hooge proposed that $1/f$ noise is essentially a bulk phenomenon. Working with metal films he extracted an empirical relation for $1/f$ noise in terms of resistance fluctuations [8-11]. In particular, the Hooge model suggests gate voltage dependence in the input-referred noise. The model is based on empirical experimental observations of the noise in homogeneous samples, and the input referred noise shows strong gate bias dependence.

1.2.2. Combined Models

A lot of research has focused on resolving the above two theories regarding the $1/f$ noise in CMOS technology. However, it has not provided any satisfactory resolution. Inconsistent experimental results were proposed for both n- and p- type channel devices.

Specifically, in an effort to fit as much published data as possible, some authors have combined the number fluctuation and mobility fluctuation theories into a single model. The main idea is that charges, when trapped, cause correlated surface mobility fluctuations due to inversion layer carriers scattering on the trapped charges. Jayaraman and Sodini [14] were the first to propose this model. They also showed that oxide band bending due to a non-uniform oxide trap distribution gives rise to a gate voltage dependence of the flicker noise coefficient, and of α. However, from their data, they were unable to determine if the gate voltage dependence they observed was due to correlated mobility fluctuations, or to non-uniform oxide trap distributions (with a corresponding higher trap density), or both.

Trying to improve the flicker noise model in circuit simulators such as SPICE, the unified model idea was developed further by Hung et al [15, 16]. By examining the channel current modulation due to the trapping of a single electron (popcorn noise in a very small device), they were able to quantify the contribution of the correlated surface mobility fluctuations. However, their model is geared towards numerical simulation.

Literature supports number fluctuations as the primary source of flicker noise in MOSFETs. Responding to this evidence, Hooge [17] has suggested that in semiconductors there is always a mobility-correlated 1/*f* noise but that other types of 1/*f* noise may be present and may be dominant.

Other authors have suggested that 1/*f* noise in NMOS devices is caused by carriers number fluctuations while in PMOS devices the noise is due to mobility fluctuations [18, 19]. This dichotomy makes sense if the NMOS transistors are surface channel devices where the carriers interact more readily with Si-SiO_2 surface states, and the PMOS transistors are buried channel devices where the carriers interact only with the lattice (however in modern deep-submicron technologies (0.35μm and below), both NMOS and PMOS transistors are surface channel devices, so that this dichotomy may no longer exist). Furthermore, differences in the energy trap density between NMOS and PMOS transistors could also account for the differences in observed bias dependence [21].

1.2.3. Noise Simulation Using the BSIM MOSFET Model in Software Packages

Flicker Noise Simulation Models

While conclusions and design criteria concerning noise are extracted using simulations in different types of typical analog integrated circuits and while other issues like the MOS transient response were studied in respect to the simulation model and Level, few extended research has been performed on the validity of the available 1/*f* noise simulation models in relation to the MOS transistor model that is developed in the simulator. In particular BSIM4 model, which is one of the most commonly used Spice MOS transistor simulation models, develops two distinct models, SPICE-Flicker and BSIM-Flicker, to calculate flicker noise.

In particular, when 1/*f* noise performance of MOS transistors or MOSFET circuits has to be simulated there are different flicker noise models that might be used. The simulation of MOSFET 1/*f* noise performance is differentiated in relation to the used flicker noise model. HSPICE simulator incorporates in its algorithms specific flicker noise models. On the other hand BSIM4.X (and BSIM3V3.X) develops its own flicker noise models. The above model selector parameters and the respective equations are listed in Table I.

For all flicker noise models incorporated either in HSPICE or in BSIM algorithms, there are serious inaccuracies in terms of MOSFET noise estimation. It seems that one single model can not simulate accurately flicker noise performances of MOSFETs in all operating regions and for both transistor types, P and N.

As it can be seen in Table I, in the HSPICE simulator there are three different equations and so three different models that can be used for flicker noise estimation depending on the value of the selector parameter NLEV. Also, in case of BSIM3/BSIM4 two different equations can also be chosen depending on the selector parameters fnoimod/noimod.

Compatibility between the BSIM3/BSIM4 flicker noise models and the HSPICE flicker noise models is presumed when fnoimod is 0 in BSIM4 (noimod 1,4 in BSIM3) and when NLEV is 0 in HSPICE (Level 49).

Table I. Flicker noise models

Selector simulation parameter	Noise model	
NLEV = 2 or 3 (HSPICE)	$\frac{\overline{i_{1/f}^2}}{\Delta f} = \frac{KF \cdot g_m^2}{C_{ox} W_{eff} L_{eff} f^{AF}}$	(3)
NLEV = 1 (HSPICE)	$\frac{\overline{i_{1/f}^2}}{\Delta f} = \frac{KF \cdot I_{DS}^{AF}}{C_{ox} W_{eff} L_{eff} f}$	(4)
NLEV = 0 (HSPICE) fnoimod = 0 (BSIM4.X) noimod = 1 or 4 (BSIM3V3.X)	SPICE-Flicker in BSIM4 (Spice2 in BSIM3): $\frac{\overline{i_{1/f}^2}}{\Delta f} = \frac{KF \cdot I_{DS}^{AF}}{C_{ox} L_{eff}^2 f^{EF}}$	(5)
fnoimod = 1 (BSIM4.X) noimod = 2 or 3 (BSIM3V3.X)	BSIM-Flicker in BSIM4 (BSIM in BSIM3): $\frac{\overline{i_{1/f}^2}}{\Delta f} \approx h(NOIA, NOIB, NOIC)\left[\frac{I_{DS}}{C_{ox} L_{eff}^2 f^{EF}} + \frac{g(EM) I_{DS}}{W_{eff} L_{eff}^2 f^{EF}}\right]$	(6)

Regarding noise parameters, they are listed in Table I and have the following meaning: *F*licker noise coefficient *KF*, is a fitting parameter and depends critically on processing, g_m is the transconductance, W_{eff} and L_{eff} are the effective channel width and length, respectively, *AF* (flicker frequency exponent) and *EF* (flicker exponent) characterize the power dependence of the measured $\overline{i_{1/f}^2}$ on I_{DS} and frequency respectively. *h* is a function of the SPICE parameters *NOIA, NOIB, NOIC* and g is a function of the SPICE parameter *EM*. *NOIA, NOIB* and *NOIC* can be treated as fitting parameters, like *KF* in the SPICE-Flicker model. *NOIA*, *NOIB*, and *NOIC* have different default values in NMOS and PMOS transistors [24]. In addition these parameters are unitless in BSIM3V3, but have some rather complicated units in BSIM4. *EM* represents the field at which the carrier velocity saturates. *EM* and *KF* retain the same meaning in both BSIM3V3 and BSIM4. *EM* is a parameter that affects only the noise models but not the I-V or the C-V characteristics [24].

The unified BSIM-Flicker noise model of BSIM4 also follows the framework of BSIM3V3, but with the slight improvements of using smoothing functions and considering the bulk charge effect [24] – [27].

Until recently two works that examine which flicker noise simulation model is more appropriate have been published [28], [29]. They concern NMOS devices operating from subthreshold to strong inversion and PMOS transistors only in the saturation region. Noise comparisons refer to old versions of the BSIM MOSFET model. 1/*f* noise performance is examined in relation to the gate-source bias, the transconductance and its linearity, and the MOSFET channel length. Different 1/*f* noise models are considered suitable for different

MOS operating regions and the accuracy of each 1/*f* model is examined at only two particular frequencies (1 Hz and 1 kHz).

According to the above works, for long channel NMOS devices (L=5 μm), the input-referred gate voltage noise is independent of gate bias, which shows the surface state model and the noise model (NLEV=2,3) in SPICE are more appropriate. SPICE can predict the long channel NMOS noise performance very well in saturation regime, but not in the sub-threshold region, and also cannot be used to simulate short-channel NMOS noise performance. For short channel NMOS devices (L=0.6 μm) the input-referred gate voltage noise is independent of the gate-source voltage V_{gs}, increasing proportionally as V_{gs} increases which is due to the non linearity of the transconductance g_m variation with gate bias. These devices can be modeled with the BSIM3.3 model and the same model is also appropriate. However these models are applicable to NMOS devices only [28]. Regarding PMOSFETs, for long channel devices, the flicker noise is due to the bulk effect [29]. For short channel PMOS devices, however it is due to the surface state effect. NLEV=0 is the most appropriate value for the simulation of long channel devices and NLEV=2,3 for short channel devices.

Contrary to the above results the authors have extracted different selection criteria considering the accuracy of the available 1/f noise simulation models [30],[31]. Particularly, in order to assess the flicker noise models described above, both n- and p-type MOS transistors were used. MOSFETs were fabricated in the 0.6 μm CMOS process by Austria Mikro Systeme (AMS) and their dimensions were 400μm/1.2μm for the NMOS and 1000μm/1.2μm for the PMOS transistor. Bias conditions varied as to measure MOS flicker noise performance from the threshold of linear operating region to saturation. The measurements were obtained for a frequency range of 10 Hz to 10 kHz. Respective simulations were performed using built-in noise models in the HSPICE simulator and the BSIM4 MOS transistor model. Comparison between measurements and simulations was made in relation to the same MOSFET drain current as to achieve accuracy. Temperature was constant at 295 K.

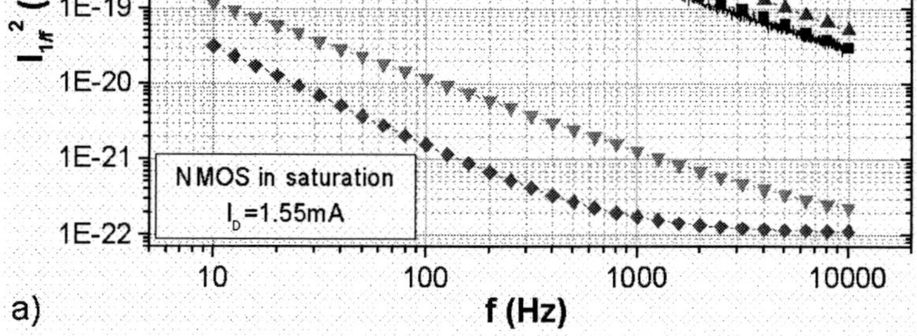

Figure 2. Continued on next page.

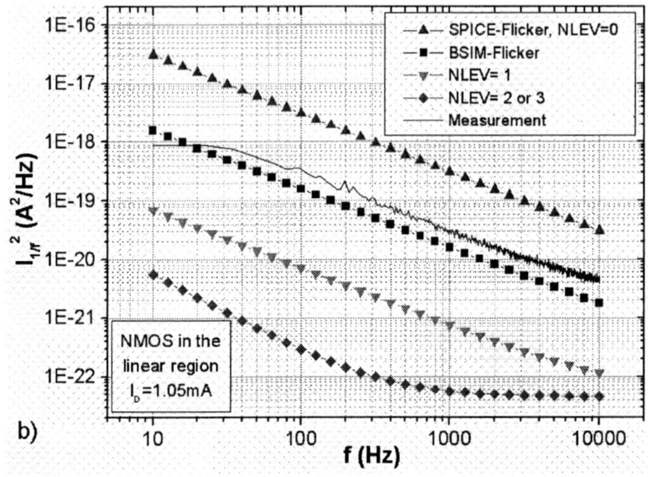

Figure 2. Noise measurement and simulations of an NMOS transistor operating (From [30] with permission) a) in the saturation region for a drain current of 1.55 mA, b) in the linear region for a drain current of 1.05mA.

In particular, for an NMOS transistor operating in the linear or saturation region, BSIM-Flicker is the best simulation model, since it appears to be closer to the experimental results. In both operating regions, and especially in saturation, the selected model provides an accurate description of the NMOS flicker noise behavior. A low frequency measurement and the respective simulations using all the available $1/f$ noise models in the saturation and in the triode regime are given in fig.2

Additionally, for a p-type MOS transistor, SPICE-Flicker noise model (and NLEV=0 for HSPICE) is more preferable in the saturation region and BSIM-Flicker in the triode region. In both operating regions, the selected models overestimate flicker noise. However, the best noise model for a PMOS transistor is not as accurate as it is for an NMOS transistor. In all

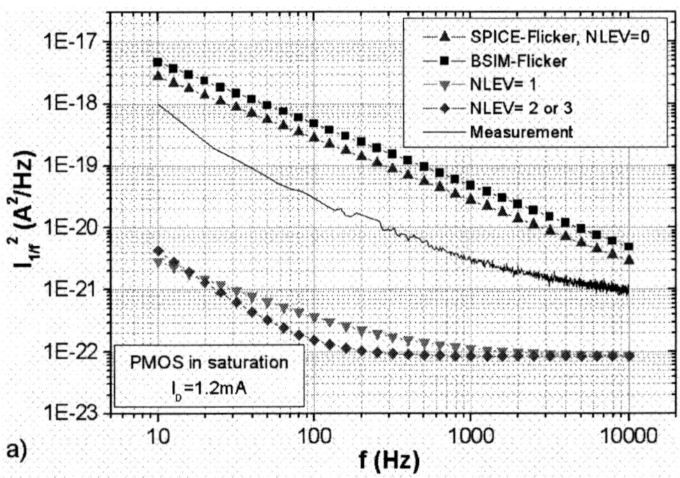

Figure 3. Continued on next page.

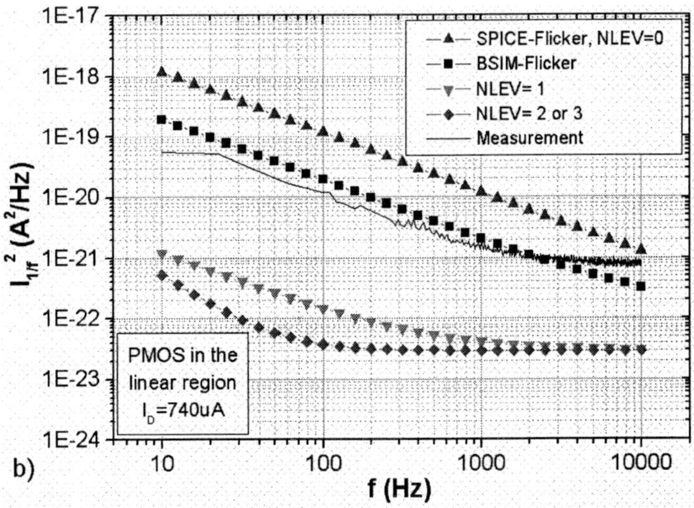

Figure 3. Noise measurement and simulations of a PMOS transistor operating a) in the saturation region for a drain current of 1.2 mA, b) in the linear region for a drain current of 740 μA (From [30] with permission).

operating regions, for an NMOS and a PMOS transistor, simulation noise models used by HSPICE for NLEV = 1 and NLEV = 2 or 3 underestimate largely 1/*f* noise and appear to be inaccurate. A low frequency measurement and the respective simulations using all the available 1/*f* noise models in the saturation and in the triode regime are given in fig.3

The above study implies that NMOS transistor flicker noise performance can be simulated with reliability using the BSIM-flicker noise model of BSIM4.X instead of SPICE-flicker or HSPICE 1/*f* models. However, all the above noise models appear to be quite problematic for a PMOS transistor. This suggests the necessity to develop a new model to calculate PMOS flicker noise performance and to implement it in the simulator.

Table II. Usability of all models for a particular device type under particular bias conditions

Device type	Operating Region	Optimum model	Accuracy
NMOS	Saturation	BSIM-Flicker	Good
	Linear region	BSIM-Flicker	Good
	Threshold	BSIM-Flicker	Good
PMOS	Saturation	SPICE-Flicker	Medium
	Linear region	BSIM-Flicker	Medium
	Threshold	-	-

An overall view of the usability of all models for a particular device type under particular bias conditions is given in Table II. Finally, the results about the reliability of the specific 1/*f* noise simulation models can be considered as valid for other typical analog CMOS scaled processes since, relatively long channel MOSFETS in analog circuit applications are used. However the accuracy of the noise magnitude each model provides is process dependent, since the majority of the model parameters depend critically on processing. Therefore the models reliability is also conditioned by an effective noise characterization of the used CMOS process.

Analysis was also performed in terms of the *1/f* noise exponent [31], [32]. In order to find out the exact flicker noise dependence on frequency variations and compare it to the respective simulation parameter (*EF*), which is common to both models, *1/f* noise measurements were exponentially fitted. The flicker noise measurements of an NMOS and a PMOS FET in the linear and saturation region are shown in Fig.4 and Fig.5.

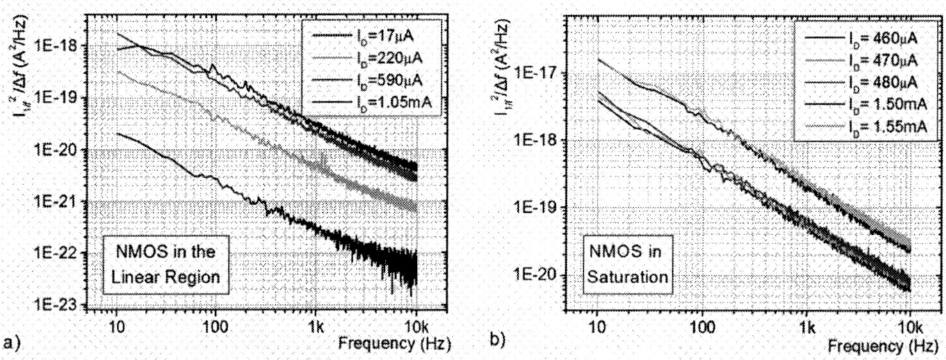

Figure 4. NMOS transistor noise performance a) in linear and b) in saturation region for various values of drain current.

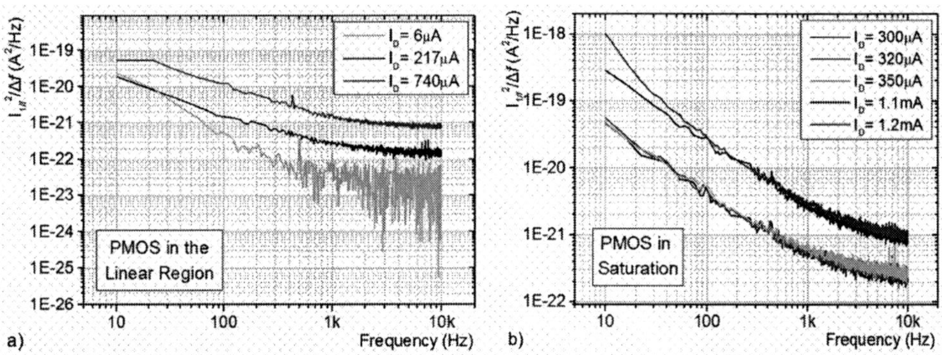

Figure 5. PMOS transistor noise performance a) in linear and b) in saturation region for various values of drain current.

Regarding the description of an NMOS transistor noise behavior in linear and saturation regions, the experimental values of the flicker noise exponent, varied from 0.91 to 0.98, quite close to that of the simulation parameter *EF*, which is unity. Therefore, both models can describe with accuracy the noise spectral density dependence on frequency. The flicker noise exponent of an N-MOSFET at the threshold of the linear region, was an exception, since it was found to be 0.85 due to harmonics in the kHz range.

Contrary, P-MOSFET *1/f* noise measurements showed a different dependence on frequency. Particularly, in the saturation region, the noise parameter *EF* ranged from 0.63 to 0.79, whereas in the linear region, from 0.53 to 0.58. These values, in all operating regions and especially in the triode region, appear to be quite lower than unity. This is due to the fact the generation-recombination (GR) noise sources also contribute to the total noise performance. The respective error on the flicker noise exponent in the PMOS transistor is much higher than that on the NMOS. This implies even more difficulty in the PMOS noise estimation using the above simulation models in the specific frequency region.

J. C. Vildeuil et al [33] proposed an appropriate noise parameters (*NOIA, NOIB, NOIC*) extraction methodology in order to optimise the BSIM3 1/*f* noise model. Specifically, n and p-channel LDD MOSFETs issued from a 0.8 μm CMOS technology and noise measurements were analyzed versus gate bias to identify the origin of the 1/*f* noise and then BSIM3 noise parameters were deduced using an appropriate extraction methodology.

Particularly, they showed that BSIM3 model provided good agreement between measurements and simulations for 1/*f* intrinsic channel noise developing an extraction methodology for the respective noise parameters. For P-MOSFETs 1/*f* noise agreed with the Hooge model and in that case *NOIA, NOIB* were constant and provided relatively accurate simulation from subthreshold to strong inversion. For N-MOSFETs 1/*f* noise agreed with Mc Whorter's model. *NOIA* was still constant whereas in strong inversion good agreement between the measurements and modelling was obtained only if the *NOIB* parameter varied as $(V_{gs}-V_T)^{-1}$. For both n- and p-channel transistors the *NOIC* parameter was negligible.

Critical Issues on the Unified 1/*f* Noise Models

The above unified models that are available in SPICE, assumes implicitly that the mobility, limited by Coulomb scattering at trapped interface charges, does not depend on the inversion carrier density. However recent experimental results indicate that the mobility, limited by Coulomb scattering, is proportional to the square root of the carrier inversion density [34]-[36]. These results are in agreement with a theoretical model based on the work of Nicollian and Brews [37], [38]. E.P. Vandamme and L. K. J. Vandamme [39] showed that the correlated mobility fluctuations can be neglected compared to the noise contribution from the carrier number fluctuations, if the correct dependence of the Coulomb scattering limited mobility on inversion carrier density is taken into account. As a result the unified 1/*f* noise model can not predict the experimentally observed 1/*f* noise as a function of gate bias in p-type MOSFETs. Specifically the unified 1/*f* noise model predict the same gate bias dependence of the noise parameter α as the Δn model (α is defined as the noise parameter to compare the noise modeling results with experimental ones from different types of materials [17] and is given by: $a = fN\frac{S_{I_d}}{I^2{}_d}$ where N is the total number of free carriers), which is not in agreement with the experimental results on PMOSFETs that show α values independent of $\left|V_{gs}-V_T\right|$ at low gate bias. Besides the unified 1/*f* noise models show weaker dependence of the current noise and input referred noise on gate bias than experimental results, and largely underestimate the noise level in strong inversion. Thus the present unified noise models cannot predict the 1/*f* noise in p-type MOS transistors.

1.2.4. Impact of Scaling Down Technologies on 1/*f* Noise

Low frequency noise (LFN) in submicron MOSFETs is an important parameter for analog and RF applications. In complex mixed-signal and low-power integrated devices maintaining a high signal-to-noise ratio becomes more difficult as the power supply voltage and signal amplitude decrease. Assessing the impact of process scaling on 1/*f* noise is therefore of great interest. As LFN is strongly technology dependent, the introduction of advanced processing steps may lead to drastic changes in performances that are not easily predicted. As a result, opposite trends of the 1/*f* noise behaviour with downscaling in CMOS technology were reported [40–44]. The basic reason is that the noise is mainly governed by interface or border traps density related to the gate oxide thickness t_{ox} [42]. Marin et al [45] focused their study only on the impact of silicon dioxide thickness, leaving out the influence of any additional change in the oxidation process steps introduced into advanced CMOS technology.

Particularly, in their experiments, they studied three sets of surface channel p-MOS devices originated from splits processed in 0.25, 0.18 and 0.13 μm dual-gate CMOS technology nodes from STMicroelectronics (Crolles-France). The essential features of the device fabrication were: electric gate oxide thickness t_{ox} = 6, 3.5 and 2 nm respectively for the 0.25, 0.18 and 0.13 μm technologies. Transistors were designed for a 2.5 V (0.25 μm technology), 1.8 V (0.18 μm) and 1.2 V (0.13 μm) power supply voltage. The channel width was equal to W =12.5 μm for 0.25 μm technology and 10 μm for 0.18 and 0.13 μm technologies. Several gate lengths were available.

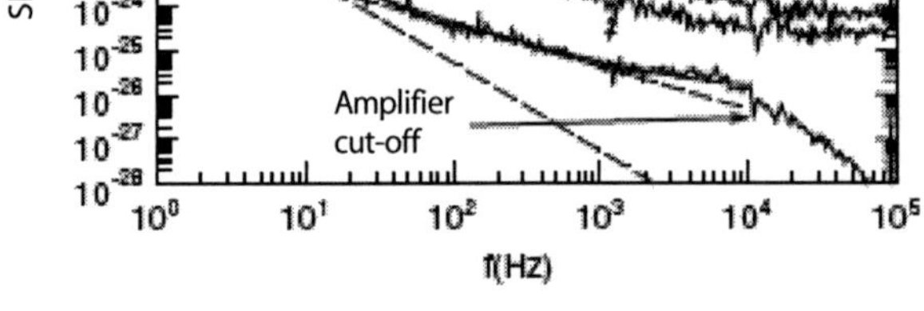

Figure 6. Typical variations of the drain current noise S_{ID} spectra in p-MOSFETs (10 × 0.22 μm^2), measured from weak to strong inversion at V_{DS}= -25 mV. Dotted lines represent the 1/*f* and G/R extracted noise components, solid line represents the total noise at V_{GS} < V_T. (From [45] with permission).

Measured spectra are predominantly $1/f^{\gamma}$ like for large area devices. An overall average frequency index of $0.96 < \gamma < 1.06$ was obtained by least-squares regression of the log–log plot of S_{ID} versus *f*. Moreover, it should be mentioned that when the drain current noise spectral density includes a Lorentzian spectrum behavior in the subthreshold regime or for small area devices, the 1/*f* noise component was extracted assuming that $S_{ID}(f)$ results in a superposition of 1/*f* noise and G/R noise components. Typical drain current noise spectra taken on L =0.22 μm devices are depicted in Fig. 6. The signal limitation of noise power spectral density above 10 kHz was determined by the parasitic RC constant of the low noise current amplifier set-up in the subthreshold regime [45].

According to their work, the 1/*f* channel noise studied in the subthreshold, linear and saturation regimes can be attributed to the carrier number with correlated mobility fluctuations. The slow oxide trap density N_t and the Coulomb scattering parameter have been deduced. At sufficient high gate overdrive voltages, the increase in the noise level observed for the longer transistors is explained by the impact of the series resistance noise. The theoretical curves of S_{ID}/I^2_D calculated by considering the contribution of the access resistance noise to the intrinsic channel noise have shown a good agreement in relation to experimental data (see Figure 7).

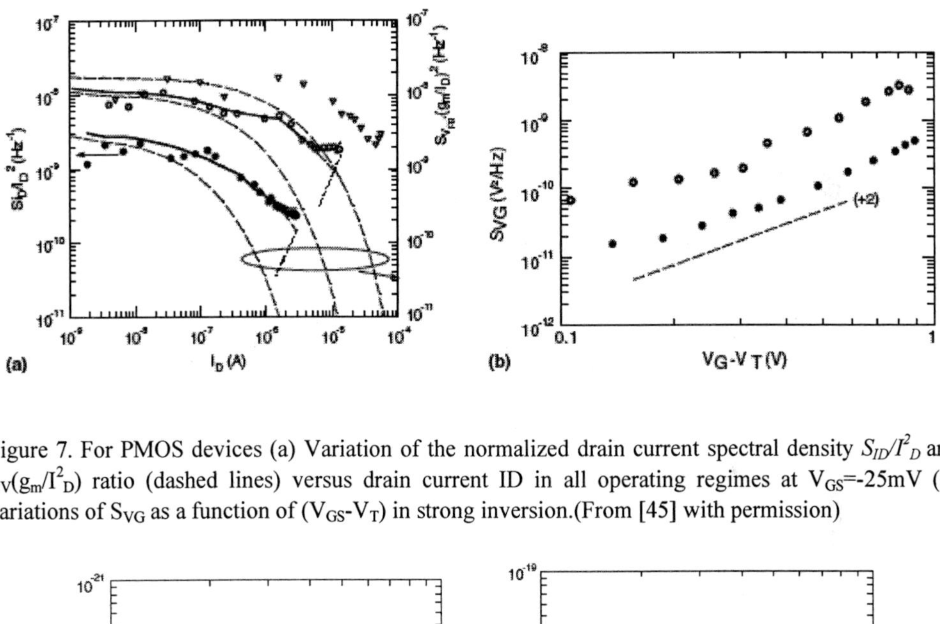

Figure 7. For PMOS devices (a) Variation of the normalized drain current spectral density S_{ID}/I^2_D and $S_V(g_m/I^2_D)$ ratio (dashed lines) versus drain current ID in all operating regimes at V_{GS}=-25mV (b) variations of S_{VG} as a function of (V_{GS}-V_T) in strong inversion.(From [45] with permission)

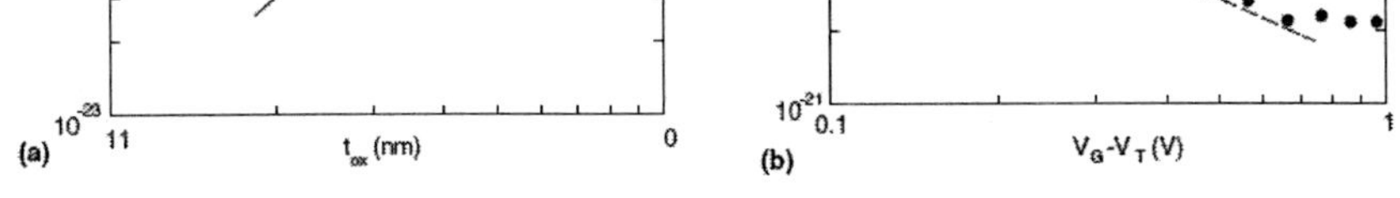

Figure 8. Input-referred noise spectral density S_{VG} vs t_{ox} for three pMOS device generations (0.13, 0.18, 0.25 μm) biased in subthreshold region at V_{DS}=-50mV and f=1 Hz. Dashed line represents $S_{VG} \propto t^2_{ox}$ dependence as predicted by the ITRS roadmap (b) Normalized current noise spectral density WLS_{ID}/I^2_D vs the gate overdrive voltage for t_{ox}=6 nm, 3.5 nm and 2nm respectively (PMOS devices in linear region at f=1 Hz and V_{DS}=-50 mV) (From [45] with permission).

In addition, three PMOS device technology generations (0.25, 0.18 and 0.13 μm) were compared with respect to noise performances. It is observed that as the oxide thickness is reduced, the interface trap density N_t increases, the Coulomb scattering parameter α_s stays almost constant and the Hooge parameter α_H increases. These parameter variations contribute to an input-referred noise $S_{VG}(t_{ox})$dependence lower than t^2_{ox} in weak inversion as previously predicted by the 1999 International Technology Roadmap for Semiconductors and to a non $S_{ID}/I^2_D \propto t_{ox}$ dependence in the linear regime (fig. 8 [45]).

Their study clearly demonstrated that to leave out a number of important noise parameters and considering only gate oxide thickness in order to predict trends in S_{VG} versus technology scaling, leads to inaccurate or too optimistic results.

1.3. Noise Reduction

1.3.1. 1/*f* Noise under Switched Bias Conditions

A. P. van der Wel et al performed low frequency measurements with switched biased MOSFETs (fig. 9) [46]. The aim was to compare switched bias and constant bias noise power for the same measurement bandwidth. For 50% duty cycle, a low-frequency noise power that is reduced by 6 dB compared to the constant-bias situation was expected and indeed, a gap of about 6 dB was observed. Further noise reduction was observed when the gate-source voltage in the "off" state was decreased. A more or less flat plateau below the switched frequency, which gradually changes to a noise spectrum with reduced power density was also observed at low frequencies. The switching frequency was varied from 100 Hz up to 1 MHz (Fig. 9b). Even at 1 MHz, where the settling of the output voltages becomes incomplete, this noise reduction was also found. Below, a quasi-plateau has been observed, while a noise spectrum

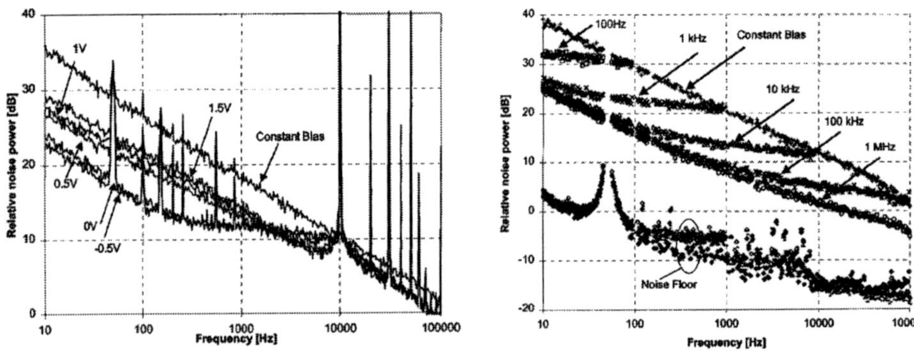

a) b)

Figure 9 a) Noise reduction as a function of the "off" voltage: HEF 4007 NMOS V_{gs}=2.5V, V_t=1.9V f_{switch} = 10 kHz duty cycle = 50% (taken from [46] with permission ©2000 IEEE) b) Noise reduction while switching at different frequencies (indicated in the figure with different symbols): HEF4007 NMOS, V_{GSon} = 2.5 V, $V_{GS,off}$ = 0V, duty cycle = 50%, all spectra are shown up to 0.9 * fswitch and points within 5 Hz of multiples of 50 Hz are omitted (taken from [46] with permission ©2000 IEEE).

appeared again at more than two decades lower frequencies. This can be explained by assuming different noise generating processes, as suggested in [47]: 1) a process that *is* affected by switching resulting in the plateau (e.g., oxide trap noise with associated RTS's, that effectively disappears when switching between inversion and accumulation [47]); and 2) other processes that are not affected by the switching and render the residual low frequency noise (e.g., mobility fluctuations) [46].

In order to study the noise reduction for different CMOS processes, devices from six different manufacturers were tested by A. P. van der Wel et al [46]. An intrinsic noise reduction was found in all cases, with values of 5–8 dB. The effect occurs in all devices, and was also observed for submicron MOS transistors [47]. It was also shown that phase noise in CMOS oscillators [48]–[50], operating at switching frequencies in the lower MHz region, benefits from switching.

Applications of the Switched Bias Technique

1/*f* noise is of increasing interest as minimum size transistors in newer CMOS processes tend to have higher corner frequencies, typically well above 1 MHz. Moreover, as shown in Fig. 10, the effect of noise is not limited to low frequencies: noise is also upconverted to high frequencies [51], e.g., in active mixers, VCO's [52], and frequency dividers [53]. This is one of the important difficulties in implementing analog CMOS circuits with similar noise performance than their bipolar counterparts. Circuit techniques like chopping and correlated double sampling reduce the effect noise has on circuits. In contrast, Eric A. M. Klumperink at al. [54] have proposed a technique that tackles the problem at its physical roots: it reduces the intrinsic noise itself. It exploits an intriguing physical effect: Cycling a MOS transistor between strong inversion and accumulation reduces its 1/*f* noise.

From a designer's perspective, a pragmatic view is usually taken, concentrating on the question: given a certain IC process, what can be done to reduce the 1/*f* noise? The main controllable parameters are *W* and *L* of the transistor and its biasing.

Changing the biasing only helps for transistors showing significant bias dependence (as for the mobility fluctuation model). However, biasing also affects many other important circuit performance aspects like transfer function, signal swing, speed, linearity and current efficiency.

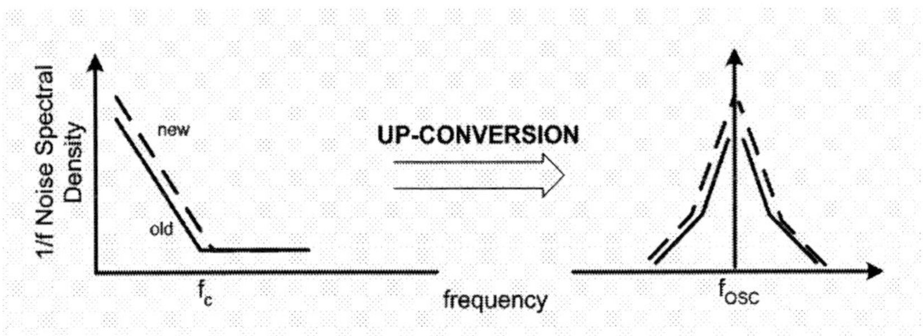

Figure 10. The 1/*f* corner frequency *f* of minimum size transistors in newer CMOS processes tends to increase. Upconversion, e.g., in oscillators renders also additional noise at high frequencies.

Due to the many tradeoffs, changing the bias is rarely practical. Increasing the gate area *WL* at constant *W/L* and constant bias has much less impact on most of the performance aspects mentioned. The main disadvantage in this case is the increase in gate-capacitance resulting in speed penalty. Lowering the impedance level, the loss in speed can be compensated. However, a lower impedance level requires higher transconductance values of the transistor, resulting in more power consumption. It is concluded that the only practical way to reduce $1/f$ noise is by increasing device gate area. However, this comes at the cost of speed or power consumption [54].

1.3.2. Available Methods

Known circuit techniques for noise reduction will be reviewed briefly, in order to make a distinction between these techniques and the switched bias technique [54].

A. $V_{DS} = 0$:

$1/f$ current becomes zero for a MOS operating in the triode region at V_{DS}=0V. This property is useful in resistive circuits, for instance in current dividers [55] or transconductors [56]. However, if V_{DS} is increased, the $1/f$ noise comes up. Furthermore, almost all transistor circuits require voltage amplification, while a MOS transistor in the triode region has a maximum “intrinsic voltage gain” of 1. The majority of MOS circuits exploit MOS transistors operating in saturation, i.e., $V_{DS} \neq 0$.

B. DC Offset and Drift Reduction Techniques

DC offset and drift are basic problems in analog circuits, and several techniques have been developed to reduce the resulting low frequency errors. Examples of such techniques are chopper stabilization [57], auto-zeroing techniques (correlated double sampling [57]), dynamic element matching [58], dynamic current mirrors [59] and current copiers [60]. As $1/f$ noise at low frequencies is indistinguishable from drift, a reduction of low frequency $1/f$ noise is also achieved. The reduction factor is typically limited by device mismatch, timing errors and charge injection. All these techniques are limited to use at low frequencies. They do not help to reduce up-converted $1/f$ noise.

C. Reduce the Upconversion of 1/f Noise

Hajimiri proposed a theory that is useful to quantify the upconversion effect of $1/f$ noise in oscillators [51]. The oscillator is modeled as a linear time variant system, and an impulse sensitivity function is defined that characterizes the sensitivity to upconversion. Based on this theory, it can be shown that symmetry in the oscillator waveform helps to minimize the upconversion [61]. However, the achievable symmetry is limited especially in cases where complementary devices are used.

D. Phase-Locked Loop (PLL)

In a PLL the phase of a voltage controlled oscillator (VCO) is locked to the phase of a reference signal by means of a phase detector with feedback loop. If the reference is clean, the PLL suppresses the phase noise of the VCO by an amount determined by the loop gain [24]. Although this is a very useful commonly used technique, it has its limitations. For frequencies larger than the loop bandwidth, the phase noise of the (effectively free running) VCO still determines the phase noise of the PLL. Hence, 1/*f* noise can be a problem, especially for applications requiring a low loop bandwidth. Within the loop bandwidth, the noise contribution of the phase detector and the frequency divider, that is often present between the VCO and the phase detector, is a source of phase noise contributions that are not attenuated by the loop gain [61]. Hence, noise improvements are wanted in these blocks.

E. The switched biasing approach

Cycling between Strong Inversion and Accumulation

Fig.11 illustrates the basic idea of cycling a transistor between strong inversion and accumulation for an NMOS transistor. A voltage source with a square wave signal switches the gate-source voltage of the NMOS between two bias values. The high level, V_{gson}, is larger than the threshold voltage, so that the transistor is biased at a constant voltage in strong inversion (NMOSFET is "on"). The low level, V_{gsoff} , is equal or lower than the threshold voltage (NMOSFET "off") and is variable. Depending on V_{gsoff}, the low state of the square wave corresponds to biasing in moderate inversion, weak inversion or finally accumulation (holes in the p-bulk accumulate under the gate-oxide).

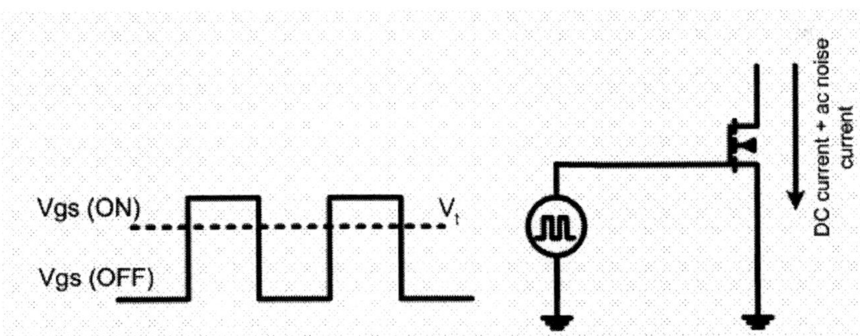

Figure 11. Varying V_{gs} cycles a MOSFET between strong inversion and accumulation.

Having a switched bias the noise power is roughly 6 dB lower. This is what we can expect for 50% duty cycle, as the switching operation can be represented as a multiplication of the *1/f* noise current with a square-wave signal with 50% duty cycle, *m(t)* , as follows:

$$m(t) = \frac{1}{2} + \frac{2}{\pi}\sin\omega_{sw}t + \frac{2}{3p}\sin 3\omega_{sw}t + \frac{2}{5p}\sin 5\omega_{sw}t + ..., \quad (7)$$

In the frequency domain this corresponds to a convolution of the *PSD* of the *1/f* noise with a spectrum with delta functions at dc, ω, 3ω, 5ω, etc. The dc-term determines the resulting noise power in baseband, which is -6 dB (or $(1/2)^{1/2}$) compared to the original *1/f* noise power.

The general concept of the switched biasing technique is demonstrated below:

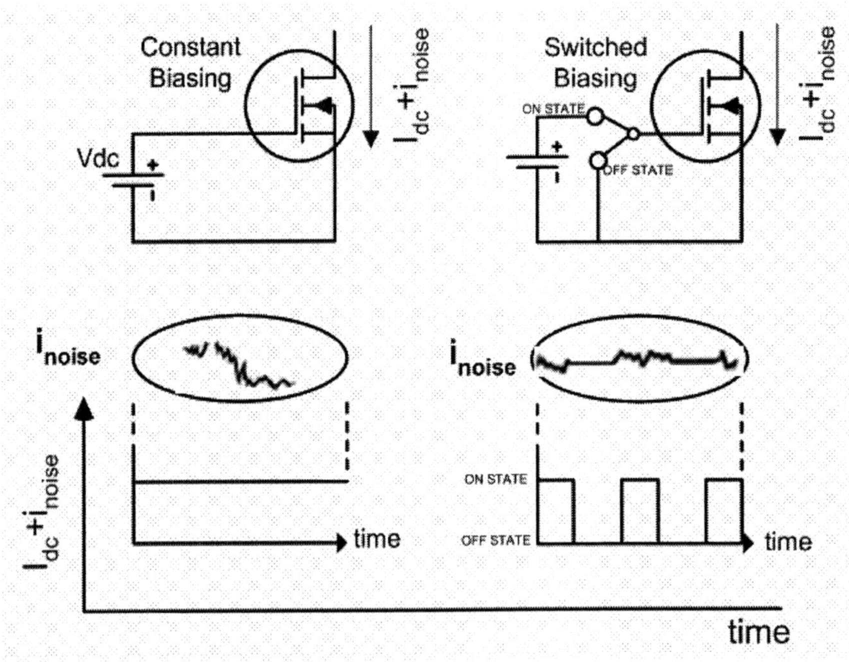

Figure 12. Concept of the switched biasing technique (V_T>0).

Many applications where the specific technique was applied in order to reduce the 1/*f* noise, are given in [54].

1.4. Measurement of Low Frequency Noise in MOSFETs

1.4.1. Low Frequency Noise Measurement Technique under Constant Bias

It is well known that the extrinsic noise behaviour of a transistor is fully described by two noise generators, together with their correlation. Two different characterization techniques can be used to measure these noise sources: the first one, called "multi-impedance technique" consists to measure the noise sources with only one low noise voltage amplifier and several different resistors at the input impedance of the device under test (DUT. Then the two input noise generators, voltage v_G and current i, and their coherence γ' (Fig. 13b) are extracted with a numerical procedure [62]. The second one, called "direct technique", is based on two low noise amplifiers at the input and the output of the DUT in order to directly measure all the noise sources. This configuration usually required two low noise current amplifiers (transimpedances) [63] or one voltage amplifier associated to a transimpendance amplifier

[64]. Thus by using two transimpedance amplifiers, the noise sources are directly measured like on Fig. 13c: the noise generators are at the input (i_G) and at the output (i_D) terminals of the DUT, together with their coherence γ. It is possible from one noise representation to derive to another one due to the equivalence between all the noise representations.

During some previous studies of bipolar transistors, we already had the opportunity to compare between two LFN set-ups based on these two different techniques [63]: the direct technique is rather utilized than the multi-impedance one due to a smaller characterization time without a numerical procedure.

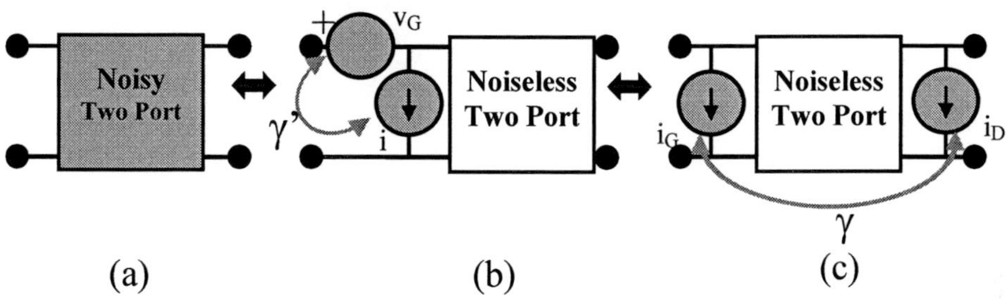

Figure 13. Equivalent LFN noise representations.

A very simple view of the set-up is on Fig.14: the current noise sources (i_G and i_D) are directly AC short-circuited at the transistor input and output terminals with the low input impedance of two transimpedance amplifiers.

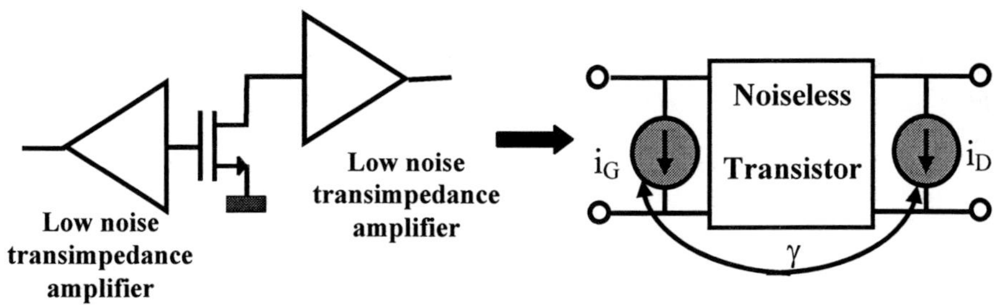

Figure 14. LFN set-up based on transimpedance amplifiers.

The study of the gate noise source i_G and of the coherence can give some important physical information [64]. However, as the gate leakage from the transistor is small and as the input impedance of the transistor is typically much more than the driving source impedance generally used, the LFN measurement can be reduced to the study of the noise source i_D as a first approximation in order to obtain the LFN behavior from a circuit simulation [65, 66]. A more complete view of the set-up is given on Fig. 15 for the drain LFN measurements.

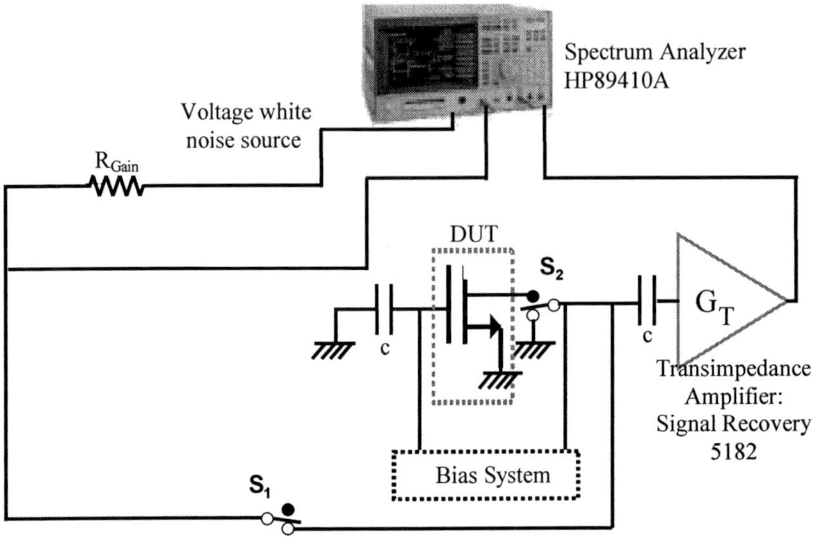

Figure 15. LFN set-up based on transimpedance amplifier for the S_{ID} measurement.

The Several S_1 and S_2 are used in order to configure the set-up: to measure either the noise of the set-up or the amplifier gain or the drain current spectral density S_{ID} respectively (Table III).

The voltage white noise source and the resistance R_{Gain} are applied in order to obtain the gain G_T of the transimpedance. The capacitors *c* have been introduced to separate the DC part from the LFN part.

In order to obtain S_{ID}, it is necessary to divide the spectrum density measured at the output of the amplifier by G_T and to substrate the noise of the set-up.

Table III. The different measurement configurations, the position of each switch is given by an open or filled dot, as shown in Fig.15

Switch / Measurement configuration	S_1	S_2
Set-up noise	● (like a floating position)	O
G_{TD} Gain	O	●
S_{ID}	●(like a floating position)	●

The design of the set-up could be more direct by connecting the amplifier and the bias system in serial instead of the transimpedance with a capacitor in parallel of the bias system and of the DUT (Fig. 15). Two points prevented this last configuration. The first and main point is that the amplifier that we utilized for the set-up has a virtual ground at the input:

without a physical access of the ground, the serial connection is not possible. The second one is due to that the transimpedance is easily saturated at the drain terminal by the amplitude of the DC current amplitude.

By using the Eq.8, it is possible to obtain from the S_{ID} measurement the input noise voltage spectrum density S_{VG}.

$$S_{VG} = \frac{S_{ID}}{g_m^2} \tag{8}$$

It is important to notice that to characterize only one LFN source, S_{ID} or S_{VG}, a voltage amplifier can be used instead of the current amplifier at the output of the DUT: the measurement consists of dividing the voltage noise measured at the output of the amplifier by the gain of the system "DUT and Amplifier" and to take into account the noise of the set-up.

1.4.2. Low-Frequency Noise Measurement Technique under Switched Bias

In order to measure MOSFET noise under constant and switched bias conditions, the measurement setup of Fig. 16 was proposed [46] . This setup exploits the high common mode rejection ratio of a differential probe to measure small differential noise currents superimposed

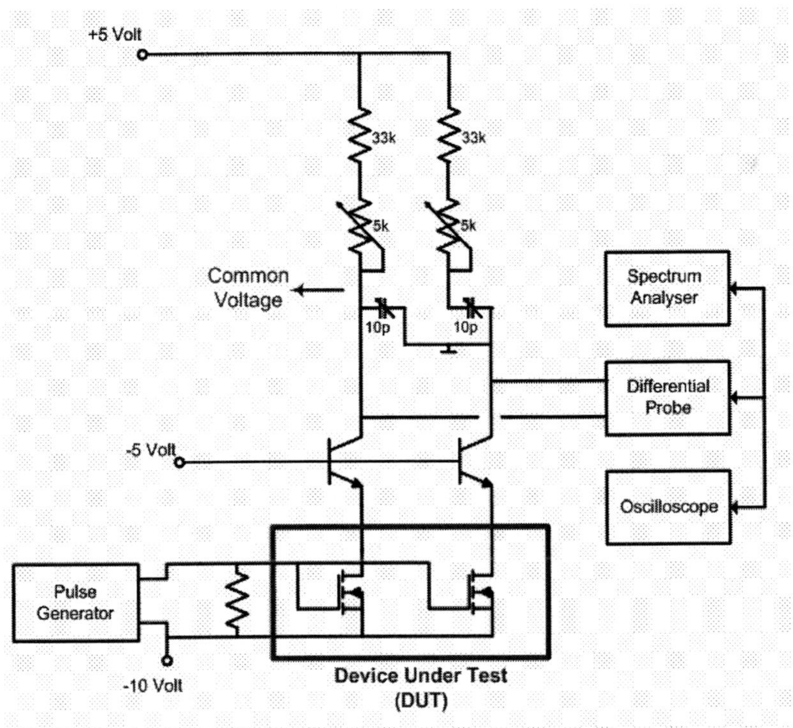

Figure 16. Low-Frequency noise measurement setup for MOSFETs under switched bias conditions. The large common signal is largely rejected by the differential probe.

on much larger common mode switched bias currents of two equal (DUT large-signal behaviour should be equal for minimal common to differential conversion) MOSFET devices under test. The bipolar cascode transistors supply the drains of the MOSFET's with an almost constant voltage and convey the MOSFET drain currents to resistors R_{D1} and R_{D2}, which act as I-V converters. To establish switched bias conditions, the gate-source bias voltages of the two DUT's are driven by a common square-wave signal. As a result, the input signal of the differential probe corresponds to two distinct bias conditions: in the first, both MOSFET's are "off" (below threshold), and in the second, they are "on" in strong inversion and saturation. The probe ideally only senses the differential voltage that is dominated by the noise current of the two DUT's in the "on" state. However, a large common signal also occurs at the probe input due to the switched biasing. Mismatches in the large-signal DUT characteristics and the large, but finite common mode rejection of the probe (>80 dB from 10 Hz to 50 kHz), result in a residual switching signal at the output of the probe. Resistive and capacitive trimmers are provided to equalise the common signals and thus minimize the differential residual to sufficiently low values so as not to exceed the input dynamic range of the spectrum analyzer. Furthermore, the periodic nature of the switching residual results in spectral peaks that can easily be distinguished from noise contributions.

References

[1] Halford D. A general model for f^a spectral density random noise with special reference to flicker noise *1/f*. *IEEE Proceeding* 1968; 56(3): 251–258.

[2] McWhorter AL. *1/f* noise and germanium surface properties. In *Semiconductor surface physics*, R.H. Kingston, Ed.,University of Pennsylvania Press: Philadelphia, 1957; 207–228.

[3] Christensson S, Lundström I, Svensson C. Low frequency noise in MOS transistors – I. Theory. *Solid-State Electronics* 1968; 11(9): 797–812.

[4] Christensson S, Lundström I. Low frequency noise in MOS transistors – II. Experiments. *Solid-State Electronics* 1968; 11(9): 813–820.

[5] Das Makunda B, Moore JM. Measurements and interpretation of low-frequency noise in FETs. *IEEE Transactions on Electron Devices* 1974; 21(4): 247–257.

[6] Reimbold G. Modified *1/f* trapping noise theory and experiments in MOS transistors biased from weak to strong inversion – influence of interface states. *IEEE Transactions on Electron Devices* 1984; 31(9): 1190–1198.

[7] Ghibaudo G. A simple derivation of Reimbold's drain current spectrum formula for flicker noise in MOSFETs. *Solid-State Electronics* 1987; 30(10): 1037–1038.

[8] Hooge FN. *1/f* noise is no surface effect. *Physics Letters* 1969; 29(3): 139–140.

[9] Hooge FN. Discussion of recent experiments on *1/f* noise. *Physica* 1972; 60(1): 130–144.

[10] Hooge FN, Vandamme LKJ. Lattice scattering causes *1/f* noise. *Physics Letters* 1978; 66(4): 315–316.

[11] Hooge FN. *1/f* noise. *Physica B* 1976; 83(1): 14–23.

[12] Jindal RP, Van Der Ziel A. Phonon fluctuation model for flicker noise in elemental semiconductors. *Journal of Applied Physics* 1981; 52(4): 2884–2888.

[13] Vandamme LKJ, de Werd HMM. 1/*f* noise model for MOSTs biased in the nonohmic region. *Solid-State Electronics* 1980; 23(4): 325–329.

[14] Jayaraman R, Sodini CG. A 1/*f* noise technique to extract the oxide trap density near the conduction band edge of silicon. *IEEE Transactions on Electron Devices* 1989; 36(9): 1773–1782.

[15] Hung KK, Ko PK, Hu C, Cheng YC. A unified model for the flicker noise in metal-oxide-semiconductor field-effect transistors. *IEEE Transactions on Electron Devices* 1990; 37(3): 654–665.

[16] Hung KK, Ko PK, Hu C, Cheng YC. A physics-based MOSFET noise model for circuit simulators. *IEEE Transactions on Electron Devices* 1990; 37(5): 1323–1333.

[17] Hooge FN. 1/*f* noise sources. *IEEE Transactions on Electron Devices*1994; 41(11): 1926–1935.

[18] Vandamme LKJ, Xiaosong Li, Rigaud D. 1/*f* noise in MOS devices, mobility or number fluctuations? *IEEE Trans Electron Devices* 1994; 41(11) : 1936–1945.

[19] Chang J, Abidi AA, Viswanathan CR. Flicker noise in CMOS transistors from subthreshold to strong inversion at various temperatures. *IEEE Trans Electron Devices* 1994; 41(11): 1965–1971.

[20] Yannis P. Tsividis. Operation and Modeling of the MOS Transistor. McGraw-Hill, New York, 1987 pp. 343.

[21] John H. Scofield, Nick Borland, and Daniel M. Fleetwood. Reconciliation of different gate- voltage dependencies of 1/f noise in n-MOS and p-MOS transistors. *IEEE Transactions on Electron Devices*, 41(11):1946-1952, November 1994

[22] Xie D, Cheng M, Forbes L. SPICE models for flicker noise in n-MOSFETs from subthreshold to strong inversion. *IEEE Trans Computer-Aided Design* 2000; 19(11): 1293–1303.

[23] Zhou J, Cheng M, Forbes L. SPICE models for flicker noise in p-MOSFETs in the saturation region. *IEEE Trans Computer-Aided Design* 2001; 20(6): 763-767.

[24] Liu W. MOSFET models for SPICE simulation including BSIM3V3 and BSIM4, New York: Wiley–Interscience; 2001

[25] BSIM4.3.0 MOSFET- User's manual. Berkeley: The Regents of the University of California; 2003.

[26] BSIM3v3.2.2 Manual. Berkeley: UC Berkeley; 1999.

[27] Star-HSPICE manual. Fremont: Avant! Corporation and Avant! Subsidiary; 2001.

[28] Xie D, Cheng Mengzhang, Forbes L. SPICE models for flicker noise in n-MOSFETs from subthreshold to strong inversion. *IEEE Trans Computer-Aided Design* 2000; 19(11): 1293-303.

[29] Zhou J, Cheng Mark, Forbes L. SPICE models for flicker noise in p-MOSFETs in the saturation region. *IEEE Trans Computer-Aided Design* 2001; 20(6): 763-7.

[30] Noulis, T., Siskos, S., and Sarrabayrouse, G.: 'Comparison between BSIM4.X and HSPICE flicker noise models in NMOS and PMOS transistors in all operating regions'. *Microelectronics Reliability,* 2007, 47, pp. 1222-1227.

[31] Noulis, T., Siskos, S., and Sarrabayrouse, G., 'Analysis and selection criteria of BSIM4 flicker noise simulation models', *Intenational Journal of Circuit Theory and Applications* (accepted for publication).

[32] Noulis, T., Siskos, S., Sarrabayrouse, G., Bary, L. and Hatzopoulos, A.A.,: 'Detailed study of BSIM3v3 flicker noise models in NMOS and PMOS transistors from threshold to saturation', *Proc. 20th DCIS,* Portugal, Lisboa, 2005

[33] Vildeuil J.C., Valenza M., Rigaud D., 'Extraction of the BSIM3 1/f nosie parameters in CMOS transistors', *Microelectronics Journal,* 1999, 30, pp. 199-205

[34] Koga J., Takagi S., and Toriumi A., ' A comprehensive study of MOSFETelectron mobility in both weak and strong inversion regimes', *IEDM Tech. Dig.*, 1994, pp. 475-478.

[35] Takagi S., Toriumi A., Iwase M., and Tango H., 'On the universality of inversion layer mobility in Si MOSFETs: Part I-effects on substrate impurity concentration' *IEEE Trans Electron Devices* 1994; 41, pp. 2357–2362.

[36] Mujtaba A., Takagi S., Dutton R., 'Accurate modelling of coulombic scattering and its impact on scaled MOSFETs' in Symp. *VLSI Tech. Dig.*, 1995, pp. 99-100.

[37] Soppa W. M., and Waagemann H. G., 'Investigation and modelling of the surface mobility of MOSFETs from -25 to +150°C' *IEEE Trans Electron Devices* 1988; 35, pp. 970–977.

[38] Nicollian E. H., and Brews J. R., MOS Metal oxide Semiconductor Physics and Technology, New York: Wiley, 1982.

[39] Vandamme E. P., and Vandamme L. K. J., 'Critical discussion on unified 1/f noise models of MOSFETs', *IEEE Trans Electron Devices* 2000; 47, pp.2146–2152.

[40] Simoen E., Claeys C., 'On the flicker noise in submicron silicon MOSFETs' *Solid-State Electronics* 1999, 43, pp.865-882

[41] Ghibaudo G., Roux-dit-Buisson O., Brini J.,. *Phys Status Solidi* (a) 1992;132:501.

[42] Kolarova R., Skotnicki T., Chroboczek JA., 'Low frequency noise in thin gate oxide MOSFETs', *Microelectronics Reliability* 2001;41:579.

[43] Brederlow R, Weber W, Schmitt-Landsiedel D, Thewes R. 'Fluctuations of the low frequency noise of MOS transistors andtheir modeling in analog and RF-circuits' International Electron Devices Meeting (IEDM), 1999, 159-162.

[44] Knitel MJ, Woerlee PH, Scholten AJ, Zegers-Van Duijnhoven ATA. In: TechnicalDigest International Electron Devices Meeting

[45] Marin M., Akue Allogo Y., De Murcia M., Llinares P., Vildeuil J. C., 'Low frequency noise characterization in 0.13 μm p-MOSFETs. Impact of scaled-down 0.25, 0.18 and 0.13 μm technologies on 1/f noise', *Microelectronics Reliability* 44, 2004, pp. 1077-1085

[46] Van der Wel A. P., Klumperink E. A. M., Gierkink S. L. J, Wassenaar R. F., and Wallinga H, 'MOSFET 1/f Noise Measurement Under Switched Bias Conditions', *IEEE Electron Devices Letter*, 2000; 21(1), pp. 43–46.

[47] Dierickx B. and Simoen E., "The decrease of 'random telegraph signal' noise in metal-oxide-semiconductor field-effect transistors when cycled from inversion to accumulation," *J. Appl. Phys.*, vol. 71, pp. 2028–2029, Feb. 1992

[48] Gierkink S. L. J., Van der. Wel A., Hoogzaad G., Klumperink E. A. M., and V. Tuijl A. J. M., "Reduction of the 1/f noise induced phase noise in a CMOS ring oscillator by increasing the amplitude of oscillation," in *Proc. 1998 Int. Symp. Circuits and Systems (ISCAS'98)*, CD-ROM Paper MPA9-8, Monterey, CA, May 31–June 3 1998.

[49] Gierkink S. L. J., Klumperink E. A. M., Van derWel A. P., Hoogzaad G., Van Tuijl A. J. M., and Nauta B., "Intrinsic 1/f device noise reduction and its effect on phase noise in

CMOS ring oscillators," *IEEE J. Solid- State Circuits*, vol. 34, pp. 1022–1025, July 1999.

[50] Gierkink S. L. J., Klumperink E. A. M., Van Tuijl E., and Nauta B., "Reducing MOSFET 1/f noise and power consumption by "Switched Biasing"," in *Proc. 25th Europ. Solid-State Circuits Conf.*, Duisburg, Germany, Sept. 21–23, 1999, pp. 154–157.

[51] Hajimiri A. and Lee T. H., "A general theory of phase noise in electrical oscillators," *IEEE J. Solid-State Circuits*, vol. 33, pp. 179–194, Feb. 1998.

[52] Craninckx J. and Steyaert M. S. J., "A fully integrated CMOS DCS-1800 frequency synthesizer," *IEEE J. Solid-State Circuits*, vol. 33, pp. 2054–2065, Dec. 1998.

[53] Krishnapura N. and Kinget P., "A 5.3 GHz programmable divider for HiPerLAN in 0.25 _m CMOS," in *Proc. 25th Eur. Solid-State Circuits Conf.*, Duisburg, Germany, Sept. 21–23, 1999, pp. 142–145.

[54] Klumperink Eric A. M., Gierkink Sander L. J., Van der Wel Arnoud P.,, and Nauta Bram, 'Reducing MOSFET 1/f Noise and Power Consumption by Switched Biasing' *IEEE J. OF Solid-State Circuits*, vol. 35, No. 7, July 2000

[55] Bult K. and Geelen G. J. G. M., "An inherently linear and compact MOST-only current division technique," *IEEE J. Solid-State Circuits*, vol. 27, pp. 1730–1735, Dec. 1992.

[56] Mensink C. H. J., Nauta B., and Wallinga H., "A CMOS "soft-switched" transconductor and its application in gain control and filters," *IEEE J. Solid-State Circuits*, vol. 32, pp. 989–998, July 1997.

[57] Gregorian R. and Temes G. C., *Analog MOS Integrated Circuits for Signal Processing.* New York, NY: Wiley, 1986.

[58] Plassche R. J. V. D., "Dynamic element matching for high accuracy monolithic D/A converters," *IEEE J. Solid-State Circuits*, vol. SSC-11, pp. 795–800, Dec. 1976.

[59] Wegmann G. and Vittoz,E. A. "Very accurate dynamic current mirrors," *Electron. Lett.*, vol. 25, pp. 644–646, May 1989.

[60] Daubert S. J. *et al.*, "Current copier cell," *Electron. Lett.*, vol. 24, pp. 1560–1562, Dec. 1988.

[61] Hajimiri A., Limotyrakis S., and Lee T. H., "Jitter and phase noise in ring oscillators," *IEEE J. Solid-State Circuits*, vol. 34, pp. 790–804, June 1999.

[62] Escotte L., Plana R., Graffeuil,J. "Evaluation of noise parameter methods," *IEEE Trans. Microwave Theory Tech.*, vol. 41, pp. 369-381, Mar 1993.

[63] Bary L., Borgarino, Plana M. R., Kovacic S., Lafontaine H., Feng Zhou Z. and Graffeuil J., "Transimpedance amplifier based full low noise characterization set-up for Si/SiGe HBT's," *IEEE Trans. Electron Devices,* vol. 48, no. 4, pp. 767-773, April 2001.

[64] Giusi G., Crupi F., Ciofi C. and Pace C., "Instrumentation Design for Gate and Drain Low Frequency Noise Measurements," *Instrumentation and Measurement Technology Conference (IMTC) 2006, Pro. of the IEEE,* pp. 1747-1750, April 2007.

[65] Van der Wel A. P., Klumperink E. A., Kolhatkar J. S., Hoekstra E., Snoeij M. F., Salm C., Wallinga H. and Nauta B.., "Low-Frequency Noise Phenomena in Switched MOSFETs," *IEEE Journal of Solid-State Circuits*, vol. 42, pp. 540-550, March 2007.

[66] Scholten A. J., and Klaassen D. B. M., "New 1/f noise model in MOS Model , Level 903," Philips Research Nat. Lab., Unclassified Report 816/98, 1998 [Online], Available: http://www.nxp.com/Philips_Models/mos_models/model9/

In: MOSFETs: Properties, Preparations and Performance
Editors: N.T. Andre et al. pp. 381-400 ISBN: 978-1-60456-762-5

Chapter 11

A Rigorous Analysis of the Parameters which Govern the SILC in MOSFET's with Oxide Thickness in the 1-2 Nanometer Range

D. Bauza, G. Ghibaudo and F. Rahmoune
Institut de Microélectronique, Electromagnétisme et Photonique (IMEP),
(INP Grenoble, UJF, CNRS), Minatec, 3, parvis Louis Neel, BP 257,
38016 Grenoble Cedex 1, FRANCE

Abstract

The experimental conditions allowing the extraction of reliable SILC characteristics in n-channel MOSFET's with ultrathin oxides are discussed first. These characteristics are often studied using the relative variations in the gate current density. This approach is critically reviewed. It is shown that the variations (instead of the relative variations) in the gate current density, allow the extraction of reliable SILC properties and information on the SILC mechanism to be obtained. These curves are used to discuss the model proposed in [1]. In this paper and at negative bias, the gate electrons tunnel from the metal Fermi level into Si-SiO_2 interface states from where they recombine with the substrate majority carriers. It is shown that the SILC mechanism in those devices cannot be that one. Then, results confirming that conclusion and allowing information on the actual SILC mechanism to be obtained are discussed. This is done through the comparison between the trap density relative variations extracted from the SILC and from charge pumping, and through the characteristics of the traps required in $I_g(V_g)$ curves simulation with regard to those measured usually in MOS devices. The information obtained on the actual SILC mechanism is finally gathered.

1. Introduction

In state-of-the-art MOS devices, the ultrathin oxides and the high doping concentrations induce quantum mechanical effects which strongly impact device operation. This is the reason why, in recent years, numerous papers have dealt with the simulation of gate current vs. gate voltage, $I_g(V_g)$, or of capacitance vs. gate voltage, $C(V_g)$, characteristics of such devices. In one of these papers, Ghetti et al. [1] have shown that at small negative bias in n-channel MOSFET $I_g(V_g)$ curves, the magnitude of the current measured is much larger than that expected from direct tunneling. An additional current, a trap assisted tunneling current (TATC), also crosses the oxide. After measuring the magnitude of this current at different temperatures, the authors concluded that it results from gate electrons tunneling elastically from the metal Fermi level, E_{fM}, into Si-SiO_2 interface states, which recombine then with the majority carriers of the substrate. As this current is therefore proportional to the Si-SiO_2 interface trap density, D_{it}, at E_{FM}, the authors proposed to use $I_g(V_g)$ curve fitting to extract the D_{it} vs. energy, $D_{it}(E)$, profiles in the upper part of the silicon bandgap. At positive gate biases, a TATC flowing in the reverse direction also crosses the oxides [2]. It may be due to electrons tunneling now from the Si-SiO_2 interface states to the gate Fermi level [2]. After electrical stress, these TATC's, rather called stress induced leakage currents (SILC's), should therefore enable the extraction of the D_{it} relative variations, $\Delta D_{it}/D_{it}$ [1-3]. As D_{it} extraction in MOSFET's with large gate leakage currents is feasible using charge pumping (CP) [4], SILC and CP $\Delta D_{it}/D_{it}$ results can be compared.

In a recent paper, devices with oxide thickness from 2.3 to 1.2 nm have been studied [5]. Depending on the device doping concentration, it has been shown that the electrons entering the TATC/SILC at positive gate bias are supplied either by the inversion layer or by the substrate [5]. Anomalous CP curves have also been reported. A good correlation between the CP curve distortions and the reservoir supplying the carriers entering the TATC/SILC (substrate or inversion layer) has been found [5].

In this chapter, the TATC/SILC properties are discussed further. The conditions that allow the extraction of reliable SILC characteristics are dealt with first. Then SILC properties are studied. It is found that the results obtained contradict the model presented in [1]. Information on the actual SILC mechanism is extracted and gathered.

2. Band Diagram and Ghetti's TATC/SILC Mechanism

For an n-channel device with an n^+ polysilicon gate, the metal Fermi level, E_{fM}, which is close to the silicon conduction band edge at flat band, V_{fb}, scans the upper part of the silicon bandgap in the direction of the intrinsic level, E_i, when V_g is varied from V_{fb} to the device threshold voltage, V_{th} (Fig. 1a). According to Ghetti et al. [1], the SILC stems at negative V_g's, from electrons tunneling from E_{fM} into Si-SiO_2 interface states, from where they recombine with the majority carriers of the substrate (Fig. 1a). At positive gate biases, the tunneling process may be simply reversed [2], the electrons being provided either by the inversion layer or resulting from electron-hole pair generation into Si-SiO_2 interface states [5].

Generally speaking, in the TATC/SILC models, the trap capture cross section, σ, and the trap density, i.e. N_t, in $eV^{-1}.cm^{-3}$, when the TATC/SILC flows through bulk oxide traps or D_{it}, in eV^{-1} cm^{-2} when it flows through interface traps, have the same impact on the gate current. In the frame of the model proposed in [1], at $V_g < 0$ the TATC/SILC results from a series process [2, 6, 7]:

$$J_{SILC} = J_e \, J_{int}/(J_e + J_{int}) \quad (1)$$

where J_e and J_{int} are the oxide and the substrate current densities [1].

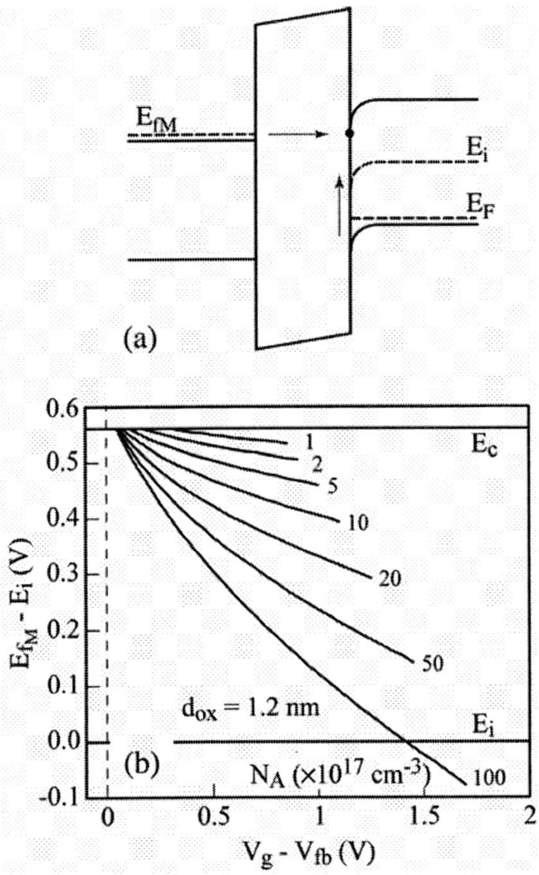

Figure 1. a) Band diagram showing the TATC/SILC mechanism proposed in [1]; b) region scanned at the Si-SiO_2 interface by E_{fM} when V_g is varied from V_{fb} to V_{th}, for different values of N_A (d_{ox} = 1.2 nm).

If tunneling is the TATC/SILC limiting step [1], $J_{SILC} \approx J_e$ is given by:

$$J_{SILC} \approx J_e \approx q.n_g.v_{th}.T_R.(\sigma.D_{it}).(1 - \gamma) \tag{2}$$

where q is the absolute electron charge, n_g is the gate electron concentration, v_{th} is the average carrier thermal velocity, T_R is the Si-SiO_2 barrier transparency and γ is the fraction of occupied traps at the Si-SiO_2 interface ($\gamma \ll 1$).

Therefore, after electrical stress, the relative variations of ($\sigma.D_{it}$) at energy E in the silicon bandgap corresponding to that of E_{fM} (Fig. 1a), are obtained from those of the SILC, $\Delta J_{SILC}/J_{SILC0}$:

$$\Delta J_{SILC}/J_{SILC0} = \Delta(\sigma.D_{it})/(\sigma_0.D_{it0}) = \Delta D_{it}/D_{it0} + \Delta\sigma/\sigma_0 \tag{3}$$

If the variations of σ with stress are negligible, Eq. 1 reduces to [1-3]:

$$\Delta J_{SILC}/J_{SILC0} = \Delta D_{it}/D_{it0} \tag{4}$$

At positives V_g's, if the TATC/SILC mechanism is simply reversed and when the electrons are provided by the inversion layer, n_s, the electron concentration at the Si-SiO_2 interface replaces n_g in Eq. 2 [2]. In the present work, the electrons are provided by the inversion layer in the devices with d_{ox} = 1.5, 1.8 and 2.3 nm [5]. For the devices with d_{ox} = 1.2 nm, between 0 and 0.7 V, the electrons result from electron/hole pairs generation into interface states [5].

The energy region E_{fM} scans at the Si-SiO_2 interface is shown in Fig. 1b for different values of the substrate doping concentration, N_A, in the case of a device with d_{ox} = 1.2 nm. In this figure, the position of E_{fM} with regard to E_i is represented as a function of $(V_g - V_{fb})$. V_g varies between V_{fb} and V_{th} and the energy region scanned is calculated for N_A values ranging from 10^{17} cm^{-3} to 10^{19} cm^{-3}. E_{fM} has been taken at 60 meV above E_c. This corresponds to an n^+ doping concentration around $5x10^{19}$ cm^{-3}.

3. SILC Characteristics Extraction

The n-channel transistors with n^+ polysilicon gates used in this work have been fabricated at LETI, Grenoble-France. The oxide thicknesses are $d_{ox} \approx$ 2.3, 1.8, 1.5 and 1.2 nm. N_A = 1.5-3 10^{18} cm^{-3} for the devices with the first three oxide thicknesses and $N_A \approx 10^{19}$ cm^{-3} for the last one [5]. All $I_g(V_g)$ measurements have been carried out on 300x100 μm^2 transistors. The devices have been stressed using constant voltage stress (CVS) at -2.5, -2.7, -3, and -3.5 V when d_{ox} = 1.2, 1.5, 1.8, and 2.3 nm, respectively (the electric field in the oxide is around 10 $MV.cm^{-1}$ in all cases). The stressing time, t_{stress}, was increased up to 10^4 s.

3.1. $\Delta J_g/J_{g0}$ Curves

The SILC characteristics are often studied using the gate current relative variations $\Delta J_{gn}/J_{g0}$ where J_{gn} is the gate current at stress level n, and J_{g0}, the gate current before stress [8-12]. At $V_g < 0$, if the SILC has the form given in Eq. 2, it is clear that ΔJ_{gn} only accounts for the SILC. This is also true at $V_g > 0$ as long as ΔD_{it} is not large enough to modify significantly the surface potential, i.e. n_s. However, the $\Delta J_{gn}/J_{g0}$ approach is justified only when $\Delta J_{gn}/J_{g0} = \Delta J_{SILC}/J_{SILC0}$, i.e. in the region where the SILC is much larger than J_T, the direct tunneling current so that $J_g >> J_T$.

Fig. 2a shows $J_{gn}(V_g)$ curves recorded for all stress levels from a device with d_{ox} = 1.2 nm. The resulting gate current relative variations, $\Delta J_{gn}/J_{g0}$, are shown in Fig. 2b. The simple reversal of the SILC mechanism from $V_g < 0$ to $V_g > 0$ has been suggested by the continuous variations of $\Delta J_{gn}/J_{g0}$ around $V_g = 0$, well visible in Fig. 2b [2].

The curves in Fig. 2b may suggest a rapid decrease of the SILC outside the -0.8 V /0.8 V interval with a peak in the degradation at $V_g \approx$ -0.8 V, i.e. close to $V_{fb} \approx$ - 1 V. However, Fig. 2a shows that $J_{g0} \geq 10\ J_T$ only between -0.7 V and 0.67 V (vertical dashed lines in Fig. 2b). Outside this region, J_T strongly increases (Fig. 2a) and swamps the SILC, if any, so that $J_T \approx J_{gn}$. As a result, $\Delta J_{gn}/J_{g0} \approx \Delta J_{gn}/J_T$ decreases abruptly and produces the peak at $V_g \approx$ - 0.8 V in Fig. 2b. None of the two decreases of $\Delta J_{gn}/J_{g0}$ outside the -0.8 V/0.8 V interval result from the

SILC. Also, the region outside –1/1 V, where $\Delta J_{gn}/J_{g0}$ is small due to the large contribution of J_T to the gate current, cannot be interpreted in terms of SILC mechanism [10, 11].

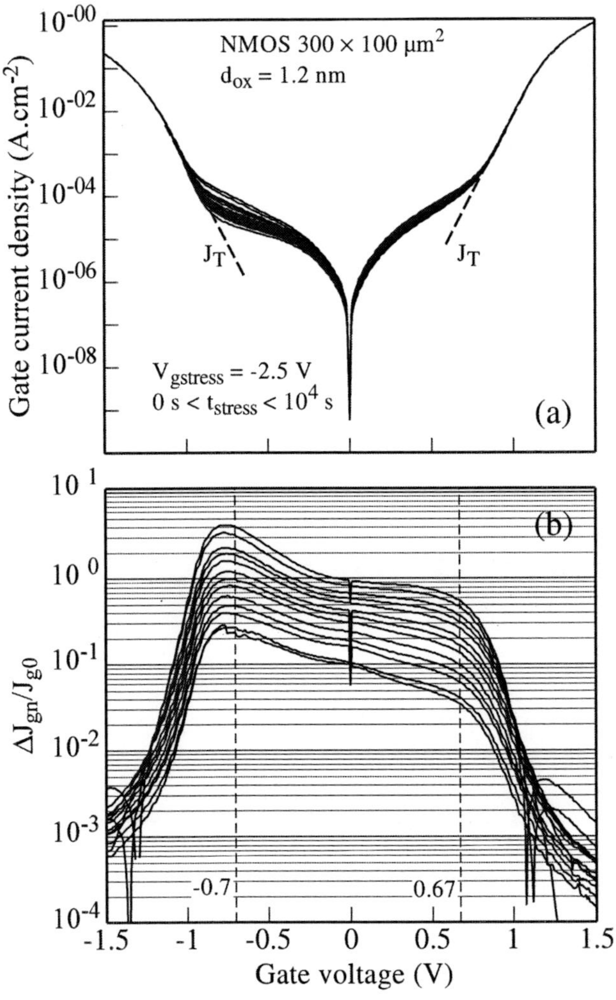

Figure 2. a) $I_g(V_g)$ curves recorded for a device with d_{ox} = 1.2 nm stressed under CVS during stressing times, t_{stress}, between 0 and 10^4 s (J_T: direct tunneling current); b) resulting $\Delta J_{gn}/J_{g0}$ curves. n is the stress step number. The region -0.7 V/0.67 V where $J_{g0} >> J_T$ is marked by vertical dashed lines.

Therefore, $\Delta J_{gn}/J_{g0} = \Delta J_{SILC}/J_{SILC0}$ in small gate bias ranges. This is especially true for devices such as that in Fig. 3. For this device with d_{ox} = 1.5 nm, the TATC/SILC dominates J_{gn} only in a small region at $V_g > 0$. In this region, the electrons entering J_{gn} are provided by the inversion layer [5]. It seems that this is the reason why the continuous behavior of $\Delta J_{gn}/J_{g0}$, from $V_g < 0$ to $V_g > 0$ observed in Fig. 2b [2, 5], is not so well obtained (Fig. 3b). As a result, in this device, $\Delta J_{gn}/J_{g0} \approx \Delta J_{SILC}/J_{SILC0}$ only for V_g values only between - 0.15 V and - 0.7 V (vertical dashed lines in the figure).

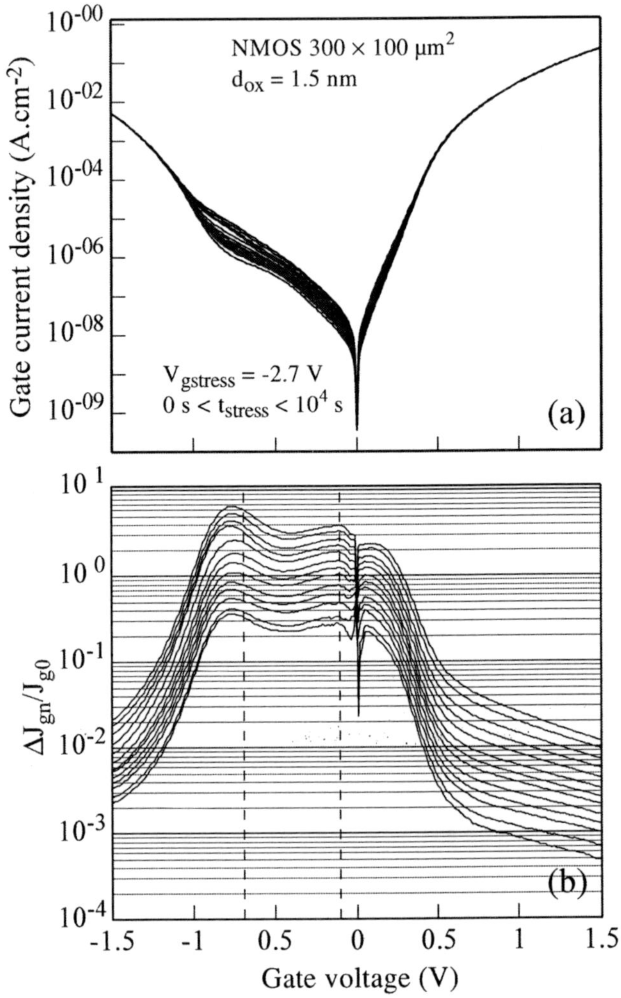

Figure 3. Same measurements as in Fig. 2 but recorded from a device with d_{ox} = 1.5 nm.

3.2. $\Delta J_{gn} = (J_{gn} - J_{g0})$ Curves

The $\Delta J_{gn} = (J_{gn} - J_{g0})$ curves deduced from the $J_g(V_g)$ data in Figs. 2a and 3a, are shown in Figs 4a and 4b, respectively. In theses figures, V_{th} of the corresponding device has been noted. It is around 0.6 V for the devices with d_{ox} = 2.3, 1.8, 1.5 nm and close to 1 V for the device with d_{ox} = 1.2 nm. With regard to the V_{th} values measured experimentally those indicated in Figs. 4a and 4b are shifted by ΔV_{th} = - 0.35 V (d_{ox} = 1.2 nm) and ΔV_{th} = - 0.17 V (d_{ox} = 2.3, 1.8, 1.5 nm), to account for the effect of inversion layers quantization [13, 14].

Whatever V_g, the $\Delta J_{gn} = (J_{gn} - J_{g0})$ curves in Figs. 4a and 4b only account for the SILC. Therefore, with regard to the $\Delta J_g/J_{g0}$ curves in Figs. 2b and 3b, it is clear from these figures that the SILC exists in the whole gate bias ranges used. Also, the greater the V_g value, the greater the SILC is.

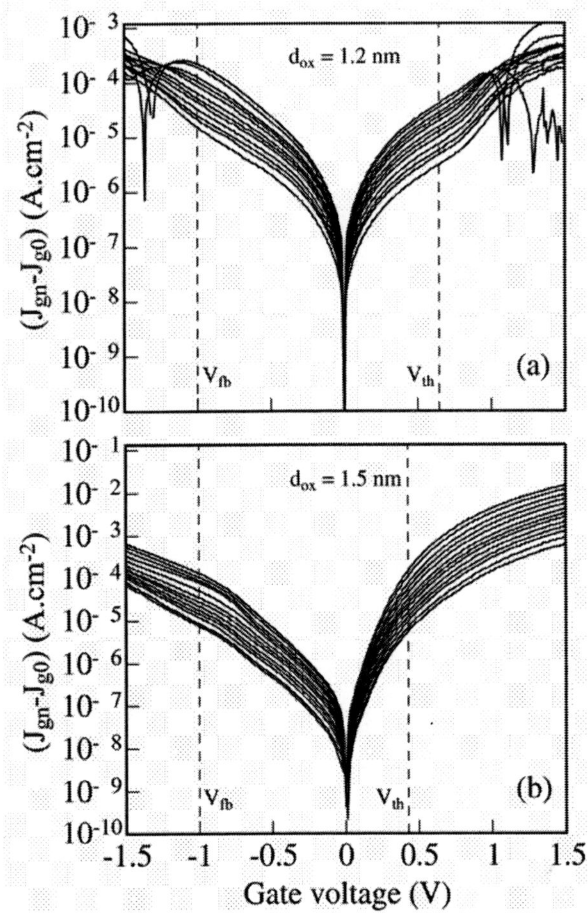

Figure 4. $\Delta J_{gn} = (J_{gn} - J_{g0})$ vs. V_g curves extracted from the data in a) Fig. 2a and b) Fig. 3a.

In the frame of the TATC/SILC mechanism proposed in [1] (Fig. 1a), these curves should reflect the variations with stress and with energy of D_{it} at E_{fM}. Three points contradict this fact.

First, at $V_g < V_{fb}$, E_{fM} is above the silicon conduction band (Fig. 1a), so that there are no Si-SiO_2 interface traps available anymore at E_{fM}. The fact that a SILC still exists in this region means that bulk oxide traps are involved. Then, the carriers may tunnel from the traps to the silicon conduction band or loose their energy in the oxide.

Second, in Fig. 4a and 4b, $(J_{gn} - J_{g0})$ increases continuously from $V_g = 0$ towards both $V_g > 0$ and $V_g < 0$ values. At $V_g < 0$, T, the transparency of the Si-SiO_2 barrier, varies weakly with V_g. Therefore, in this region, the curves in Figs. 4a and 4b are supposed reveal to the variation of D_{it}, i.e. ΔD_{it}, with stress and with V_g.

However, there is no reason for ΔD_{it} to tend to zero at $V_g = 0$, except if one assumes that D_{it} is much smaller there. But the devices in Figs. 4a and 4b have different N_A values. This strongly impacts the device characteristics at $V_g > 0$, as can be seen in Figs. 2a and 3a, so that either the substrate or the source and drain provide the carriers used by the TATC/SILC [5]. As a result, $V_g = 0$ V corresponds to different energy positions of E_{fM} with regard to E_i (Fig.

1b). In other words, $E_{fM} = E_i$ at $V_g = 0.27$ V for the device with $d_{ox} = 1.2$ nm in Fig. 4a, while $E_{fM} = E_i$ at $V_g = -0.18$ V for that with $d_{ox} = 1.5$ nm in Fig. 4b. Once again, there is no reason for $(J_{gn} - J_{g0})$ to be nil at $V_g = 0$ in both cases. Instead, one should observe a minimum in the $(J_n - J_0)$ curves corresponding to the minimum in trap density, if any, with a continuous behavior of $(J_n - J_0)$ from $V_g < 0$ to $V_g > 0$, as in Fig. 2b.

Finally, when the model proposed in [1] is assumed, exponential variations of D_{it} from the intrinsic level E_i towards the conduction bans edge, E_c, are required for $I_g(V_g)$ curve fitting (Figs. 2a and 3a and [1, 6, 13, 15]). But looking at Figs. 4a and 4b, one can conclude that exponential variations of ΔD_{it} with E should also be required for explaining the curves in Fig. 4. From $V_g \approx 0$ up to V_{fb}, the degradation increases by two (Fig. 4a) or three (Fig. 4b) orders of magnitude. Assuming that $E_{FM} = E_i$ at $V_g = 0$, there is no reason again to obtain such large increases of ΔD_{it} from E_i towards E_c.

The same observations hold for the $I_g(V_g)$ curves in Figs. 2a and 3a. Indeed, between $V_g \approx 0$ and $V_g = V_{fb}$ the TATC/SILC increases by more than two (Fig. 2a) and three (Fig. 3a) orders of magnitude. This corresponds to the exponential increase in D_{it} required in $I_g(V_g)$ curves simulation [1, 6, 13, 15] but seems unrealistic with regard to actual $D_{it}(E)$ variations (see section 4.2.). Finally, in the case of the device in Fig. 1a, where the SILC dominates $I_g(V_g)$ on both side of $V_g = 0$, one may expect again a continuous evolution in $|I_g(V_g)|$ on both sides of $V_g = 0$ with $I_g \neq 0$ at $V_g \approx 0$.

3.3. Limitation of the SILC by Majority Carrier Concentration, by Channel Debiasing or by the Series Resistance during Stress

Before discussing further TATC/SILC characteristics, three points deserve to be dealt with. Indeed, the gate current may be limited by other factors than the TATC/SILC component through the oxide.

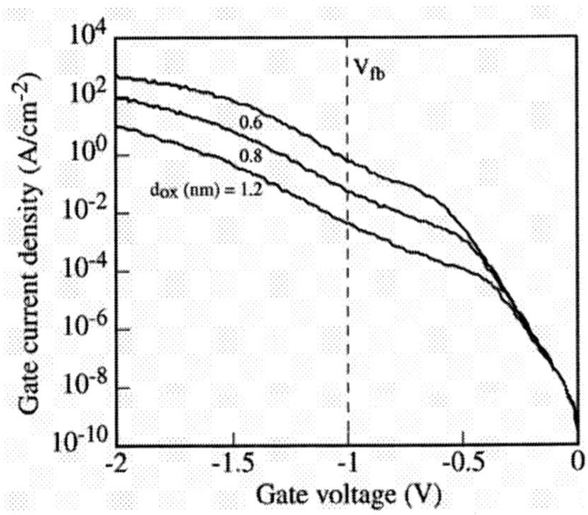

Figure 5. $I_g(V_g)$ characteristics recorded from devices with different oxide thickness (reprinted from [17] with permission). When the TATC/SILC is limited by the majority carrier concentration, I_g is independent of d_{ox} at small negative V_g's.

First, at small negative V_g's, I_g may be limited by the majority carrier concentration at the Si-SiO_2 interface [6]. As can be seen in Fig. 5, where $I_g(V_g)$ curves recorded from devices with different thicknesses are presented, at small negative V_g's, the $I_g(V_g)$ curves overlap regardless of d_{ox}, as expected in that case [15]. At larger negative V_g values, the $I_g(V_g)$ curves depend on d_{ox}, as expected when I_g limits the current [15]. These two regions are separated by a breaking point.

Neither our devices nor those used in [1] show such a behavior. Therefore, in Figs. 2a and 3a, the regions at small negative V_g where the SILC dominates J_g can be fully interpreted in terms of TATC/SILC mechanism through the oxide.

Second, in MOSFET's with so thin oxides, debiasing of the channel at $V_g > 0$ may exist. This occurs when the gate electrons are provided by the inversion layer [5, 16-18]. Then, for small enough d_{ox} and long devices, gate leakage can induce non-uniform inversion layers and impacts $I_g(V_g)$ curves [16-18].

For the devices with d_{ox}= 1.2 nm (Fig. 2), between 0 V < V_g < 0.67 V, the source and drain currents are much smaller than the gate current, i.e. the minority carrier density is too low to contribute to the TATC/SILC [5]. In this bias region, the gate current is equal to the substrate current and the TATC/SILC results from electron-hole pair generation into interface states [5]. Therefore, up to V_g = 0.67 V surface potential is uniform along the channel. Beyond 0.87 V, the source and drain currents become greater than the substrate current [5]. Surface potential should therefore become non uniform between 0.67 V and 0.87 V.

This is confirmed in Fig. 6. This figure shows I_g measured when the source and drain are connected to the ground divided by the same current when one of these terminals is disconnected. From the debiasing point of view, disconnecting the source or the drain is equivalent to multiply the length of the device by two. If the current which crosses the gate in the centre of the channel is negligible with regard to that close to source and drain, surface potential decreases from these regions towards the centre of the channel where it becomes constant [16-18]. Then, when disconnecting the drain or the source, the gate current is divided by two. Of course, if the gate current does not impact the inversion layer, removing one of these connections does not change I_g.

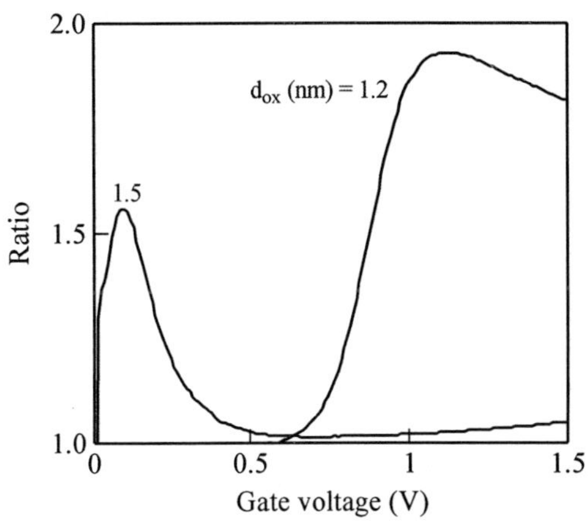

Figure 6. Ratio between the gate current measured when recording $I_g(V_g)$ curves with the source and drain connected to the ground and when one of them is disconnected.

Fig. 6 shows that for the device with d_{ox} = 1.2 nm, I_g is not significantly modified up to $V_g \approx 0.7$ V. At $V_g \approx 1$ V, the ratio is larger than 1.8 V, as expected from a strong debiasing of the channel. Therefore, the channel is not uniform beyond 0.7 V, in accord with the previous discussion on the respective magnitude of I_{SILC} with regard to I_g in Fig. 2. Note however, that beyond $V_g = 0.7$ V, what is measured in Fig. 2 is still a SILC.

Concerning the device with d_{ox} = 1.5 nm, the curves of which are shown in Fig. 3, the electrons which cross the gate are provided by the inversion layer in the whole $V_g > 0$ V region [5]. As can be seen in Fig. 6, between 0.5 V and 1.5 V, disconnecting the drain or the source does not change the gate current significantly so that debiasing is negligible in this region. In weak inversion, I_g is more strongly modified. This is not due to debiasing but results from the contribution of the gate overlap which can significantly impact I_g in this region [18]. However, with regard to the variations of several orders of magnitude of the SILC measured in Fig. 4b, a contribution of 50 % of the gate overlap at $V_g \approx 0.1$ V, does not significantly impact the results and their meaning.

As large area devices have been used, the last point that must be addressed is the impact of series resistance during CVS. It may have limited the gate current and therefore reduced the trap creation kinetics. For given stressing bias conditions, this effect depends on the technology [19]. Series resistance only impacts $I_g(V_g)$ curves at high current densities. Such an impact can be seen at high current level for the device with d_{ox} = 0.6 nm in Fig. 5. There, the slope of the $I_g(V_g)$ curve is smaller than that of the devices with larger d_{ox}'s. Therefore, only the devices with the thinnest oxides (d_{ox} = 1.2 nm), may have suffered from this effect [15]. As exactly the same tendencies of the SILC and between the SILC and CP (see section 4.2) have been observed in the whole paper regardless of d_{ox}, the results are physically meaningful.

Now that it has been shown that the results in Figs. 2 to 4 are reliable, from sections 3.1 and 3.2, it can be concluded from the curves in Figs. 2a, 3a, 4a, and 4b, that the TATC/SILC is not controlled by a single tunneling step across the oxide and by D_{it}. The fact that J_{gn} and $(J_{gn} - J_{g0})$ are equal to zero at $V_g = 0$ and strongly increase with V_g demonstrates the TATC/SILC is controlled by the Fermi level gradient or difference between the metal and the semiconductor. The existence of a SILC beyond V_{fb} as observed in Figs. 4a and 4b but also for the devices with the other oxide thicknesses, implies that bulk oxide traps are involved. But the continuous behavior of $(J_{gn} - J_{g0})$, from $V_g < V_{fb}$ to $V_g > V_{fb}$ in these figures suggests that the same mechanism operates regardless of the position of E_{fM} with regard to the semiconductor conduction band. This supports an inelastic mechanism in the oxide rather than tunneling from the traps to the substrate conduction band.

3.4. $(J_{gn} - J_{g0})/(J_{g1} - J_{g0})$ Curves

In Figs. 7a and 7b, the $(J_{gn} - J_{g0})/(J_{g1} - J_{g0})$ curves obtained using again the data in Figs. 2a and 3a are drawn. In the frame of the model proposed in [1], these curves should highlight the D_{it} variations in the gap after the different stress, normalized with regard to the degradation during the first stress. One can see that at high enough stress level, the degradation follows the same trends regardless of d_{ox}: it increases on both sides of $V_g = 0$ V, exhibits a maximum around V_{fb} and V_{th}, and decreases beyond these maxima (the same behavior is also observed for the other oxide thicknesses). Here again, there is no reason to observe a minimum at V_g =

0 V regardless of N_A and d_{ox}. This also supports that the Fermi level difference between the metal and the semiconductor is the driving force for the TATC/SILC.

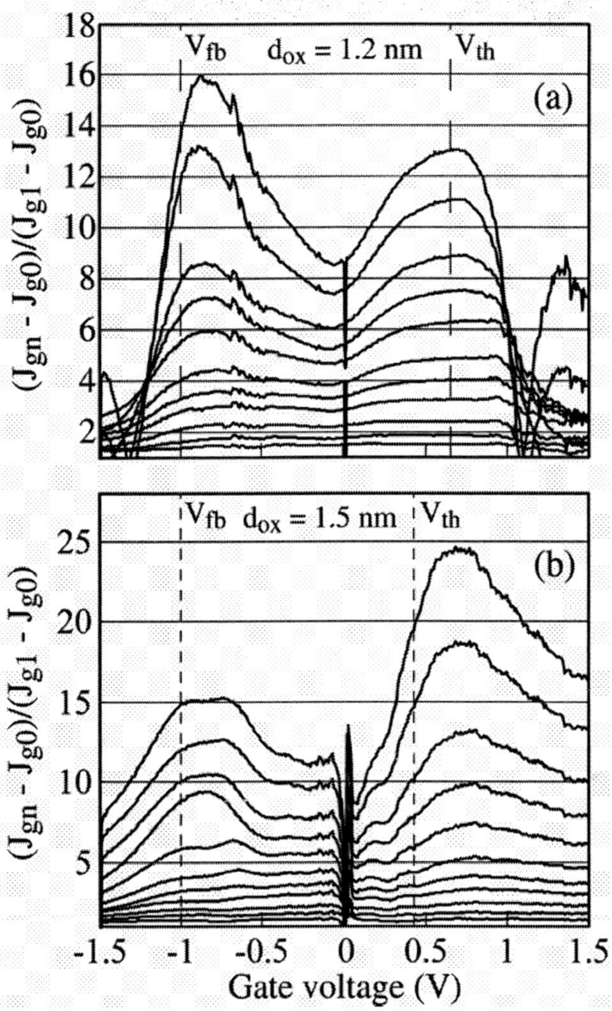

Figure 7. $(J_{gn} - J_{g0})/(J_{g1} - J_{g0})$ vs. V_g curves obtained from the data in a) Fig. 2a and b) Fig. 3a.

Note that $(J_{gn} - J_{g0})/(J_{g1} - J_{g0})$ does not vary continuously around $V_g = 0$ V in Fig. 5b with regard to Fig. 5a. As in Fig. 3b with regard to Fig. 2b, this behavior may be attributed to the change in the reservoir providing the carriers involved in the TATC/SILC.

3.5. $d(\log\Delta J_n/dV_g)$ Curves

A further comparison of the curves in Figs. 4a and 4b can be carried out by plotting $d(\log\Delta J_{gn})/dV_g = (1/\Delta J_{gn}).(d\Delta J_{gn}/dV_g)$, i.e. the slope of the $(d\Delta J_{gn}/dV_g)$ curves normalized with regard to ΔJ_{gn}. In the frame of the model proposed in [1], at a given V_g, and for a given stress level, these curves should reflect the variations of D_{it} with energy E, normalized by the increase of D_{it} at E that results from this stress.

In Eq. 2, T, the transparency of the Si-SiO_2 barrier expresses as [13]:

$$T_R = \exp\left[\frac{-4d_{ox}\sqrt{2m_0 q}}{3\hbar V_{ox}}\left[\Phi^{1.5} - (\Phi - V_{ox})^{1.5}\right]\right] \quad (5),$$

where m_0 is the electron effective mass in the oxide, ħ is the reduced Plank constant, V_{ox} and Φ being the voltage drop across the oxide and the Si-SiO_2 barrier height, respectively.

At $V_g < 0$, from Eqs. 2 and 5, $d(\log\Delta J_{gn})/dV_g$ reads:

$$\frac{1}{\Delta J_{gn}}\frac{d\Delta J_{gn}}{dV_g} = \frac{d\left[\log\Delta J_{gn}\right]}{dV_g} = \frac{-4d_{ox}\sqrt{2m_0 q}}{3\hbar V_{ox}}\left[B(\Phi, V_{ox})\right]\frac{dV_{ox}}{dV_g} + \frac{d(\log\Delta D_{it})}{dV_g} \quad (6),$$

with

$$B(\Phi, V_{ox}) = \frac{1}{V_{ox}}\left[\Phi^{3/2} - (\Phi - V_{ox})^{3/2}\right] + \frac{3}{2}(\Phi - V_{ox})^{1/2} \quad (7).$$

It is easy to show that the magnitude of the right hand side of Eq. 6 is essentially determined by the second term, i.e. that which depends on ΔD_{it}.

At $V_g > 0$, and for the devices in which the electrons of the TATC/SILC are provided by the inversion layer (d_{ox} = 1.5, 1.8, 2.3 nm) [5], n_s in Eq. 2 induces an additional term, $d(\log n_s)/dV_g$, in the right hand side of Eq. 6. Therefore, the impact of ΔD_{it} is expected to be swamped in that term.

The $d(\log\Delta J_{gn})/dV_g = (1/\Delta J_{gn}).d(\Delta J_{gn}/dV_g) \approx [1/\Delta D_{it}\ d(\Delta D_{it}/dV_g)]$ curves recorded from the devices with d_{ox} = 1.2 nm and d_{ox} = 1.5 nm are shown in Figs. 8a and 8b for all stress levels. We focus on the device with d_{ox} = 1.5 nm first (Fig. 8b). It is clear in Fig. 8b and at first order, that at $V_g < 0$, $[1/\Delta D_{it}\ d(\Delta D_{it}/dV_g)]$ does not depend on stress level. This means that the slope of the curves in Fig. 4a is proportional to the degradation ΔJ_{gn} or ΔD_{it}: at a given V_g, $d(\Delta D_{it}/dV_g) = \alpha.\Delta D_{it}$ with α = Cste.

For this device, at $V_g > 0$, $[1/\Delta J_{gn}.(d\Delta J_{gn}/dV_g)]$ is significantly larger than at $V_g < 0$. This agrees with the fact that at $V_g > 0$, the additional term, $d(\log n_s)/dV_g$, dominates the right hand side of Eq. 6. As expected also from that term, the curves do not depend on stress level.

The same curves recorded from the device with d_{ox} = 1.2 nm are shown in Fig. 8a. $[1/\Delta D_{it}\ d(\Delta D_{it}/dV_g)]$ is again independent on stress level at $V_g < 0$. The large variations beyond V_{fb} mainly result from the resolution limit of the measurement apparatus which induces some negative $(J_{gn} - J_{g0})$ values at large V_g's in Fig. 4a.

Note that at $0 < V_g < 0.7$ V, in the region where the TATC/SILC results from minority carriers generated into interface states [5], $[1/\Delta D_{it}.(d\Delta D_{it}/dV_g)]$ is also independent on stress level.

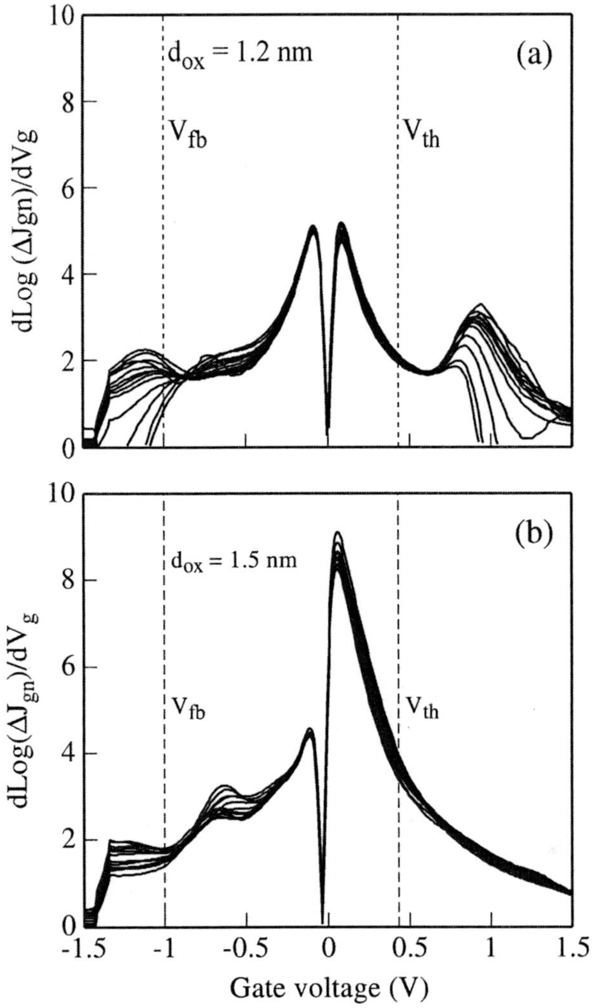

Figure 8. $d[\log(\Delta J_{gn}/dV_g)]$ vs. V_g curves plotted for all stress levels for devices with a) $d_{ox} = 1.2$ nm and b) $d_{ox} = 1.5$ nm.

$d(\log\Delta J_{gn})/dV_g$ curves recorded from the different devices used are compared in Fig. 9. As expected from devices which have V_{th} values close together (d_{ox} = 1.5, 1.8 and 2.3 nm), the curves nearly superimpose at $V_g > 0$. At $V_g < 0$ the shape of the curves is also independent on the device characteristics (The curve recorded from the device with $d_{ox} = 1.5$ nm, deviates from the general shape at $V_g \approx - 0.6$ V). As a result, Fig. 8 and 9 show that when the carriers of the TATC/SILC are provided by the gate, $[1/\Delta J_{gn}.(d\Delta J_{gn}/dV_g)]$ is independent of stress level and of the device, i.e. the slope of the $\Delta D_{it} = f(V_g)$ curves is, at first order, proportional to ΔD_{it}. In the frame of the mechanism proposed in [1], there is no reason at all for that. This supports again that the TATC/SILC does not depend only defect density but also on Fermi level difference between the gate and the semiconductor. Such curves may be useful tools for identifying the actual SILC mechanism.

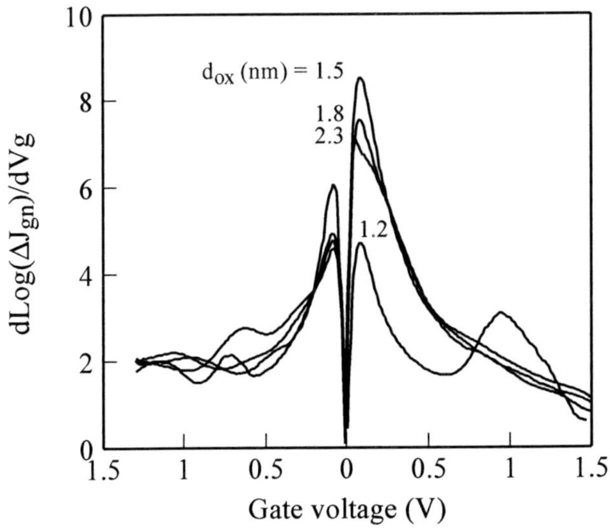

Figure 9. Comparison between $d[\log(\Delta J_{gn}/dV_g)]$ vs. V_g curves recorded from devices with all oxide thicknesses.

4. TATC/SILC and Trap Parameters. Comparison with Those Measured by CP

In section 3, the $J_{gn}(V_g)$ data have been used in different ways in order to extract reliable TATC/SILC characteristics. From this study it results that TATC/SILC mechanism cannot be that proposed in [1]. By doing so important characteristics of the actual mechanism have been pointed out. In this section, the trap characteristics required in $I_g(V_g)$ curve fitting to account for the TATC/SILC are discussed. The D_{it} variations under stress extracted from the SILC and from CP are also compared. Both further confirm previous results and provide additional information on the actual TATC/SILC mechanism.

4.1. Trap parameters Required for $I_g(V_g)$ Curve Fitting

($\sigma.D_{it}$) and $I_g(V_g)$ curve fitting

In the TATC/SILC tunneling models and regardless of d_{ox}, N_t and σ have the same impact on the gate current [1, 7, 20-22]. Then, values of σ in the order of 10^{-13} cm^2 are required in $I_g(V_g)$ curve simulation to get large enough ($\sigma.N_t$) (or ($\sigma.D_{it}$)) with acceptable N_t [1, 22-24]. Such σ values are large and should correspond to Coulombic attractive traps. This led Ielmini et al. [22] to assume in their devices with d_{ox} = 7.6 nm, that the traps responsible for the SILC were Coulombic attractive bulk oxide traps. But the SILC is widely accepted to occur through neutral traps in such oxides [25, 27]. Such values of σ are therefore too large with regard to those expected.

This is also true for the devices with ultrathin oxides. If the TATC/SILC flows through interface traps, the traps implied are essentially those in the upper half of the silicon bandgap (Fig. 1b) [1]. It is generally accepted these traps are acceptor-like traps, i.e. neutral when empty, those in the lower half of the gap being donor-like traps. For neutral traps, σ should be in the 10^{-14}-10^{-18} cm^2 range. Therefore, $\sigma \approx 10^{-13}$ cm^2, as also required in that case to account for the TATC/SILC in $I_g(V_g)$ curve, is again a too large value [1].

($\sigma.D_{it}$) versus energy variations and $I_g(V_g)$ curve fitting

The Si-SiO_2 barrier transparency varies slowly in weak inversion (Eq. 5). In the frame of the TATC/SILC mechanism proposed in [1], the strong increase in the gate current in the -1/0 V bias region therefore mainly results from ($\sigma.D_{it}$) variations. As noted above, exponential increases of ($\sigma.D_{it}$) from E_i to E_c are necessary to fit the $I_g(V_g)$ characteristics [1, 6, 13, 15]. Besides, it is well-known that $D_{it}(E)$ exhibits a U-shape. For a reason which is not clear yet, $D_{it}(E)$ is found to increase faster towards band edges when measured using capacitance-voltage techniques with regard to DLTS or CP [28]. But in most cases, $D_{it}(E)$ is quite flat in the centre of the gap and strongly increases only close to band edges [29, 30], so that an exponential increase of D_{it} from midgap towards band edge does not seem realistic with regard to experimental data.

With the CP method used here for measuring D_{it} [4], Si-SiO_2 interface trap depth concentration profiles such as those reported in [31] can be recorded. The parts of the profiles obtained using different gate pulse heights overlap well when the trap parameters, N_t and σ, do not vary with energy [31]. By doing so, neither the trap densities nor the trap cross sections have been found to vary significantly with energy in the region of the silicon bandgap probed, that is $\pm$ 0.3 eV to $\pm$ 0.4 eV around midgap [31]. In no case, the small variations of these parameters measured can be compared with the large $D_{it}(E)$ (or $\sigma(E)$) variations required in [1, 13] (or [15]) to fit the $I_g(V_g)$ curves. Note that exponential variations of D_{it} with E increase the average ($\sigma.D_{it}$) with regard to the values measured experimentally.

Therefore, from the large D_{it} and σ required for $I_g(V_g)$ curve fitting, one can conclude that the ($\sigma.D_{it}$) values used in that case are at least ten times larger than the those expected from actual trap characteristics. This strongly suggests that a much larger number of traps than those around the energy corresponding to the metal Fermi level [1, 3] are involved in the TATC/SILC. For instance, all the traps between both Fermi levels may be implied in the process. Alternately, the efficiency of the traps in generating the SILC may be greater than that expected from $\Delta J/J_0 = \Delta D_{it}/D_{it0}$ or $\Delta N_t/N_{t0}$ (see section 4.2).

4.2. Comparison between SILC and CP $\Delta D_{it}/D_{it0}$

Implication of the Si-SiO_2 interface traps in the TATC/SILC

In [5], anomalous CP curves have been reported. They correlate well with the fact that either the substrate or the source and drain provide the carriers entering the TATC/SILC. Therefore, CP curves suggest that the TATC/SILC flows through interface traps.

$I_g(V_g)$ curves may also be put forward to support that the TATC/SILC flows through Si-SiO_2 interface traps. Indeed, when recorded from devices with smaller and smaller d_{ox}, they show parallel upward shifts regardless of V_g, as can be seen in Fig. 5 [15]. In this figure, this is true in both the TATC/SILC (-1 V < V_g < -0.5 V) and the direct tunneling (V_g < -1 V) regions, suggesting that what is viewed in these two V_g ranges is the full oxide thickness. This feature may be put forward to supports the mechanism proposed in [1]. As all the preceding contradicts such a conclusion, this shift should be explained in the frame of the actual TATC/SILC process.

As a result, at present time, only CP results support that the TATC/SILC flows through Si-SiO_2 interface states. This point deserves to be confirmed.

Comparison between SILC and CP $\Delta D_{it}/D_{it0}$ results

If one assumes that the TACT/SILC flows through Si-SiO_2 interface states, and according to the mechanism proposed in [1], the variations with stress of $\Delta D_{it}/D_{it0}$ obtained from the SILC and from CP should be the same. A CP technique proposed for extracting D_{it} in MOSFET's with ultrathin oxides has been described in [4]. It also allows the measurement of σ. For the devices studied, σ does not vary significantly with stress in the stress conditions used. This is in accord with the small D_{it} variations measured by CP [32]. Therefore, for the extraction of $\Delta D_{it}/D_{it0}$ from the gate current, Eq. 4 can be used.

CP measurements have been carried out on devices of either 300×0.15 μm^2 or 10×10 μm^2. $\Delta D_{it}/D_{it0}$ results obtained that way are shown in Fig. 10 as a function of t_{stress} for a devices with d_{ox} = 1.5 nm (see [5] for d_{ox} = 1.2 nm). The corresponding SILC results measured at different V_g's in the region of reliable extraction in Fig. 3b are also shown in the figure.

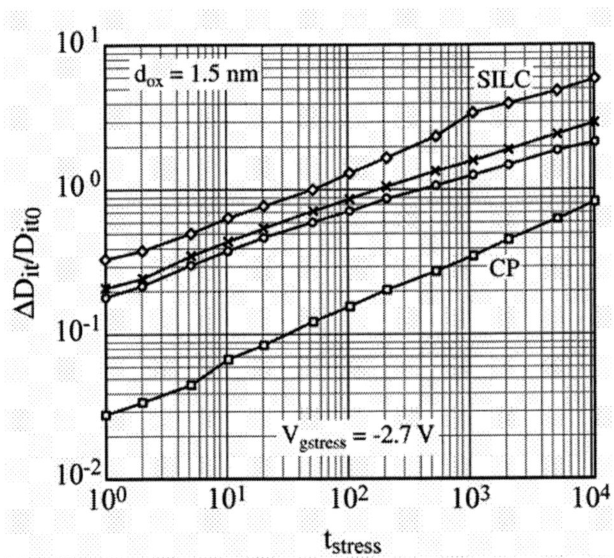

Figure 10. $\Delta D_{it}/D_{it0}$ values as a function of the stressing time obtained from the SILC using Eq. 4. The $\Delta J_{gn}/J_{g0}$ data have been measured at different V_g's in the region of reliable extraction in Fig. 3b. Comparison with CP results.

As expected from the SILC [1, 2, 9, 12, 33], the curves follow a t_{stress}^{α} dependence. Regardless of d_{ox} and of the sensing bias, α is between 0.27 and 0.34, i.e. comparable with the values obtained by other groups in similar stressing conditions [1, 2, 9, 12, 33]. The CP $\Delta D_{it}/D_{it0}$ curve also follow a power law with α between 0.36 and 0.46, in accord with results obtained on devices with d_{ox} = 2.3 nm using conventional CP [34]. The α values corresponding to the CP results are therefore significantly larger than those extracted from the SILC. This suggests that at least partly, the SILC does not flow through the same traps as those measured by CP. However, the main feature in Fig. 10 is that the SILC $\Delta D_{it}/D_{it0}$ values are much larger than those extracted from CP. Depending on the device and on the bias where the SILC $\Delta D_{it}/D_{it0}$ curves are measured, the ratio between them is comprised between 3 and 50 (Fig. 10 and Ref. 5, Fig. 4).

It is well known CP averages the interface trap density measured over the region probed. Depending on the gate voltage swing and on trap capture cross section, this region can reach ± 0.3 to ± 0.4 eV around midgap [35]. On the contrary, in the frame of the model proposed in [1], the TATC/SILC results from traps at E_{FM}, E_{FM} scanning a more or less large region of the Si-SiO_2 interface depending on the doping concentration (Fig. 1a) [1]. This point has been put forward to explain the very different $\Delta D_{it}/D_{it0}$ values obtained using the SILC and CP. However, it has been shown that the energy regions where the SILC and CP probe the Si-SiO_2 interface overlap [5]. This is especially true for the devices with high N_A concentrations, i.e. for the device with d_{ox} = 1.2 nm (Fig. 1a and [5]). Consequently, it seems difficult to explain the large differences in the $\Delta D_{it}/D_{it0}$ values measured by the two techniques by differences in the energy region probed.

Therefore, the results in Fig. 10 contradict again the TATC/SILC mechanism proposed in [1]. As there is no doubt that CP probes interface or near interface states, two situations may explain these results: either the traps involved in the TATC/SILC are different from those probed by CP, i.e. are not interface states, and are such that their concentration increases much faster with stress than those at the interface, or the TATC/SILC mechanism is such that the traps created are much more efficient than expected in creating the SILC, i.e. $\Delta J/J_0 >> \Delta D_{it}/D_{it0}$ or $\Delta N_t/N_{t0}$. As so large differences in the trap creation rate with stress from some traps to some others seems unrealistic (first possible explanation), the results in Fig. 10 suggest that the impact of the traps in creating the SILC, is greater than expected. Such a feature has already be noted [25].

As a result and similarly as ($\sigma.D_{it0}$) studied in section 4.1, $\Delta D_{it}/D_{it0}$ is also always extremely large when evaluated using the TATC/SILC with regard to the trap characteristics measured. In sections 4.1 this conclusion was obtained from the comparison between experimental data and simulation. Here, it only results from experimental data, and therefore excludes any hypothesis on the trap characteristics. In section 4.1, it was concluded that either the number of traps involved in the TATC/SILC is much larger than that at E_{FM}, or that the efficiency of the traps in producing the TATC/SILC is larger than expected from Eq. 4. Here, and as so large differences in the generation rates experienced by different traps are unlikely, the second explanation necessarily enters the results obtained.

Recombination in the oxide has been proposed in [22] for explaining the TATC/SILC in devices with ultrathin oxides. Then, and as soon as the traps are not located in the vicinity of the Si-SiO_2 interface anymore, their filling depends on the carrier concentration at the Si-SiO_2 interface [1, 22]. Also, and contrary to the conclusion drawn in [8], TATC/SILC beyond V_{fb} is expected in that case, as observed experimentally in Fig. 4.

Then, on the one hand, according to the results presented in [1], the TATC/SILC should become excessively dependent on temperature with regard to the experimental results. But on the other hand, accounting for a volumic trap distribution throughout the oxide and that the traps between the metal and semiconductor Fermi levels may contribute to the TATC/SILC, the curves in Fig. 4 along with the large (σ Dit) values required in [1] may be explained. The contribution, throughout the oxide, of a large part of the traps created, providing a large number of recombination paths, may also explain the efficiency of the degradation in the oxide with regard to that predicted at the Si-SiO_2 interface by Eq. 4.

5. Actual TATC/SILC Mechanism

The TATC/SILC properties can be gathered as follows. The TATC/SILC results from a bulk and inelastic mechanism governed by Fermi level gradient between the metal and the semiconductor. It should involve a much larger number of traps than those located close the metal Fermi level. The efficiency of the traps created for generating the SILC should be greater than expected from Eq. 4. These requirements may be fulfilled if a large number of traps between both Fermi levels and throughout the oxide are involved.

The TATC/SILC mechanism should also be such that $[1/\Delta J_n.(d\Delta J_n/dV_g)] \approx$ Cste and should produce parallel shifts of the $I_g(V_g)$ characteristics when reducing d_{ox}. The Si-SiO_2 interface states may enter the TATC/SILC, but this point deserves to be confirmed.

6. Conclusion

In this chapter, the TATC/SILC in n-channel MOSFET's with n^+ polysilicon gates and oxides in the 1-2 nm range has been studied thoroughly. First, the use $J_g(V_g)$ characteristics for extracting information on the SILC has been discussed. The $\Delta J_{gn}/J_{g0}$ curves yield reliable $\Delta J_{SILC}/J_{SILC0}$ values only in small gate bias regions. The reliability of the results obtained in terms of TATC/SILC properties, has been verified. Then, it has been shown that plots such as $(J_{gn} - J_0)$, $(J_{gn} - J_{g0})/(J_{g1} - J_{g0})$, or $d(\log(\Delta J_{gn}))/dV_g$, which only account for the SILC, allow to obtain more information, including important insights on the SILC mechanism. From these results it has been concluded that the TATC/SILC mechanism is not that proposed in [1]. The TATC/SILC is governed by Fermi level gradient between the metal and the semiconductor. It likely implies both interface and oxide traps as well as inelastic transitions. From the comparison between the trap characteristics required for $J_g(V_g)$ curve fitting and those generally measured it has been concluded that more traps at the Si-SiO_2 interface than those at the energy of the metal Fermi level are involved. Finally, the results strongly suggest that the efficiency of the traps in generating the TATC/SILC is greater than that expected from the SILC models.

References

[1] Ghetti, A.; Sangiorgi, E.; Bude, J.; Sorsch, T. W.; Weber, G. *IEEE Trans. Electron. Devices* 2000, 47, 2358-2365.

[2] Lime, F.; Ghibaudo, G.; Guégan, G. *Microelectron. Reliab.* 2001, 41, 1421-1425.
[3] DiMaria D. J.; Stathi J. H. *Appl. Phys. Lett.* 1997, 71, 3230-3232.
[4] Bauza, D. *Solid-St Electron.* 2003, 47, 1677-1683.
[5] Bauza, D.; Rahmoune, F.; Laqli, R.; Ghibaudo G. *Microelectron. Reliab.* 2005, 45, 849-852.
[6] Clerc, R.; Spinelli, A. S.; Ghibaudo, G.; Leroux, C.; Pananakakis, G. *Microelectron. Reliab.* 2001, 41, 1027-1030.
[7] Endoh, T.; Chiba, T.; Sakuraba, H. ; Lenski, M.; Masuoka, F. *J. Appl. Phys.* 1999, 86, 2095-2099.
[8] Nicollian, P. E.; Rodder, M.; Grider, D. T.; Chen, P.; Wallace, R. M.; Hattangady, S. V. 1999 *Inter. Reliab. Phys. Symposium (IRPS)*, San Diego, California, Proceedings pp. 400-404.
[9] Meinertzhagen, A.; Petit, C.; Zander, D.; Simonetti, O.; Maurel, T.; Jourdain, M.; *J. Appl. Phys.* 2002, 91, 2123-2132.
[10] Loh, W. Y.; Cho, B. J.; Li, M. F. *Appl. Phys. Lett.* 2002, 81, 379-381.
[11] Crupi, F.; Ciofi, C.; Germanò, A.; Iannaccone, G.; Sthatis, J. H.; Lombardo, S. *Appl. Phys. Lett.* 2002, 80, 4597-4599.
[12] Goguenheim, D.; Trapes, C.; Bravaix. A. *J. Non-Crystal Solids* 2003, 322, 183-190.
[13] R. Clerc, PhD thesis. Institut National Polytechnique de Grenoble, November 2001 the 14th.
[14] van Dort, M. J.; Woerlee, P.H.; Walker, A.J.; Juffermans, C. A. H.; Lifka, H. *IEEE Trans. Electron Devices* 1992, 39, 932-938.
[15] F. Lime, PhD thesis, Institut National Polytechnique de Grenoble, January 2004 the 29th.
[16] Romanjek, K. ; Lime, F. ; Ghibaudo, G.; Leroux, C. *Solid-St. Electron.* 2003, 47, 1657-1661.
[17] Gilibert, F.; Rideau, D.; Bernardini, S.; Scheer, P.; Minondo, M.; Roy, D.; Gouget, G.; Juge, A. *Solid-St. Electron.* 2004, 48, 597-608.
[18] Shen, M. Y. C.; Jopling, J; Massoud, H. Z.; in *The Physics and Chemistry of* SiO_2 *and the Si-*SiO_2 *interface – 5* symposium, H.Z. Massoud, J.H. Stathis, T. Hattori, D. Misra, and I. Baumol, Editors, ECS transactions 2005, Vol. 1, N° 1, pp. 283-294.
[19] Leroux, C.; Mur, P.; Rochat, N.; Rouchon, D.; Truche, R.; Reimbold, G.; Ghibaudo G. *Microelectron. Eng.* 2004, 72, 121-124.
[20] Ghetti, A.; Sangiorgi, E.; Bude, J.; Sirsch, T. W.; Weber, G. *International Electron Device Meeting (IEDM)* 1999. Proceedings pp. 731-734.
[21] Ghetti, A.; Bude, J.; Silverman, P.; Hamad, A.; Vaida, H. *IEICE Trans. Electron.* 2000, 83-C, 1175-1152.
[22] Ielmini, D.; Spinelli, A. S.; Rigamonti, M. A.; Lacaita, A. L. *IEEE Trans. Electron Devices* 2000, 47, 1266-1272.
[23] Ricco, B.; Gozzi, G.; Lanzoni, M. *IEEE Trans. Electron. Devices* 1998, 45, 1554-1560.
[24] Lee, J.; Boshman, G.; Green, K. R.; Ladwig, D. *IEEE Trans. Electron Devices* 2002, 49, 1232-1241.
[25] DiMaria, D. J.; Cartier E. *J. Appl. Phys.* 1995, 78, 3883-3864.
[26] Yoon, G. W.; Joshi, A. B.; Kim, J.; Kwong, D.L. *IEEE Electron Device Lett.* 1993, 14, 231-233.

[27] De Blauwe, J.; Van Houdt, J.; Wellekens, D.; Groeseneken, G.; Maes, H. E. *IEEE Trans. Electron Devices* 1998, 45, 1745-1760.

[28] Autran, J. L.; Seigneur, F.; Plossu C.; Balland, B. *J. Appl. Phys.* 1993, 74, 3932-3935.

[29] Trombetta, L. P.; Feigh, F. J.; Zeto, R. J. *J. Appl. Phys.* 1991, 69, 2512-2521.

[30] Bhattacharyya, A; Vorst, C.; Carim, A. H. *J. Electrochem. Soc.* 1985, 132, 1900-1908.

[31] Manéglia, Y.; Rahmoune, F.; Bauza, D.; *J. Appl. Phys.* 2004, 97, 0145021-0145028.

[32] Y. Maneglia and D. Bauza, in *Silicon Nitride and Silicon Dioxide Thin Insulating Films VII* symposium, K.B. Sundaram, M.J. Deen, W.D. Brown, R.E. Sah, E. Poindexter, and D. Misra, Editors. The Electrochemical Society Proceedings 1999, vol. 99-6, pp. 59- 74.

[33] Zander, D.; Saigne, F.; Petit, C.; Meinertzhagen, A. J. *Non-Cryst Solids* 2001, 280, 86-91.

[34] Zander, D. ; Petit, C. ; Saigne, F. ; Meinertzhagen, A. *Microelectron. Reliab.* 2001, 41, 1023-1026.

[35] Bauza, D. *J. Appl. Phys.* 2003, 94, 3239-3248.

This work has been supported by Ultimox/RMNT and RA-Nanodiel projects. The authors wish to thank Martine Gri from IMEP for technical assistance.

In: MOSFETs: Properties, Preparations and Performance ISBN: 978-1-60456-762-5
Editors: N.T. Andre et al. pp. 401-419

Chapter 12

Analog and Digital Circuit Functionality under the Influence of Gate Oxide Degradation and Breakdown

R. Rodríguez*[*], *J. Martín-Martínez, R. Fernández, M. Nafría and X. Aymerich

Dept. d'Enginyeria Electrònica. ETSE. Universitat Autònoma de Barcelona. 08193 Bellaterra. Barcelona, Spain

Abstract

To investigate the impact of the gate oxide degradation and breakdown (BD) on CMOS circuits functionality, an accurate description of the electrical characteristics of the stressed devices is presented, which can be included in circuit simulators. The proposed picture considers, on the one hand, the variation of the channel current and, on the other, the increase of the gate current due to the oxide degradation and BD and is able to account for different levels of oxide damage. The effect of gate oxide degradation and BD on simple analog (current mirror) and digital (RS latches, SRAM memory cells, CMOS inverters) circuits has been evaluated.

Index Terms: CMOS, dielectric breakdown, hard breakdown, oxide reliability.

I. Introduction

Gate oxide thickness and MOSFET channel dimensions have rapidly decreased in the last years as a consequence of the continuous scaling of devices [1], leading to the appearance of several failure mechanisms which limit the IC reliability. In the case of the gate oxide, the dielectric breakdown (BD), that is, the loss of its dielectric properties is one of the most relevant. However, it seems that the oxide BD effect is overestimated and that devices and

[*] E-mail address: Rosana.Rodriguez@uab.es. Tel.: +34 93 581 3217; Fax.:+34 93 581 26 00. (Corresponding author.)

circuits can still work after an important level of oxide degradation [2-4]. Therefore, the analysis of the impact of dielectric BD (and of its previous degradation) on MOS devices and circuits functionality is critical for the evaluation of present CMOS reliability. In this direction, the development of models for broken down devices that can be included in circuit simulators is needed [5]. Customary, the effect of BD on device performance has been studied through the variation of some device parameter (off current, threshold voltage, transconductance...) [6-8], but less attention has been paid to the modification of the complete output characteristics of the device (I_D-V_{DS}). In addition, equivalent electrical circuits for FETs broken down at different positions of the channel have been already presented, which can be included in SPICE simulators [2,9]. In these cases, different equivalent circuits are needed, depending on the BD hardness and BD position along the channel, but the possible effects of BD on other device parameters [6,7] are not considered. An alternative is the description of the broken down MOSFET using models like, for example, BSIM, whose parameters are extracted after the oxide BD [10]. However, when this is done, the gate BD current is not usually considered [11]. Independently of the approach, BD is the main focus, but much less attention has been paid to the effects of the degradation previous to the device BD [12], which can have a large impact in analog circuit performance.

In this work, a unified approach for the description of the electrical characteristics of worn-out (before BD) and broken down MOSFETs is presented, which can be easily introduced in circuit simulators and simultaneously considers the increase of the gate current and the modification of the conduction through the channel due to the oxide degradation. In this approach, the I-V curves of the device in all the operation regions are described using an equivalent circuit for the post-BD gate current and BSIM4 SPICE model to account for the variation of the current along the channel. The data obtained on nMOSFETs that have suffered different levels of oxide degradation and/or BD at different locations along the channel have been correctly described using this approach. They have been introduced in a circuit simulator to evaluate the impact of degradation and BD on the functionality of simple analog and digital circuits. In addition, the adequacy of the description has been checked on experimental data at circuit level.

II. Experimental

nMOSFETs of 1.5V bulk technology with SiO_2 or SiON as gate dielectric have been studied. For SiO_2 devices (tox=2.2nm) two different aspect ratios, W/L = 10μm/10μm and W/L=10μm/0.175μm, have been considered. SiON devices (1.6nm EOT) were much narrower, with a W/L ratio of 0.35μm/0.13μm. Constant voltage stresses (3.7V and 3V for the SiO_2 and SiON samples, respectively) were applied to the gate with the other terminals grounded to provoke the degradation and/or BD of the oxide. The stress was stopped periodically to measure the transistor I_G-V_G, I_D-V_{DS}, I_D-V_G characteristics after different degradation levels of the oxide, from the fresh case until their dielectric breakdown. After BD, in addition to these curves, the I_G, I_D, I_S and I_B currents were also measured when a ramp voltage was applied to the gate with the other terminals grounded, to determine the position of the BD path along the channel, following the procedure described in [13]. SBD events were measured in the SiO_2 samples, whereas HBD events were mostly observed in the SiON devices, in agreement with the transistor geometry dependence observed in other works [14].

The extracted parameters of the damaged bulk transistors characteristics have been included in a circuit simulator to evaluate the effect of the oxide degradation and breakdown on a current mirror and RS latches.

On the other hand, the impact of gate oxide breakdown was also analyzed on circuits fabricated with SOI technology. A CMOS 6T SRAM cell was investigated using a 0.13μm /1.2V partially depleted SOI technology [15]. In this case, cell transfer ratios (width ratio of n-FET/pass-gate) were approximately 2-2.4. The effect of the oxide SBD and HBD on the cell functionality was evaluated through simulations. Only the gate-to-diffusion (source or drain) breakdown was considered, since these represent the worst case situations. In order to quantify the cell stability, the static noise margin (SNM), i.e., the maximum DC noise voltage necessary to flip the state of the cell [16, 17], was obtained. A butterfly plot, which consists of drawing and mirroring the transfer curve of the inverters of the cell (Fig. 10), has been used to obtain the SNM. Additionally, the influence of the dielectric degradation and BD was also experimentally investigated on CMOS inverters with SOI technology. In this case, inverters fabricated with 0.13μm/1.2V partially depleted SOI technology with W=0.175μm L=0.125μm and tox=1.5 nm for both NFETs and PFETs were analyzed [18]. Constant voltage stresses (positive and negative) were applied to the inverter input to provoke the oxide breakdown with inverter output at ground and the other terminals floating. After the application of a constant stress until a determined current limit was reached, the I-V characteristics (applying a ramp voltage at the inverter input with the other terminals grounded) and the inverter transfer curve were measured. The BD location and the conduction characteristics are obtained from the I-V curves and the effect of the stress on the inverter performance was evaluated by the changes in the inverter transfer curve. The stress was continued by increasing the stress voltage or current limit to increase the oxide damage.

III. Modeling Worn-Out and Broken Down Transistors

Figures 1a and 1b show the effect of different levels of oxide damage (before and after a SBD event close to the source of the device) on the I_D-V_D and I_D-V_G characteristics of short channel transistors (SiO_2 devices, L=0.175μm). Figure 1c shows the corresponding gate leakage characteristics. In this case (a source SBD) only a small variation of the I_D-V_D characteristics is measured after BD, although they can be strongly affected in other cases (especially after a HBD at the drain [19]). In addition to the increase of the gate current (Figure 1c), the stress leads to a variation in the drain current of the transistors, which indicates changes in the conduction along the channel. This effect is visible even before BD. This variation should be related to the generation of defects in the oxide during the stress, which will finally end in the creation of the percolation path leading to BD. As a result of the stress, a decrease of the saturation current, an increase in the threshold voltage and a decrease in the transconductance are observed [6, 20]. The same qualitative results were obtained for the longer (L=10μm) channel SiO_2 transistors (not shown here) although the variations of the transistor characteristics are more pronounced in this case [20]. Before BD, the observed changes have been related to the generation of defects at the SiO_2/Si interface [21]. The experimental curves of the worn-out device (before BD) in Figures 1a and 1b have been fitted to the BSIM4 model [22]; as observed, good fittings have been obtained (continuous lines in figures 1a and 1b). Of course, a modified parameter set (different from the one used to fit the fresh data) is needed. The I_G-V_G curve of the stressed devices prior to BD (figure 1c) has been also fitted using BSIM4, pointing out that BSIM4 model is also able to account for the additional gate leakage due to oxide degradation (before BD).

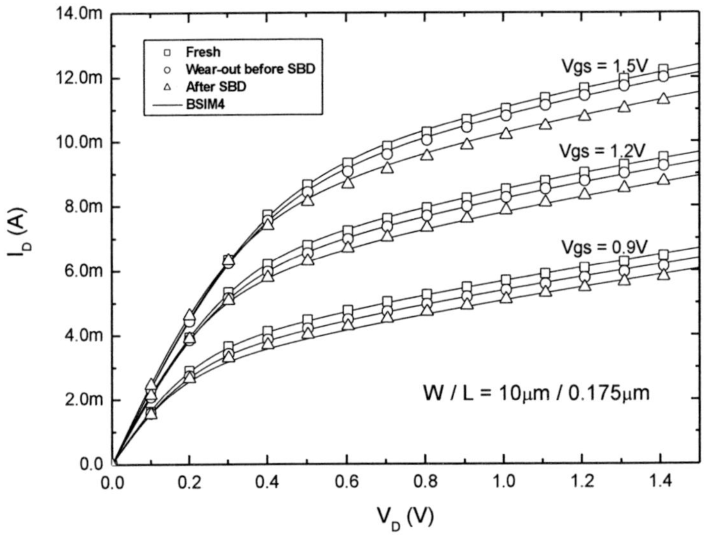

(a)

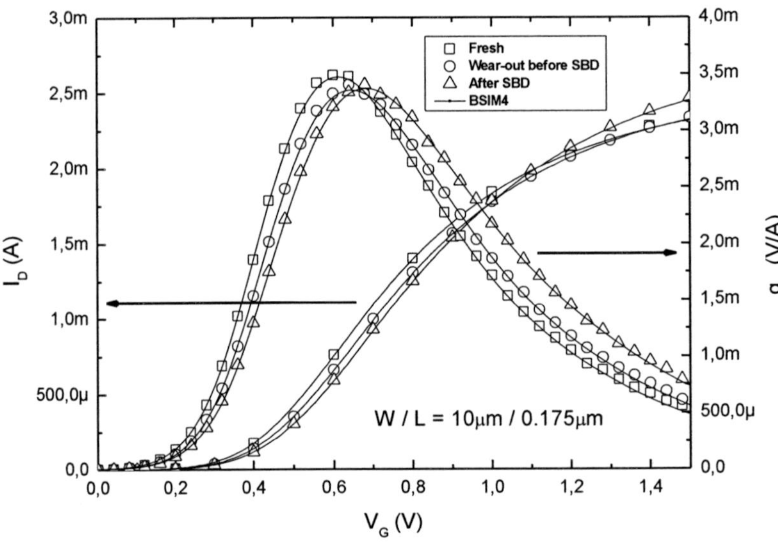

(b)

Figure 1. Continued on next page.

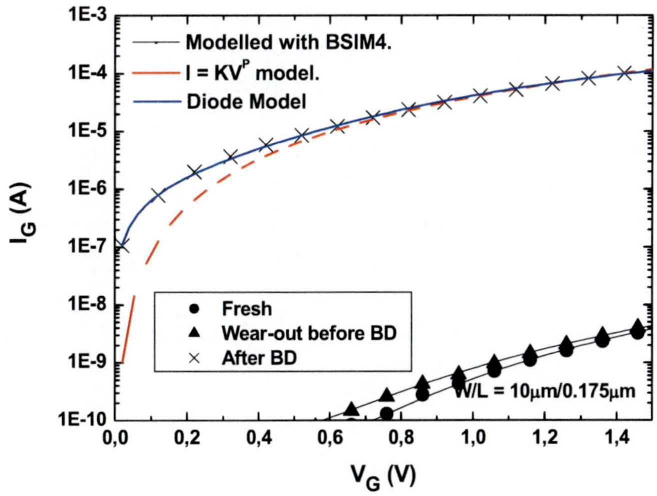

(c)

Figure 1. a). I_D-V_{DS} curves measured for a fresh SiO_2 transistor and after different levels of oxide degradation. Continuous lines correspond to the fitting of the data to the BSIM4 model. A source BD (BD in the overlap region between the gate and the source) took place. b). I_D-V_{GS} curves and transconductance measured for V_D=0.1V, for the transistors whose I_D-V_{DS} curves are shown in Figure 1a. Continuous lines correspond to the fitting of the data to the BSIM4 model. .c). I_G-V_G curves for the transistors whose I_D-V_{DS} and I_D-V_{GS} are shown in Fig 1a and 1b. Fresh and worn-out gate currents have been correctly described by the BSIM4 model.The post-BD gate current has been fitted to the power and exponential BD laws. The last one is more accurate in the low voltage range.

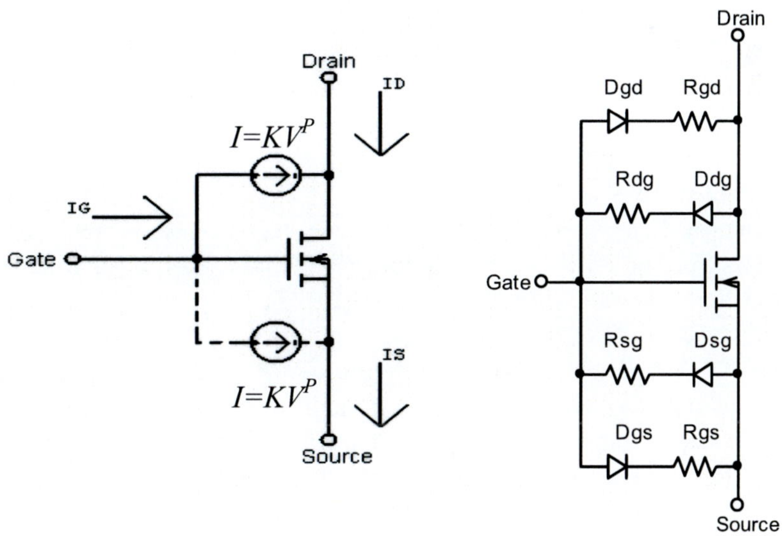

(a) (b)

Figure 2. Equivalent model for the post-breakdown currents in MOSFETs. The current through the gate is described using an equivalent circuit a) based in voltage controlled current sources or b) based in a diodelike description. A modified BSIM4 parameter set is used to describe the conduction through the channel in the stressed device (this point has been indicated by the asterisk).

After BD, BSIM4 model is still able to describe the channel current (see the fittings in Figures 1a and 1b) but, due to the local nature of the BD phenomenon, it is not well suited for the description of the BD gate leakage current. To include the post-BD gate current in the MOSFET performance, the connection of voltage controlled current sources between gate and/or source and drain (depending on the location of the BD path along the channel) has been proposed (Fig. 2a) [4]. These sources obey the well-known power law dependence of the BD gate current on the gate voltage [23], $I_G=KV_G^P$ so that a maximum of 4 parameters are needed for the description of the BD current (8 if the asymmetry in the BD current with the voltage polarity is considered, so that a double branch should be included for BD location). However, it has been recently shown that in some cases (SBD and HBD in the very low voltage range), the post-BD gate current can be better described by an exponential dependence on voltage, so that the use of a diode plus a series resistance has been proposed to describe the BD gate current [24]. As an example, Figure 1c shows that the power-law fails to correctly describe the gate current at low voltages, whereas the exponential law is more accurate in this voltage range. In [24] this dependence is discussed in detail but the main idea is that the diode-like behaviour can be associated with the contact effect between the electrodes, whereas the series resistance with the confinement of the electron wave function. In this way, the equivalent circuit model shown in Figure 2b can be used to represent the post-BD current in the device (which also accounts for the voltage polarity asymmetry). This equivalent circuit represents the SBD case (if the series resistance is zero) and also the HBD one (even at very low biases [24]), so that a unique equivalent circuit is considered for SBD and HBD in the entire voltage range. In all cases, current through the diode is described by the implicit equation (1), where V_T corresponds to the thermal voltage (KT/q) and η to the ideality factor; the ***i*** index indicates the different voltage polarities and BD locations (GD, DG, SG or GS).

$$I_i = I_{0i}\left(e^{\frac{V_i - I_i R_i}{\eta_i V_T}} - 1 \right) \tag{1}$$

Then, 3 parameters are needed to describe one of the diode+resistance branches (I_0, R and η), so that a complete description of the after-BD source and drain currents requires a maximum of 12 parameters (6 when only one voltage polarity is considered). These parameters can be easily obtained using the procedure described in [24]. The adequacy of the diode + series resistance representation of the BD path , independently of the BD hardness (SBD and HBD), location along the channel and voltage polarity, in the whole bias range, has been checked through the fitting of all the currents (I_D, I_S and I_G) in the broken down device for different BD events. As an example, Figure 3a shows all the currents in two SiON transistors during a voltage sweep in the gate (with the other terminals grounded) after their BD. The data correspond to a HBD in the channel (but closer to the drain) and a channel SBD. Note the good matching between the experimental data (symbols) and the fitting (continuous lines). From these data, the parameters of the BD current equivalent circuit (Figure 2b) are extracted. In any case, the exponential or power-law dependences for the gate current can be equivalently used, being the only advantage of the first one the more accurate description of the BD current at low gate voltages [24]. Therefore, the exponential (and consequently the diode+resistance description of the BD gate current) can better describe the sub-threshold behaviour of

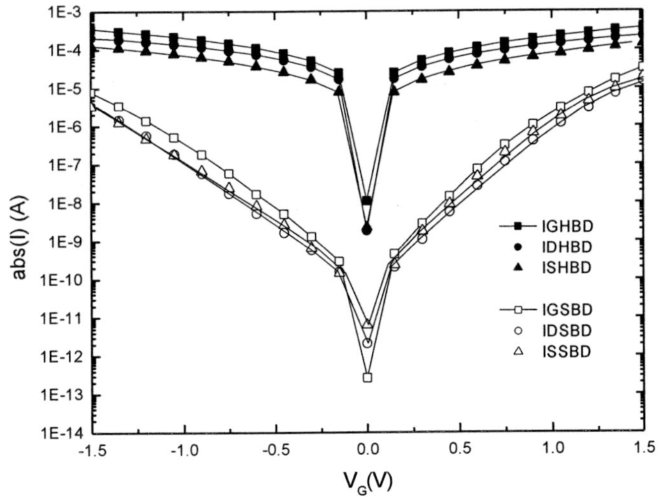

(a)

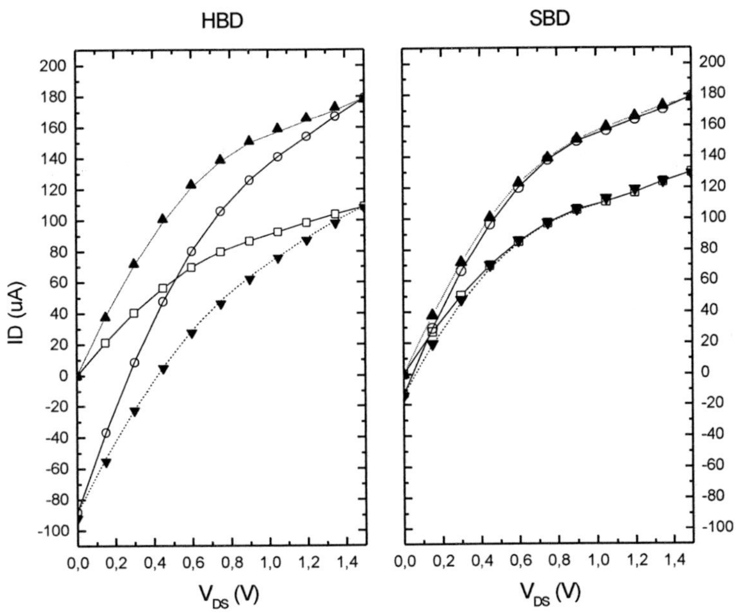

(b)

Figure 3. a) Experimental abs(I)-V_{GS} characteristics for a HBD and a SBD. SBD took place in the channel and HBD close to the drain. Circles correspond to measured I_D current, up triangles to I_S current and squares to the total I_G current. The HBD and SBD data are plotted using solid and open symbols, respectively. The continuous lines correspond to the fitting of the data to the model in Figure2b. b) Solid symbols: Experimental I_D-V_{DS} curves measured before BD (up triangles) and after BD (down triangles). The figure on the left (right) corresponds to the HBD (SBD) event whose I_G-V_{GS} curves are included in Figure 3a. Open symbols: Calculated post-BD I_D-V_{DS} curves when only the variation of channel (squares) or gate current (circles) is considered. The gate voltage was 1.5V in all cases.

MOSFETs, which can be relevant in some circuits. The use of the current-source (Figure 2a) or diode + resistance (Figure 2b) equivalent circuit will depend on the application. Whatever the equivalent circuit for the gate current is chosen, the channel current of the stressed devices can be described using the BSIM4 model with a modified parameter set. This point has been indicated in Figure 2 (a and b) by marking the MOSFET with an asterisk.

The presented description of the post-BD MOSFET behavior allows to separately analyse the contributions of the post-BD gate current and of the variation of the current along the channel to the post-BD electrical characteristics of the device. Figure 3b shows the I_D-V_{DS} curves measured on fresh (up triangles) and on the broken down transistors (down triangles). HBD data are shown on the left and SBD data on the right. The corresponding post-BD I_G-V_G characteristics are plotted in Figure 3a. The BSIM4 model has been used to fit the fresh curves (dotted lines in Fig. 3b). To describe the post-BD I_D-V_D curves (down triangles), BSIM4 model and the contribution of I_G to I_D are needed (dashed line). Note that the post-BD BSIM4 model parameters are obtained from the I_D-V_G and I_D-V_D curves, once the contribution to I_D of the BD gate current has been subtracted, for each of the V_D values. When this is done, the open squared curves in Figure 3b are obtained (which would correspond to the MOSFET characteristics after BD, if only the conduction along the channel would have been affected). The comparison of these curves to the fresh ones allows to conclude that an important modification of the MOSFET behavior has occurred (even when the gate current is not considered), and, consequently, a new BSIM4 parameter set is needed to describe the post-BD MOSFET performance. In addition, Figure 3b includes the I_D-V_D characteristics that would have been obtained if BD only led to an increase of the gate current (open circles), i.e., fresh BSIM4 parameters and the contribution of the post-BD gate current to drain current have been used to calculate the curve. Again, when comparing these curves to the post-BD experimental behavior, one can conclude that the inclusion of the post-BD gate current is not enough to explain the post-BD MOSFET I_D-V_D characteristics. If the curves obtained when only the modification of the channel (open squares) or gate current (open circles) are considered are compared to the experimental post-BD curve, one can conclude that, when the transistor is "on", the low-V_D region (linear region) of the post-BD curve is mainly affected by the gate current, but in the high-V_D region (saturation region) the post-BD behavior is mainly determined by the variation of the channel conduction and the gate current has a smaller influence. This observation is applicable to HBD and SBD. However, in the case of SBD (i.e. when BD I_G current is less important, when compared to the fresh I_D current), the variation of the channel current has a larger effect on the post-BD I_D curve and the SBD gate current has only an appreciable effect in the low-V_D region of the curve (<0.5V in this case).

IV. Impact of Oxide Degradation and BD on Circuit Performance

In this section the above presented description of the stressed MOSFETs has been used to study the impact of oxide wear-out and BD on circuit functionality of simple analog (a current mirror) and digital (RS latches, SRAM memory cells and CMOS inverters) circuits, through simulated and measured data.

A. Current Mirror

The performance of a current mirror (Figure 4) after the failure of one of its devices has been simulated. Since analog circuits are more sensitive to transistor parameter variations, degradation and SBD effects will be considered in this case, so that the SiO_2 devices will be our focus. The experimental data corresponding to the long and short channel SiO_2 MOSFETs have been used to extract the model parameters (BSIM4 and BD equivalent circuit) for fresh, degraded and soft-broken down devices. BSIM4 model parameters have been extracted using Aurora software [25]. Due to the characteristics of the model, the influence of the variation of the channel and gate currents on the circuit performance will be separately considered. Typical worn-out and broken down curves measured on these devices are shown in Figures 1a, 1b and 1c, which have been considered for the circuit simulation. Before BD, the device electrical characteristics have been fitted to the BSIM4 model, with an appropriate set of parameters. After BD, in addition to BSIM4 model for the channel current, one of the equivalent circuits for the post-BD gate current is needed. For simplicity, since the devices will be always operating in the saturation region (they will never enter the subthreshold region), the post-BD gate current has been modeled using the current-source representation (Figure 2a). Arbitrarily, T2 has been chosen as the degraded device (while T1 remains fresh), but equivalent results would have been got if T1 was degraded instead. Note that the degradation of the device implies a decrease of its saturation current (Fig. 1a). For the post-BD simulations, a source BD event has been considered. Source BD is expected to have a larger impact on the current mirror response, because gate and drain are at the same or nearly at the same potential, so that the contribution of the gate BD current to the drain current will be negligible. Two different cases have been taken into account: (i) wear-out of T2 transistor, without including the gate current in any of the two transistors and (ii) wear-out of T2, with gate current in both, T1 and T2. For the two cases, the figure of merit of the circuit performance, I_{out} / I_{in}, has been studied as a function of the stress time (Fig. 5). Figure 5 shows that, in short channel devices, a small degradation of I_{out}/I_{in} is observed, but the inclusion of the gate current does not make any difference. For the longer channel devices, a small decrease in the figure of merit is observed when the tunnel current is considered before SBD. These observations can be explained taking into account that, before SBD, current through the gate due to oxide degradation is very low compared to the drain current operation values and consequently, the decrease of I_{out}/I_{in} is due mainly to the reduction of the saturation current of the damaged transistor, which is larger for the longer transistors. Therefore, before SBD, the variation of the channel conduction as a consequence of oxide degradation has a larger effect on the functionality of the current mirror than the tunnel current through the gate. After the SBD of the oxide, the performance of the circuit is only remarkably affected in the case of long channel transistors. This behavior can be explained taking into account that for long channel devices the saturation currents are lower, so that BD gate current is relatively more important in I_{in} (I_{in} is the saturation current of T1 plus gate currents of both transistors), what means a larger influence in the I_{out}/I_{in} ratio. It should be noted that a more important decrease in the figure of merit would be observed in narrower devices after BD, since in this case I_{sat} is strongly reduced [6].

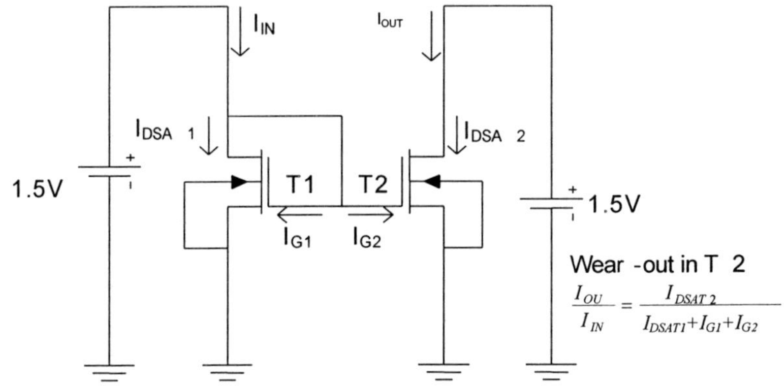

Figure 4. Schematics of the current mirror used to analyse the effect of oxide degradation on the circuit performance of an analog circuit. Short and long channel SiO_2 transistors have been considered. $I_{DSATFRESH}$ is 200μA and 12mA for long channel and short channel transistors, respectively, at 1.5V operation.

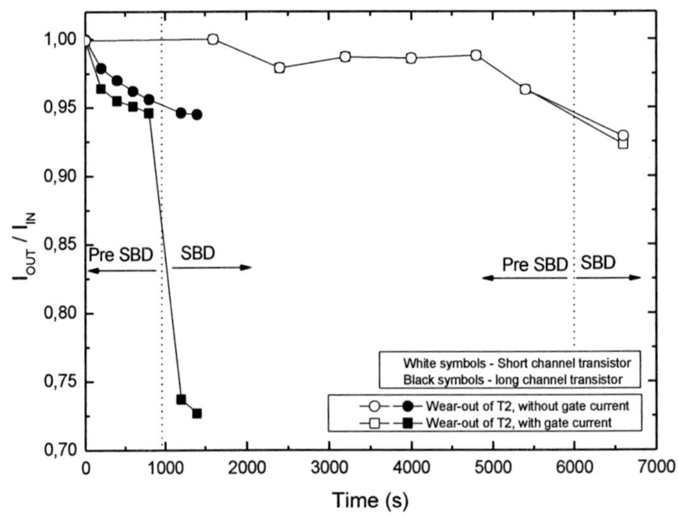

Figure 5. Evolution of the in Iout / Iin ratio in the current mirror shown in Figure 4, for different transistor lengths. Time in the y-axis of the graph means level of degradation. For short channel devices a lower influence of the tunnel current is observed.

B. RS Latches

The performance of NOR and NAND topologies of RS latches (Figure 6) has been simulated after the BD of one of their nMOSFETs. An additional load capacitance of 100fF has been also included in order to consider load effects. Due to the larger robustness to parameter variation of digital circuits, HBD has been considered in this case. Therefore, the data

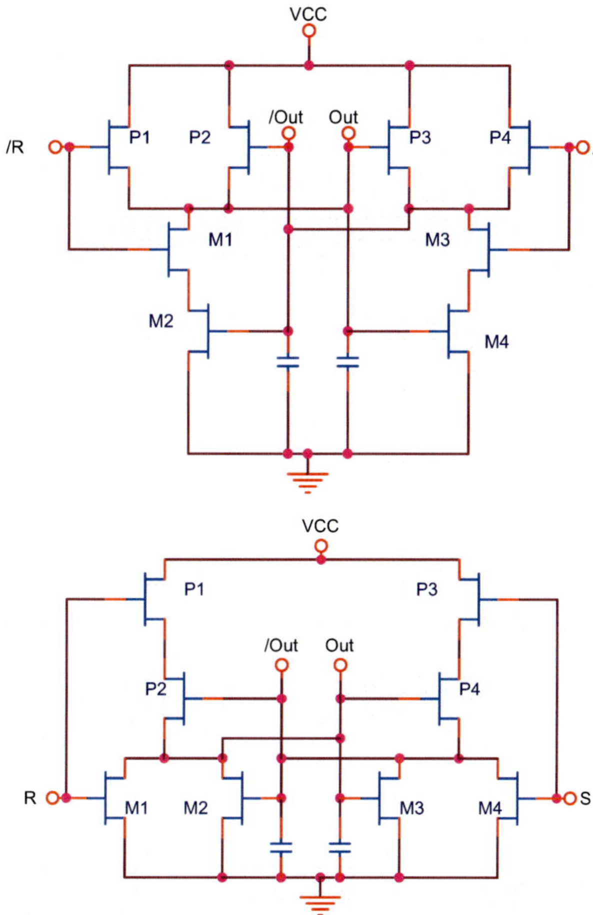

Figure 6. Schematics of the simulated RS latches based on NAND gates (top) and NOR gates (bottom).

obtained on the SiON devices of section III will be used to extract the model parameters. Typical post-BD I_D-V_D curves are shown in Figure 3b left (down triangles), which are characterized by negative values of the I_D current and a strong decrease of the saturation current [26]. The corresponding gate current is shown in Figure 3a. These data have been used during the simulations. The diode+series resistance equivalent circuit has been used in this case to describe the post-BD gate current, which better reproduces the post-HBD current at low voltages. A detailed analysis of the circuit behaviour shows that differences in the output levels, depending on the broken down transistor, can be observed, but the circuits do not lose their functionality. Only reductions in the noise margin and variations in the delay times during the level transitions are observed [26]. As for the current mirror, the effects of the variation of the channel current and the increase of the gate current have been separately analysed. To do so, a BD in transistor M3 has been considered, which has been determined to be one of the worst cases for both circuit topologies [26]. Four different cases have been simulated: fresh device (case A), fresh BSIM parameters plus additional current due to BD (case B), post-BD BSIM parameters plus additional BD current (case C, i.e., the experimental device behaviour) and only post-BD BSIM parameters (case D). Figure 7 shows the

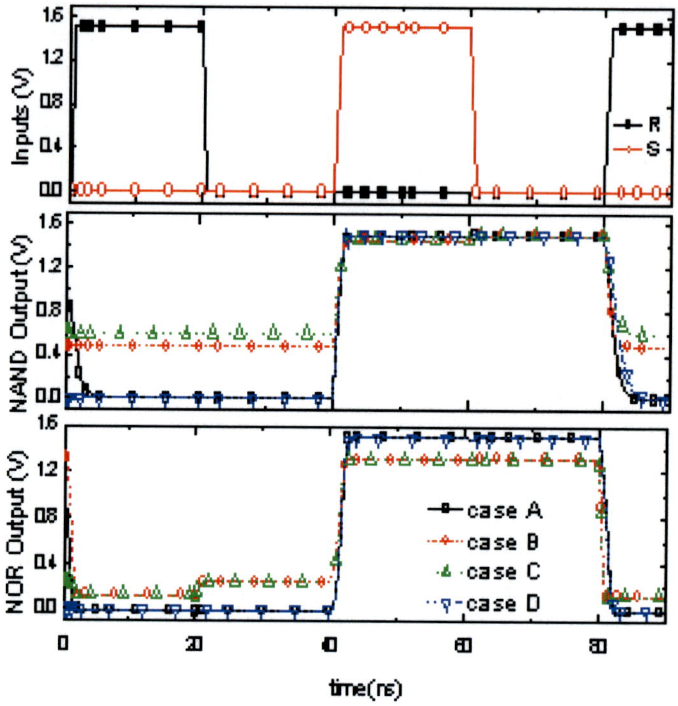

Figure 7. Responses of the NAND (middle) and NOR (bottom) RS latches. The upper figure shows the input waveform. Fresh (Case A) and broken down devices (Case C) are considered. The circuits have been also simulated when only the effects of the additional gate current (Case B) and the variation of channel current (case D) after BD are separately considered (see text).

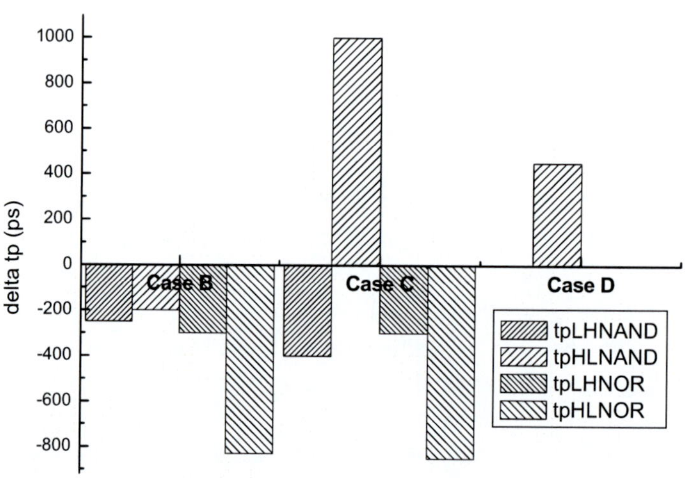

Figure 8. Switching times variation (from the fresh device case) for NAND and NOR topologies when the additional BD gate current and/or the variation of the channel current are considered.

simulation results for the four considered cases for all possible input combinations (Fig.7 top). In case of NAND topology (Fig.7 middle), for cases A (fresh device) and D (only channel conduction variation) the latch shows a similar behaviour. On the other hand, the latch performance for cases B (only post-BD gate current) and C (variation of channel + post-BD gate currents) is also similar, but a ~0.15V better noise margin is obtained for case B. The NOR simulation (Fig.7 bottom) shows similar tendencies, although in this topology the outputs for cases B and C are almost identical. Figure 8 shows the variation of the switching times in cases B to D from those measured in case A (fresh device). Note that all of them suffer important modifications when the gate current is included in the simulations (cases B and C). The results in Figures 7 and 8 indicate that, for RS latches, the circuit behaviour is mainly affected by the additional post-BD gate current and that the variation of the conduction along the channel has a negligible effect on the post-BD latch performance.

C. SRAM Memory Cell

The effect of oxide breakdown was also evaluated in a more complex circuit as a 6T SRAM memory cell. The BD current limit that the circuit is able to tolerate without losing its functionality was analyzed. There are only three topologically distinct breakdown locations in an SRAM cell, denoted drain, *p*-source, and *n*-source in Fig 9. Breakdown at *p*- or *n*-source is also equivalent to increased source-to-drain leakage (I_{off}) in the opposite *p*- or *n*-FET. For simplicity, during the simulations, gate oxide HBD was modeled as ohmic conduction I=(1/R)V and SBD following the power law model $I_G=KV_G^P$. In this case, the values of the K

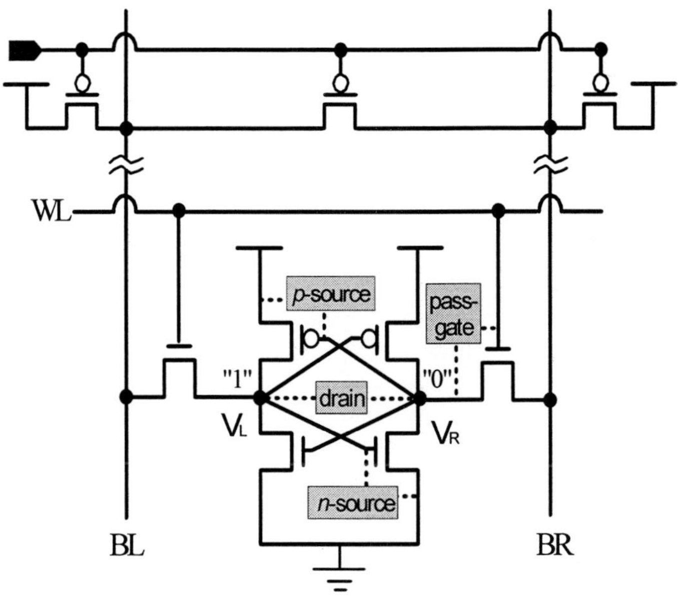

Figure 9. 6-T SRAM cell with possible BD leakage paths corresponding to the cell state shown. The precharge devices were not present in the cells studied.

and p parameters were extracted from experimental SBD I_G-V_G curves of the transistors used in the cell. As the transistors of the cell are in the subthreshold or linear region it is expected that the gate current has a greater influence than the channel current in the post-BD model. Fig. 10 illustrates schematically the effect of HBD at each of these locations. A *p*-source or an *n*-source BD as defined in figure 9 does not affect the transfer curve at the inverter with the *p*-(*n*-)source BD (dotted line in Fig. 10b and continuous line in Fig. 10d respectively). However, the *p*-(*n*-)source BD raises (lowers) the voltage at the output of the opposite inverter (continuous line in Fig. 10b and dotted line in Fig. 10d), fighting the channel resistance of the *n*-(*p*-)FET of the inverters without the BD. Consequently, the SNM and therefore the stability of the cell is modified. For fixed R, breakdown at *p*-source has less effect than *n*-source on minimum SNM, because the opposing *n*-FET is stronger (relative to the *p*-FET). Breakdown at drain (equivalent to input-output short) reduces output swing, with the largest effect being to reduce the 'high' state, again because of the weaker *p*-FET.

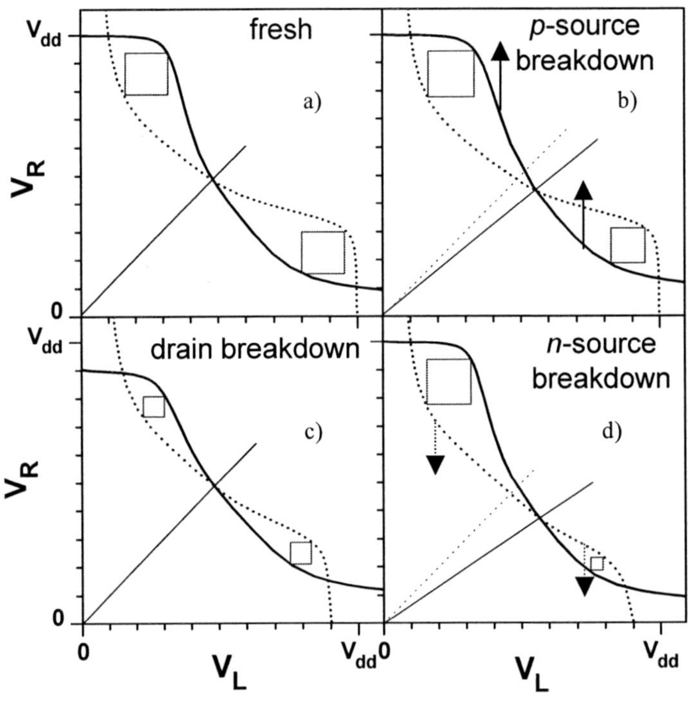

Figure 10. Typical butterfly plots illustrating the effect of breakdown at each location in the cell. HBD, R=40kΩ. SNM is the noise voltage across the cell which causes the transfer characteristics of the inverters to intersect at only one point. Graphically, it is the side of smaller of the two maximum nested squares. The higher this value, the better is the stability of the cell.

A cell is most sensitive to noise during a "read" operation. The worst-case scenario occurs in the half-selected state [17], where the word line is pulled high while the bitlines are pre-charged high before the "read" operation. In this case, breakdown at *p*-source is indistinguishable from breakdown at the inverter side of a pass gate. Fig. 11 shows the SNM in the half-selected state, normalized to the SNM of the cell without BD for HBD and SBD. For the cells considered in this paper, a 50% degradation in SNM results from oxide BD when the current through the BD spot reaches ~20-50μA at V_{dd} for the worst-case *n*-source breakdown. Pass-gate or *p*-source breakdown may tolerate higher leakage, up to ~500μA. These values are comparable with the on-currents of the fresh *p*-FET and *n*-FET respectively used in this SRAM cell, but decrease with device widths, e.g. for smaller SRAM cells or for cells with lower transfer ratios. These values give targets for tolerable values of leakage caused by gate oxide breakdown.

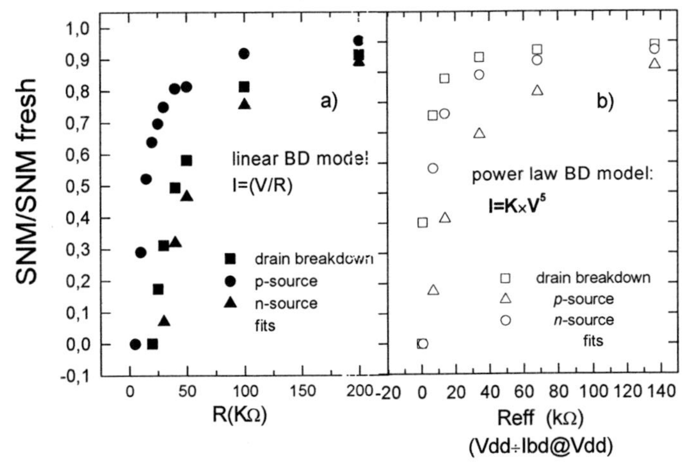

Figure 11. a) Normalized SNM for HBD as a function of ohmic breakdown resistance R at various locations in cell. b) Normalized SNM for SBD at various locations in cell as a function of the effective resistance of the breakdown spot at V_{dd}.

D. CMOS Inverters

CMOS inverters have been subjected to electrical stress to experimentally evaluate the impact of oxide BD on their performance. The effects of the stress have been measured through the I-V characteristics on the different terminals of the circuit and the transfer curve. The results show that the circuit performance and the devices degradation are affected differently depending on the stress polarity applied to the circuit input. For positive polarity stress, the evolution of the I-V characteristic (Fig 12a) shows that the leakage current through the NFET oxide is higher than through the PFET, which indicates that the NFET oxide is more damaged than the PFET oxide. Consequently, a voltage increase of the lower part of the inverter transfer curve, V_{OL}, can be observed as the NFET oxide becomes more damaged (Fig. 13a). However, for negative voltage stresses, the PFET oxide is mainly damaged (Fig.12b) and a voltage decrease of the higher part of the inverter transfer curve, V_{OH} (Fig. 13b) is produced. In addition, for positive (negative) voltage stress the transfer curves exhibit a shift to the right (left) due to a change in the threshold voltage (V_t) of the NFET (PFET), not related to the breakdown event itself.

In all the cases, the experimental transfer curves after the different stresses can be fitted by a combination of a threshold voltage shift of the transistors and a leakage current between the inverter input and output. The leakage current follows the form of the power law formula, $I_G=KV_G^p$, and was account for using the equivalent circuit shown in Fig. 2a connected between gate and drain of the inverter transistors. Different values of K and p factor are necessary to fit the high and low part of the inverter transfer curve due to the different degradation level of the transistor with the stress polarity. There is usually a difference of one order of magnitude between the K values between these two parts of the transfer curve, with lower values of K for the more distorted portion of the transfer curve (output high for negative stresses and output low for positive streses). Note the good results of the fittings (symbols in Fig. 13).

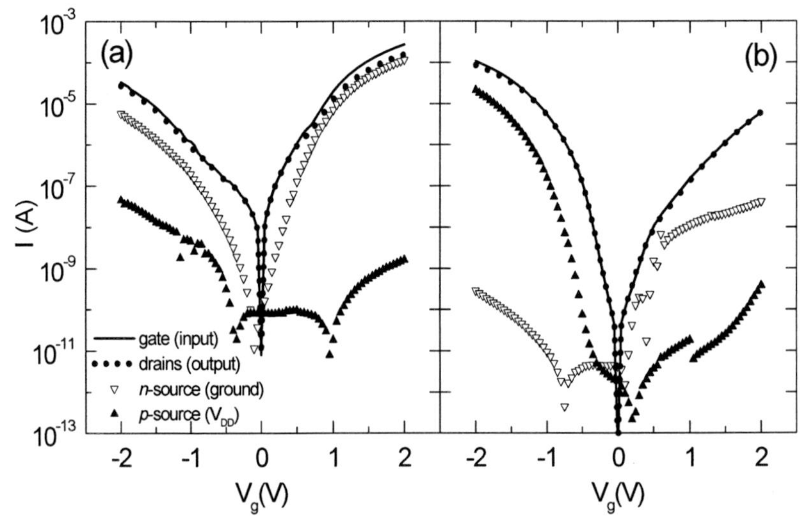

Figure 12. Post-BD IV characteristics of inverters. (a) positive stress on input, (b) negative.

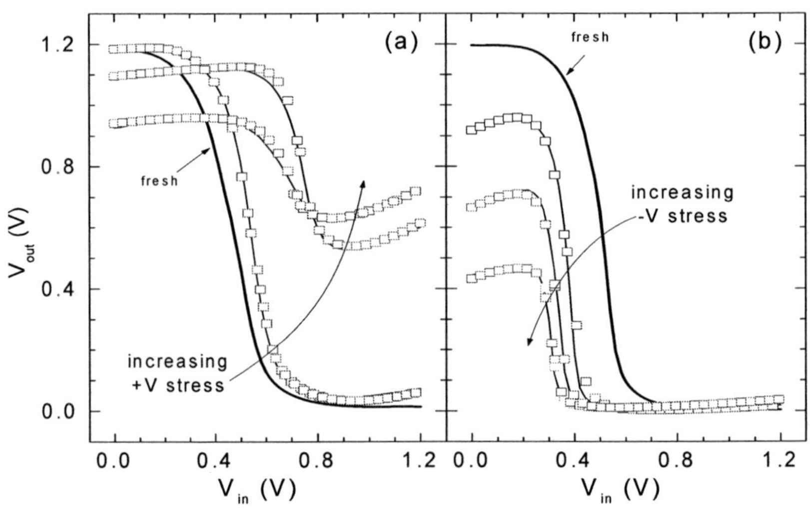

Figure 13. Transfer curves of inverters after breakdown to various levels. (a) positive stress on inverter input. (b) negative. Lines are experiment, symbols are model.

V. Conclusion

A description for the degraded (before BD) and post-BD behaviour of MOSFETs has been presented, which includes the variation of the conduction along the channel and the increase of the gate current due to the electrical stress. The variation of the channel current is accounted for through the variation of the BSIM4 parameters. BSIM4 model is also able to describe the increase of the gate current in the degraded (before BD) devices. For the post-BD gate current, the use of an equivalent circuit is proposed (current source or diode+resistance), whose complexity depends on the degree of accuracy required and on the application. This equivalent circuit is able to describe the post-BD gate current for all BD hardness and positions along the channel and device geometries by simply changing the model parameters. The model can be included in a circuit simulator to analyse the impact of degradation and BD on circuits Due to the characteristics of the adopted description of the I-V curves (channel and gate currents are separately considered), the model has allowed to determine the influence of the variation of the channel current and of the increase of the gate current on the output characteristics of the devices and on the performance of analog (current mirror) and digital (RS latches). The results point out that, at device level, the contribution of the post-BD gate current is more important in the low-V_D region and that the variation of the conduction along the channel has a larger influence in the high-V_D region. For the current mirror, the gate current influence (before and after BD) on the circuit performance is very small and it only has a strong effect in long channel devices after BD, probably due to the differences between the drain operation currents and the wear-out gate tunnel current due to the oxide degradation. Contrarily, after oxide BD on the RS latches, the variation of the circuit performance is mainly due to the HBD gate current, and the change in the conduction along the channel has a negligible effect in the circuit functionality. In addition, the model has been used to study the stability of the SRAM cell. The worst SRAM cell stability has been obtained for gate-oxide breakdown between gate and source of an n-FET of the cell. In this case, BD leakage of ~20-50 μA (comparable to the fresh p-FET current) produces a reduction of the 50% of the SNM. The model has been also able to explain data experimentally obtained on fabricated CMOS inverters. The results show that on CMOS inverters, the BD affects the inverter performance in a different way depending on the polarity of the applied stress to the inverter input. For positive (negative) stress polarity the NFET (PFET) is mainly damaged.

Acknowledgment

The authors would like to thank IBM and IMEC for sample provision. Part of the work was made during a postdoctoral stay funded by a Fulbright grant of R. Rodríguez at the Thomas J. Watson research Center of IBM in Yorktown Heights, New York, under the supervision of Dr. J. H. Stathis. This work has been partially supported by the Spanish MEC (TEC2004-00798/MIC and TEC2007-61294), the DURSI of the Generalitat de Catalunya (2005SGR-00061) and the European Commission (Marie Curie Actions, APROTHIN Project).

References

[1] *International Technology Roadmap for Semiconductors*, Semiconductor Industry Association (2007)

[2] B. Kaczer, R. Degraeve, M. Rasras, K.V.de Mieroop, P.J. Roussel, G. Groeseneken, "Impact MOSFET gate oxide breakdown on digital circuit operation and reliability". *IEEE Trans. Elect. Dev.*, **49** (3), pp. 500-506 (2002)

[3] R. Rodríguez, J.H. Stathis, B. P, Linder, S Kowalczyk, C. T. Chuang, R. V. Joshi, G. Northrop, K. Bernstein, A. J. Bhavnagarwala, S. Lombardo, " The impact of gate-oxide breakdown on SRAM stability", *IEEE Electron Device Lett.* , Vol. 23(9), pp. 559-561 (2002)

[4] R. Rodríguez, J. H. Stathis, B. P. Linder, "A model for gate oxide breakdown in CMOS inverters", *IEEE Electron Device Letters,* Vol. 24 (2), pp. 114-116 (2003)

[5] R. Fernández, R. Rodríguez, M. Nafría, X. Aymerich, "DC broken down MOSFET model for circuit reliability simulation", *IEE Electronics Lett.*, Vol. 41(6), pp. 368-370 (2005)

[6] A. Cester, S. Cimino, A. Paccagnella, G. Ghidini, G. Guegan, "Collapse of MOSFET drain current after soft breakdown and its dependence on the transistor aspect ratio W/L", *IRPS Proc.*, pp. 189-195 (2003)

[7] T. Pompl, H. Wurzer, M. Kerber, R.C.W. Wilkins, I. Eisele. "Influence of Soft Breakdown on NMOSFET Devices Characteristics". *IRPS Proc.*, pp. 82-87 (1999)

[8] G. Ghidini, A. Garavaglia, G. Giusto, R. Bottini, D. Brazzelli, "Soft and hard breakdown: impact of annealing recovery on transistor performances", *Microelec. Eng.,* Vol. 72, pp. 5-9 (2004)

[9] T. S. Yeoh, "Influence of MOS transistor gate oxide breakdown on circuit performance", *IEEE Int. Conf. on Semiconductor Electronics*, pp. 59- 63 (1998)

[10] H. Yang, J.S. Yuan, E. Xiao, "Effect of gate oxide breakdown on RF device and circuit performance", *IRPS Proc,*. pp. 1-4 (2003)

[11] Q. Li, J.Zhang, W. Li, J.S. Yuan, Y. Chen, A.S. Oates, 'RF circuit performance degradation due to softbreakdown and hot carrier effect in 0.18μ CMOS technology', *IEEE Radio Freq. Int. Circ. Symp.*, pp. 139-142 (2001)

[12] A. Cester, L.Bandiera, S.Cimino, A. Paccagnella, G. Ghiaini, "Incidence of oxide and interface degradation on MOSFET performance", *Microelec. Eng.*, Vol. 72, pp. 66-70 (2004)

[13] R. Degraeve, B. Kaczer, A. De Keersgieter , G. Groeseneken, "Relation between breakdown mode and breakdown location in short channel nMOSFETs and its impact on reliability specifications". *IRPS Proc.,* p. 360-366 (2001)

[14] E. Y. Wu and J. Suñe, "The effects of device dimension on the Post-breakdown characteristics of ultrathin gate oxides" *IEEE Electron Device Lett.,* Vol 26, pp. 401-403 (2005)

[15] P. Smeys, *et al.,* "A high performance 0.13μm SOI CMOS technology with Cu interconnects and low-k BEOL dielectric," *VLSI Tech. Symp.*, pp. 184-185 (2000)

[16] E. Seevink, F. J. List, and J. Lohstroh, " Static-noise margin analysis of MOS SRAM cells," *IEEE J. Solid State Circuits*, vol. SC-22, pp. 748-754 (1987)

[17] A. J. Bhavnagarwala, X. Tang, and J. D. Meindl, "The impact of intrinsic device fluctuations on CMOS SRAM cell stability," *IEEE J. Solid State Circuits*, vol. 36, pp. 658-665 (2001)

[18] P. Smeys, *et al.,* "A high performance 0.13μm SOI CMOS technology with Cu interconnects and low-k BEOL dielectric," *VLSI Tech. Symp.*, pp. 184-185 (2000)

[19] R. Fernández, R. Rodríguez, M. Nafría and X. Aymerich, "MOSFET output characteristics after oxide breakdown", *Microelec. Eng.,* Vol. 84, pp. 31-36 (2007)

[20] J. Martín-Martínez, R.Rodríguez, M.Nafría, X.Aymerich, J.H.Stathis, "Worn-out oxide MOSFET characteristics: Role of gate current and device parameters on a current mirror", *Microelectr. Reliab.*, Vol. 47, pp. 665-668 (2007)

[21] J. Martín-Martínez, S. Gerardin, R.Rodríguez, M.Nafría, X.Aymerich, A. Cester, A. Paccagnella, G. Ghidini, "Lifetime estimation of analog circuits from the electrical characteristics of stressed MOSFETs", *Microelectr. Reliab.* Vol. 47, pp. 1349-1352 (2007)

[22] BSIM4 webpage: www-device.eecs.berkeley.edu/~bsim3/bsim4_intro.html

[23] E. Miranda, J. Suñé, R. Rodríguez, M. Nafría and X. Aymerich, "A function-fit model for the soft-breakdown failure mode", *IEEE Elec. Dev. Lett.*, Vol. 20, No. 6, pp. 265-267 (1999)

[24] E. Miranda, "Analytic model for the post-breakdown conductance of sub-5-nm SiO_2 gate oxides", *IEEE Electron Device Lett.* , Vol. 26(9), pp.673-675 (2005)

[25] http://www.synopsys.com

[26] R. Fernández, R. Rodríguez, M. Nafría, X. Aymerich, "Effect of oxide breakdown on RS latches", Microelectr. Reliab., *Microelectr. Reliab.*, Vol. 47, pp. 581-584 (2007).

In: MOSFETs: Properties, Preparations and Performance
Editors: N.T. Andre et al. pp. 421-439
ISBN: 978-1-60456-762-5

Chapter 13

MOSFET's Programmable Conductance: The Way of VLSI Implementation for Emerging Applications from Biologically Plausible Neuromorphic Devices to Mobile Communications

I.S. Han
Institute for Information Technology Covergence Korea Advanced Institute of Science and Technology, Daejeon, Korea(South)

Abstract

This paper describes a mixed-signal VLSI design utilizing the controlled conductance properties of MOSFETs, for tunable filters, low power amplifiers, and analog-mixed signal processing. The programmable linear conductance is realized by the pair of MOSFETs in the triode region, and it produces the voltage-controlled tunable function or the analog computing element. It demonstrates the application areas of neural synaptic functions, electronic neurons with biologically plausible characteristics, and tunable analogue complex filters for RF communication. The programmable conductance by MOSFETs is analyzed by the experimentation of prepared MOSFETs and the SPICE simulation.

The amplifier based on the proposed MOSFET conductance is employed to enforce the tunable filters for RF applications. It has the tuning range of +/- 50% and the bandwidth over 1 GHz is exhibited. For applications to analog computing, the design of an analog multiplier and biologically plausible neuromorphic devices are illustrated by utilizing MOSFET's programmable conductance.

I. Introduction

The voltage-controlled conductance of MOS transistors has been widely used in the analog signal processing or VLSI circuits of programmable amplifiers, tunable filters, neural computing and others. The MOSFET's conductance in the triode region can be exploited for transconductors as in [1-4], though it is necessary to eliminate the nonlinear characteristics. Various methods were reported to improve the linearity of transconductance, as the advantages of low voltage and low power operation are exploited in the areas of analog complex filter, switched capacitor filter or continuous time filter [5-8]. Many cascode or source degeneration topologies were proposed to improve its linearity of triode-region transconductor , and its application to the tunable filter was introduced [1-9].

Recently the new improvement by a pair of MOSFETs in triode region was presented as a way of implementing the controlled linear conductance for programmable amplifier and tunable filters, while the early design based on a pair of MOS transistors in the triode region was developed for continuous time filters [10, 11]. It can be used to implement the new four quadrant analog multiplication function for modulators, and basic signal processing elements. The voltage controlled linear transconductance of MOSFETs was also adopted to produce the synaptic or neuron function of various applications from robots to regenerative medicine, which was inspired by the biological plausibility and low power supply [12-16]. The neuron circuit based on the MOSFET controlled conductance demonstrated the asynchronous spike generation with a refractory period, the biological behavior of integration-and-firing, and the synaptic computation [17-20].

In this paper, the proposed design principle of MOSFET transconductance is introduced and demonstrated its performance for various applications of a programmable amplifier, tunable complex filters of wireless system, and the neuromorphic device for bio-inspired implementation.

II. MOSFET for Analog VLSI Signal Processing

The expanding demand of modern information technology drives the rapid growth of semiconductor industry and VLSI products of MOS transistors. The analog signal processing has been an essential element for the success of many VLSI products, particularly for the telecommunication and wireless technology. For analogue VLSI signal processing, it went back to the VLSI filter for voice CODEC in the early days of digital communication. For the advantage of digital processing or advanced VLSI computing, the front-end analog filter is required to select only the wanted input signal before the A/D conversion. The integrator circuit of Fig. 1 is a general building block of analog filter, where the additional fabrication process is required for tuning the resistors' value of VLSI. It is commonly resolved by the laser trimming for the exact resistance value, as the standard VLSI process has the inherent constraints.

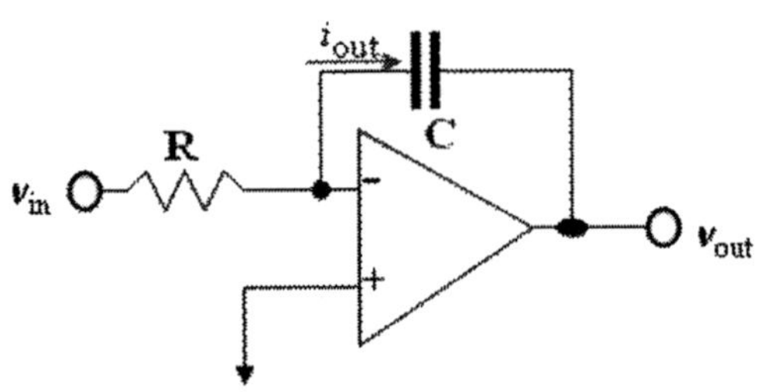

Figure 1. The basic building block of analog filer – an integrator circuit.

The switched capacitor (SC) technique was introduced to improve the problem of precise resistance value of an integrator, based on the SC integrator of Fig. 2.

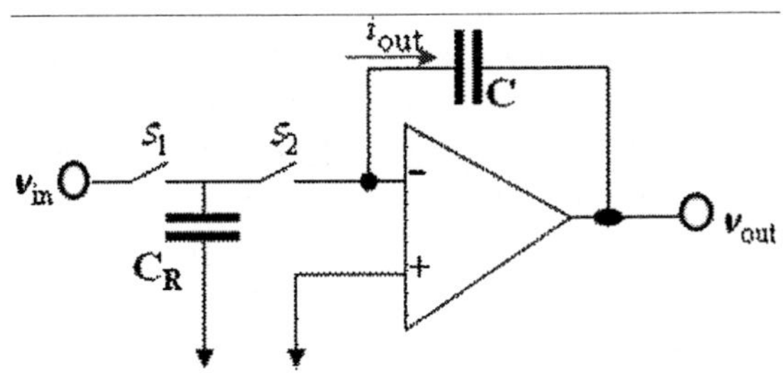

Figure 2. The switched capacitor integrator.

The integrator circuit of Fig. 1 is based on the integration of current i_{out} at capacitor C, which is represented as v_{out}. For the SC integrator of Fig. 2, the current i_{out} is equivalent to the total amount of the electric charge (Q) per second and can be represented as in the following equation (1).

$$i_{out} = f \bullet (v_{in} \bullet C_R) \tag{1}$$

where f is the switching frequency of s_1 and s_2. Switches of s_1 and s_2 are controlled by the non-overlapping clock and the clocking frequency f is assumed to be higher than the Nyquist sampling rate of the input signal v_{in}. It is based on the property that the charge of ($v_{in} \bullet C_R$) contributes to Q at each clock cycle. The switch is implemented by a single MOS transistor or a CMOS transistor pair, and the SC circuit has been widely employed for analog or analog-mixed VLSI products.

However, the SC based VLSI design becomes limited for certain applications, such as digital mobile communication of the high frequency and low power requirements. There are two constraints for the proper SC integrator of Fig. 2, which are the high clocking frequency and a CMOS operational amplifier. For an example, the filter of 1MHz bandwidth is necessary for Wifi applications. It means that the sampling clock becomes much higher compared 4 KHz bandwidth of voice CODEC. In addition to the higher clock frequency, the CMOS operational amplifier is likely prohibitive for such high frequency and low power requirements. Instead of utilizing SC type of analog sampled data type design, the integrator based on transconductance amplifier (Fig. 3) became widely used.

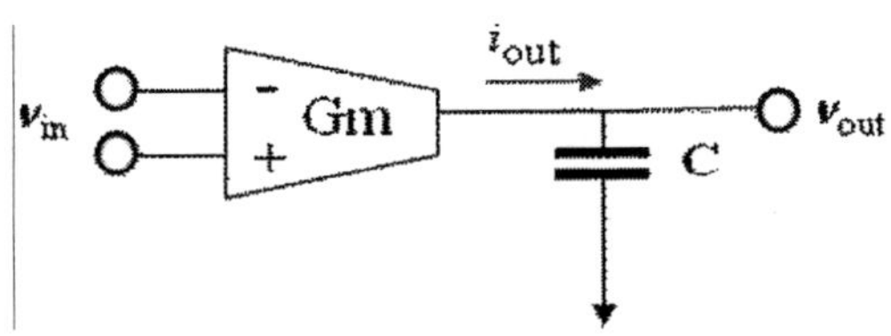

Figure 3. The integrator by a transconductance amplifier.

The advantage of transconductance amplifier is its relatively large bandwidth and the low power consumption. One of key aspects in transconductance amplifier design is the utilization of MOSFET's channel conductance, and there are a few ways of exploiting the channel conductance. The conductance in the triode region has been adopted for realizing the tunable resistance, and the equivalent circuit of Fig. 4 shows a MOSFET's channel characteristics in the triode region.

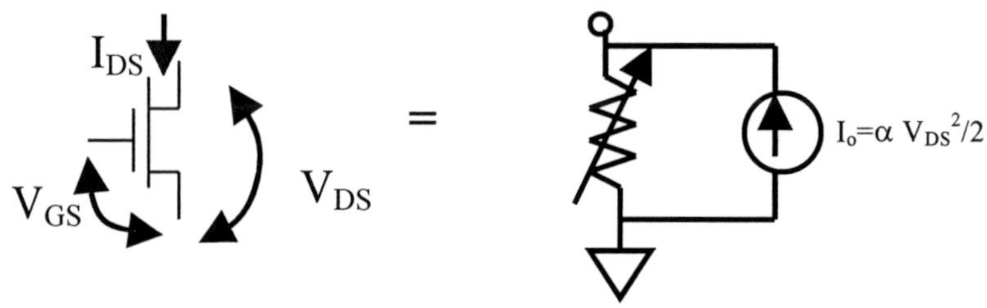

Figure 4. The equivalent circuit of a MOSFET in the triode region.

The I-V characteristics of MOSFET in the triode region can be modeled as in the equation (2).

$$I_{DS} = \alpha\ [\ (V_{GS}-V_T)V_{DS}-V_{DS}^2/2],\ \alpha=\mu C_{OX}W/L \tag{2}$$

where $V_{GS}-V_T > V_{DS}$. The second order nonlinearity well illustrated in the I-V curves of Fig. 5. The geometry of MOSFET is the [gate width (W)]/[gate length(L)]=1.

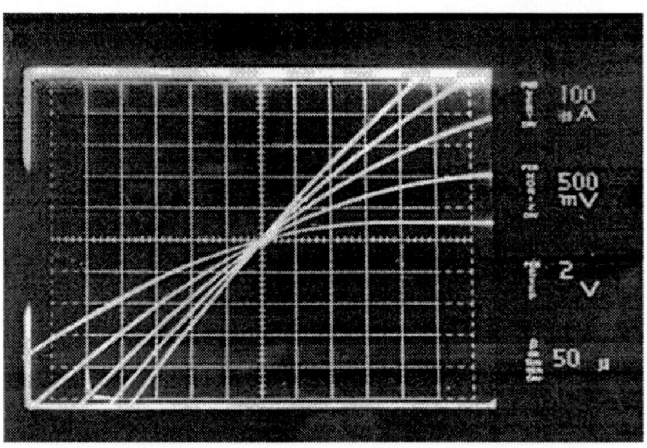

Figure 5. I-V characteristics of a MOSFET in the triode region, W/L=1.

The I-V characteristics of Fig. 5 shows the channel conductance controlled by the gate-source voltage, and it illustrates the feasibility of implementing the programmable conductance by MOSFETs in triode region. The parallel connection of two MOSFETs in Fig. 6 was proposed to improve the linearity of equivalent conductance in Fig. 5, which was motivated to challenge the limited characteristics of SC filters [10].

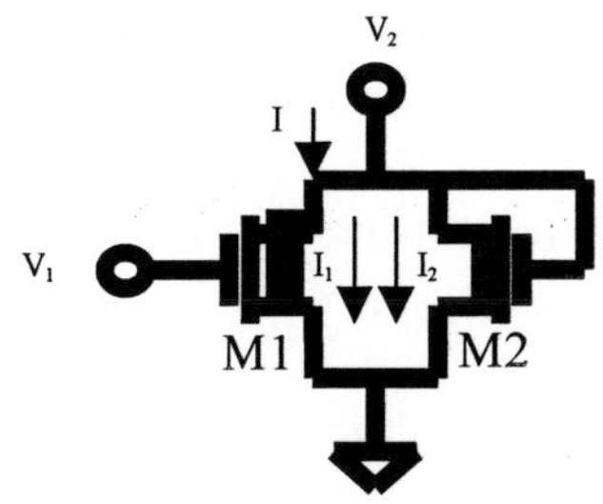

Figure 6. The voltage-controlled linear conductance by two MOSFETs in the triode region.

The operation principle of the circuit in Fig. 6 is to produce the quadratic term of V_2^2 by M2, and compensate the inherent negative quadratic terms of V_2^2 by M1 or M2. The compensation of the nonlinear component in (2) is proved by the equations of (3a), (3b) and (4). The equation (3a) represents the operation of M1, while the equation (3b) for M2. By adding I_1 and I_2, the equivalent conductance can be derived by equation (4) of total current I.

$$I_1 = \alpha[(V_1-V_T)V_2-V_2^2/2] \tag{3a}$$

$$I_2 = \alpha[(V_2-V_T)V_2-V_2^2/2] \tag{3b}$$

$$I = \alpha (V_1-2V_T) V_2 \tag{4}$$

It is necessary to satisfy the condition for the triode region operation, e.g. the threshold voltage V_T <0. The experimental result is illustrated in Fig. 7, where two MOS transistors are identical to the transistor used for the experimentation in Fig. 5. For the negative input, the swapping of drain and source does not affect the equation (4) representing I, as the change related to the gate-source bias voltage is disappeared after the summation process of I_1 and I_2. The comparison of Fig. 5 and Fig. 7 demonstrated the susbstitute to the resistors or SC-type design by the linear channel conductance of MOSFET in thertriode region. The simple parallel connection of two transistors satisfied the requirements of voice CODEC, though the slight degradation in Fig. 7 was caused by the threshold voltage modulation.

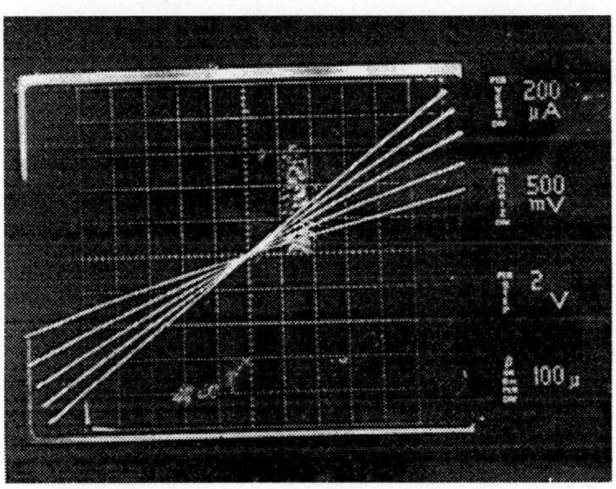

Figure 7. The voltage-controlled linear conductance by two MOSFETs.

There exists the small nonlinearity, as threshold voltage V_T in equation (4) is modulated by the input signal V_2. The change of V_2 drives the change in the source-substrate reverse-biased voltage, as the source voltage is equal to V_2 for the negative input. There is another but minor cause to the nonlinearity, which is due to the modulation of channel carrier mobility. The programmable conductance by the circuit of Fig. 6 demonstrated the improvement of issues by SC circuits, because of the tunable characteristics without relying on the discrete sampled data operation. As in the SC circuits, operational amplifies are necessary for the circuit of Fig. 6, which is still an issue for the modern mobile communication or special purpose applications like large scale analog-mixed VLSI integration.

III. Programmable Linear Conductance by MOSFET in the Triode Region – Principles and Applications

The requirement of general purpose operational amplifiers imposes the constraint on some applications, because of its power consumption, limited frequency bandwidth or etc. An example can be VLSI neural networks, which requires the large scale of multiplication and summation. Instead, the nature of biological neural networks drives the appropriate neuromorphic computation based on the large scale multiplication of neural states and synaptic connections with the excitatory-inhibitory weight value. It is a straightforward implementation to utilize the programmable conductance like I-V characteristics in Fig. 7, as it is similar to the neural networks model. With the large scale requirements like neural networks, it is realistic to implement the programmable linear conductance without using an operational amplifier.

The circuit of Fig. 8(a) was proposed for the programmable conductance without an operational amplifier. The operation principle is based on M1 in the triode region and the current mirror of M3-M5 [11]. The nonlinear term in equation (2) is eliminated by applying the positive and negative voltage of the same amplitude to the drain and the source respectively.

Assuming the drain voltage *v* and the source voltage –*v*, the drain-source current I of M1 is

$$\begin{aligned} I &= \alpha[\,(V_{in}+v-V_T)(2v)-(2v)^2/2] \\ &= \alpha(V_{in}-V_T)(2v) \qquad (5) \end{aligned}$$

The output current I_{out} is the mirrored current I by M3 and M5, which is linearly proportional to either V_{in} or $2v$. M2 and M3 are the same ones, which introduce the same voltage drop across them. With the presence of M2 and M3, the same amplitude voltage with the different polarity is applied to the drain and the source of M1 respectively. M4 acts as the offset tuning or reference generator for the bipolar current output.

The circuit of Fig. 8(a) shows the linear characteristics of voltage-controlled transconductance, without an operational amplifier. It suggested a feasible design solution for the large scale VLSI neural networks, with a switch attached to the output. The switch can multiply the synaptic weight (V_{in}) and the spike neural input, and every synapse cell of Fig. 8a can be connected in parallel without the restriction [21, 22]. The bipolar supply voltage is one of key element for the programmable linear conductance of Fig. 8(a), compared to the single voltage supply. For the analog VLSI, the differential inputs are generally preferred to the single-ended signal.

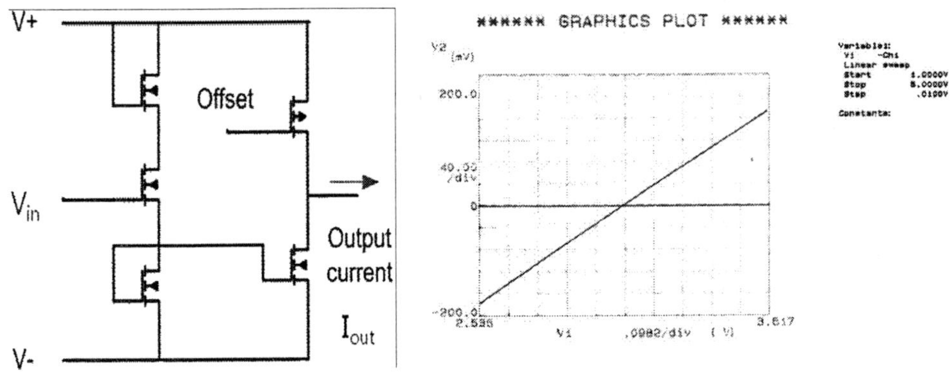

(a) (b)

Figure 8. (a) The linear transconductance circuit based on a MOSFET in the triode region (b) the measured I-V transfer characteristic.

Hence, the transconductor circuit of Fig. 9 was recently designed to accommodate the differential inputs, followed by two previous design methodologies of Fig. 6 and Fig. 8a [11]. The circuit of Fig. 9 is based on the parallel connection of two MOSFETs M1 and M2. Both M1 and M2 operate in the triode region, which are similar to MOSFETs M1 and M2 in Fig. 6. Instead of an operational amplifier, the current mirror configuration was adopted as the circuit of Fig. 8a. The role of M3 is to implement the current mirror, while it is also indispensable for the triode region operation of M2. M2 operates in the triode region with the ordinary positive threshold voltage, as the drain voltage of M2 is kept for the triode-region operation condition by V_{BIAS} of M3.

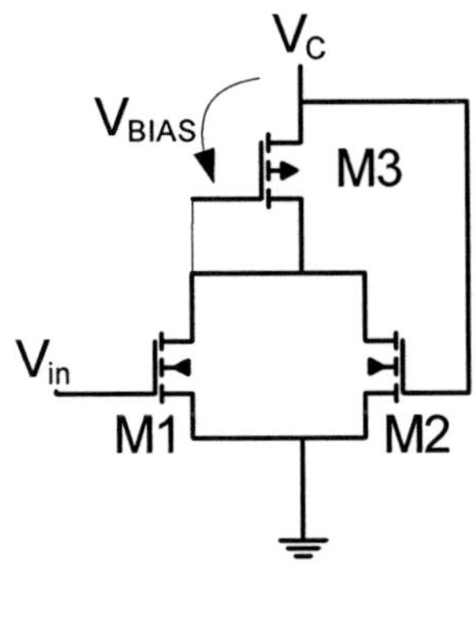

Figure 9. The voltage-controlled linear transconductance circuit based on a pair of MOS transistors in the triode region.

The drain current I_D of M1 and M2 may be expressed as

$$I_{D1} = \alpha\ [(V_{GS} - V_{TH})V_{DS} - V_{DS}^2/2] \tag{6a}$$

$$I_{D2} = \alpha\ [(V_C - V_{TH})V_{DS} - V_{DS}^2/2] \tag{6b}$$

where $V_{GS} = V_{inDC} \pm \Delta V_{in}$ is the gate-source voltage of transistor M1, $V_{DS} = V_C - V_{BIAS}$ is the drain-source voltage of transistors M1and M2, V_C is the programming voltage of transconductance, and V_{inDC} is the DC offset of input voltage.

V_{BIAS} of the diode connected M3 operates as the DC voltage, neglecting the minor non-ideal characteristics. Hence, V_{DS} is $(V_C - V_{BIAS})$ and the total current I_D is determined by

$$\begin{aligned} I_D &= I_{D1} + I_{D2} \\ &= \alpha\ (V_{GS} - V_{BIAS} - 2V_{TH})(V_C - V_{BIAS}) \end{aligned} \tag{7}$$

The analysis of equations (6) and (7) explains that the 2nd order distortion can be removed by a matched pair of M1 and M2 in the triode region. The mobility degradation or variation of bias voltage can cause the slight nonlinearity of transconductance in equation (7). The term of V_{BIAS} in (7) is the gate-source voltage of M3, while the drain current of M3 is I_D. The deviation of V_{BIAS} from the DC voltage is undesirable but exists due to the mobility reduction effect or the nonlinear characteristics of M3 in diode connection. The circuit of Fig. 10 compensates such variation by introducing the differential mode operation, as the variation of V_{BIAS} is almost common between both basic transconductance circuits of Fig. 9. In principle, the circuit of Fig. 10 is based on the parallel operation of basic transconductor circuits in Fig. 9 with the linear and programmable transconducatnce.

Repeated two circuits in Fig. 10 operate in the same condition, other than the small differential input signals applied to the gate of M1 and M1A in Fig. 10. Therefore the gate-source voltage of M3 and M3A is almost same, though the amplitude of small differential signal can cause a slight difference. The transistor pairs of (M3, M4), (M3A, M4A), and (M5A, M5) are connected as the current mirror circuit. M4 copies the drain current I_{D3} of M3 and M5 copies the drain current of M5A, which is equal to the drain current I_{D3A} of M3A. Based on equation (7), I_{D3} and I_{D3A} can be given by

$$I_{D3} = Gm(V_{GS1} - V_{BIAS} - 2V_{TH}) \tag{8a}$$

$$I_{D3A} = Gm(V_{GS1A} - V_{BIAS} - 2V_{TH}) \tag{8b}$$

Where $V_{GS1} = V_{inDC} + \Delta V_{in}/2$, $V_{GS1A} = V_{inDC} - \Delta V_{in}/2$, and $Gm = K(V_C - V_{BIAS})$.
Therefore, the output current I_{OUT} is

$$I_{OUT} = Gm\ \Delta V_{in} \tag{9}$$

where $I_{OUT} = I_{D3} - I_{D3A}$.

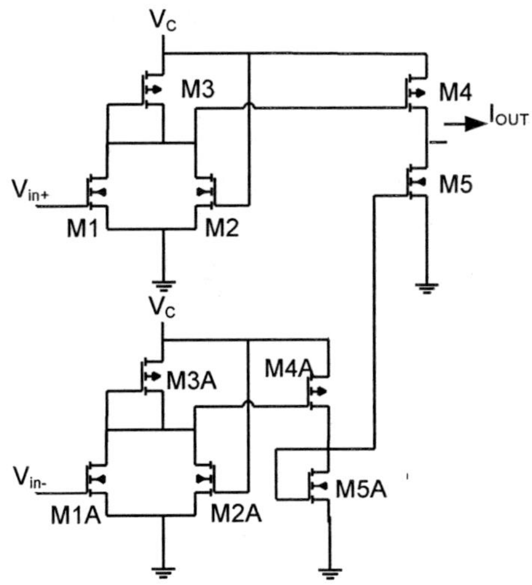

Figure 10. The transconductance amplifier based on the voltage-controlled transconductance circuit of Fig. 9.

The I-V relation of equation (9) is also based on the triode-region operation of transistors M1, M2, M1A, and M2A. By comparing the equations (7) and (9), the variation by the term of V_{BIAS} is less influential in equation (9). For transistors M1 and M1A, the triode-region condition of ($V_{GS} - V_{TH} \geq V_{DS}$) determines the input signal dynamic range. The V_{BIAS}, the gate-source voltage of M3 or M3A maintains the operation of M2 or M2A in the triode region, while M3 or M3A operates in the saturation region with the tuning voltage V_C. Hence, the circuit dynamic range is decided approximately by

$$V_{in} > V_C - V_{BIAS}$$

where V_{BIAS} is larger than V_{TH} of M3 and M3A, V_C is larger than V_{TH} of M2 and M2A. The transconductance amplifier of Fig. 10 was designed for the high frequency application and the simulated result illustrates the operation over the Giga Hertz range in Fig. 11. There are several types of transconductance amplifier with different characteristics and target applications, as there are various requirements with constraints of the linearity, the operation frequency, the tunable characteristics, power consumption, noise characteristics, and so on. The transconductance amplifier circuit of Fig. 10 has the advantage of the high frequency operation with the improved linearity, in addition to the controlled conductance by simple and flexible configuration. The gain bandwidth is observed up to 2 GHz in Fig. 11(a), while the signal dynamic characteristics are shown in the Fig. 11(b).

The equation (9) implies the analog multiplication, as the term of ($V_C \Delta V_{in}$) in equation (9) represents the mathematical multiplication. Though the digital system or signal processing is dominant nowadays, the analog multiplication is expected to bridge the gap between the high precision digital computing and the large scale special computing [23-25]. The neural networks or convolution decoder is an example of such applications. The offset components of V_C and V_{BIAS} for analog operation can be removed by a pair of same transconductance amplifiers. The differential operation for V_C can be implemented by two transconducatance amplifiers in Fig. 12. It removes the common terms of V_{BIAS} and the offset voltage of V_C from the final output current.

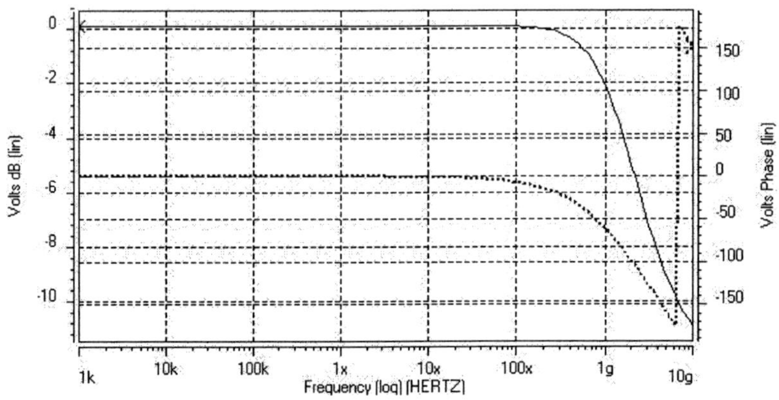

(a)

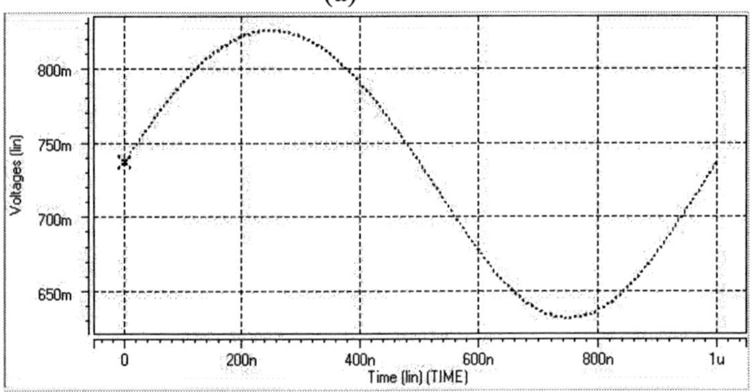

(b)

Figure 11. (a) The frequency characteristics of the transconductance amplifier in Fig. 10, (b) the output waveform dynamics.

Based on equation (9), the output currents of upper and lower transconductor circuit in Fig. 12 are described as

$$I_{OUT1} = k(v_{iny} - V_{BIAS})\, \Delta v_{inx} \tag{10a}$$

$$I_{OUT2} = k(v_{iny-} - V_{BIAS})\, \Delta v_{inx} \tag{10b}$$

where I_{OUT1} represents the output from left hand side transconductors and I_{OUT2} represents the output of right hand side transconductors. As $\Delta v_{iny} = v_{iny} - v_{iny-}$, the output current I_{OUT} of the circuit of Fig. 12 is determined by

$$I_{OUT} = I_{OUT1} - I_{OUT2} = k(\Delta v_{inx} \Delta v_{iny}) \tag{11}$$

From equation (11), the circuit of Fig. 12 works as the four quadrant analog multiplier with the dynamic range of ($\pm\Delta v_{inx}/2$. $\pm\Delta v_{iny}/2$).

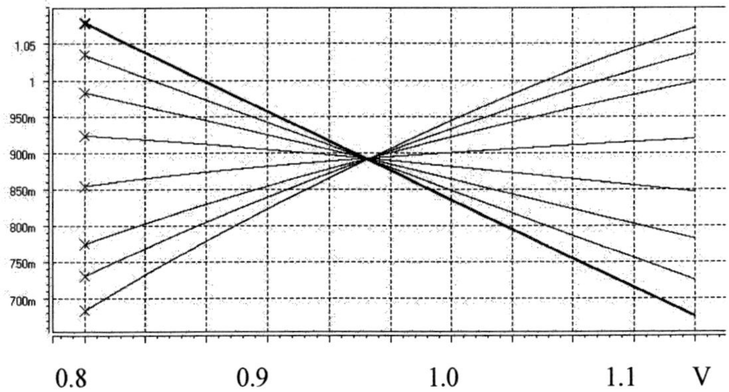

Figure 12. Analog multiplier by two transconductance amplifiers of Fig. 10.

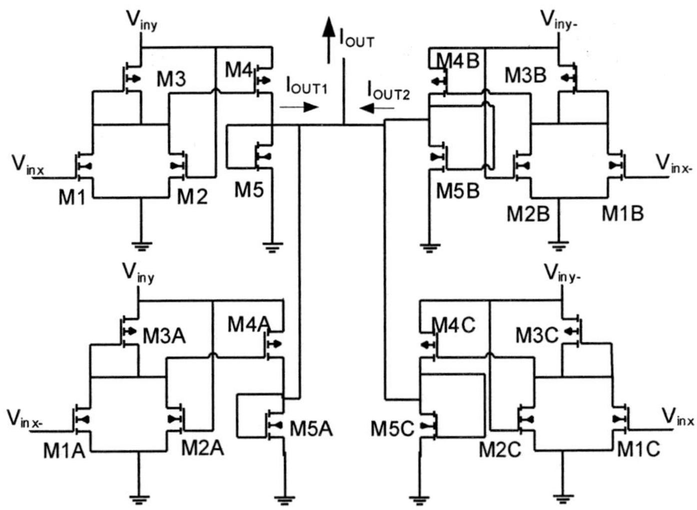

Figure 13. Linearity of analog multiplier of Fig. 12 : x-axis v_{inx} (0.8V-1.14V), y-axis multiplier output (converted to voltage via a load resistor), v_{iny} (0.9V-1.5V).

The simulated result of a proposed analog multiplier is exhibited in Fig. 13, with $\pm\Delta v_{inx}/2 = \pm 0.17$V and $\Delta v_{iny}/2 = \pm 0.3$V. The output current is measured as the voltage via a resistor connected to the output, with the MOSFETs of $W_1/L_1 = W_2/L_2 = W_{1A}/L_{1A} = W_{2A}/L_{2A} = W_{1B}/L_{1B} = W_{2B}/L_{2B} = W_{1C}/L_{1C} = W_{2C}/L_{2C} = 0.9/7.2$, $W_3/L_3 = W_4/L_4 = W_{3A}/L_{3A} = W_{4A}/L_{4A} = W_{3B}/L_{3B} = W_{4B}/L_{4B} = W_{3C}/L_{3C} = W_{4C}/L_{4C} = 3.6/0.36$, $W_5/L_5 = W_{5A}/L_{5A} = W_{5B}/L_{5B} = W_{5C}/L_{5C} = 1.8/0.36$. The overall power consumption of an analog multiplier is 250μW.

The transconductance amplifier of Fig. 14 has the output buffer circuits added to the circuit of Fig. 10, for the complete transconductance amplifier and the balanced structure of filter applications. The transistors M6-M9 in Fig. 14 are added for the output circuit to be independent of the tuning voltage V_C, and the output circuit can be flexible to the application requirements such as the robust current level or dynamic range. The balanced output current can be available from the other transconductor element, by adding the equivalent output circuit. The transistor pairs of (M5, M6) and (M7, M8) are introduced to copy the current in M3 or M4 to the output with the supply voltage of V_{DD}. The current in M3A or M4A is mirrored to the transistor M9, which contributes to the differential output current I_{OUT} as in Fig. 10. The transistor sizes used are $W_1/L_1 = W_2/L_2 = W_{1A}/L_{1A} = W_{2A}/L_{2A} = 0.125$, $W_3/L_3 = W_4/L_4 = W_{3A}/L_{3A} = W_{4A}/L_{4A} = 10$, $W_5/L_5 = W_{5A}/L_{5A} = 5$. The input V_{in} is 1.25V ± 0.2V and the tuning voltage V_C is 1.2V± 0.2V.

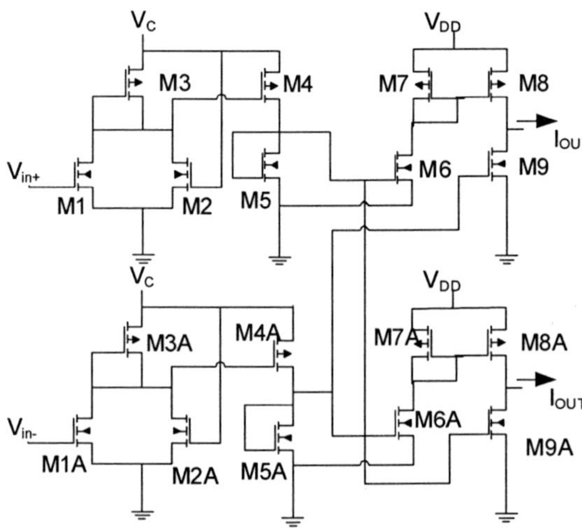

Figure 14. The transconductance amplifier with balanced outputs.

The linearity from SPICE simulation is summarized in Table I [11]. The results shown in Table I correspond to the worst case ±1% mismatch of gate geometry (W and L) for all transistors of the circuit in Fig. 14. The simulation is based on V_{DD}=1.8V, V_C=1.0 V, and V_{inDC}=1.25V. From the results in Table I, the total harmonic distortion is less than -65 dB with the input of 0.6 V_{PP}. The input with larger DC value or common-mode can increase the dynamic range. The input common-mode of DC offset of the circuit in Fig. 14 can be varied from 1V to the maximum signal level in the circuit like 1.5V. The simulation result shows the 100μW of power consumption, the tunable transconductance range of ± 50% with the control voltage V_C of 1 V ± 0.2V.

Table I.

V_{in} (V_{PP}) at10 MHz	0.2 V	0.3 V	0.4 V	0.5 V	0.6 V
Total Harmonic Distortion	-70dB	-72dB	-69dB	-68dB	-65dB

The tansconductance amplifier in Fig. 14 was evaluated for the GmC filter of 5th order low pass filter. The 1 MHz bandwidth with the Leap-Frog structure was simulated as in Fig. 15.

The linearity of the MOSFET's conductance yielded the total harmonic distortion of the filter as -68 dB with 0.5V input, and it is proved be an efficient way of implementing the programmable conductance [11].

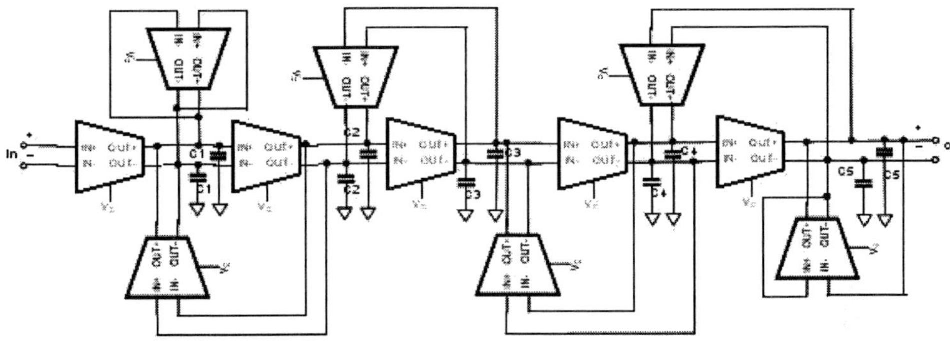

(a)

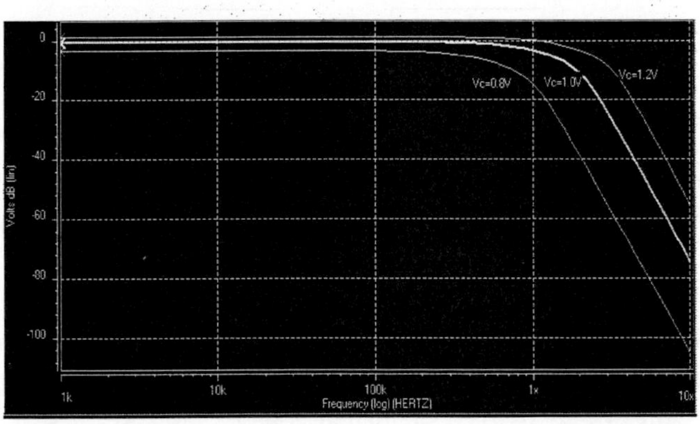

(b)

Figure 15. (a) Fifth-order Gaussian low-pass filter, C1=0.9 pF, C2=1.9 pF, C3=2.6 pF, C4=3.1 pF C5=7.6 pF, using balanced transconductance amplifiers in Fig.5, (b) Cut-off frequency vs. control voltage V_C; 0.5 MHz (0.8V), 1 MHz (1.0V), 1.6 MHz (1.2V).

The conductance by MOSFET in the triode region has other advantages, and the one is the flexible programming for the analog computing or complex analog-mixed signal processing. For an example, an analogue multiplier of Fig. 12 was implemented by the principle based on simple equations (10) and (11). An analog polyphase filter is designed to evaluate the effectiveness of utilizing the programmable conductance, as the tunable analogue filters have been emerged as key issues for advanced wireless applications of Mobile TV, digital RF, and Cognitive Radio.

For implementing an analog polyphase filter, the feedback between I and Q signals requires the higher transconductance coefficient to shift the centre frequency from dc to the range of MHz. Two transconductance amplifiers (G2) of Fig. 16(a) correspond to the feedback paths of I and Q signals, where v_{in} represents the I channel signal and jv_{in} of 90 degrees phase shifted v_{in} represents the Q channel signal. The control voltage v_{Ct} is the transconducatnce tuning voltage of G2 to tune the centre frequency of polyphase bandpass filter. The output buffer of transconductance amplifier of Fig. 14 improves the dynamic characteristic by maintaining the consistent condition, even with the dynamical change of V_C. To meet the requirement of a different transconductance value for I and Q channel feedback, a simple way of current amplification is employed at the first stage by multiplying the transistors of M4 and M4A of G2 amplifiers in Fig. 16(a). The G2 amplifier has the 6 times larger gate width of M4 and M4A than G1 amplifiers. The MOSFETs of G1 amplifier are $W_1/L_1 = W_2/L_2 = W_{1A}/L_{1A} = W_{2A}/L_{2A} = 1/8$, $W_3/L_3 = W_4/L_4 = W_{3A}/L_{3A} = W_{4A}/L_{4A} = 10$, $W_5/L_5 = W_6/L_6 = W_9/L_9 = W_{5A}/L_{5A} = W_{6A}/L_{6A} = W_{9A}/L_{9A} = 5$, $W_7/L_7 = W_8/L_8 = W_{7A}/L_{7A} = W_{8A}/L_{8A} = 10$, with the simulated analogue polyphase filter characteristics of Fig. 16(b). The capacitor C is 3.5 pF for a low pass prototype filter of 1 MHz bandwidth.

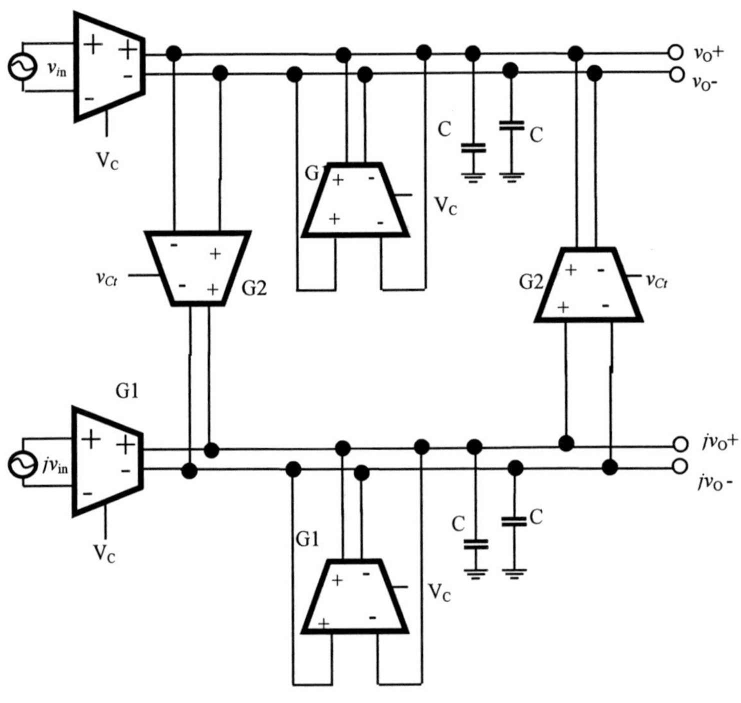

(a)

Figure 16. Continued on next page.

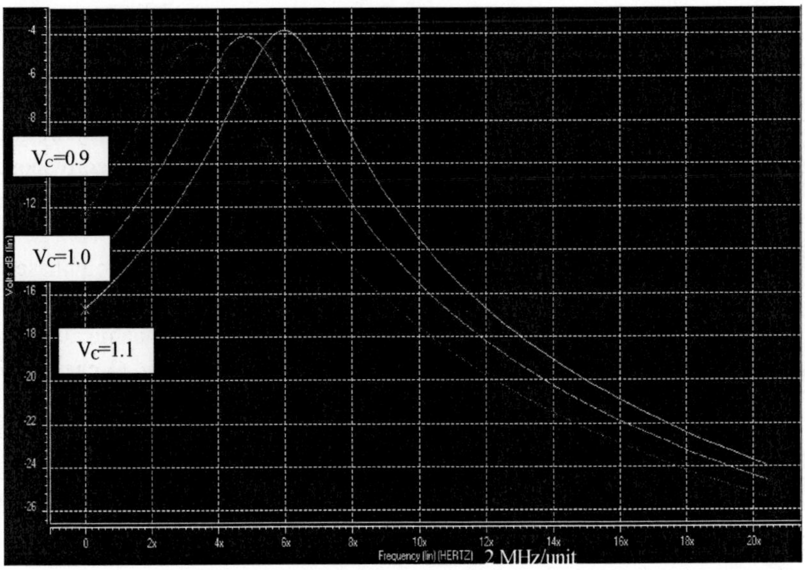

(b)

Figure 16. (a) An analog complex filter using transconductance amplifiers of Fig. 14, (b) simulated tuning of analog complex filter: the control voltage v_{Ct} of G2 is in the range of 1V ± 0.1V, the control voltage V_C of G1 is 1.0V.

The overall linear characteristic was evaluated by SPICE simulation, where the total harmonic distortion was less than 0.1 % with the input signal of 0.6 V_{PP}. The supply voltage V_{DD} of CMOS tunable analogue polyphase filter is 1.8V with the power consumption of 100 μW and the control voltage is in the range of 1V ± 0.1V. The simulated characteristics in Fig. 16(b) illustrated that the centre frequency of the pass band can be controlled linearly by the change of control voltage v_{Ct} of G2 transconductance amplifiers. The control voltage V_C of transconductance amplifiers G1 is maintained to 1.0V throughout the simulation.

The new programmable conductance by MOSFETs can be utilized for implementing the biologically inspired applications of neuromorphic signal processing. For neuromorphic applications, the transconductor of Fig. 10 has the advantage in implementing synaptic connections with excitatory and inhibitory weights, based on biologically plausible spike-based neural networks [26]. By adding a MOSFET at the output, the neural computing for synaptic function can be realized by an additional MOSFET for a switch, as the output current simulates the post synaptic state. There is another advantage of implementing the biologically plausible neuron like Hodgkin-Huxley(H-H) formalism as in Fig. 17, which has been well-known for the electrical model based on the controlled conductance for its physiological dynamics [27-31].

The circuit of Fig. 17 works on the dynamics of H-H formalism, with the simplified but simulating the membrane conductance model. The H-H formalism can include two digits of ion numbers depending on the purposes, and the objective of Fig. 17 is to implement the fully asynchronous spike neuron with the refractory period like the biological nature. The circuit operation is illustrated in Fig. 18. In Fig. 18, the membrane potential (a) of neuron (axon) produced the control voltage of the membrane conductance (b). Once the membrane potential reached the threshold of firing the spikes (d), neural input spikes (c) to the neuron won't contribute to increase the membrane potential for a certain period.

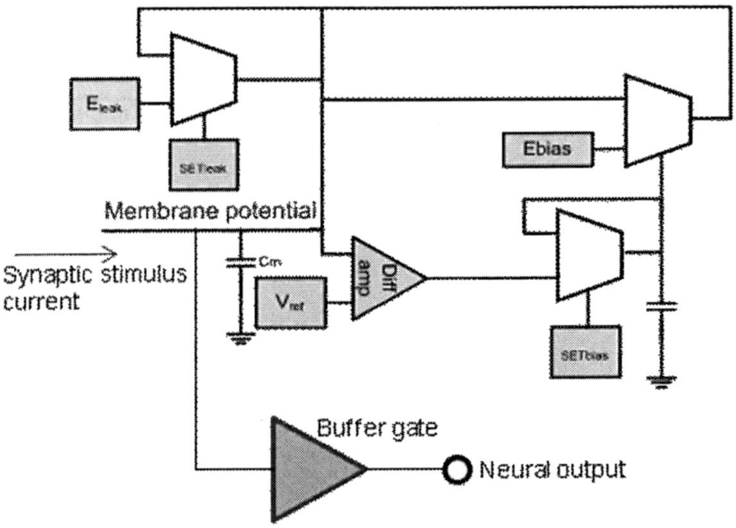

Figure 17. The neuron circuit of Hodgkin-Huxley formalism based the transconductance amplifier.

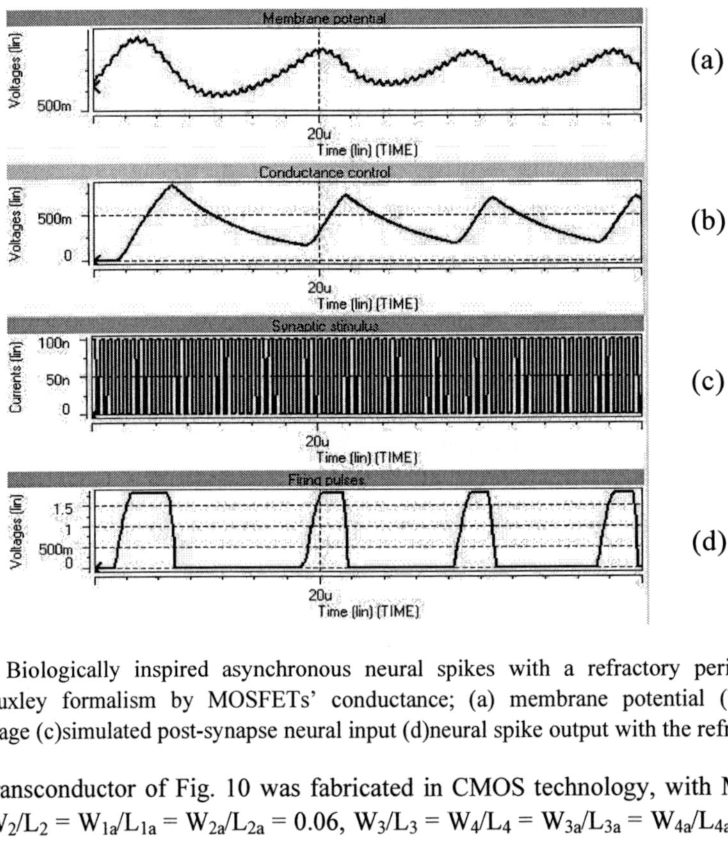

Figure 18. Biologically inspired asynchronous neural spikes with a refractory period, based on Hodgkin-Huxley formalism by MOSFETs' conductance; (a) membrane potential (b)conductance control voltage (c)simulated post-synapse neural input (d)neural spike output with the refractory period.

The transconductor of Fig. 10 was fabricated in CMOS technology, with MOSFETs of $W_1/L_1 = W_2/L_2 = W_{1a}/L_{1a} = W_{2a}/L_{2a} = 0.06$, $W_3/L_3 = W_4/L_4 = W_{3a}/L_{3a} = W_{4a}/L_{4a} = 20$, W_5/L_5

= W_{5a}/L_{5a} = 10. The modulated sinusoidal waveform in Fig. 19 shows the tuning and linear characteristics of tunable linear transconductor, where a sinusoidal signal of V_{in} and a sawtooth signal of V_C were applied as input signals. The sawtooth wavefrom in Fig. 19 represents V_C, while a resistor was added to convert the output current to voltage at the output.

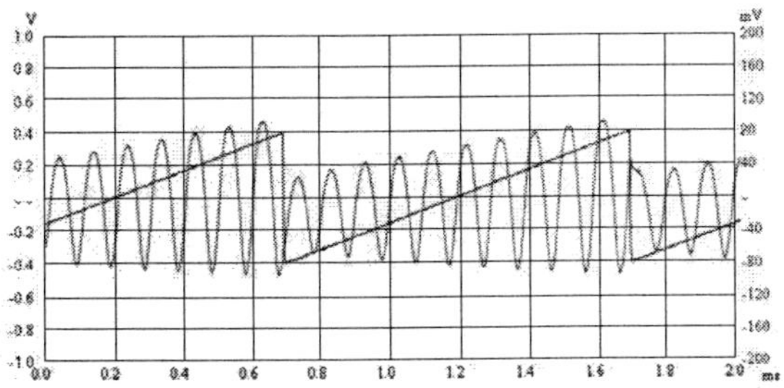

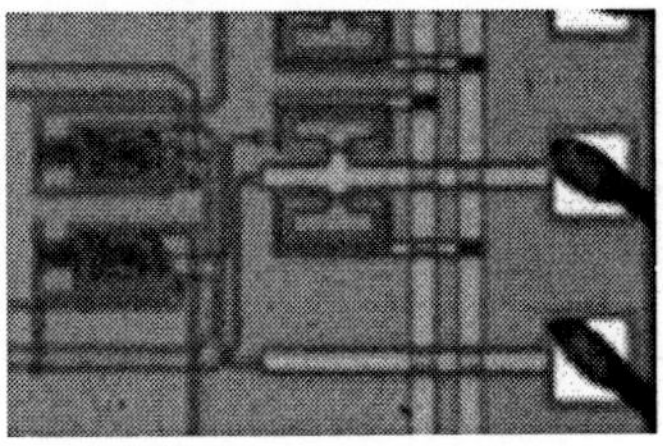

Figure 19. The experimental result and CMOS circuit of voltage-controlled conductance in Fig. 10.

IV. Conclusion

The conductance of MOSFET is proposed as a methodology of realizing analog-mixed VLSI for analog signal processing, analog front-end for innovative mobile information technology, and the biologically inspired systems. In this paper, the ways of utilizing the MOSFET in the triode region are introduced and analyzed for implementing the linear and programmable conductance from the nonlinear physical property of MOS transistors. The recent design concept is based on the pair structure of two transistors in the triode region, and the controlled conductance is realized with the flexibility of adapting to complicated applications.

The special function like analog multiplier can be implemented by repeating the basic circuit.

Example applications are investigated for amplifiers, filters, analog multiplier, and neuromorphic computing. The accuracy or the flexibility is demonstrated by the performance of filters and special analog signal processing, with the tunable range of 100% and the linearity equivalent to - 65dB in total harmonic distortion. The application to the biologically plausible signal processing is demonstrated by implementing the H-H formalism in analog-mixed MOSFET circuit, while the H-H formalism is too complex computing to be efficient by the digital VLSI. The spiking neuron output with the refractory period is simulated by MOSFETs' programmable conductance.

The linear and programmable conductance by MOSFETs is proposed as a new way of the VLSI implementation for analog-mixed applications, mobile communications to biologically inspired information processing. The principle is flexible to extend or modify depending on application specific requirements, as it is based on the simple configuration of two MOS transistors in the triode region.

References

[1] D. V. Morozov and A. S. Korotkov, A realization of Low-Distortion CMOS Transconductance Amplifier, IEEE Trans. Circuits Syst. I, 2001, vol. 48, pp. 1138-1141

[2] M. Kachare et al, A Compact Tunable CMOS Transconductor with High Linearity, *IEEE Trans. Circuits Syst. II*, 2005, vol 52, pp. 82-84

[3] S. Korotkov, D. V. Morozov and R. Unbehauen, Low-voltage continuous-time filter based on a CMOS transconductor with Enhanced Linearity, *International Journal of Electronics and Communications*, 2002, vol 56, pp.416-420

[4] E. A. M. Klumperink and B. Natua, Systematic Comparison of HF CMOS Transconducators, *IEEE Trans. Circuits Syst. II,* 2003, vol 50, pp. 728-741

[5] Q. Huang, A MOSFET-Only Continuous-Time Bandpass Filter, *IEEE J. Solid-State Circuits*, 1997, Vol. 32, pp. 147-158

[6] L. Corban and P. E. Allen, Low-voltage CMOS transconductance cell based on parallel operation of triode and saturation transconductors," *Electron lett.*, 1994, Vol. 14, pp. 1124-1126

[7] S. H. Yang et al,"A Novel CMOS Operational Transconducatance Amplifier Based on a Mobility Compensation Technique," *IEEE Trans. Circuits Syst. II,* 2005, vol 52, pp. 37-42

[8] U. Yodprasit and C. C. Enz, "A 1.5 V 75-dB dynamic range third-order Gm-C filter integrated in a 0.18-um Standard digital CMOS process," *IEEE J. Solid-State Circuits,* 2003, Vol. 7, pp. 1189-1197

[9] M. Banu and Y. Tsividis, "Fully Integrated Active RC Filters in MOS Technology," *IEEE J. Solid-State Circuits, 1983*, Vol. 6, pp. 644-651

[10] S. Han and S. B. Park, "Voltage-controlled linear resistor by two MOS transistors and its application to active RC filter MOS integration," *Proc. IEEE*, 1984, Vol. 11, pp. 1655-1657

[11] S. Han, A novel tunable transconductance amplifier based on voltage-controlled resistance by MOS Transistors, IEEE trans. Circuits Syst. II, 2006, Vol. 53, pp. 662-666

[12] Y. Horio, K. Aihara, and O. Yamamoto, Neuron–Synapse IC Chip-Set for Large-Scale Chaotic Neural Networks, *IEEE Trans on Neural Networks*, 2003, vol. 14, pp1393-1404

[13] M. Milev and M. Hristov, Analog Implementation of ANN With Inherent Quadratic Nonlinearity of the Synapses, *IEEE Trans on Neural Networks,* 2003, vol. 14, pp1187-1200

[14] G. Bugmann, Biologically plausible neural computation, *Biosystems*, 1997, Vol. 40, pp. 11-19

[15] J. Taylor, Paying attention to consciousness, *Progress in Neurobiology*, 2003, vol. 71, pp.305-335

[16] J. Taylor, Private communication, 2004

[17] Christodoulou, G. Bugmann, and T. Clarkson, A spiking neuron model: applications and learning, *Neural networks*, 15, 2002, pp. 891-908

[18] Bartolozzi and G. Indiveri, A neuromorphic selective attention architecture with dynamic synapses and integrate-and fire neurons, *Proceedings of BICS 2004*, Sterling, 2004, BIS2.2

[19] S. Han, Mixed-signal neuron-synapse implementation for large-scale neural networks, *Neurocomputing,* 2006, Vol .69, pp. 1860-1867

[20] S. Han, A pulse-based neural hardware implementation based on the controlled conductance by MOSFET circuit, *Proc. IJCNN*, 2006, pp5100-5106

[21] S. Han and R. Webb, Neural Network Switch Controller with Analogue-Digital Mixed Neural Network VLSI, *Proc. of EANN'97*, 1997, pp. 299-302

[22] S. Han, Neural network VLSI implementation and its applications, *Korea Telecom Journal*, 1997, Vol.2, pp. 12-20

[23] L. Coban, and P. E. Allen, Low-voltage, four quadrant analogue CMOS multiplier, *Electron. Lett.,* 1994, Vol. 30, pp. 1044-1045

[24] Ramirez-Angulo et al, Low-voltage CMOS analogue four quardrant multiplier based on flipped voltage followers, *Electron. Lett.*, 2003, Vol. 39, pp. 1171-1172

[25] Maundy and M. Maini, A comparison of three multipliers based on V_{gs}^2 technique for low voltage applications, *IEEE Trans. Circuit & Systems-I,* 2003, Vol. 50, pp. 937-940

[26] S. Han, Mixed-signal neuron-synapse implementation for large scale neural network, *Proceedings of BICS,* Sterling, 2004, NC4.2

[27] Linares-Barranco et al, A CMOS implementation of FizHugh-Nagumo neuron model, *IEEE JSSC*, 1991, Vol. 26, pp956-965

[28] M. Hausser, The Hodgkin-Huxley theory of action potential, *Nature neuroscience* supplement, 2000, vol. 3, pp. 1165

[29] M. Izhikevich, Which model to use for cortical spiking neurons?, *IEEE trans. Neural Networks*, 2004, vol. 15, pp1063-1070

[30] T. Asai et al, A subthreshold MOS neuron circuit based on the Volterra system, *IEEE trans Neural Networks*, 2003, Vol. 14, pp.1308-1312

[31] M. Simoni et al, A multiconductance silicon neuron with biologically matched dynamics, *IEEE Trans. Biomedical Eng*, 2004, Vol. 51, pp. 342-354

INDEX

C

D

E

F

G

H

I

N

O

P

Q

R

S

T

U

V

W

X

Y

Z